T0396556

Tea Polyphenols,

Oxidative Stress and Health Effects

Volume 2

Tea Polyphenols,

Oxidative Stress and Health Effects

Volume 2

Edited by

Baolu Zhao

Institute of Biophysics
Chinese Academy of Sciences, China

W‍‍ World Scientific

NEW JERSEY · LONDON · SINGAPORE · BEIJING · SHANGHAI · HONG KONG · TAIPEI · CHENNAI · TOKYO

Published by

World Scientific Publishing Co. Pte. Ltd.

5 Toh Tuck Link, Singapore 596224

USA office: 27 Warren Street, Suite 401-402, Hackensack, NJ 07601

UK office: 57 Shelton Street, Covent Garden, London WC2H 9HE

Library of Congress Cataloging-in-Publication Data

Names: Zhao, Baolu, editor.

Title: Tea Polyphenols, Oxidative Stress and Health Effects, In 2 Volumes / edited by
    Baolu Zhao, Institute of Biophysics, Chinese Academy of Sciences, China.

Description: New Jersey : World Scientific, [2024] | Includes bibliographical references and index.

Identifiers: LCCN 2023016990 | ISBN 9789811274206 (set) (hardcover) |
    ISBN 9789811274213 (set) (ebook) | ISBN 9789811274220 (set) (ebook for individuals) |
    ISBN 9789811285301 Volume 1 | ISBN 9789811285318 Volume 2

Subjects: LCSH: Tea--Health aspects. | Tea--Therapeutic use. | Polyphenols--Therapeutic use. |
    Oxidative stress.

Classification: LCC RM251 .T44 2024 | DDC 615.3/21--dc23/eng/20230823

LC record available at https://lccn.loc.gov/2023016990

British Library Cataloguing-in-Publication Data

A catalogue record for this book is available from the British Library.

For any available supplementary material, please visit
https://www.worldscientific.com/worldscibooks/10.1142/13350#t=suppl

Desk Editor: Vanessa Quek ZhiQin

Typeset by Stallion Press
Email: enquiries@stallionpress.com

https://doi.org/10.1142/9789811274213_fmatter

About the Editors

Professor Yuefei Wang: Director of Tea Research Institute of Zhejiang University, China.

Professor Youying Tu: Tea Research Institute of Zhejiang University, China.

Professor Yushun Gong: Department of tea, College of horticulture, Hunan Agricultural University, China.

Associate Professor Jihong Zhou: Tea Research Institute of Zhejiang University, China.

Dr. Ying Chen: Department of tea, College of horticulture, Hunan Agricultural University.

Dr. Guoliang Jie: General Manager of Huangshan Maofeng Tea Group Co., Ltd., China.

Caisheng Lan: National first-class tea appraiser, general manager of Yibaozhai Tea Industry Co., Ltd., China.

Jianping Sun: National first-class tea appraiser, chairman of Yibaozhai Tea Industry Co., Ltd., China.

Professor Baolu Zhao: State Key Laboratory of brain and cognition, Institute of Biophysics, Chinese Academy of Sciences, China.

Preface

In recent years, many research articles about tea polyphenols have been published in world academic journals. However, so far, there is no scientific, comprehensive, and objective monograph on tea polyphenols, oxidative stress, and health. The author, Professor Baolu Zhao, formerly chairman of the Asian Free Radical Society and director of the Chinese Free Radical Biology and Medicine Committee of China, was selected as the world's top scientists with global lifetime impact in 2022. The author has been engaged in the research of free radicals and tea polyphenols for decades and has published more than 300 research papers and 6 monograph books, of which more than 30 have directly studied tea polyphenols. In addition, Professor Yuefei Wang, Professor Youying Tu, and Associate Professor Jihong Zhou of Zhejiang University of China, Professor Yushun Gong of Hunan Agricultural University China, Dr Ying Chen of Hunan Agricultural University China, Dr Jie Guoliang the general manager of Huangshan Maofeng Tea Group Co., Ltd. China, national first-class tea evaluator Caisheng Lan, and Jianping Sun the chairman and general manager of Wuyishan Yibaozhai Tea Co., Ltd. China, were invited to participate in the writing of this book. They have studied tea polyphenols for many years and have published many books and papers in scientific journals. On the basis of summarizing these research results, this book refers to the literature on tea polyphenols in recent years and attempts to comprehensively introduce the research results of tea polyphenols, oxidative stress, and health. Therefore, this is not only a monograph on the research results of tea polyphenols, but also a popular science book. This book focuses on the properties and structural characteristics of tea polyphenols and systematically expounds the biological functions of tea polyphenols, especially the free radicals scavenging and antioxidant effects of tea polyphenols, the regulation of oxidative stress, the protection of nerve cells, anticancer effects, strengthening body immunity, preventing cardiovascular and cerebrovascular diseases, anti-inflammatory effects, reducing blood lipid, blood glucose and anti-arteriosclerosis, myocardial protection, especially Alzheimer's disease and Parkinson's disease, anti-aging, improving memory, anti-radiation effect of tea polyphenols, etc. In addition, the sources and safety of tea polyphenols and the precautions for drinking tea and using tea polyphenols are also included.

This book can be read and referred by the majority of scientific researchers in free radical, biology, chemistry, and medicine and teachers and students from colleges and universities of relevant majors. It can also be used as a reference for technicians engaged in the research and development of free radicals and antioxidants, tea processing and sales, and tea lovers and drinkers. At the same time, this book is popular science. Therefore, the publication of this book will serve as a good teacher and friend for the health of the people.

Most of the contents of this book come from published research papers by authors. The authors have sorted and summarized these research papers. The book has also collected hundreds of literature articles, and the list of research articles are placed at the end of the relevant chapters as references.

The author would like to express his sincere gratitude to all his colleagues who participated in writing this book! The author would like to thank to Professor Liping Du and her students at Tianjin University of Science and Technology for providing some useful and meaningful materials for the first chapter of this book. Additionally, the author would like to thank to Associate Professor Chunai Zhang for some valuable suggestions for this book. Finally, the author would like to thank to World Scientific Publishing Co. Pte. Ltd. for editing and publishing this book, so that this book can be met with a wide audience.

Contents

Chapter 10

Tea Polyphenols and Obesity

Baolu Zhao

Institute of Biophysics, Chinese Academy of Sciences, Beijing, China

10.1 Introduction

Metabolic syndrome is known as the main health killer in the 21st century. It is a state in which a variety of diseases such as hypertension, abnormal blood glucose, dyslipidemia, and obesity accumulate in the human body, which directly leads to the occurrence of serious cardiovascular diseases and death. Metabolic syndrome mainly refers to high body weight, hypertension, hyperlipidemia, hyperglycemia, high fatty acid, high uric acid, high blood viscosity, high insulin anti-body, etc. If there are two of the eight highs, it can be called metabolic syndrome. The incidence of metabolic syndrome in adults over the age of 20 in the United States is about 24%, including 60% of obese patients [Thethi *et al.*, 2006]. According to the survey of the Chinese Medical Association, the prevalence of metabolic syndrome is 14–16% among the people over 20 years old in Chinese cities. Metabolic syndrome increases with age and reaches the peak among the people aged 50–70, of which female patients are more than men. It is estimated that there will be one patient in every eight in the next seven years will die from metabolic syndrome. The pathogenesis of metabolic syndrome mainly involves the following three aspects: 1) Obesity and adipose tissue metabolic disorder; 2) Insulin resistance; 3) Other related independent risk factors, such as chronic inflammatory state, abnormal vascular function, etc. [Fonseca, 2005; Thethi *et al.*, 2006].

Obesity is a chronic disease caused by many factors, which can be defined as the increase of body fat accumulation. Adipose tissue is not only the storage organ of triglycerides, but also the role of white adipose tissue as the producer of bioactive substances called adipokines. Among adipokines, it was found to have some inflammatory functions, such as interleukin-6 (IL-6); other fat factors have the function of regulating food intake, so they have a direct impact on weight control. It acts on the limbic system by stimulating the absorption of dopamine and produces a sense of satiety. However, these fat factors induce the production of reactive oxygen species (ROS), which cause oxidative stress. Since adipose tissue is an organ that secretes adipokines, which in turn produce ROS, adipose tissue is considered to be an independent factor in the production of systemic oxidative stress. There are many mechanisms for obesity to produce oxidative stress. The first is the oxidation of mitochondria and peroxidase enzymes of fatty acids, which can produce ROS in the oxidation reaction, while the other mechanism is the excessive consumption of oxygen, which can produce free radicals in the mitochondrial respiratory chain combined with oxidative phosphorylation in mitochondria. High fat diets can also produce ROS because they can change oxygen metabolism. With the increase of adipose tissue, the activities of antioxidant enzymes such as superoxide dismutase (SOD), catalase (CAT), and glutathione peroxidase (GPX) decreased significantly. Finally, the production of highly ROS and the reduction of antioxidant capacity led to various abnormalities. Among them, it was found that endothelial dysfunction, which is characterized by the bioavailability of vasodilators (especially nitric oxide), decreases and the endothelium-derived contraction factor increases, which is conducive to atherosclerotic diseases.

Obesity has become a social health problem. It must be taken seriously. Studies have found that drinking tea can reduce body weight and natural antioxidant tea polyphenols can inhibit oxidative stress caused by obesity and reduce a series of health risks caused by obesity. This chapter will discuss the relationship between oxidative stress and obesity, and the inhibition effects of tea polyphenols on oxidative stress caused by obesity and obesity syndrome.

10.2 Obesity syndrome

Obesity refers to a certain degree of obvious overweight and overly thick fat layer, which is a state caused by excessive accumulation of body fat, especially triglycerides. It does not mean simple weight gain, but the excess accumulation of adipose tissue in the body. Excessive fat accumulation in the body due to excessive food intake or changes in body metabolism leads to excessive weight gain and human pathological and physiological changes or latent. If the obesity degree is within ± 10%, it is called normal and moderate. If obesity is more than 10%, it is called overweight. If obesity is more than 20–30%, it is called mild obesity. If obesity is more than 30–50%, which is called moderate obesity. If obesity is more than 50%, it is called severe obesity. If obesity is less than −10%, it is called lean and if obesity is less than −20%, which is called emaciation. Obesity and adipose tissue metabolic disorder is a disease symptom caused by excessive fat accumulation in the body due to the long-term excess of energy intake. At present, it is considered that when a person's weight exceeds more than 20% of the standard weight, or body mass index, referred to as body mass index BML, if BML >25 kg/m^2 in China, it is called obesity. The World Health Organization (WHO) limit value of body mass index is usually used internationally (BML, men >27kg/m^2 and women >25 kg/m^2 as indicators).

10.2.1 *Predisposing factors of obesity syndrome*

Although there are many causes of obesity, the most basic one is the result of the imbalance of energy metabolism in the body, that is, too much intake and too little consumption. Many factors can lead to disorders of energy metabolism in the patients, such as over nutrition, reduced physical activity, endocrine and metabolic disorders, hypothalamic injury, genetic factors, or emotional disorders can lead to obesity. According to the location of fat accumulation, obesity can be divided into central obesity and peripheral obesity. Central obesity is closely related to metabolic syndrome, and the harm is much greater than peripheral obesity [Arner, 2001]. The adipose tissue of the body is mainly divided into visceral fat and peripheral fat; visceral fat is mainly distributed in the abdominal mesangium and around the viscera. Central obesity is mainly caused by the excessive accumulation of visceral fat. Central obesity is mainly characterized by increased waist circumference. Epidemiological investigation also confirmed that central obesity is more related to metabolic syndrome than peripheral obesity.

Genetic factors are identified as "multi factor inheritance". For example, non-insulin-dependent diabetes mellitus and obesity belong to this kind of inheritance. If one of the parents is obese, the child has a 40% chance of obesity. If both parents are obese, the probability of obesity increases to 70–80%. Psychological factors include being too orderly to relieve emotional troubles and emotional instability, causing many people also use "eating" as a way to vent. This is the cause of excessive diet and obesity.

Exercise helps to consume fat in daily life, with the development of transportation, the mechanization of work, and the reduction of housework, the human body has fewer opportunities to consume calories. Since the energy intake has not been reduced, obesity is formed. Obesity leads to slower and lazier daily activities and reduces the consumption of calories again, leading to a vicious circle and contributing to the occurrence of obesity. The quality of sleep will affect your weight. Sleeping less than 6.5 hours or more than 8.5 hours can lead to weight gain.

10.2.2 *Hazards of obesity syndrome*

Obesity not only affects physical beauty, but also causes human pathological changes, produces a variety of diseases harmful to human health, and brings trouble to healthy life. Obesity can increase the risk of cardiovascular disease and is closely related to many other diseases such as hypertension, high cholesterol, sleep apnea, type II diabetes, and so on [Caballero, 2019]. Recent studies have reported that obesity has weakened lung function and caused the decline of immune function. Therefore, once obese people are infected with COVID-19, the hospitalization time of the infected patients was longer than that of the normal weight group. Moreover, obesity exacerbates the cytokine storm and becomes an important risk factor for the severity of pneumonia [Yu *et al.*, 2020]. Obesity is a chronic metabolic disease. As the pathogenesis of obesity is complex, it brings great challenges to the prevention of obesity and how to lose weight, which may also be the main reason why monotherapy cannot achieve good weight loss effect at present. Therefore, it is an urgent task for basic and clinical research to develop weight loss and health products and drugs with multiple components. To sum up, obesity can increase the risk of cardiovascular disease, affect the function of digestive system, affect the function of endocrine system, and increase the risk of cancer. In addition, there are joint soft tissue injury, decreased reproductive ability, psychological disorders, heart disease, diabetes, atherosclerosis, fatty liver, gallstones, edema, gout, etc.

To reduce the impact of obesity on health, the obesity patients need to lose weight. There are many ways to lose weight at present. However, the most important thing is diet control, limit the total energy intake to 1000–1500 kcal/ day, and reduce fat intake, which should be 25–35% of the total energy. At the same time, more exercise is needed to increase the utilization of fatty acids and glucose by muscle tissue, so as to achieve the balance between input and output. Those that cannot be controlled by themselves need to bear the pain of medicine and surgery. Research has found that drinking tea can reduce weight without any pain, and it is also a kind of enjoyment. This chapter discusses obesity and oxidative stress, and how tea polyphenols help to reduce weight and prevent oxidative stress damage.

10.3 Obesity and oxidative stress

More and more evidence show that oxidative stress is a key factor linking obesity and its related complications. Systemic and tissue-specific chronic inflammation and oxidative stress are common characteristics of obesity. Obesity itself can induce systemic oxidative stress through various biochemical mechanisms, such as superoxide production by NADPH oxidase, oxidative phosphorylation, automatic oxidation of glyceraldehyde, activation of protein kinase C, and polyols and hexosamine pathway. Other factors of oxidative stress in obese patients include hyperleptinemia, low antioxidant defense, chronic inflammation, and postprandial

ROS production. Oxidative stress and inflammation are central mediators of obesity. With the induction of inflammation, pro-inflammatory molecules may interfere with insulin signal transduction, resulting in insulin resistance in obesity patients. In addition, other molecules promote atherosclerotic inflammation of obesity patients.

10.3.1 *Obesity, inflammation, and oxidative stress*

A large number of studies have shown that obesity leads to atherosclerosis, cardiovascular and cerebrovascular diseases, and inflammation. The oxidative stress which accompanies feeding, particularly when there is excessive ingestion of fat and/or other macronutrients without concomitant ingestion of antioxidant-rich foods/beverages, may contribute to inflammation attributed to obesity. Recent studies indicate that the accumulation of biologically-active lipids in adipose tissue may regulate the synthesis/secretion of adipokines and pro-inflammatory cytokines. A research also showed that some foods are non-inflammatory or anti-inflammatory, so they will not interfere with the insulin signal. Obesity is associated with the adipocyte dysfunction, macrophage infiltration, and low-grade inflammation, which probably contributes to the induction of insulin resistance. Adipose tissue synthesizes and secretes numerous bioactive molecules, namely adipokines and cytokines, which affect the metabolism of both lipids and glucose. Once the body fat accumulates too much, it will secrete a large number of inflammation cytokines that inhibit the function of fat and muscle tissue. These cytokines mainly include free fatty acids (FFA) and inflammatory factors (such as TNF-α), ROS etc. [Kojta *et al.*, 2020; Grundy, 2004]. In addition, high level of anxiety was independently associated with low NO levels in the overweight/obesity [Chung *et al.*, 2020]. Cytokines can act directly on adipocytes in the form of autocrine and paracrine to produce insulin resistance and disorder glucose and lipid metabolism; they can also enter the blood to act on muscle cells to produce insulin resistance and reduce their energy storage and consumption capacity, or act on organs such as pancreas and damage their functions [Holguin *et al.*, 2019]. Adipokines entail the functions of regulating food intake, therefore exerting a direct effect on weight control. Obese patients secrete more free fatty acids (FFA) than normal people.

These adipokines induce the production of ROS, generating a process known as oxidative stress (OS). Since adipose tissue is the organ that secretes adipokines and these in turn generate ROS, adipose tissue is considered an independent factor for the generation of systemic oxidative stress [Fernández-Sánchez *et al.*, 2011]. Oxidative stress is the product of triglyceride decomposition by adipocytes, so the body fat content is directly proportional to the secretion of FFA. Once the fat is stored too much, the visceral adipose tissue secretes more FFA than the peripheral adipose tissue, which is an important reason why the harm of central obesity is much greater than that of peripheral obesity [Flashner *et al.*, 2020]. Although FFA is the basic energy supply form of fat cells to other tissues during starvation, it is very dangerous for fat cells in obese patients to secrete excessive FFA. FFA will lead to the inhibition of carbohydrate oxidation in muscle tissue [Fernández-Sánchez *et al.*, 2011]. FFA can inhibit the phosphorylation of insulin receptor, thereby inhibiting the insulin signaling pathway, leading to insulin resistance and inducing type 2 diabetes [Becerril *et al.*, 2019]. The secretion of inflammatory factors such as IL-6 and TNF-α also increased significantly in obese patients. They can induce muscle insulin resistance, damage vascular endothelial cells and lead to atherosclerosis [Ridker & Morrow, 2003]. More importantly, inflammatory factors can act directly on adipocytes in the form of autocrine and paracrine, inhibiting insulin signaling pathway and PPARγ [Miles *et al.*, 1997; Yan *et al.*, 2013]. Inflammatory factors can also increase the level of ROS in adipocytes, leading to oxidative stress [Furukawa *et al.*, 2004]. Adipocytes that accumulate excess fat increase the activity of NADPH oxidase and synthesize a large amount of ROS [Sonta *et al.*, 2004]. Oxidative stress of adipocytes not only leads to increased P-JNK levels, inhibits AKT phosphorylation, the insulin signaling pathway, and type 2 diabetes mellitus, but also secretes ROS into the blood by adipocytes, leading to systemic oxidative stress and damage to islets β-cells [Houstis *et al.*, 2006]. The functional pleiotropism of adipose tissue relies on its ability to synthesize and release a large number of hormones, cytokines, extracellular matrix proteins, and growth and vasoactive factors, collectively termed adipokines that influence a variety of physiological and pathophysiological processes. In the obese state, excessive visceral fat accumulation causes adipose tissue dysfunctionality that strongly contributes to the onset of obesity-related comorbidities. The mechanisms underlying adipose tissue dysfunction include adipocyte hypertrophy and hyperplasia, increased inflammation, impaired extracellular matrix remodeling, and fibrosis together with an altered secretion of adipokines [Unamuno *et al.*, 2018]. TNFα is mainly produced by activated macrophages — NK cells and T lymphocytes. Glycerol release into the medium was significantly increased (50%, $p = 0.027$) by exposure of the adipose-tissue fragments to TNF-alpha (4 nmol/L) for 24 hours [Porter *et al.*, 2002]. Adipose tissue secretes a large amount of the secretion of inflammatory factors, such as TNF-α and IL-6. The inflammatory factors is also increased significantly in obese patients. It can induce muscle insulin resistance, damage vascular endothelial cells, and lead to atherosclerosis. These cytokines mainly include free fatty acids (FFA), inflammatory factors (such as TNF-α, Resistin, etc.), and ROS [Rondinone, 2006]. Cytokines can act directly on adipocytes in the form of autocrine and paracrine, causing them to produce insulin resistance and disorder glucose, inflammation, and oxidative stress [Paoletti *et al.*, 2006]. More importantly, inflammatory factors can act directly on adipocytes in the form of autocrine and paracrine, inhibit insulin signaling pathway and expression of transcription factors of peroxisome proliferator activated receptor (PPAγ), regulate adipocyte function, reduce the plasma membrane localization of glucose transporter glucose transporter (GLUT-4), thereby leading to insulin resistance. Inflammatory factors can also increase the level of ROS in adipocytes, leading to oxidative stress, reduce the secretion of factors beneficial to glucose metabolism, such as adiponectin, and increase the secretion of resistin [Furukawa *et al.*, 2004].

The growing recognition that obesity and atrial fibrillation are closely intertwined disorders has spurred major interest in uncovering their mechanistic links. There is growing evidence linking oxidative stress and inflammation to adverse atrial structural and electrical remodeling that leads to the onset and maintenance of abdominal fat (AF) in the diabetic heart [Karam, 2017]. Obesity affects children and adults differentially. In those with early onset asthma, it associated with increased eosinophilic inflammation, whereas in late onset, it correlates with lower nitric oxide (NO) and predominantly non-T2 inflammation. There are probably multiple pathways by which obesity impacts asthma, airway and systemic oxidative stress has been proposed as a mechanism that could potentially explain the obesity mediated increased comorbidity and poor response to treatment. More likely than not, oxidative stress is an epiphenomenon of a very diverse set of processes driven by complex changes in airway and systemic metabolism [Grasemann & Holguin, 2021]. Periodontal disease is a chronic inflammatory disease of the jaws and is more prevalent in obesity. Local and systemic oxidative stress may be an early link between periodontal disease and obesity. A study was done to detect whether increased periodontal disease susceptibility in obese individuals is associated with local and systemic oxidative stress. Oxidant status analyses revealed lower gingival crevicular fluid and serum total antioxidant capacity, and higher gingival crevicular fluid and serum oxidative stress index values in the obese women ($p < 0.05$ for all). Gingival crevicular fluid total antioxidant capacity values showed a negative correlation with body mass index, whereas gingival crevicular fluid oxidative stress index was positively correlated with fasting insulin and low-density lipoprotein-cholesterol levels ($p < 0.05$). Clinical periodontal indices showed significant correlations with body mass index, insulin, and lipid levels, and also oxidant status markers. The results suggest that young obese, otherwise healthy women, show findings of early periodontal disease (gingival inflammation) compared with age-matched healthy lean women, and that local/periodontal oxidative stress generated by obesity seems to be associated with periodontal disease. Accordingly, it analyzed periodontal status and systemic (serum) and local (gingival crevicular fluid, gingival crevicular fluid) oxidative status markers in young obese women in comparison with age-matched lean women [Dursun *et al.*, 2016].

It can be seen from the above results that it may not only cause inflammation caused by obesity itself, but also cause inflammation and oxidative stress in obesity related diseases.

10.3.2 *High cholesterol obesity and oxidative stress*

A large number of studies have shown that obese people with high cholesterol, especially oxidized low-density lipoprotein (LDL, i.e. bad cholesterol) can lead to serious diseases such as atherosclerosis and cardiovascular and cerebrovascular diseases. High low-density lipoprotein (LDL)-cholesterol is a recognized pathogenic factor for atherosclerosis and causes heart diseases by increasing the formation of atherosclerotic plaques. Cholesterol metabolism disorder is obvious in obesity, which may be directly related to oxidative stress caused by metabolic inflammation. A number of clinical and experimental studies concur to identify endothelial dysfunction as a primary step in the development of atherosclerosis, as well as a risk factor for subsequent clinical events. Oxidant stress resulting from chronic elevation of plasma LDL-cholesterol is a major contributor to both endothelial dysfunction and its complications, for example, through alterations of endothelial nitric oxide signaling [Hermida & Balligand, 2014]. A study found that hyper-cholesterolemic vessels (1 mol cholesterol-fed rabbits) produced threefold more $\bullet O_2^-$ than compared to non-hyper-cholesterolemic vessels ($p < 0.001$). Another study found that vessels with endothelium from cholesterol-fed rabbits produced 4.5-fold more $\bullet O_2^-$ than vessels from normal animals. Endothelial removal increased $\bullet O_2^-$ production in non-hyper-cholesterolemic vessels ($p < 0.05$), while decreasing it in hyper-cholesterolemic vessels ($p < 0.05$). Oxypurinol, a noncompetitive inhibitor of xanthine oxidase, normalized $\bullet O_2^-$ production in hyper-cholesterolemic vessels but had no effect in non-hyper-cholesterolemic vessels. Thus, the endothelium is a source of $\bullet O_2^-$ in hypercholesterolemia probably via xanthine oxidase activation. Increased endothelial $\bullet O_2^-$ production in hyper-cholesterolemic vessels may inactivate endothelium-derived nitric oxide and provide a source for other oxygen radicals, contributing to the early atherosclerotic process [Ohara *et al.*, 1993, 1995]. Xanthine oxidase activity was elevated more than two-fold in plasma of hyper-cholesterolemic rabbits. Incubation of vascular rings from rabbits on a normal diet with purified xanthine oxidase (10 milliunits/ml) was also impaired.

The above results indicate the continuous production of ROS and the simultaneous production of NO and $\bullet O_2^-$, suggesting that peroxynitrite (ONOO$^-$), which oxidized cholesterol to produce oxidized LDL, may irreversibly damage vascular function. It plays a central role in vascular inflammation and atherosclerosis. The increase of superoxide ($\bullet O_2^-$) production significantly impairs the nitric oxide-dependent vasodilation function of cholesterol feeding.

10.3.3 *Obesity, fat accumulation, and oxidative stress*

A large number of studies have shown that fat accumulation in obese people can lead to atherosclerosis and cardiovascular and cerebrovascular diseases. Obese patients secrete more free fatty acids than normal people, which is the product of triglyceride decomposition by adipocytes, so the body fat content is directly proportional to the secretion of free fatty acids. Once too much fat is stored, visceral adipose tissue will secrete more free fatty acids than peripheral adipose tissue. Adipocyte hypertrophy, through physical reasons, facilitates cell rupture that will evoke an inflammatory reaction. Inability of adipose tissue development to engulf incoming fat leads to deposition in other organs, mainly in the liver, with consequences on insulin resistance and oxidative stress. Central obesity leads to metabolic syndrome for two reasons: 1) The visceral adipose tissue itself has too much fat accumulation, which leads to the decline of adipocyte storage capacity and can-not store more excess lipids and sugars, resulting in the increase of blood lipid and fat content of other organs, which is harmful to health; 2) The increase of fat attached to mesentery and viscera will also affect the function of visceral organs. Due to the existence of adipose tissue as a secretory organ, especially visceral adipose tissue, once it accumulates too much fat, it will secrete a large number of cytokines that inhibit the function of fat and muscle tissue. Although FFA is the basic energy supply form of fat cells to other tissues under starvation, it is very dangerous for fat cells in obese patients to secrete excessive free fatty acids. Free fatty acids can lead to the inhibition of carbohydrate oxidation in muscle tissue. Free fatty acids can inhibit the production of insulin receptors phosphorylation, thereby inhibiting the insulin signaling pathway, leading to insulin resistance and inducing type 2 diabetes. Free fatty acids can also act directly on islets β-cells, which damage their function, also cause type 2 diabetes [Paoletti *et al.*, 2006]. Free fatty acids can increase the content of triglycerides in liver, lead to fatty liver, reduce high-density lipoprotein synthesis, and increase low-density lipoprotein synthesis. In addition, free fatty acids can also lead to hypertension. We use dexamethasone (DEX) (20 nm) or TNF-α long-time treatment (six days)-induced oxidative stress in mature and differentiated adipocytes. It is found that ROS is produced by activating NADPH oxidase, which leads to oxidative stress and increases JNK phosphorylation to activate JNK pathway. The ROS results measured by ESR show that long-term high-fat diet can significantly increase the ROS level of rat adipose tissue (52.3%) [Yan *et al.*, 2013].

Adipocytes that accumulate excess fat increase the activity of NADPH oxidase and synthesize a large amount of ROS. The oxidative stress of adipocytes not only leads to increased P-JNK levels, but also inhibits AKT phosphorylation, inhibits insulin signaling pathway, and induces type 2 diabetes. Moreover, adipocytes can secrete ROS into the blood, leading to systemic oxidative stress and damage to islets β-Cells cause type 1 diabetes. ROS in blood can also damage vascular endothelial cells and lead to atherosclerosis. In addition, resistin produced by adipocytes can also directly affect adipocytes and inhibit insulin activity, leading to type 2 diabetes [Gonzalez *et al.*, 2006].

Therefore, obesity caused by excessive accumulation of lipids, on the one hand, will lead to hyperglycemia and hyperlipidemia due to insufficient sugar and lipid storage capacity of adipocytes, leaving excess sugar and lipids in other parts of the body, on the other hand, adipocytes will secrete a large number of harmful factors, ROS, leading to hyperglycemia, hyperlipidemia, fatty liver, hypertension, and atherosclerosis, thereby resulting oxidative stress injury.

10.3.4 *Obesity, atherosclerosis, and oxidative stress*

A large number of studies have shown that the prevalence of atherosclerosis in obese people is very high. A primary event in atherogenesis is the infiltration of activated inflammatory cells into the arterial wall. There, they secrete reactive oxygen species and oxidize lipoproteins, inducing foam cell formation and endothelial cell apoptosis, which in turn lead to plaque growth, erosion, and rupture. In addition, there is evidence that this vicious circle between oxidative stress and inflammation occurs not only in the diseased arterial wall but also in adipose tissues in obesity. In this condition, oxidative stress and inflammation impair adipocyte maturation, resulting in defective insulin action and adipocytokine signaling. Studies have provided compelling evidence demonstrating the roles of vascular oxidative stress and NO in atherosclerosis. Atherosclerosis is now considered a chronic inflammatory disease. Oxidative stress induced by generation of excess ROS has emerged as a critical, final common mechanism in atherosclerosis. ROS and oxidized low-density lipoprotein cholesterol play a key role in the development of atherosclerosis. The atherosclerotic process is accelerated by a myriad of factors, such as the release of inflammatory chemokines and cytokines, the generation of ROS, growth factors, and the proliferation of vascular smooth muscle cells. Inflammation and immunity are key factors for the development and complications of atherosclerosis [Marchio *et al.*, 2019]. Key molecular events in atherosclerosis such as oxidative modification of lipoproteins and phospholipids, endothelial cell activation, and macrophage infiltration/activation, are facilitated by vascular oxidative stress and inhibited by endothelial NO. All established cardiovascular risk factors such as hypercholesterolemia, hypertension, diabetes mellitus, and smoking, enhance ROS generation and decrease endothelial NO production. ROS at moderate concentrations have important signaling roles under physiological conditions. Excessive or sustained ROS production, however, when exceeding the available antioxidant defense systems, leads to oxidative stress [Förstermann *et al.*, 2017]. The ROS and RNS are the most important endogenous sources produced by non-enzymatic and enzymatic (myeloperoxidase (MPO), nicotinamide adenine dinucleotide phosphate (NADH) oxidase, and lipoxygenase (LO)) reactions that may be balanced with anti-oxidative compounds (glutathione (GSH), polyphenols and vitamins), and enzymes (glutathione peroxidase (Gpx), peroxiredoxins (Prdx), superoxide dismutase (SOD) and paraoxonase (PON)). However, the oxidative and antioxidative imbalance causes the involvement of cellular proliferation and migration signaling pathways and macrophage polarization leads to the formation of atherosclerosis plaques [Khosravi *et al.*, 2019]. Adipose tissue represents an important source of ROS; ROS may contribute to the development of obesity-associated insulin resistance and type 2 diabetes. Moreover, the levels of oxidative stress present in several other types of cells or tissues, including those in the brain, arterial walls, and tumors, have been implicated in the pathogenesis associated with hypertension, atherosclerosis, and cancer. The increased levels of systemic oxidative stress that occur in obesity may contribute to the obesity-associated development of these diseases [Matsuda & Shimomura, 2013].

Atherosclerosis oxidative stress is the main cause of cardiovascular disease. Atherosclerotic oxidative stress is the main cause of cardiovascular disease. Atherosclerosis is formed by the deposition of blood lipids under endothelial cells. The formation of atherosclerosis is related to the increase of blood lipid. When high cholesterol and high saturated fatty acids are taken for a long time, the blood lipid level will increase, especially the total cholesterol and low-density lipoprotein cholesterol, which will damage the vascular endothelium and penetrate the endothelium through the sub-endothelial space. The lipid components invading the endothelium will attract macrophages, monocytes, etc., and gather around. After the aggregation, macrophages will phagocytose lipids and form foam cells. Macrophages die, and more inflammatory factors will be released after the death of macrophages. Inflammatory factors stimulate smooth cell proliferation and migration, and finally form lipids under the endothelium, acting as the core in the middle, and surrounding smooth muscle cells surround lipid plaques. Chronic inflammation, ROS, RNS, oxidative stress, low-density lipoprotein (LDL) particles, and fiber components gradually accumulate in large and medium-sized arterial lesions. These fibrous fatty lesions in the arterial wall gradually become unstable and thrombosis, leading to heart attack, stroke, or other severe cardiac ischemia syndrome.

10.3.5 *Obesity, mitochondria, and oxidative stress*

Mitochondrial dysfunction contributes to the oxidative stress and systemic inflammation seen in metabolic syndrome. Brown adipocytes have a large number of mitochondria, which provide a place for lipid oxidation and heat production. In brown adipocytes, peroxisome proliferators activate receptors, PPARδ and PPARα, which are important regulators of oxidation and heat production. PPARδ regulates the downstream expression of key enzymes such as acetyl CoA oxidase (AOX) and lipoacyl carnitine transferase-1 (CPT-1). It can regulate the expression of uncoupling protein-1 (UCP-1), so that the proton dynamic potential generated by fat oxidation on both sides of mitochondrial inner membrane will generate heat energy through uncoupling protein [Barish *et al.*, 2006]. PGC-1α can also increase the number of mitochondria and make the oxidative function of brown adipocytes stronger. PPARδ and PGC-1α expression is very low in white adipocytes under normal conditions. These are the mitochondrial and peroxisomal oxidation of fatty acids, which can produce ROS in oxidation reactions, while another mechanism is the overconsumption of oxygen, which generates free radicals in the mitochondrial respiratory chain that is found coupled with oxidative phosphorylation in mitochondria. Lipid-rich diets are also capable of generating ROS because they can alter oxygen metabolism. Upon the increase of adipose tissue, the activity of antioxidant enzymes such as superoxide dismutase (SOD), catalase (CAT), and glutathione peroxidase (GPx), was found to be significantly diminished. High ROS production and the decrease in antioxidant capacity leads to various abnormalities, such as endothelial dysfunction, which is characterized by a reduction in the bioavailability of vasodilators, particularly nitric oxide (NO), and an increase in endothelium-derived contractile factors, favoring atherosclerotic disease [Fernández-Sánchez *et al.*,2017]. Adipocytes undergo intense energetic stress in obesity resulting in loss of mitochondrial mass and function. We have found that adipocytes respond to mitochondrial stress by rapidly and robustly releasing small extracellular vesicles (sEVs). These sEVs contain respiration-competent but oxidatively-damaged mitochondrial particles, which enter circulation and are taken up by cardiomyocytes, where they trigger a burst of ROS [Crewe *et al.*, 2021].

Autophagy is a lysosomal degradation pathway recycling intracellular long-lived proteins and damaged organelles, thereby maintaining cellular homeostasis. Autophagy is a highly-regulated process that has an important role in the control of a wide range of cellular functions, such as organelle recycling, nutrient availability, and tissue differentiation. A recent study has shown that an increased autophagic activity in the adipose tissue of obese subjects and a role for autophagy in obesity-associated insulin resistance was proposed. In addition to inflammatory processes, autophagy has been implicated in the regulation of adipose tissue and beta cell functions. In obesity and type 2 diabetes, autophagic activity is modulated in a tissue-dependent manner. A study explore the connections of autophagy with mitochondria in obesity. Mitophagy eliminates this vicious cycle of oxidative stress and mitochondrial damage, and thus counteracts pathogenic processes. On one hand, autophagy mediates exercise-induced increases in muscle glucose uptake and protects β cells against endoplasmic reticulum stress in diabetogenic conditions. On the other hand, adipose tissue autophagy promotes adipocyte differentiation, possibly through its role in mitochondrial clearance. Being involved in many aspects, autophagy appears to be an attractive target for therapeutic interventions against obesity and diabetes [Sarparanta *et al.*, 2017].

Inflammation, oxidized low density lipoprotein (oxLDL), atherosclerosis and fat accumulation, autophagy, and mitochondrial associated with obesity, all of which can lead to oxidative stress, damage to body tissue cells, and lead to cardiovascular and cerebrovascular diseases, etc.

10.4   Effects of tea polyphenols on obesity

Epidemiological investigation shows that drinking green tea can significantly reduce obesity and metabolic syndrome caused by abnormal lipid metabolism. As a worldwide beverage, green tea has many beneficial effects [Mitscher *et al.*, 1997], especially green tea polyphenols (GTC), which can significantly reduce body weight and body fat content. It is mainly that EGCG can reduce the incidence rate of obesity and diabetes [Lin *et al.*, 1996] and prevent neurodegenerative diseases induced by obesity oxidative stress [Haqqi *et al.*, 1999]. Green tea polyphenols catechin are not only free radical scavengers, but also can activate a series of cellular mechanisms related to their neuroprotective activities. These include pharmacological activities such as iron chelation, scavenging free radicals, activating survival genes and cellular signaling pathways, regulating mitochondrial function, and ubiquitin proteasome system [Mendel & Youdim, 2004]. Tea polyphenols can significantly reduce the content of triglycerides and cholesterol in the blood, increase the level of high-density lipoprotein, and reduce the release capacity of liver lipoprotein. The mechanism of tea polyphenols improving obesity and metabolic syndrome is multifaceted.

Recent studies have shown the role of dietary polyphenols in the prevention of obesity and obesity-related chronic diseases. The prevalence and incidence of overweight and obesity worldwide continues to increase, as well as diseases related to these conditions. This is attributed to an increase in energy intake and a decrease in energy expenditure. Consumption of green tea has been linked to a reduction in body fat and body weight. There are many reported clinical results and animal studies that strongly suggest that commonly consumed polyphenols have a pronounced effect on obesity as shown by lower body weight, fat mass, and triglycerides through enhancing energy expenditure and fat utilization and modulating glucose hemostasis. Limited human studies have been conducted in this area and are inconsistent about the anti-obesity impact of dietary polyphenols probably due to the various study designs and lengths, variation among subjects (age, gender, ethnicity), chemical forms of the dietary polyphenols used, and confounding factors such as other weight-reducing agents about the consumption of green tea is related to the reduction of body fat and weight. However, the research on green tea has been diversified. A number of studies have evaluated the content of green tea and its epigallocatechin gallate (EGCG) and evaluated its impact on human body fat and weight. Although the research results vary greatly, the conclusions are consistent.

10.4.1 *Effect of drinking tea and tea polyphenols on body weight*

Weight is an important indicator of obesity. If the body mass index (BMI) exceeds the standard value, many diseases may occur. Epidemiological investigation, clinical results, and animal experiments have shown that tea and tea polyphenols can help obesity to reduce weight [Zhao, 2020].

1. Epidemiological investigation results
Consumption of green tea has been linked to a reduction in body fat and body weight. However, research on green tea has been very diverse. A paper assesses the investigations that have been made with green tea and its EGCG content, evaluating its effect on body fat and body weight

in humans. A search was made in the PubMed and Web of the Science databases that gave a first total result of 424 potential articles; 409 were excluded and 15 articles were used for this systematic paper. It has found that daily consumption of green tea with doses of EGCG between 100–460 mg/day has shown greater effectiveness on body fat and body weight reduction in intervention periods of 12 weeks or more [Vazquez Cisneros *et al.*, 2017]. A systematic literature search of MEDLINE, EMBASE, CENTRAL, and the Natural Medicines Comprehensive Database was conducted through April 2009. This includes evaluating the impact of GTC on BMI and weight. A randomized controlled trial (RCT) was conducted to study the effects of green tea polyphenols (GTP) on anthropometry, and contradictory results were obtained. Taking GTC and caffeine can significantly reduce BMI and weight, however, the clinical significance of these reductions is not significant. Current data do not suggest that GTC alone positively alter anthropometric measurements [Phung *et al.*, 2010].

There is another meta-analysis about the effects of green tea on weight loss and weight maintenance. Different outcomes of the effect of green tea on weight loss (WL) and weight maintenance (WM) have been reported in studies with subjects differing in ethnicity and habitual caffeine intake. The purpose is to elucidate by meta-analysis whether green tea indeed has a function in body weight regulation. English-language studies about weight loss and WM after green tea supplementation were identified through PubMed and based on the references from retrieved articles. Out of the 49 studies initially identified, a total of 11 articles fitted the inclusion criteria and provided useful information for the meta-analysis. Effect sizes (mean weight change in treatment versus control group) were computed and aggregated based on a random-effects model. The influence of several moderators on the effect sizes was examined. Results showed catechins significantly decreased body weight and significantly maintained body weight after a period of weight loss ($p < 0.001$). So, they got conclusions: catechins or EGCG-caffeine mixture have a small positive effect on weight loss and weight maintenance [Hurse *et al.*, 2009]. A study conducted a randomized, double-blind trial registered under Clinical Trials. A total of 115 women with central obesity were screened at our clinic; 102 of them with a body mass index (BMI) ≥ 27 kg/m^2 and a waist circumference (WC) ≥ 80 cm were eligible for the study. These women were randomly assigned to either a high-dose green tea group or placebo group. The total treatment time was 12 weeks. Significant weight loss, from 76.8 ± 11.3 kg to 75.7 ± 11.5 kg ($p = 0.025$), as well as decreases in BMI ($p = 0.018$) and waist circumference ($p = 0.023$) were observed in the treatment group after 12 weeks of high-dose EGCG treatment. This study also demonstrated a consistent trend of decreased total cholesterol, reaching 5.33%, and decreased LDL plasma levels. There was good tolerance of the treatment among subjects without any side effects or adverse events. Significantly lower ghrelin levels and elevated adiponectin levels were detected in the study group than in the placebo group [Chen *et al.*, 2016]. To test the efficacy of theaflavin administration on body fat and muscle, a randomized, double-blind, placebo-controlled was studied and the effect of theaflavins administration on the body composition using of healthy subjects was investigated. In this study, 30 male and female Japanese were enrolled, and participants were randomly allocated to receive placebo, theaflavin (50 mg/day or 100 mg/day), or catechin (400 mg/ml) for 10 weeks. Theaflavin administration significantly improved body fat percentage, subcutaneous fat percentage, and skeletal muscle percentage when compared to with the placebo. In contrast, there was no significant difference in all measured outcomes between the catechin and the placebo groups [Aizawa *et al.*, 2017].

The above epidemiological investigation and clinical experiments show that drinking tea or tea polyphenols can really reduce weight. In particular, the habit of theaflavins is very obvious, but the effect of catechins is not consistent. In the future, strict epidemiological investigation and clinical trials should be carried out to verify the true effect of tea polyphenols on weight loss.

2. Results of investigation in animals

A large number of animal experiments have proved that tea polyphenols can reduce weight. In a paper, the canines were fed to dogs either with low (0.48% g/kg), medium (0.96% g/kg), or high (1.92% g/kg), doses of green tea polyphenols (GTP) for 18 weeks. The results showed that green tea polyphenols decrease weight gain, ameliorate alteration of gut microbiota, and mitigate intestinal inflammation in canines with high-fat diet-induced obesity [Li *et al.*, 2020; Cheng *et al.*, 2020]. In another experiment with SD female rats, the rats fed with high-fat diet group increased body weight as compared to the control group. Supplementation of GTP in the drinking water in the rats fed with high-fat diet+GTP group reduced body weight as compared to the high-fat diet group [Lu *et al.*, 2012]. Black tea, commonly referred to as "fermented tea", has shown a positive effect on reducing body weight in animal models. Black tea polyphenols are the major components in black tea which reduce body weight. Black tea polyphenols are more effective than green tea polyphenols. Black tea polyphenols exert a positive effect on inhibiting obesity [Pan *et al.*, 2016]. The main bioactive components of Pu-erh tea, such as theabrownin, polysaccharides, polyphenols, and statins, may downregulate the biosynthesis of fat and upregulate the oxidation of fat to cut weight and reduce the content of lipids in blood [Zou *et al.*, 2012]. Theafuscin increases the level of ileal conjugated bile acids (BAs), thereby inhibiting the intestinal FXR-FGF15 signal pathway, leading to increased liver production and fecal excretion of BAs, decreased liver cholesterol, and reduced fat production [Huang *et al.*, 2019].

Another experiment also shows that the daily dose of tea polyphenols has no significant effect on appetite in the short term, but at the same time, the body weight has decreased significantly [Yan *et al.*, 2013]. Treatment with GTPs, black tea (BT) and oolong tea (OT), and an LF/HS diet led to significantly lower body weight, total visceral fat volume by MRI, and liver lipid weight compared with mice in the high-fat/high-sucrose (HF/HS) control group. Only GTPs reduced food intake significantly by ~10%. GTP, BTP, and low-fat/high-sucrose (LF/HS)-diet treatments significantly reduced serum monocyte chemotactic protein-1 (MCP-1) compared with HF/HS controls [Heber *et al.*, 2014].

Our experiments have proved that tea polyphenols can reduce weight. The rats fed with high HF had significant gain in weight compared to the rats fed with chow diet CHOW and GTCs significantly decreased the body weight (Figs. 10-1a, 10-1b). Feeding GTCs to the rats caused significant body weight reduction (about 9.4% and 6.3% compared to the corresponding control groups) in both HF and CHOW groups within 30 days (Fig 10-1c).

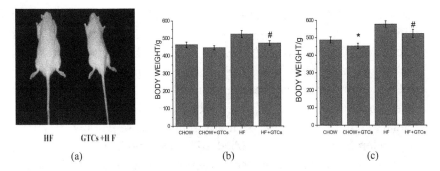

Figure 10-1. Effect of GTCs on SD rat body weight. (a) Photographs of representative rats fed with or without GTCs for 45 days in the high fat diet group. Photographs of representative rats fed with or without GTCs for (b) 30 days and (c) 45 days in the high fat diet group.

The above epidemiological investigation, clinical results, and animal experiments show that tea and tea polyphenols can indeed help obesity to reduce body weight.

10.4.2 *Effect of drinking tea and tea polyphenols on visceral fat*

Epidemiological investigation, clinical results, and animal experiments have shown that tea and tea polyphenols can help obesity to reduce visceral fat.

1. Epidemiological investigation results

Six human trials pooled the efficacy of tea polyphenols-rich beverages to reduce abdominal adiposity and metabolic syndrome risks in obese and overweight subjects. This post-hoc pooled analysis assessed the effectiveness of GTP to reduce the risk of metabolic syndrome (MetS) associated with abdominal fat reduction, because previous findings are unclear. Data were pooled from six human trials (n = 921, 505 men) comparing the effects of GTC-containing beverages (540–588 mg GTC/beverage) and a placebo beverage. Outcome measures were abdominal fat (total fat area (TFA), visceral fat area (VFA), subcutaneous fat area (SFA)) and metabolic syndrome risk. Consumption of GTC-containing beverages for 12 weeks significantly reduced TFA (-17.7 cm^2, 95%CI: -20.9 to -14.4), VFA ($-7.5cm^2$, 95%CI: -9.3 to -5.7), SFA (-10.2 cm^2, 95%CI: -12.5 to -7.8), body weight, body mass index, and waist circumference, and improved blood pressure. Continual consumption of GTP-containing beverages reduced abdominal fat and improved metabolic syndrome, suggesting its potential to prevent diabetes and cardiovascular disease [Hibi *et al.*, 2018]. In a randomized placebo-controlled trial, 182 moderately overweight Chinese subjects, consumed either two servings of a control drink (30 mg tea polyphenols /day) for 90 days. Results showed a decrease in estimated intra-abdominal fat (IAF) area of 5.6 cm^2 in the green tea 3 group. In addition, we found decreases of 1.9 cm in waist circumference and 1.2 kg body weight in the GT3 group versus control group ($p < 0.05$). We also observed reductions in total body fat (green tea 2 group, 0.7 kg, $p < 0.05$) and body fat % (green tea 1 group, 0.6%, $p < 0.05$) [Wang *et al.*, 2010]. In a randomized, placebo-controlled, double-blind, crossover study, in the first test period participants were asked to drink either a beverage containing 55 mg black tea polyphenols (BTP) or a control beverage without BTP three times a day for 10 days. After an 11-day interval, for the second test period, they then drank the alternate test beverage three times a day for 10 days. It was found that postprandial elevation of triacylglycerol was suppressed by the intake of black tea polyphenol (BTP) [Ashigai *et al.*, 2016].

2. Results of investigation in animals

Data from laboratory studies have shown that green tea has important roles in fat metabolism by reducing food intake, interrupting lipid emulsification and absorption, suppressing adipogenesis and lipid synthesis, and increasing energy expenditure via thermogenesis, fat oxidation, and fecal lipid excretion [Huang *et al.*, 2014]. GTP treatment attenuated weight gain ($p < 0.05$) and visceral fat accumulation (27.6%, $p < 0.05$), and significantly reduced fasting serum glucose ($p < 0.05$) and insulin ($p < 0.01$) levels [Cheng *et al.*, 2020]. Oxidized tea polyphenols alleviated the accumulation of lipids in the rat liver tissue and changed the expression levels of the regulators of lipid metabolism, i.e., peroxisome proliferation-activated receptors (PPARs), compared with the rats fed with a high-fat diet alone [Wang *et al.*, 2017].

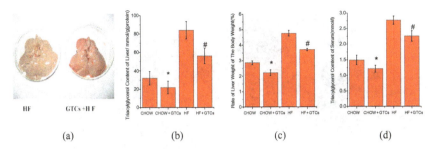

(a)   (b)   (c)   (d)

Figure 10-2. Effect of GTCs on SD rat body index. (a) Photographs of the representative livers from rats fed with or without GTCs for 45 days. (b) GTCs reduced liver triglyceride content in both HF ($p < 0.05$) and CHOW ($p < 0.05$) groups. (c) GTCs reduced the ratio of liver weight/body weight and the triglyceride content in the liver and serum in both HF ($p < 0.05$) and CHOW ($p < 0.05$) groups. (d) GTCs reduced liver MDA (TBARS) level in both HF ($p < 0.05$) and CHOW ($p < 0.05$) groups.

We also found that tea polyphenols can significantly reduce the body fat of rats fed with high fat [Yan, 2013].

(1) GTCs showed anti-obesity and hypolipidemic effects
The effect of GTPs on body weight and triglyceride content became more obvious (about 11.8% and 8.2%) after 45 days of GTCs feeding. GTCs feeding significantly reduced the ratio of liver weight/body weight and the triglyceride content in the liver and serum (Figs. 10-2a, 10-2b, 10-2c). GTCs feeding also decreased MDA levels in the liver (Fig. 10-2d).

(2) EGCG decreased the lipid accumulation in adipocyte 3T3-L1 cell line
The effect of EGCG on the accumulation of lipid was investigated in adipocyte 3T3-L1 cell line and it was found that EGCG (0–50 μM) significantly decreased the lipid accumulation in the adipocytes as shown in Figure 10-3(a) (using Oil Red O staining). The result from the quantitative triglyceride assay showed that EGCG decreased the lipid accumulation in a concentration dependent manner (about 8% with 2.5 μM, 20% with 25 μM, and 35% with 50 μM, respectively) (Fig. 10-3b). EGCG also increased the glycerol content in medium (Fig. 10-3c, but did not significantly changed the free fatty acid in the medium (Fig. 10-3d).

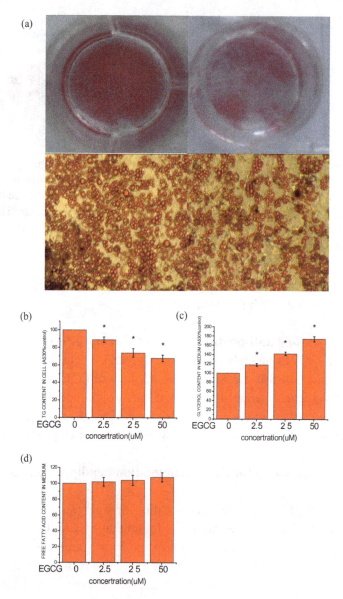

Figure 10-3. EGCG decreased the lipid accumulation in adipocyte 3T3-L1 cell line. (a) Oil Red O staining of cells treated with or without EGCG. Two-day post-confluent 3T3L1 pre-adipocytes (day 0) were stimulated for 48 hours with differentiation medium. 5 μM EGCG were added on day 8 and incubated for 48 hours and Oil Red O staining was performed on day 10. Only differentiated adipocytes can be stained by Oil Red O and visualized. After fully differentiation, EGCG were added and incubated for 48 hours. Quantification of triglyceride content in the cell by (b) triglyceride assay kit, (c) glycerol in the medium, and (d) free fatty acid level in the medium.

The above epidemiological investigation, clinical results, and animal experiments show that tea and tea polyphenols can help obesity to reduce visceral fat.

10.4.3 *Effect of drinking tea on level of low-density lipoprotein*

Cholesterol exists in lipoprotein in blood, and its existing forms include high-density lipoprotein cholesterol, low-density lipoprotein cholesterol, and very low-density lipoprotein cholesterol. The vast majority of cholesterol in blood is cholesterol ester combined with fatty acid, and less than 10% of cholesterol is free. High density lipoprotein (HDL) can help clear cholesterol in cells, while excessive LDL is generally considered as a precursor of cardiovascular disease.

Cholesterol often exists in the blood in the form of lipoprotein, while low density lipoprotein (LDL) in plasma is the main carrier for transporting endogenous cholesterol, which is degraded and transformed by binding to the low-density lipoprotein receptor (LDL-R) on its cell membrane. The deficiency of LDL-R function will reduce the clearance ability of plasma LDL-C, and eventually lead to the formation of atherosclerotic plaque in arterial intima. Therefore, the content of LDL-C is related to the incidence rate and lesion degree of cardiovascular diseases and is considered as the main pathogenic factor of atherosclerosis. Its concentration is obviously positively related to the incidence rate of coronary heart disease and is also an important index to evaluate the risk factors of coronary heart disease. Epidemiological investigation, clinical results, and animal experiments have shown that tea and tea polyphenols can help obesity to reduce low density lipoprotein.

1. Epidemiological investigation results

There was a meta-analysis about the effects of green tea extract on overweight and obese women with high levels of low density-lipoprotein-cholesterol (LDL-C); a randomized, double-blind, crossover placebo-controlled clinical trial was conducted. This study aims to examine the effects of green tea extract (GTE) supplement on overweight and obese women with high levels of low-density lipoprotein-cholesterol (LDL-C). The randomized, double-blind, crossover placebo-controlled clinical trial was conducted from August 2012 to December 2013. Seventy-three out of 90 subjects aged between 18–65 years, with BMI ≥ 27 kg/m2 and LDL-C ≥ 130 mg/dl were included in the analysis. The subjects were randomly divided into groups A and B. Group A received green tea extract supplement treatment for the first six weeks, while group B received placebo daily. After six weeks of treatment and 14 days of washout period, group A switched to placebo and group B switched to green tea extract treatment for six weeks. Results showed that subjects treated with GTE (n = 73) for six weeks showed significant differences, with 4.8% ($p = 0.048$) reduction in LDL-C and 25.7% ($p = 0.046$) increase in leptin. However, there was no statistical difference in the levels of total cholesterol, triglyceride, and high-density lipoprotein between the green tea extract and placebo groups after treatments. This study shows that green tea extract effectively increases leptin and reduces LDL in overweight and obese women after six weeks of treatment even though there were no significant changes in other biochemical markers related to overweight [Huang *et al.*, 2018]. An open-label, randomized, controlled trial was performed among 79 patients aged 20–55 years with grade I or grade II systemic arterial hypertension. In the age- and gender-matched case control observation, it found that individuals with frequent tea consumption (n = 224) had the lower plasma pro-protein convertase subtilisin/kexin type 9 (PCSK9) and low-density lipoprotein cholesterol (LDL-C) levels compared with ones without tea consumption (n = 224, $p < 0.05$) [Cui *et al.*, 2020]. Double-blind, randomized, placebo-controlled, parallel-group trial set in outpatient clinics in six urban hospitals in China. A total of 240 men and women 18 years or older on a low-fat diet with mild to moderate hypercholesterolemia were randomly assigned to receive a daily capsule containing theaflavin enriched green tea extract (375 mg) or placebo for 12 weeks. After 12 weeks, the mean +/– SEM changes from baseline in total cholesterol, LDL-C, HDL-C, and triglyceride levels were –11.3% ± 0.9% ($p = 0.01$), –16.4% ± 1.1% ($p = 0.01$), 2.3% ± 2.1% ($p = 0.27$), and 2.6% ± 3.5% ($p = 0.047$), respectively, in the tea extract group. The mean levels of total cholesterol, LDL-C, HDL-C, and triglycerides did not change significantly in the placebo group. No significant adverse events were observed. The theaflavin-enriched green tea extract studied is an effective adjunct to a low-saturated-fat diet to reduce LDL-C in hypercholesterolemic adults and is well tolerated [Maron *et al.*, 2003].

2. Results of investigation in animals

Crude tea polyphenols prepared from green tea powder were supplemented at a 1% and 2% of the lard diet and cholesterol diet, respectively. Tea polyphenols decreased plasma total cholesterol, cholesterol ester, total cholesterol — HDL-cholesterol (VLDL-+LDL-cholesterol) and atherogenic index (VLDL-+LDL-cholesterol/HDL-cholesterol). Hematocrit and plasma glucose were not altered by the addition of tea polyphenols [Muramatsu *et al.*, 1986]. Administration of green tea polyphenol effectively inhibited LDL oxidation and elevated serum anti-oxidative activity to the same degree as probucol. Green tea polyphenol may exert an anti-atherosclerotic action by virtue of its antioxidant properties and by increasing HDL cholesterol levels [Yokozawa *et al.*, 2002]. These experiments in rats have also proved that tea polyphenols can reduce the accumulation of cholesterol in the body [Yan, 2013]. The influence of black tea polyphenols on plasma lipid levels was investigated in rats fed a 15% lard and 1% cholesterol diet. The diet was supplemented with 1% black tea polyphenols extracted and condensed from black tea. Rats fed with the lard-cholesterol diet showed an increase in plasma cholesterol and liver lipids compared to rats fed with a basal diet. The supplementation of black tea polyphenols in this lard-cholesterol diet decreased the lipid levels in the plasma and increased the fecal excretion of total lipids and cholesterol [Matsumoto *et al.*, 1998].

The above epidemiological investigation, clinical results, and animal experiments show that tea and tea polyphenols can indeed help obesity to reduce cholesterol, low density-lipoprotein-cholesterol (LDL-C).

10.4.4 *Tea polyphenols prevent nonalcoholic fatty liver by regulating intestinal function*

Obesity can lead to the accumulation of fat in the liver to form fatty liver, further developing into liver cirrhosis and liver cancer, and seriously damage human health. Non-alcoholic fatty liver disease (NAFLD) is a constellation of progressive liver disorders that are closely related to obesity and diabetes. Studies have found that tea polyphenols can prevent fatty liver. NAFLD may occur as relatively benign, non-progressive liver steatosis, but in many individuals, it may progress in severity to nonalcoholic steatohepatitis, fibrosis, cirrhosis, and liver failure or hepatocellular carcinoma. No validated treatments currently exist for NAFLD except for weight loss, which has a poor long-term success rate. Thus, dietary strategies that prevent the development of liver steatosis or its progression to non-alcoholic steatohepatitis are critically needed. Green tea is rich in polyphenolic polyphenols that have hypo-lipidemic, thermogenic, antioxidant, and anti-inflammatory activities that may mitigate the occurrence and progression of NAFLD. One study established a model of non-alcoholic fatty liver disease in C57BL/6N mice. A high-fat diet-induced C57BL/6N mouse model of NAFLD was established. The results showed that tea polyphenols could effectively reduce the body weight, liver weight, and liver index of NAFLD mice. The serum effects of tea polyphenols can decrease in alanine aminotransferase (ALT), aspartate aminotransferase (AST), alkaline phosphatase (AKP), total cholesterol (TC), triglyceride (TG), low density lipoprotein cholesterol (LDL-C), D-lactate (D-LA), diamine oxidase (DAO), lipopolysaccharide (LPS), and an increase of high density lipoprotein cholesterol (HDL-C) levels; decrease of inflammatory cytokines such as interleukin 1 beta (IL-1β), interleukin 4 (IL-4), interleukin 6 (IL-6), interleukin 10 (IL-10), tumor necrosis factor alpha (TNF-α), and interferon gamma (INF-γ); decrease the ROS level in liver tissue; alleviate pathological injuries of liver, epididymis, and small intestinal tissues caused by NAFLD and protection of body tissues. The results also showed that tea polyphenols could upregulate the mRNA and protein expressions of LPL, PPAR-α, CYP7A1, and CPT1, and downregulate PPAR-γ and C/EBP-α in the liver of NAFLD mice. In addition, tea polyphenols can down-regulated the expression of CD36 and TNF-α in the small intestines of NAFLD mice. Studies on mice feces showed that tea polyphenols reduced the level of *Firmicutes* and increased the minimum levels of *Bacteroides* and *Akkermansia*, as well as reduced the proportion of *Firmicutes/Bacteroides* in the feces of NAFLD mice, which play a role in regulating intestinal microecology. Tea polyphenols improved the intestinal environment of NAFLD mice with the contained active ingredients, thus playing a role in preventing NAFLD. The effect was positively correlated with the dose of 100 mg/kg, which was even better than that of the clinical drug bezafibrate [Liu *et al.*, 2019; Cheng *et al.*, 2020]. Green tea contains many polyphenolic constitutes, which might prevent NAFLD. A study investigated whether green tea extract given at doses reflecting habitual consumption of green tea beverages prevents the development of NAFLD in rats fed a high-fat diet (HFD). The study groups received a HFD (approximately 50% energy from fat), enriched with 1.1% and 2.0% green tea extract E, respectively, for a total of 56 days. The control groups were fed a HFD alone and normal standardized diet (low-fat diet), respectively, for the same period. The percentage of hepatocytes affected by steatosis in the HFD group (median: 25%) was higher ($p < 0.033$ and $p < 0.050$, respectively) than in the HFD-2.0% green tea extract group (9%) and normal diet group (10%). No significant differences were observed for the group consuming HFD-1.1% green tea extract, in which intermediate results were observed (15%). This finding points towards the hepatoprotective potential of green tea extract in preventing dietary-induced liver steatosis. In view of the increasing incidence of overweight and obesity, a simple and cheap dietary modification, such as green tea extract supplementation, could prove to be useful clinically [Karolczak *et al.*, 2019].

Compared with the mice fed with high fat/high sugar control group, the weight, total visceral fat volume and liver fat weight of the mice fed with GTPs, oolong tea OTPs, black tea BTPs, and low-fat/high sugar LF/HS diet significantly decreased. Hallmark features of non-alcoholic steatohepatitis (NASH) are fatty hepatocytes and inflammatory cell infiltrates in association with increased activation of hepatic nuclear factor kappa-B (NFκB) that exacerbates liver injury. The anti-inflammatory activities of green tea extract on gut barrier function as well as prebiotic and antimicrobial effects on gut microbial ecology that help to limit the translocation of gut-derived endotoxins (e.g., lipopolysaccharides) to the liver, where they otherwise upregulate NFκB activation by Toll-like receptor-4 signaling [Hodges *et al.*, 2020]. The protective effect of tea polyphenols on chronic alcoholic liver injury in rats compared with the control group rats, the fat/body weight ratio, SOD/MDA, T-AOC, and GSH-Px activity of chronic liver injury rats decreased significantly ($p < 0.05$, $p < 0.01$). Meanwhile the liver index, FAT/CD36 protein level, and lipid deposition in liver of chronic liver injury rats were increased ($p < 0.01$). Compared with chronic liver injury rats, the tea polyphenols intervention increased fat/body weight ratio ($p < 0.05$) and significantly increased SOD/MDA, T-AOC, and GSH-Px activity ($p < 0.01$). Meanwhile the tea polyphenols intervention reduced liver index ($p < 0.01$), FAT/CD36 protein level ($p < 0.01$), and lipid deposition in liver [Zhang *et al.*, 2018]. Our results show that GTC plays an anti-obesity role in weight loss by regulating PPAR related pathways, such as PPAR's different regulatory mechanisms. Moreover, in BAT, SWAT, and VWAT, GTC results in an increase in the net lipid flow of fat cells and the oxidation of fat in fat cells, leading to weight loss. Based on these results, it seems that regular drinking of green tea may be a good strategy to prevent overweight and obesity [Yan *et al.*, 2013].

Another study found that green tea extract could prevent nonalcoholic hepatic steatosis in rats fed a high-fat diet and tea polyphenols can effectively prevent and inhibit nonalcoholic fatty liver and seminal fatty liver in obese people.

10.4.5 *Effect of tea polyphenols on food intake and absorption*

Obesity is caused by eating more food and absorbing more energy than energy consumption. If tea polyphenols can inhibit the intake food and absorption of energy, they can play a role in weight loss from the source. Studies have found that tea polyphenols can indeed inhibit the intake and absorption of food energy. Recent findings indicate that the major effect of tannins in tea was not due to their inhibition on food consumption or digestion but rather the decreased efficiency in converting the absorbed nutrients to new body substances [Chung *et al.*, 1998]. Data from laboratory studies have shown that green tea has important roles in fat metabolism by reducing food intake, interrupting lipid emulsification and absorption, suppressing adipogenesis and lipid synthesis and increasing energy expenditure via thermogenesis, fat oxidation, and fecal lipid excretion [Huang *et al.*, 2014]. Tea polyphenols can interact with food in the digestive tract, interfere with the digestion and absorption of lipids, and reduce energy intake. Although the effects of tea polyphenols on appetite have been reported, most of them make the subjects take tea polyphenols much higher than the daily dose of tea, and there is no significant change in the value of food intake/body weight. In particular, the secretion of leptin, an important factor (reducing appetite), decreased significantly after taking tea polyphenols. These mechanisms may be related to certain pathways, such as through the modulations of energy balance, endocrine systems, food intake, lipid and carbohydrate metabolism, the redox status, and activities of different types of [Kao *et al.*, 2006]. Our experiment also shows that the daily dose of tea polyphenols has no significant effect on appetite in the short term, but at the same time, the body weight has decreased significantly [Yan *et al.*, 2013]. Tea polyphenols can decrease absorption of lipids and proteins in the intestine, thus reducing calorie intake and activating (adenosine monophosphate) AMP-activated protein kinase by tea polyphenols that are bioavailable in the liver, skeletal muscle, and adipose tissues [Yang *et al.*, 2016].

Therefore, the effect of tea polyphenols on food absorption can partly explain the improvement effect of tea polyphenols on obesity and metabolic syndrome, but the effect of tea polyphenols on appetite has no obvious relationship. In addition, tea can prevent obesity via reduction of appetite, food consumption, and food absorption in gastrointestinal system and through the changes in fat metabolism.

10.5 The mechanism of tea polyphenols for weight loss and lipid reduction

On the basis of epidemiological investigation, clinical and animal experiments that tea polyphenols can reduce body weight, the accumulation of fat and cholesterol in the body and liver, inhibit appetite, and increase the oxidation of energy in the body. Many studies have carried out various studies on the mechanism of tea polyphenols to reduce weight. These include inhibition of pre-adipocyte synthesis and adipocytes, and stimulation of adipocyte oxidation.

10.5.1 *Tea polyphenols have significant inhibitive effects on pre-adipocytes*

The first is the inhibition of tea polyphenols on pre-adipocyte proliferation. The increase of body weight is not only the increase of fat storage in adipocytes, but also the increase of the number of adipocytes [Wolfram *et al.*, 2006]. Adipocytes themselves cannot proliferate, only front adipocytes proliferate and then differentiate into adipocytes extracellular regulated protein kinase (ERK) and cyclin (CDK) controlled pathways play an important role in pre-adipocyte proliferation. Pre-adipocyte proliferation as indicated by an increased number of cells and greater incorporation of bromodeoxyuridine (BrdU) was inhibited by EGCG in dose-, time-, and growth phase-dependent manners. Also, EGCG dose- and time-dependently decreased levels of phospho-ERK1/2, Cdk2, and cyclin D(1) proteins, reduced Cdk2 activity, and increased levels of G(0)/G(1) growth arrest, p21(waf/cip), and p27(kip1), but not p18(ink), proteins and their associations to Cdk2. Increased phospho-ERK1/2 content and Cdk2 activity, respectively, via the transfection of MEK1 and Cdk2 cDNA into pre-adipocytes prevented EGCG from reducing cell numbers. These data demonstrate the ERK- and Cdk2-dependent antimitogenic effects of EGCG. Moreover, EGCG was more effective than epicatechin, epicatechin gallate, and epigallocatechin in changing the mitogenic signals. The signal of EGCG in reducing growth of 3T3-L1 pre-adipocytes differed from that of 3T3 fibroblasts. *In vitro* studies showed that 10–50 µM EGCG could significantly inhibit the phosphorylation of ERK1/2 in 3T3-L1 pre-adipocytes, but the protein level of ERK1/2 did not change. At the same time, the phosphorylation of MEK1 and p38-MAPK did not change, indicating that EGCG specifically inhibited the phosphorylation of ERK1/2 and thus inhibited the proliferation of pre-adipocytes [Hung *et al.*, 2005]. Secondly, tea polyphenols inhibit the differentiation of pre-adipocytes into adipocytes. At present, two signal pathways are mainly inhibited. The first is cell experiment, which shows that 10–50 µM EGCG can obviously activate AMPK pathway during differentiation, which may be related to the abnormal changes of ROS [Qanungo *et al.*, 2005]. Additionally, 50 µM EGCG can significantly inhibit C/EBPαand PPARγ expression in 3T3-L1 pre-adipocytes during differentiation. The specific protein required for adipocyte differentiation is also C/EBPα and PPARγ expression of downstream-regulated specific proteins, AP-2, Fas, fat, and GLUT-4, inhibited fat accumulation, carbohydrate absorption, and adipocyte differentiation [Moon *et al.*, 2006].

It was confirmed that green tea reduced adipose tissue weight without any change in body weight, other tissue weights, and food and water intakes. Green tea also significantly reduced the plasma levels of cholesterols and free fatty acids. For mechanisms of the anti-obesity actions, green tea significantly reduced glucose uptake accompanied by a decrease in translocation of glucose transporter 4 (GLUT4) in adipose tissue, while it significantly stimulated the glucose uptake with GLUT4 translocation in skeletal muscle. Moreover, green tea suppressed the expression of peroxisome proliferator-activated receptor gamma and the activation of sterol regulatory element binding protein-1 in adipose tissue. In conclusion, green tea modulates the glucose uptake system in adipose tissue and skeletal muscle and suppresses the expression and/or activation of adipogenesis-related transcription factors, as the possible mechanisms of its anti-obesity actions. The *in vivo* results also showed that tea polyphenols could reduce the expression of PPARγ and GLUT-4 in visceral adipose tissue [Ashida *et al.*, 2004]. However, this may make the body's normal need for mature adipocytes unsatisfied. Tea polyphenols can also induce apoptosis of mature and differentiated adipocytes. Greater than 100 μM EGCG could significantly induce apoptosis of mature and differentiated 3T3-L1 cells for a long time [Lin *et al.*, 2005]. Another important aspect of the effect of tea polyphenols on mature and differentiated adipocytes is to improve the ability of adipocytes to oxidize fatty acids and decouple heat. Tea polyphenols can significantly increase the energy consumption and heat production of human body within 24 hours [Dulloo *et al.*, 1999]. EGCG can increase the content of brown adipose tissue in animals [Choo, 2003]. EGCG can improve the thermogenic capacity of brown adipose tissue. This is due to the specific presence of a receptor adipocytes β3-adrenoceptor for the sympathetic nervous system transmitter norepinephrine [Dulloo *et al.*, 2000]. EGCG can inhibit the activity of Catechol-O-methyltransferase (COMT), an enzyme that decomposes norepinephrine, so it can delay the action of norepinephrine [Borchardt, 1975]. β3-adrenoceptor also exists in white adipocytes, but the effect of tea polyphenols on the thermogenic capacity of white adipocytes has hardly been reported. However, the improvement of oxidative heat production of white fat is considered to be a very effective method to treat excessive fat accumulation [Chen, 2001]. A study revealed that oxidized tea polyphenols failed to decrease the serum concentrations of triglyceride and total cholesterol. Oxidized tea polyphenols alleviated the accumulation of lipids in the liver tissue and changed the expression levels of the regulators of lipid metabolism, i.e., peroxisome proliferation-activated receptors (PPARs), compared with the rats fed a high-fat diet alone. It also observed a significantly decreased reduction of weight in the visceral white adipose, enhanced regulation of fatty acid β-oxidation by PPARα, and enhanced biosynthesis of mitochondria in the visceral white adipose of the oxidized tea polyphenols rats compared with the high-fat diet rats. Additionally, oxidized tea polyphenols promoted the excretion of lipids [Wang *et al.*, 2017].

Above results indicate that tea polyphenols can significantly inhibit obesity by affecting the physiological functions of pre-adipocytes.

10.5.2 *Tea polyphenols have an inhibitive effects on synthesizing fatty acid*

Tea polyphenols can reduce the body weight by inhibit synthesizing fatty acids. The existing experimental and clinical results show that tea polyphenols can effectively inhibit obesity and metabolic syndrome and affect many aspects of fat metabolism and physiological function of adipocytes, but there are still many aspects to be explained, new mechanisms need to be proposed, and the existing mechanisms need to be improved, so as to provide a basis for the prevention and treatment of obesity and metabolic syndrome. It has been reported that inhibition of fatty-acid synthase (FAS) is selectively cytotoxic to human cancer cells. Considerable interest has developed in identifying novel inhibitors of this enzyme complex. Previous work showed that green tea EGCG could inhibit FAS *in vitro*. To elucidate the structure-activity relationship of the inhibitory effects of tea polyphenols, a paper investigated the inhibition kinetics of the major catechins and analogues on FAS. Ungallated catechins from green tea do not show obvious inhibition compared with gallated catechins. Another gallated catechin, (-)-epicatechin gallate (ECG), was also found as a potent inhibitor of FAS and its inhibition characteristics are similar to EGCG. Our atomic orbital energy analyses suggest that the positive charge is more distinctly distributed on the carbon atom of ester bond of galloyl moiety of gallate catechins, and that gallated forms are more susceptible for a nucleophilic attack than other catechins [Zhao *et al.*, 1992]. Furthermore, the analogues of galloyl moiety without the catechin skeleton such as propyl gallate also showed obvious slow-binding inhibition, whereas the green tea ungallated catechin did not. For the first time, it was identified that the galloyl moiety of green tea catechins is critical in the inactivation of the ketoacyl reductase activity of FAS. [Wang *et al.*, 2003]. EGCG and ECG inhibit FAS with IC (50) values of 52 microM and 42 microM mainly by reacting on the beta-ketoacyl reductase (KR) domain of FAS. The inhibitory ability of catechin gallate (CG) is 15 folds and 12 folds higher than that of EGCG and ECG, respectively. The inhibition kinetics show that they inhibit FAS competitively with acetyl CoA and most likely react mainly on acyl transferase domain [Tian, 2006]. By analyses of the inhibitory kinetics and the structure of the gallated catechins, it was found that the acyl transferase domain may be the main site reacting with (-)-CG, the structure consisting of a B ring, a C ring, and a gallate ring, which is possibly essential for its inhibitory efficacy. The polyphenols rather than the alkaloids are the main fractions contributing to the inhibitory effect of green tea extract on FAS. During separation we also found that the total ability of this portion to inhibit FAS increases by 15-fold, and this may be due to some novel potent inhibitor of FAS other than (-)-CG being formed [Zhang *et al.*, 2006]. Dietary supplementation of EGCG resulted in a dose-dependent attenuation of body fat accumulation. Food intake was not affected but feces energy content was slightly increased by EGCG, indicating a reduced food digestibility and thus reduced long-term energy absorption. Leptin and stearoyl-coenzyme A desaturase 1 (SCD1) gene expression in white fat was reduced but SCD1 and uncoupling proteins1 (UCP1) expression in brown fat was not changed. In liver, gene expression of SCD1, malic enzyme (ME), and glucokinase (GK) was reduced and that of UCP2 increased. Acute oral administration of EGCG over three days had no effect on body temperature, activity, and energy expenditure, whereas respiratory quotient during night (activity phase) was decreased, supportive of a decreased lipogenesis, and increased fat oxidation [Klaus *et al.*, 2005].

These results show that tea polyphenols can reduce weight by inhibiting fatty acid synthase and fatty acid synthesis. Tea polyphenols with different structures showed different inhibitory effects.

10.5.3 *GTCs have opposite effects on PPARγ expression in SWAT and VWAT*

Our study found that PPARγ is one of the central regulators of adipogenesis that promotes lipid storage in adipocytes. GTCs increased the PPARγ level in subcutaneous white adipose tissue (SWAT) in both the HF (about 50.2%) and CHOW (about 39.8%) groups (Fig. 10-4a, 10-4a1). However,

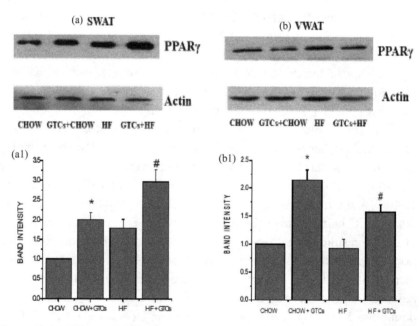

Figure 10-4. Effects of GTCs on the expression of PPARγ in (a) subcutaneous white adipose tissue (SWAT) and (b) visceral white adipose tissue (VWAT). Rats were fed different diets for 45 days and the adipose tissue samples were analyzed by Western blot. The intensities of the bands were quantified by densitometric analysis and the results (a1 and b1) are expressed as fold changes by normalizing the data to the values obtained from the CHOW diet group, which were set to 1.0 (n = 6; #, $p < 0.001$ *vs* CHOW).

GTCs decreased the levels of PPARγ in visceral (mesentery) white adipose tissue (VWAT) in both the HF (about 23.1%) and CHOW (about 18.7%) groups (Figure 10-4b, 10-4b1) [Yan *et al.*, 2013].

10.5.4 *Anti-obesity effect of green tea polyphenols by stimulation of lipids oxidation and mitochondrial biogenesis in adipocyte*

Obesity has become a worldwide problem resulting in metabolic syndrome, typified by type 2 diabetes mellitus (T2DM), cardiovascular disease, hypertension, and hyperlipidemia [Liu *et al.*, 2019]. Epidemiological studies as well as experimental data in animals have shown that drinking green tea or administration of its most effective components green tea polyphenols (GTPs) is associated with lower risk of obesity [Kao *et al.*, 2000]. GTP participates in fat metabolism and has anti-obesity effects by stimulating lipid oxidation and mitochondrial biogenesis in adipocytes [Yan *et al.*, 2012]. Increasing evidence shows that mitochondrial number and function are reduced in various tissues in human diabetes and obesity [Ritov *et al.*, 2005]. A high-fat diet downregulates genes required for mitochondrial oxidative phosphorylation in skeletal muscle. Acute lipid infusion can decrease the expression of mitochondrial proteins in man [Kuroda *et al.*, 2006]. Mechanistic study showed that hyperleptinemia upregulated PGC-1a, increased uncoupling protein uncoupling protein-1 (UCP-1) and -2, downregulated lipogenic enzymes, and increased phosphorylation of both acetyl CoA carboxylase and AMP-activated protein kinase. Lipid infusion decreases the expression of nuclear encoded mitochondrial genes and increases the expression of extracellular matrix genes in human skeletal muscle [Richardson *et al.*, 2005]. Therefore, by regulating PPAR pathway, stimulating mitochondrial biogenesis could be a novel strategy for reducing fat accumulation in obesity.

One family of transcription factors that are critical to the fat metabolism are PPARs (PPARα, γ and δ), which are ligand-activated transcription factors belonging to the nuclear receptor superfamily [Kersten *et al.*, 2000]. Each member of PPARs displays a tissue-selective expression pattern and has distinct role in lipid metabolism. Green tea components have been shown to affect the PPARs signaling pathways. It was previously observed that green tea polyphenols and hypolipidemic agent fenofibrate significantly increased PPARa expression in HepG2 cell line [Jiao *et al.*, 2003] and experimental data in animals have shown that administration of green tea catechins (GTCs) is associated with lower risk of obesity through PPARs Pathway. EGCG, the major component of GTCs, was also found to suppress adipocyte differentiation accompanied by downregulation of PPARγ and C/EBPa in high concentrations (more than 50 μM). However, the top concentration of EGCG in serum after drinking tea is not more than 2.5 μM [Lee *et al.*, 2002]. So, the mechanism of anti-obesity effects of green tea in physiological condition is still not clear. In this study, we investigated the roles of PPARs and related pathways in the anti-obesity mechanisms of green tea catechin EGCG in 3T3-L1 adipocytes and mitochondrial system. The following will discuss about PPARs and related pathways in the anti-obesity mechanisms and the main results are from the work of our laboratory [Yan *et al.*, 2013].

1. EGCG promoted b-oxidation of lipids by PPAR-related pathway in adipocyte 3T3-L1 cell line

β-oxidation of lipids causes decrease of fatty acid in adipocyte. We investigated the β-oxidation of lipids by PPAR-related pathway in adipocyte 3T3-L1 cell line. It was found that EGCG in physiological concentration upregulated the expression of C/EBP (Fig. 10-5Aa) and C/EBPβ (Fig. 10-5Ab), which is responsible in stimulating cell differentiation, while decreasing CHOP-10 (Fig. 10-5Ad). Since CHOP-10 negatively regulates the activity of C/EBPb, the expression of the transcription targets of C/EBPα, PPARγ (Fig. 10-5Ac), and PPARδ (Fig. 10-5Ae) were upregulated. We also found that EGCG upregulated the expression of the enzymes involved in energy expenditure, including AOX (acyl-CoA oxidase) (Fig. 10-5Ba) and UCP-1 (Fig. 10-5Bb) in physiological concentration. EGCG increased the expression of fatty acid transporter (FAT) (Fig. 10-4Bc) and decreased fatty acid synthase (FAS) (Fig. 10-5Bd) in very low concentration.

All these results indicate that EGCE treatment stimulated activation of PPAR related pathways for lipids oxidation in 3T3-L1 adipocyte.

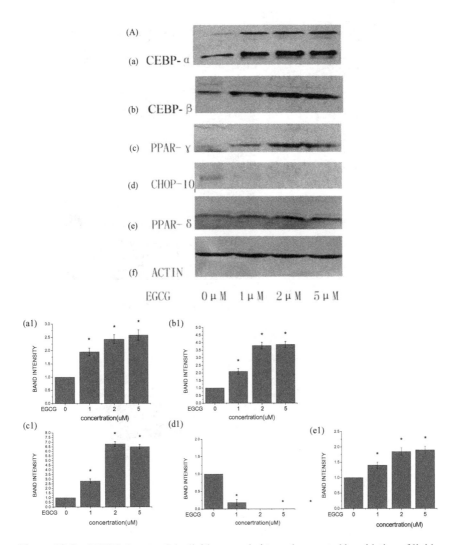

Figure 10-5. EGCG decreased the lipid accumulation and promoted b-oxidation of lipids by PPAR- related pathway in adipocyte 3T3-L1 cell line. (A) EGCG regulated PPARγ, C/EBP, and CHOP10 in 3T3L1 pre-adipocytes. Levels of (a) C/EBPα, (b) C/EBPβ, (c) PPARγ, (d) CHOP-10, and (e) PPARγ. Protein levels in 3T3L1 cells treated with or without EGCG were analyzed by Western Blot. The column graphs (a1, b1, c1, d1, e1) are the statistical results of band intensities of a, b, c, d, e; (B) Levels of (a) AOX and (b) UCP-1, (c) FAT, and (d) FAS. Protein levels in 3T3L1 cells treated with or without EGCG were analyzed by RT-PCR. The column graphs (a1, b1, c1, d1) are the statistical results of band intensities of a, b, c, d. *, $p < 0.05$ compared with the control (untreated group) (n = 8).

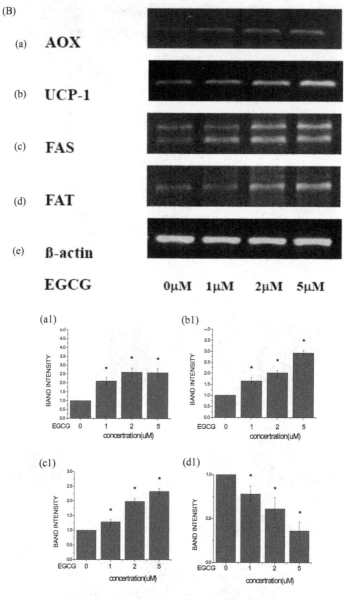

Figure 10-5. (*Continued*)

2. EGCG stimulated mitochondrial biogenesis in 3T3-L1 adipocytes

Mitochondrial is the main place for oxidation of lipid and generation of energy. We investigated the effects of EGCG on the most important makers for oxidation of lipid, mitochondrial mass, and oxygen consumption in adipocytes. It was found that as shown in Figure 10-6(a), EGCG increased the relative fluorescence intensity significantly at 1.0 μM and 5.0 μM, indicating that EGCG increased mitochondrial mass. As shown in Figure 10-6(b), the basal rate of oxygen consumption was significantly increased in adipocytes treated with EGCG at 1.0–5.0 μM.0. EGCG increased complex I, II, V, but not III protein expression in adipocytes. As shown in Figure 10-6(c), EGCG significantly increased mitochondrial complex I protein at 5.0 μM (131%), complex II protein expression at 5.0 μM (172%), complex V at 5.0 μM (276%), and at complex III 10.0 μM (227%). EGCG showed no effect on the expression of mitochondrial complex III in adipocytes. EGCG increased the mitochondrial DNA express shown as an increase in D-loop, the major site of transcription initiation on both the heavy and light strands of mitochondrial DNA. As shown in Figure 10-6(d), the ratio of mt D-loop/18SRNA was significantly increased in adipocytes treated with EGCG at 2.0–5.0 μM. EGCG increased PGC-1α protein level in adipocytes. As shown in Figure 10-6(e), EGCG showed an increasing effect on PGC-1α from 0.1–10.0 μM with a maximum protein expression at 2.0 μM (154%) and PPAR-γ2 in 5–10.0 μM (Fig. 10-6f) in mitochondria.

3. Discussion and analysis

Obesity is characterized by increase of adipocyte volume (hypertrophy) and population (hyperplasia), which depend on the balance between lipolysis and lipogenesis [Lowell & Spiegelman, 2000]. Our results showed that EGCG significantly decreased the accumulation of lipid in 3T3-L1 adipocytes. It is worth to mention that EGCG was able to improve the adipocyte function even at 1 μM, which is physiologically achievable according to the report of Lee *et al.* [2002]. PPARγ serves as an essential regulator for promoting lipid and glucose storage in mature adipocytes. Tea catechin suppresses adipocyte differentiation accompanied by the downregulation of PPARgamma2 and C/EBP alpha in 3T3-L1 cells [Furukawa *et al.*, 2004]. Therefore, the raising expression of PPARγ in adipocytes may be one of the

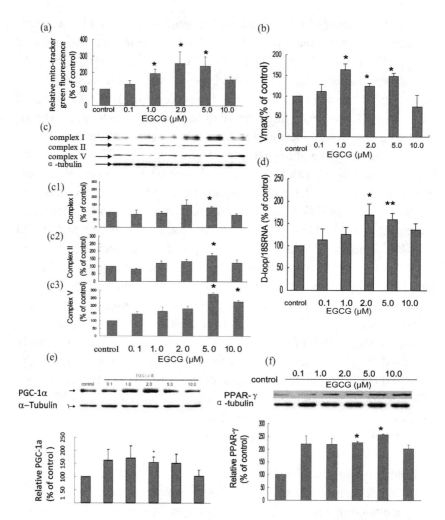

Figure 10-6. EGCG stimulated mitochondrial biogenesis in 3T3-L1 adipocytes. (a) Mitochondrial mass estimated by MitoTracker staining using fluorescence microplate spectrophotometer. Results are expressed as fold increase of the fluorescence intensity over untreated control cells. (b) The basal rate of oxygen consumption in adipocytes treated with different EGCG concentrations. (c) Expression of mitochondrial proteins of complex I, II, V. Representative immuno-blots for steady-state levels of proteins. The quantitative analyses of the bands by densitometry are shown in (c1), (c2), and (c3) for mitochondrial complex I, complex II, and complex V, respectively. (d) Expression of mito-chondrial DNA expressed as N-fold differences in mitochondrial D-loop expression rela-tive to the 18sRNA gene. Effects of EGCG stimulation on (e) mitochondrial PGC-1α and (f) PPARγ-2 in 3T3L1 adipocytes. Values are mean ± SE of five experiments, n = 5. *, $p < 0.05$ *vs* control; **, $p < 0.01$ *vs* control.

mechanisms how EGCG decreases the serum lipid level. The expression of PPARγ gene is regulated by transcription factors C/EBPα and β [Lee *et al.*, 2002]. Our data suggested that EGCG upregulates PPARγ, C/EBPβ, and C/EBPα by inhibiting CHOP-10 and upregulating C/EBPβ in 3T3-L1 cell. EGCG upregulated the C/EBPα expression in the CHOP-10-deficient keratinocyte [Furuyashiki *et al.*, 2004]. PPARγ causes a net flux of fatty acids from circulation and other organs into adipocytes. In addition, EGCG increase mitochondrial biogenesis will lead to an increase in fatty acid oxidation that will further protect against adipocyte hypertrophy. Therefore, the key point is whether the adipocytes can work efficiently to get circulation fat into them and burn the fats. We then studied the effects of EGCG on mitochondrial biogenesis.

We detected whether EGCG could induce 3T3-L1 adipocyte apoptosis. The level of intracellular triglyceride directly correlates with the volume of adipocyte. The intracellular triglyceride level of mature adipocyte decreased after EGCG treatment, whereas the glycerol level in the media increased with EGCG treatment in a concentration dependent manner, indicating that EGCG promoted triglyceride hydrolysis or lipids oxidation. However, EGCG treatment had no significant effect on the quantity of free fatty acid in the medium. So, the reduction of lipids content in adipocyte was caused by lipids oxidation. The upregulated PPARd and expression of the genes involved in fatty acid β-oxidation enzyme such as acetyl CoA oxidase (AOX), uncoupling protein-1 (UCP-1), suggests that EGCG promoted fatty acid metabolism and energy expenditure. The expression of PGC-1α, the up-regulator of mitochondrial biogenesis, is reduced in the adipose tissue of morbidly obese subjects. Thiazolidenediones, such as pioglitazone, are shown to increase PGC-1α expression and mitochondrial DNA copy number and enhance the oxidative capacity of white adipose tissue leading to insulin sensitization [Bogacka *et al.*, 2004]. The transcriptional coactivator PGC-1α plays an important role in the stimulation of the mitochondrial biogenesis program in fat and muscle cells and the interaction between PPARδ and PGC-1α has been thought to promote the oxidation of lipids in adipocytes [Rosen & Spiegelman, 2001]. Our results shown that EGCG at 2–5μM significantly increased mitochondrial biogenesis as evidenced by the enhanced mitochondrial mass, increased oxygen consumption, increased complex I, II, V expression, and increased mitochondrial DNA expression in 3T3-L1 adipocytes. EGCG enhanced the expression of both PPARδ and PGC-1α, suggesting that EGCG may promote the oxidation of lipids through the interaction of PPARd and PGC-1α.

Studies with adipocyte cell lines and animal models have demonstrated that EGCG inhibits extracellular signal-related kinases (ERK), activates AMP-activated protein kinase (AMPK), modulates adipocyte marker proteins, and downregulates lipogenic enzymes as well as other potential targets [Moon *et al.*, 2007].

These results suggest that tea polyphenols, EGCG, could control adipose tissue associated with obesity through mitochondrial remodeling, leading to an increase in fatty acid oxidation that protects against adipocyte hypertrophy due to PPAR-activation-caused net flux of fatty acids from circulation and other organs into adipocytes.

10.6 Conclusion

There are several proposed mechanisms whereby GTC may influence body weight and composition. The predominating hypothesis is that GTC influences sympathetic nervous system (SNS) activity, increases energy expenditure, and promotes the oxidation of fat. Caffeine, naturally present in green tea, also influences SNS activity, and may act synergistically with GTC to increase energy expenditure and fat oxidation. Other potential mechanisms include modifications in appetite, upregulation of enzymes involved in hepatic fat oxidation, and decreased nutrient absorption [Rains *et al.*, 2011]. Epidemiological investigation, clinical and animal experiments of green tea and black tea, tea polyphenols can reduce body weight and prevent obesity, the accumulation of fat and cholesterol in the body and liver, inhibit appetite, and increase the oxidation of energy in the body. Tea polyphenols may inhibit lipid and saccharide digestion and absorption and reduce calorie intake. Tea polyphenols exert a positive effect on inhibiting obesity involved in three major mechanisms: 1) Inhibiting lipid and saccharide digestion, absorption and intake, thus reducing calorie intake; 2) Promoting lipid metabolism by activating AMP-activated protein kinase to attenuate lipogenesis and enhance lipolysis, and decreasing lipid accumulation by inhibiting the differentiation and proliferation of pre-adipocytes; 3) Tea polyphenols also could promote lipid metabolism by activating AMPK, PGC-1,

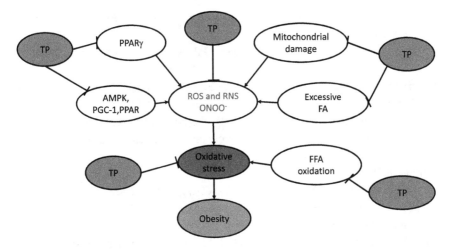

Figure 10-7. The pathway and mechanism of tea polyphenols in prevention and treatment of obesity. TP: tea polyphenols.

PPAR signal pathways, attenuating lipogenesis, and enhancing lipolysis. They would lower lipid accumulation by suppressing the differentiation and proliferation of pre-adipocytes and by reducing oxidative stress; blocking the pathological processes of obesity and comorbidities of obesity by reducing oxidative stress. In summary, the pathway and the mechanism of tea polyphenols in prevention and treatment of obesity are shown in Figure 10-7.

References

Aizawa T, Yamamoto A, Ueno T. (2017) Effect of oral theaflavin administration on body weight, fat, and muscle in healthy subjects: a randomized pilot study. *Biosci Biotechnol Biochem*, **81**(2), 311–315.

Arner P. (2001) Regional differences in protein production by human adipose tissue. *Biochem Soc Trans*, **29**, 72–75.

Ashida H, Furuyashiki T, Nagayasu H, *et al.* (2004) Anti-obesity actions of green tea: possible involvements in modulation of the glucose uptake system and suppression of the adipogenesis-related transcription factors. *Biofactors*, **22**, 135–140.

Ashigai H, Taniguchi Y, Suzuki M, *et al.* (2016) Fecal lipid excretion after consumption of a black tea polyphenol containing beverage-randomized, placebo-controlled, double-blind, crossover study. *Biol Pharm Bull*, **39**(5), 699–704.

Barish GD, Narkar VA, Evans RM. (2006) PPAR delta: a dagger in the heart of the metabolic syndrome. *J Clin Invest*, **116**, 590–597.

Becerril S, Rodríguez A, Catalán V, *et al.* (2019) Functional relationship between leptin and nitric oxide in metabolism. *Nutrients*, **11**(9), 2129.

Bogacka I, Xie H, Bray GA, *et al.* (2004) The effect of pioglitazone on peroxisome proliferator-activated receptor-gamma target genes related to lipid storage in vivo. *Diabetes Care*, **27**, 1660–1667.

Borchardt RT. (1975) Catechol-o-methyl-transferase: structure-activity relationship for inhibition by flavonoids. *Med Chem*, **8**, 120–122.

Caballero B. (2019) Humans against obesity: who will win? *Adv Nutr*, **10** (Suppl 1):S4–S9.

Chen HC FRJ. (2001) Turning WAT into BAT gets rid of fat. *Nature Medicine*, **7**, 1102–1103.

Chen IJ, Liu CY, Chiu JP, Hsu CH. (2016) Therapeutic effect of high-dose green tea extract on weight reduction: A randomized, double-blind, placebo-controlled clinical trial. *Clin Nutr*, **35**(3), 592–599.

Cheng J, Tan Y, Zhou J, Xiao L, Johnson M, Qu X. (2020) Green tea polyphenols ameliorate metabolic abnormalities and insulin resistance by enhancing insulin signalling in skeletal muscle of Zucker fatty rats. *Clin Sci (Lond)*, **134**(10), 1167–1180.

Choo JJ. (2003) Green tea reduces body fat accretion caused by high-fat diet in rats through beta-adrenoceptor activation of thermogenesis in brown adipose tissue. *J Nutr Biochem*, **14**, 671–676.

Chung KH, Chiou HY, Chang JS, Chen YH. (2020) Associations of nitric oxide with obesity and psychological traits among children and adolescents in Taiwan. *Pediatr Obes*, **15**(3), e12593.

Chung KT, Wong TY, Wei CI, Huang YW, Lin Y. (1998) Tannins and human health: a review. *Crit Rev Food Sci Nutr*, **38**(6), 421–464.

Crewe C, Funcke JB, Li S, Joffin N, Gliniak CM, Ghaben AL, An YA, Sadek HA, Gordillo R, Akgul Y, Chen S, Samovski D, Fischer-Posovszky P, Kusminski CM, Klein S, Scherer PE. (2021) Extracellular vesicle-based interorgan transport of mitochondria from energetically stressed adipocytes. *Cell Metab*, **33**(9), 1853–1868.

Cui CJ, Jin JL, Guo LN, Sun J, Wu NQ, Guo YL, Liu G, Dong Q, Li JJ. (2020) Beneficial impact of epigallocatechingallate on LDL-C through PCSK9/LDLR pathway by blocking HNF1α and activating FoxO3a. *J Transl Med*, **18**(1), 195.

Dulloo AG, Duret C, Rohrer D, *et al.* (1999) Efficacy of a green tea extract rich in catechin polyphenols and caffeine in increasing 24-h energy expenditure and fat oxidation in humans. *Am J Clin Nutr*, **70**, 1040–1045.

Dulloo AG, Seydoux J, Girardier L, Chantre P, Vandermander J. (2000) Green tea and thermogenesis: interactions between catechin-polyphenols, caffeine and sympathetic activity. *Int J Obes Relat Metab Disord*, **24**, 252–258.

Dursun E, Akalin FA, Genc T, *et al.* (2016) Oxidative stress and periodontal disease in obesity. *Medicine (Baltimore)*, **95**(12), e3136.

Fernández-Sánchez A, Madrigal-Santillán E, Bautista M, *et al.* (2011) Inflammation, oxidative stress, and obesity. *Int J Mol Sci*, **12**(5), 3117–3132.

Flashner BM, Rifas-Shiman SL, Oken E, *et al.* (2020) Obesity, sedentary lifestyle, and exhaled nitric oxide in an early adolescent cohort. *Pediatr Pulmonol*, **55**(2), 503–509.

Förstermann U, Xia N, Li H. (2017) roles of vascular oxidative stress and nitric oxide in the pathogenesis of atherosclerosis. *Circ Res*, **120**(4), 713–735.

Fonseca VA. (2005) The metabolic syndrome hyperlipidemia and insulin resistance. *Clin Cornerstone*, **7**, 61–72.

Furukawa S, Fujita T, Shimabukuro M, *et al.* (2004) Increased oxidative stress in obesity and its impact on metabolic syndrome. *J Clin Invest*, **114**, 1752–1761.

Furuyashiki T, Nagayasu H, Aoki Y, *et.al.* (2004) Tea catechin suppresses adipocyte differentiation accompanied by down-regulation of PPARgamma2 and C/EBPalpha in 3T3-L1 cells. *Biosci Biotechnol Biochem*, **68**, 2353–2359.

Gonzalez F, Rote NS, Minium J, Kirwan JP. (2006) Reactive oxygen species-induced oxidative stress in the development of insulin resistance and hyperandrogenism in polycystic ovary syndrome. *J Clin Endocrinol Metab*, **91**(1), 336–340.

Grasemann H, Holguin F. (2021) Oxidative stress and obesity-related asthma. *Paediatr Respir Rev*, **37**, 18–21.

Grundy SM. (2004) Obesity, metabolic syndrome, and cardiovascular disease. *J Clin Endocrinol Metab*, **89**, 2595–2600.

Haqqi TM, Anthony DD, Gupta S, *et al.* (1999) Prevention of collagen-induced arthritis in mice by a polyphenolic fraction from green tea. *Proc Natl Acad Sci U S A*, **96**, 4524–4529.

Heber D, Zhang Y, Yang J, Ma JE, Henning SM, Li Z. (2014) Green tea, black tea, and oolong tea polyphenols reduce visceral fat and inflammation in mice fed high-fat, high-sucrose obesogenic diets. *J Nutr*, **144**(9), 1385–1393.

Hermida N, Balligand JL. (2014) Low-density lipoprotein-cholesterol-induced endothelial dysfunction and oxidative stress: the role of statins. *Antioxid Redox Signal*, **20**(8), 1216–1237.

Hibi M, Takase H, Iwasaki M, Osaki N, Katsuragi Y. (2018) Efficacy of tea cate-chin-rich beverages to reduce abdominal adiposity and metabolic syndrome risks in obese and overweight subjects: a pooled analysis of 6 human trials. *Nutr Res*, **55**, 1–10.

Hodges JK, Sasaki GY, Bruno RS. (2020) Anti-inflammatory activities of green tea catechins along the gut-liver axis in nonalcoholic fatty liver disease: les-sons learned from preclinical and human studies. *J Nutr Biochem*, **85**, 108478.

Holguin F, Grasemann H, Sharma S, *et al.* (2019) L-Citrulline increases nitric oxide and improves control in obese asthmatics. *JCI Insight*, **4**(24), e131733.

Houstis N, Rosen ED, Lander ES. (2006) Reactive oxygen species have a causal role in multiple forms of insulin resistance. *Nature*, **440**, 944–948.

Huang F, Zheng X, Ma X, Jiang R, Zhou W, Zhou S, Zhang Y, Lei S, Wang S, Kuang J, Han X, Wei M, You Y, Li M, Li Y, Liang D, Liu J, Chen T, Yan C, Wei R, Rajani C, Shen C, Xie G, Bian Z, Li H, Zhao A, Jia W. (2019) Theabrownin from Pu-erh tea attenuates hypercholesterolemia via modula-tion of gut microbiota and bile acid metabolism. *Nat Commun*, **10**(1), 4971.

Huang L-H, Liu C-Y, Wang L-Y, Huang C-J, Hsu C-H (2018) Effects of green tea extract on overweight and obese women with high levels of low density-lipoprotein-cholesterol (LDL-C): a randomised, double-blind, and cross-over placebo-controlled clinical trial. *BMC Complement Altern Med*, **18**(1), 294.

Huang J, Wang Y, Xie Z, Zhou Y, Zhang Y, Wan X. (2014) The anti-obesity effects of green tea in human intervention and basic molecular studies. *Eur J Clin Nutr*, **68**(10), 1075–1087.

Hung PF, Wu BT, Chen HC, *et al.* (2005) Antimitogenic effect of green tea (-)-epigallocatechin gallate on 3T3-L1 preadipocytes depends on the ERK and Cdk2 pathways. *Am J Physiol Cell Physiol*, **288**, C1094–C1108.

Jiao HL, Ye P, Zhao B.L. (2003) Protective effects of green tea polyphenols on human HepG2 cells against oxidative damage of fenofibrate. *Free Radic Biol Med*, **35**, 1121–1128.

Kao YH, Hiipakka RA, Liao S. (2000) Modulation of obesity by a green tea cat-echin. *Am J Clin Nutr*, **72**, 1232–1234.

Kao YH, Chang HH, Lee MJ, Chen CL. (2006) Tea, obesity, and diabetes. *Mol Nutr Food Res*, **50**, 188–210.

Karam BS, Chavez-Moreno A, Koh W, Akar JG, Akar FG. (2017) Oxidative stress and inflammation as central mediators of atrial fibrillation in obesity and diabetes. *Cardiovasc Diabetol*, **16**(1), 120.

Karolczak D, Seget M, Bajerska J, Błaszczyk A, Drzymała-Czyż S, Walkowiak J, Marszałek A. (2019) Green tea extract prevents the development of nonal-coholic liver steatosis in rats fed a high-fat diet. *Pol J Pathol*, **70**(4), 295–303.

Kersten S, Desvergne B, Wahli W. (2000) Roles of PPARs in health and disease. *Nature*, **405**, 421–424.

Khosravi M, Poursaleh A, Ghasempour G, Farhad S, Najafi M. (2019) The effects of oxidative stress on the development of atherosclerosis. *Biol Chem*, **400**(6), 711–732.

Klaus S, Pültz S, Thöne-Reineke C, Wolfram S. (2005) Epigallocatechin gallate attenuates diet-induced obesity in mice by decreasing energy absorption and increasing fat oxidation. *Int J Obes (London)*, **29**(6), 615–623.

Kojta I, Chacińska M, Błachnio-Zabielska A. (2020) obesity, bioactive lipids, and adipose tissue inflammation in insulin resistance. *Nutrients*, **12**(5), 1305.

Kuroda Y, Mitsui T, Kunishige M, *et al.* (2006) Parkin enhances mitochondrial biogenesis in proliferating cells. *Hum Mol Genet*, **15**, 883–895.

Lee MJ, Maliakal P, Chen L, *et al.* (2002) Pharmacokinetics of tea catechins after ingestion of green tea and (-)-epigallocatechin-3-gallate by humans: formation of different metabolites and individual variability. *Cancer Epidemiol Biomarkers Prev*, **11**, 1025–1032.

Li Y, Rahman SU, Huang Y, Zhang Y, Ming P, Zhu L, Chu X, Li J, Feng S, Wang X, Wu J. (2020) Green tea polyphenols decrease weight gain, ameliorate alteration of gut microbiota, and mitigate intestinal inflammation in canines with high-fat-diet-induced obesity. *J Nutr Biochem*, **78**, 108324.

Liu B, Zhang J, Sun P, Yi R, Han X, Zhao X. Raw (2019) Bowl tea (Tuocha) polyphenol prevention of nonalcoholic fatty liver disease by regulating intestinal function in mice. *Biomolecules*, **9**(9), 435.

Lowell BB, Spiegelman BM. (2000) Towards a molecular understanding of adaptive thermogenesis. *Nature*, **404**, 652–660.

Lu C, Zhu W, Shen CL, Gao W. (2012) Green tea polyphenols reduce body weight in rats by modulating obesity-related genes. *PLoS One*, **7**(6), e38332.

Marchio P, Guerra-Ojeda S, Vila JM, Aldasoro M, Victor VM, Mauricio MD. (2019) Targeting early atherosclerosis: a focus on oxidative stress and inflammation. *Oxid Med Cell Longev*, **2019**, 8563845.

Maron DJ, Lu GP, Cai NS, Wu ZG, Li YH, Chen H, Zhu JQ, Jin XJ, Wouters BC, Zhao J. (2003) Cholesterol-lowering effect of a theaflavin-enriched green tea extract: a randomized controlled trial. *Arch Intern Med*, **163**(12), 1448–1453.

Matsumoto N, Okushio K, Hara Y. (1998) Effect of black tea polyphenols on plasma lipids in cholesterol-fed rats. *J Nutr Sci Vitaminol (Tokyo)*, **44**(2), 337–342.

Matsuda M, Shimomura I. (2013) Increased oxidative stress in obesity: implications for metabolic syndrome, diabetes, hypertension, dyslipidemia, atherosclerosis, and cancer. *Obes Res Clin Pract*, **7**(5), e330-41.

Mendel S, Youdim MB. (2004) Green tea and cancer Catechin polyphenols: neurodegeneration and neuroprotection in neurodegenerative diseases. *Free Radic Biol Med*, **37**, 304–317.

Miles PD, Romeo OM, Higo K, Cohen A, Rafaat K, Olefsky JM. (1997) TNF-alpha-induced insulin resistance in vivo and its prevention by troglitazone. *Diabetes*, **46**, 1678–1683.

Mitscher LA, Jung M, Shankel D, Dou JH, Steele L, Pillai SP. (1997) Chemoprotection: a review of the potential therapeutic antioxidant properties of green tea (Camellia sinensis) and certain of its constituents. *Med Res Rev*, **17**, 327–365.

Moon HS, Lee HG, Seo JH, *et al.* (2006) Down-regulation of PPARgamma2-induced adipogenesis by PEGylated conjugated linoleic acid as the pro-drug: attenuation of lipid accumulation and reduction of apoptosis. *Arch Biochem Biophys*, **456**, 19–29.

Moon HS, Lee HG, Choi YJ, Kim TG, Cho CS. (2007) Proposed mechanisms of (-)-epigallocatechin-3-gallate for anti-obesity. *Chem Biol Interact*, **167**(2), 85–98.

Muramatsu K, Fukuyo M, Hara Y. (1986) Effect of green tea catechins on plasma cholesterol level in cholesterol-fed rats. *J Nutr Sci Vitaminol (Tokyo)*, **32**(6), 613–622.

Ohara Y, Peterson TE, Harrison DG. (1993) Hypercholesterolemia increases endothelial superoxide anion production. *J Clin Invest*, **91**, 2546–2551.

Ohara Y, Peterson TE, Sayegh HS, (1995) Dietary correction of hypercholesterolemia in the rabbit normalizes endothelial superoxide anion production. Subramanian RR, Wilcox JN, Harrison DG. *Circulation*, **92**(4), 898–903.

Pan H, Gao Y, Tu Y. (2016) Mechanisms of body weight reduction by black tea polyphenols. *Molecules*, **21**(12), 1659.

Paoletti R, Bolego C, Poli A, Cignarella A. (2006) Metabolic syndrome, inflammation and atherosclerosis. *Vasc Health Risk Manag*, **2**, 145–152.

Phung OJ, Baker WL, Matthews LJ, Lanosa M, Thorne A, Coleman CI. (2010) Effect of green tea catechins with or without caffeine on anthropometric measures: a systematic review and meta-analysis. *Am J Clin Nutr*, **91**, 73–81.

Porter MH, Cutchins A, Fine JB, Bai Y, DiGirolamo M. (2002) Effects of TNF-alpha on glucose metabolism and lipolysis in adipose tissue and isolated fat-cell preparations. *J Lab Clin Med*, **139**, 140–146.

Qanungo S, Das M, Haldar S, Basu A. (2005) Epigallocatechin-3-gallate induces mitochondrial membrane depolarization and caspase-dependent apoptosis in pancreatic cancer cells. *Carcinogenesis*, **26**, 958–967.

Rains TM, Agarwal S, Maki KC. (2011) Antiobesity effects of green tea catechins: a mechanistic review. *J Nutr Biochem*, **22**(1), 1–7.

Richardson DK, Kashyap S, Bajaj M, *et al.* (2005) Lipid infusion decreases the expression of nuclear encoded mitochondrial genes and increases the expression of extracellular matrix genes in human skeletal muscle. *J Biol Chem*, **280**, 10290–10297.

Ridker PM, Morrow DA. (2003) C-reactive protein, inflammation, and coronary risk. *Cardiol Clin*, **21**, 315–325.

Ritov VB, Menshikova EV, He J, *et al.* (2005) Deficiency of subsarcolemmal mitochondria in obesity and type 2 diabetes. *Diabetes*, **54**, 8–14.

Rondinone CM. (2006) Adipocyte-derived hormones, cytokines, and mediators. *Endocrine*, **29**, 81–90.

Rosen ED, Spiegelman BM. (2001) PPARgamma: a nuclear regulator of metabolism, differentiation, and cell growth. *J Biol Chem*, **276**, 37731–37734.

Sarparanta J, García-Macia M, Singh R. (2017) autophagy and mitochondria in obesity and type 2 diabetes. *Curr Diabetes Rev*, **13**(4), 352–369.

Sonta T, Inoguchi T, Tsubouchi H, *et al.* (2004) Evidence for contribution of vascular NAD(P)H oxidase to increased oxidative stress in animal models of diabetes and obesity. *Free Radic Biol Med*, **37**, 115–123.

Thethi T, Bratcher C, Fonseca V. (2006) Metabolic syndrome and heart-failure. *Heart Fail Clin*, **2**, 1–11.

Tian WX. (2006) Inhibition of fatty acid synthase by polyphenols. *Curr Med Chem*, **13**(8), 967–977.

Unamuno X, Gómez-Ambrosi J, Rodríguez A, Becerril S, Frühbeck G, Catalán V. (2018) Adipokine dysregulation and adipose tissue inflammation in human obesity. *Eur J Clin Invest*, **48**(9), e12997.

Vazquez Cisneros LC, Lopez-Uriarte P, Lopez-Espinoza A, Navarro Meza M, Espinoza-Gallardo AC, Guzman Aburto MB. (2017) Effects of green tea and its epigallocatechin (EGCG) content on body weight and fat mass in humans: a systematic review. *Nutr Hosp*, **34**, 731–737.

Wang H, Wen Y, Du Y, Yan X, Guo H, Rycroft JA, Boon N, Kovacs EM, Mela DJ. (2010) Effects of catechin enriched green tea on body composition. *Obesity (Silver Spring)*, **18**(4), 773–779.

Wang S, Huang Y, Xu H, Zhu Q, Lu H, Zhang M, Hao S, Fang C, Zhang D, Wu X, Wang X, Sheng J. (2017) Oxidized tea polyphenols prevent lipid accumulation in liver and visceral white adipose tissue in rats. *Eur J Nutr*, **56**(6), 2037–2048.

Wang X, Song KS, Guo QX, Tian WX. (2003) The galloyl moiety of green tea catechins is the critical structural feature to inhibit fatty-acid synthase. *Biochem Pharmacol*, **66**, 2039–2047.

Wolfram S, Wang Y, Thielecke F. (2006) Anti-obesity effects of green tea: from bedside to bench. *Mol Nutr Food Res*, **50**, 176–187.

Yan J, Zhao Y, Zhao B. (2013) Green tea catechins prevent obesity through modulation of peroxisome proliferator-activated receptors. *Sci China Life Sci*, **56**, 804–810.

Yang CS, Zhang J, Zhang L, Huang J, Wang Y. (2016) Mechanisms of body weight reduction and metabolic syndrome alleviation by tea. *Mol Nutr Food Res*, **60**(1), 160–174.

Yokozawa T, Nakagawa T, Kitani K. (2002) Antioxidative activity of green tea polyphenol in cholesterol-fed rats. *J Agric Food Chem*, **50**(12), 3549–3552.

Yu W, Rohli KE, Yang S, Jia P. (2020) Impact of obesity on COVID-19 patients. *J Diabetes Complications*, **26**, 107817.

Zhang R, Xiao W, Wang X, Wu X, Tian W. (2006) Novel inhibitors of fatty-acid synthase from green tea (Camellia sinensis Xihu Longjing) with high activity and a new reacting site. *Biotechnol Appl Biochem*, **43**(Pt 1), 1–7.

Zhang Y, Li MM, Hua TM, Sun QY. (2018) The protective effect of tea polyphenols on chronic alcoholic liver injury in rats. *Zhongguo Ying Yong Sheng Li Xue Za Zhi*, **34**(6), 481–484.

Zhao B-L. (2020) The pros and cons of drinking tea. *Tradit Med Mod Med*, **3**(3), 1–12.

Zhao B-L, Liu S-L, Chen R-S, Xin W-J. (1992) Scavenging effect of catechin on free radicals studied by molecular orbital calculation. *Acta Pharmacol Sinica*, **13**, 9.

Zou XJ, Ding YH, Liang B. (2012) The mechanisms of weight-cutting effect and bioactive components in Pu-erh tea. *Dongwuxue Yanjiu*, **33**(4), 421–426.

https://doi.org/10.1142/9789811274213_0011

Chapter 11

Tea Polyphenols and Diabetes

Baolu Zhao

Institute of Biophysics, Chinese Academy of Sciences, Beijing, China

11.1 Introduction

Metabolic syndrome is known as the main health killer in the 21st century. It is a state in which a variety of diseases such as hypertension, abnormal blood glucose, dyslipidemia, and obesity accumulate in the human body, which directly leads to the occurrence of serious cardiovascular diseases and death. Metabolic syndrome mainly refers to high body weight, hypertension, hyper-lipidemia, hyperglycemia, high fatty acid, high uric acid, high blood viscosity, high insulin anti-body, etc. If there are two of the eight highs, it can be called metabolic syndrome. Diabetes is the most common metabolic disease. With its worldwide popularity, diabetes has become an important health problem. At present, there are 463 million people in the world (9.3% of 20–79 years old adults) suffering from diabetes. There are also 1 million children under 20 years old and 100,000 teenagers suffering from type 1 diabetes. This figure is expected to increase to 578 million by 2030. The incidence of metabolic syndrome in adults over the age of 20 in the United States is 24%, which includes 60% of obese patients. The US Centers for Disease Control (CDC) estimates that 29 million Americans have diabetes and 70% of the diabetic patients will develop to diabetic peripheral neuropathy. The economic burden of diabetes is also shocking. The American Diabetes Association's direct medical costed for diabetes was US $176 billion in 2012. The total cost of diabetes-related healthcare is estimated at about US $760 billion and is expected to reach US $825 billion over the next 10 years. According to the survey of the Chinese Medical Association, the prevalence of metabolic syndrome is 14–16% among the people over 20 years old in Chinese cities. Metabolic syndrome increases with age and reaches the peak among the people aged 50–70, with more female patients than men. It is estimated that there will be one patient in every eight in the next seven years.

Diabetes is a chronic disease with impaired insulin secretion, inappropriate insulin action or both. Insulin deficiency can lead to high blood sugar levels. In addition, diabetes can also lead to bone fragility and bone healing. Diabetes and obesity are so closely intertwined that scientists now call them a global metabolic syndrome. Diabetes is the main component of metabolic syndrome. Studies have shown that diabetes is closely related to reactive oxygen species (ROS) and reactive nitrogen species (ROS) free radical. The metabolic syndrome is characterized by a group of metabolic risk factors in one person, including abdominal obesity, atherogenic dyslipidemia, elevated blood pressure, and insulin resistance or glucose intolerance [Flier, 2004]. The dominant underlying risk factors for metabolic syndrome are abdominal obesity and insulin resistance [Freidenberg *et al.*, 1988; Spiegelman & Flier, 2001]. Obesity is the causal component in this syndrome [Montague & O'Rahilly, 2000], but the mechanistic role of obesity has not been fully elucidated. The oxidative stress in adipocyte induced by fat overload might be the origin of metabolic syndrome caused by obesity [Furukawa *et al.*, 2004]. Oxidative stress in adipocyte impaired insulin signals and decreased insulin-stimulating glucose uptake [Houstis *et al.*, 2006]. As adipocyte is not only a glucose and fat storage position, it also could secrete cytokines to affect glucose and lipids homeostasis [Wang *et al.*, 2008]. The fat over-loaded adipocyte secreted ROS, and RNS, TNF-α, resistin, and free fatty acids to cause insulin resistance in muscles and liver [Guilherme *et al.*, 2008]. ROS and RNS secreted by adipocyte could change the whole body redox system through transportation of blood. Insulin deficiency can lead to high levels of blood glucose. If not strictly controlled, it can lead to disability and life-threatening health complications, including cardiovascular disease, retinopathy, neuropathy, nephropathy, and prolonged/incomplete wound healing.

Drinking of green tea has been found to have an anti-diabetes effect for long time [Mackenzie *et al.*, 2007; Iso *et al.*, 2006; Ryu *et al.*, 2006]. It was found that green tea polyphenols (GTPs) could protect pancreas from oxidative damage [Kim *et al.*, 2003]. Recent findings show that GTPs could increase insulin sensitivity in diabetic animals [Wu *et al.*, 2004; Serisier *et al.*, 2008]. GTPs could enhance glut-4 expression, increase glucose tolerance, and promote glucose uptake in adipocyte and muscles [Ashida *et al.*, 2004; Janle *et al.*, 2005]. Also, GTPs could decrease oxidative stress in diabetic rat [Sabu *et al*, 2002; Babu *et al.*, 2006]. However, how GTPs improved the impaired insulin resistance is still not fully understood. As GTPs is one of the most used antioxidants, and epidemiological evidence also show that oxidative stress is associated with insulin resistance [Fukino *et al.*, 2005], especially oxidative stress in adipocyte might be the connection between obesity and type 2 diabetes, we propose that one mechanism of GTPs anti-diabetes effect is that GTPs decrease oxidative stress in adipocyte and improve insulin sensitivity. GTPs could directly scavenge ROS and RNS secreted by adipocyte and could act on ROS and RNS from other tissues, thus the whole body, especially muscle and liver, would have decreased oxidative stress. Since oxidative stress in adipocyte plays a central role in obesity causing insulin resistance and type 2 diabetes, we conducted our research in obese and diabetic KK-ay mouse, diet induced obese rat, and 3T3-L1 adipocyte to detect the improvement of insulin signals and the enhancement of glucose uptake by GTPs through ameliorate of oxidative stress. This chapter discusses the pathogenesis of diabetes, oxidative stress, and the prevention and treatment of diabetes by tea polyphenols.

11.2 Syndrome of diabetes

Diabetes is a metabolic disease characterized by hyperglycemia. Hyperglycemia is caused by defective insulin secretion or impaired biological effects, or both. Chronic hyperglycemia leads to chronic damage and dysfunction of various tissues, especially eyes, kidneys, heart, blood vessels, and nerves.

11.2.1 *Types of diabetes*

The main types of diabetes are type 1 diabetes, type 2 diabetes and gestational diabetes mellitus. Among these three types, type 1 diabetes is the most common type, accounting for 90% of all diabetes in the world. Type 1 diabetes is an autoimmune, multi-gene disease, wherebt the body cannot produce insulin, or cannot produce enough insulin, or cannot make full use of insulin. The cause is very complicated. Environmental factors also play an important role in the development of type 1 diabetes, such as toxins, infections, eating habits and geographical differences. In general, researchers think that type 1 diabetes is autoimmune disorder, and cytokines secreted by immune cells, such as T cells, selectively attack insulin secretion resulting apoptosis of β cell. Some researchers believe that type 1 diabetes is caused by loss of immune tolerance. Type 2 diabetes is a metabolic syndrome, which is closely related to obesity. It affects multiple organs, and metabolic and hormonal disorders gradually lead to diabetes. Type 2 diabetes is a very serious international health problem. It is estimated that type 2 diabetes will increase from 135 million in 1995 to 300 million of 2025 worldwide in China, and which will mainly occur in developing countries. Gene and environmental factors play an important role in the formation and development of type 2 diabetes. General researchers believe that over-nutrition of adipocytes and glycolipid toxicity are the underlying causes of type 2 diabetes. The over-nutrition of fat cells not only reduces their ability to store carbohydrates, but also secretes a large number of cytokines, inhibiting the function of fat and muscle, leading to type 2 diabetes. Excessive sugar and lipids produces oxidative stress, damage insulin signal channels, and damage the function of β cells. High glucose and high fat lead to oxidative stress, which activates serine/threonine kinase signal channels, such as IKK-β. Once activation, these kinases can phosphorylate insulin receptors and insulin receptor substrates (IRS-1 and IRS-2) [Charles *et al.*, 2004; Permutt *et al.*, 2001; Camp *et al.*, 2002].

11.2.2 *Etiology of diabetes*

The etiology of diabetes is very complex. There are many factors that can lead to diabetes. There are mainly three important factors, namely family genetic factors, environmental factors, and personal habits.

Family heredity is an important factor of diabetes. There is obvious genetic heterogeneity in both type 1 and type 2 diabetes. Diabetes has a family tendency, and 1/4–1/2 patients have a family history of diabetes. There are at least 60 genetic syndromes associated with diabetes. There are many DNA loci involved in the pathogenesis of type 1 diabetes, among which, the polymorphism of DQ locus in human leukocyte antigen (HLA) antigen gene is the most closely related. Many specific gene mutations have been found in type 2 diabetes, such as insulin gene, insulin receptor gene, glucokinase gene, mitochondrial gene, etc. The studies found 69 distinct diabetes syndromes. Thirty (43.5%) syndromes included diabetes mellitus as a cardinal clinical feature and 56 (81.2%) were fully genetically elucidated. Sixty-three syndromes (91.3%) were described more than once in independent case reports, of which 59 (93.7%) demonstrated clinical heterogeneity. Syndromes associated with diabetes mellitus are more numerous and diverse than previously anticipated [Shi *et al.*, 2021].

Human living environment is an important factor of diabetes. The factors contributing to diabetes risk includes aspects of diet quality and quantity, little physical activity, increased monitor viewing time or sitting in general, exposure to noise or fine dust, short or disturbed sleep, smoking, stress and depression, and a low socioeconomic status. In general, these factors promote an increase in body mass index. Since loss of β-cell function is the ultimate cause of developing overt type 2 diabetes, environmental and lifestyle changes must have resulted in a higher risk of β-cell damage in those at genetic risk. Multiple mechanistic pathways may come into play. Obesity caused by excessive eating and reduced physical activity is the most important environmental factor of type 2 diabetes, which makes individuals with genetic susceptibility to type 2 diabetes prone to disease. Patients with type 1 diabetes have abnormal immune system. After infection with some viruses such as Coxsackie virus, rubella virus, and parotid adenovirus, they will cause autoimmune reaction and destroy β cells-released insulin [Kolb & Martin, 2017].

Epidemiological investigation shows that obesity is closely related to diabetes. Obesity is an important factor inducing diabetes. About 60–80% of adult diabetes patients are obese before the onset of the disease. The degree of obesity is in direct proportion to the incidence rate of diabetes. Basic research show that with aging and the gradual reduction of physical activity, the proportion of human muscle and fat is also changing. From the age of 25–75, muscle tissue gradually decreased from 47% of body weight to 36%, while fat increased from 20% to 36%. This is one of the main reasons for the significant increase of diabetes in the elderly, especially the obese and fat elderly. Excessive diet, excess nutrition, and little exercise induces diabetes by overburdening β cell.

11.2.3 *Clinical manifestation of diabetes*

Diabetes mellitus has become a global epidemic and presents many complications, usually proportional to the degree and duration of hyperglycemia. Periodontal disease, periapical lesions, xerostomia, and taste disturbance were more prevalent among diabetic patients. The increase of blood sugar and urine sugar can cause osmotic diuresis, which can cause polyuria. With the increase of blood glucose and the loss of a large amount of water, the blood osmolality will also increase correspondingly. The high blood osmolality can stimulate the thirst center of the hypothalamus, resulting in the symptoms of thirst and polydipsia. Due to the relative or absolute lack of insulin, the glucose in the body cannot be used and the consumption of protein and fat increases, resulting in fatigue and weight loss. In order to compensate for the loss of sugar and maintain body activity, more food is needed. This has formed the typical "three more and one less" symptoms — eat more, drink more, urinate more, and lose body weight. The symptoms of polydipsia and polyuria in patients with diabetes are direct proportion to the severity of the disease. In addition, it is worth noting that the more patients eat, the higher their blood sugar will be, the more sugar in their urine will be lost, and the more severe their hunger will be, eventually leading to a vicious circle. This may eventually lead to diabetes insipidus, a disorder characterized by a high hypotonic urinary output of more than 50 ml/kg body weight per 24 hours, with associated polydipsia of more than three liters a day. Besides central DI, further underlying etiologies of DI can be due to other primary forms (renal origin) or secondary forms of polyuria (resulting from primary polydipsia) [Christ-Crain & Gaisl, 2021].

11.2.4 *Hazards of diabetes*

Long term hyperglycemia will cause pathological changes in various tissues and organs of the body, leading to acute and chronic complications, such as water loss, electrolyte disorder, nutritional deficiency, decreased resistance, impaired renal function, neuropathy, fundus diseases, cardiovascular and cerebrovascular diseases, diabetes foot, etc. It can seriously damage large blood vessels and micro-vessels and endanger the heart, brain, kidney, peripheral nerves, eyes, feet, etc. According to the statistics of the World Health Organization, diabetes has more than 100 complications, which is a disease with the most known complications. More than half of diabetes deaths are caused by cardiovascular and cerebrovascular diseases and 10% are caused by renal diseases. Amputation due to diabetes is 10–20 times more than that due to non-diabetes. Clinical data show that about 10 years after the onset of diabetes, 30–40% of patients will have at least one complication, and once the complication occurs, drug treatment is difficult to reverse. Studies also found the associations of serum folate and vitamin B12 levels with cardiovascular outcomes among patients with type 2 diabetes (T2D) [Liu *et al.*, 2022].

From the above discussion, we can see that diabetes is extremely harmful to human health, so it is imperative to control hyperglycemia and diabetes. To control diabetes, it is very important to understand the pathogenesis of diabetes and find effective methods. Studies have found that diabetes is closely related to oxidative stress injury, and natural antioxidant tea polyphenols may be an important method to prevent diabetes. This chapter will discuss this in depth.

11.3 Diabetes and oxidative stress

Increasing evidence from research on several diseases show that oxidative stress is associated with the pathogenesis of diabetes. Oxidative stress is defined as a disturbance between the production of reactive oxygen species (ROS) and reactive nitrogen species (RNS) of free radicals, and antioxidant defenses. Oxidative stress is the most relevant mechanism in the pathophysiology of type 1 and type 2 diabetes mellitus and the pathogenesis of diabetic vascular complications. It is also the main cause of death of diabetic patients caused by diabetic complications. Examples of the possible consequences of free radical damage are provided with special emphasis on lipid peroxidation. The injuries of important free radicals are described and biological sources of origin, together with the major antioxidant defense mechanisms in diabetes mellitus, are discussed in this part. Based on this research, the emerging concept is that oxidative stress is the risk factor of diabetes that exerts its deleterious effects. Oxidative stress causes a complex dysregulation of cell metabolism and cell-cell homeostasis. The role of oxidative stress in the pathogenesis of insulin resistance and β-cell dysfunction need to be discussed. The question of whether oxidative stress is increased in diabetes mellitus will also be discussed in this part.

11.3.1 *Reactive oxygen species, oxidative stress, and diabetes*

From the molecular process of biological oxidation reaction, oxygen, as an essential substance, has duality. On the one hand, oxygen, as the terminal electron receptor of respiratory chain, participates in the oxidative phosphorylation reaction of ATP, and maintains the important energy metabolism process of life. On the other hand, oxygen can generate harmful oxygen free radicals, superoxide anion, hydrogen peroxide, and hydroxyl free radicals through a series of chemical reactions, destroy the dynamic balance between oxides and antioxidants in the cell system, and trigger the cell oxidative stress response [Maximo *et al.*, 2002]. Fructose-rich diet induced an early pro-oxidative state and metabolic dysfunction in abdominal adipose tissue that would favor the overall development of ischemia reperfusion (IR) and ROS and further development of pancreatic beta-cell failure [Rebolledo *et al.*, 2008].

High concentrations of H_2O_2 activate insulin signaling and induce typical metabolic actions of insulin. This result is the first experimental documentation on the link between ROS and insulin [Czech *et al.*, 1974]. H_2O_2 employs the same pathway as insulin and causes downstream propagation of the signal yielding typical metabolic actions of insulin. H_2O_2 increases glucose uptake in adipocytes and muscles [Higaki *et al.*, 2008]. In obese mice, an increased H_2O_2 generation by adipose tissue could be observed prior to diabetes onset [Furukawa *et al.*, 2004]. Increased numbers of lipid peroxidation markers have been observed in the liver, in animal models of diabetes, and obesity [Svegliati-Baroni *et al.*, 2006]. Recent evidence for systemic oxidative stress includes the detection of increased circulating and urinary levels of the lipid peroxidation product F2-isoprostane (8-epi-prostaglandin F2α) in both type 1 and type 2 diabetic patients [Davi *et al.*, 2003].

ROS and oxidative stress plays a key role in the occurrence, development, and chronic complications of diabetes. Hyperglycemia induced ROS can reduce islets β cell insulin secretion, lowers insulin sensitivity and signal transduction in insulin sensitive tissues, and change endothelial cell function in type 1 and type 2 diabetes. Recent studies have shown that oxidative stress and excessive fatty acids can reduce glucose stimulated insulin secretion, inhibit insulin gene expression, and lead to β-cell death causes type 2 diabetes to transform into type 1 diabetes. Adipose tissue not only stores fat and energy, but also secretes many bioactive proteins, such as leptin and many inflammatory factors, such as TNF-α, IL-6, IL-8, IL-10, MCP-1, and adipose tissue can also secrete some proteins. Islets isolated from male Sprague-Dawley rats were exposed to palmitate (0.125 mM or 0.25 mM), oleate (0.125 mM), or octanoate (2.0 mM) during culture. Insulin responses were subsequently tested in the absence of free fatty ac (FFA). After a 48-hour exposure to FFA, insulin secretion during basal glucose (3.3 mM) was several-fold increased. However, during stimulation with 27 mM glucose, secretion was inhibited by 30–50% and proinsulin biosynthesis by 30–40%. In the rat that had long-term exposure to fatty acids, β-cell function was inhibited *in vivo* and *in vitro*. To further assess the clinical significance of these findings, the effects of fatty acids on glucose-induced insulin release and biosynthesis and on pyruvate dehydrogenase (PDH) activity was tested in human islets. Fatty acids, as well as ketone bodies, diminish β-cell responsiveness to glucose in human islets by the way of a glucose-fatty acid cycle. Increased plasma concentrations of fatty acids and ketones are likely to be important factors behind the negative influences on β-cell function exerted by a diabetic state in both type 1 and type 2 diabetes. Inflammatory factors can directly act on adipose tissue and muscle tissue to produce insulin resistance or induce oxidative stress in adipose and muscle cells to produce insulin resistance.

The over-nourished adipocytes themselves will be in a state of oxidative stress, produce insulin resistance, and secrete a large amount of ROS into the blood. Inflammatory factors and ROS can cause direct damage on β cells, which cause apoptosis, transforming type 2 diabetes to type 1 diabetes [Zhou & Grill, 1994, 1995; Mason *et al.*, 1999].

Long-term exposure of pancreatic β cells to palmitate decreases insulin gene expression only in the presence of elevated glucose concentrations, in part through inhibition of insulin gene promoter activity. The occurrence of diabetes and its complications is related to oxidative stress, ROS, and low chronic inflammation. ROS is an imbalance between cell oxidant and antioxidant system. It is the result of excessive production of free radicals and related ROS, which results in the promotion of chronic inflammation by high glucose, and promotes the formation of ROS, which eventually leads to diabetic complications, including vascular dysfunction. In addition, the increase of ROS level will reduce insulin secretion and damage the signal transduction of insulin sensitivity and insulin responsive tissues. Therefore, the appropriate treatment of high glucose and inhibition of excessive production of ROS is crucial for delaying the occurrence and development of diabetes and preventing subsequent complications. Gene expression analysis suggests that ROS levels are increased in both models, it was confirmed through measures of cellular redox state. ROS have previously been proposed to be involved in insulin resistance, although evidence for a causal role has been scant. In cell culture using six treatments designed to alter ROS levels, including two small molecules antioxidants and four transgenes, all treatments ameliorated insulin resistance to varying degrees. One of these treatments was tested in obese, insulin-resistant mice and was shown to improve insulin sensitivity and glucose homeostasis. ROS have a causal role in multiple forms of insulin resistance. Insulin resistance is a cardinal feature of type 2 diabetes and is characteristic of a wide range of other clinical and experimental settings. Dexamethasone (DEX) and TNF-α caused insulin resistance by inducing oxidative stress in adipocyte. A genomic analysis of two cellular models of insulin resistance was reported, one induced by treatment with the cytokine tumor-necrosis factor-alpha and the other with the glucocorticoid DEX. [Jacqueminet *et al.*, 2000; Briaud *et al.*, 2001; Houstis *et al.*, 2006].

In a diabetic milieu, high levels of ROS are induced. ROS are generated in macrophage activation and function in diabetes. This contributes to the vascular complications of diabetes. Recent studies have shown that ROS formation is exacerbated in diabetic monocytes and macrophages due to a glycolytic metabolic shift. Macrophages are important players in the progression of diabetes and promote inflammation through the release of pro-inflammatory cytokines and proteases. Since ROS is an important mediator for the activation of pro-inflammatory signaling pathways, obesity and hyperglycemia-induced ROS production may favor induction of M1-like pro-inflammatory macrophages during diabetes onset and progression. ROS induces mitogen-activated protein kinase (MAPK), signal transducer and activator of transcription (STAT), STAT6 and NFκB signaling, and interferes with macrophage differentiation via epigenetic (re) programming. ROS production links metabolism and inflammation in diabetes and its complications. ROS contributes to the crosstalk between macrophages and endothelial cells in diabetic complications [Rendra *et al.*, 2019]. ROS are the sources, consequences, and targeted therapy in type 2 diabetes. Oxidative stress has been considered as a central mediator in the progression of diabetic complication. The intracellular ROS leads to oxidative stress and it is raised from the mitochondria as well as by activation of five major pathways: increased polyol pathway flux, activation of protein kinase C (PKC) pathway, increased formation of advanced glycation end products (AGEs), overactivity of hexosamine pathway, and increased production of angiotensin II. The increased ROS through these pathways leads to β-cell dysfunction and insulin resistance, responsible for cell damage and death. The sources of ROS production and their involvement in the progression of diabetes are the pharmacological interventions and targeting of ROS in type 2 diabetes. ROS activation in type 2 diabetes, as well as ROS, is a possible target for its treatment. It would be a promising target for various strategies and drugs to modulate ROS levels in diabetes [Panigrahy *et al.*, 2017].

Two key observations that were made more than 30 years ago — oxidants can facilitate or mimic insulin action and H_2O_2 is generated in response to insulin stimulation of its target cells, have led to the hypothesis that ROS may serve as second messengers in the insulin action cascade. Specific molecular targets of insulin-induced ROS include enzymes whose signaling activity is modified via oxidative biochemical reactions, leading to enhanced insulin signal transduction, because chronic exposure to relatively high levels of ROS have also been associated with functional β-cell impairment and the chronic complications of diabetes. The best-characterized molecular targets of ROS are the protein-tyrosine phos-phatases (PTPs) because these important signaling enzymes require a reduced form of a critical cysteine residue for catalytic activity. PTPs normally serve as negative regulators of insulin action via the dephospho-rylation of the insulin receptor and its tyrosine-phosphorylated cellular substrates. However, ROS can rapidly oxidize the catalytic cysteine of target PTPs, effectively blocking their enzyme activity and reversing their inhibitory effect on insulin signaling [Goldstein *et al.*, 2005].

The level of ROS is closely related to insulin resistance. Application of antioxidants to improve insulin resistance by reducing the level of ROS has been confirmed [McClung *et al.*, 2004; Schulz *et al.*, 2007; Gomez-Cabrera *et al.*, 2008; Zhao, 2020]. It was found that tumor necrosis factor-α, both TNF-α and glucocorticoid, can induce insulin resistance but through different signal transduction mechanisms. It was found that 18% of genes had the same regulation for different stimuli and these genes were involved in the regulation of ROS. The level of ROS increased at different levels in both models. The application of different antioxidants or the expression of high-level antioxidant enzyme activity can improve the level of insulin resistance to varying degrees. Furthermore, antioxidant therapy can improve insulin sensitivity and glucose homeostasis in insulin resistant mice. These data suggest that the level of ROS is an important bioactive substance leading to insulin resistance. The role of ROS in the progression of aging and chronic diseases depends on their total amount. A certain amount of ROS plays an important role in maintaining normal physiological activities. ROS with important physiological functions include cytochrome P-450 and peroxidase activated transition compounds, hydroxyl radicals, produced by ethanol metabolism *in vivo*, RNA reduc-tion reaction, oxidation reaction, carboxylation, and hydroxylation reac-tion. Peroxidase and NADPH oxidase in macrophages react and oxidize unsaturated fatty acids to synthesize eicosanoic acid (precursor of prosta-glandins and leukotrienes) [Halliwell & Gutteridge, 1999]. However, excessive production of ROS will lead to aging and disease progression. Another pathway how oxidative stress in adipose tissue could induce whole-body insulin resistance is by secreting cytokines. Based on their effects on lipids and glucose metabolism, adipokines were generally divided into two groups: the positive group, including leptin, adiponectin etc., and the negative group, including TNF-α, resistin. Adiponectin could stimulate adipose, muscles, and other tissues to uptake glucose while resistin could induce inflammatory reactions in many glucose metabolism related tissues. Obesity-induced oxidative stress in adipocyte could reduce adiponectin but increase resistin secretion in adipocyte, which in turn resulted in adipose, liver, and muscles insulin resistance. In our experiments, dexamethasone (DEX) and TNF-α increased ROS production in adipocyte, thus reduced adiponectin synthesis and increased resistin synthesis. Tea polyphenols EGCG could weaken these changes caused by DEX and TNF-α as the results of its ROS scavenging effect [Yan *et al.*, 2012].

Major ROS-producing systems in vascular wall include NADPH (reduced form of nicotinamide adenine dinucleotide phosphate) oxidase, xanthine oxidase, the mitochondrial electron transport chain, and uncoupled endothelial nitric oxide (NO) synthase. All established cardiovascular risk factors such as hypercholesterolemia, hypertension, diabetes mellitus, and smoking enhance ROS generation and decrease endothelial NO production. Key molecular events in atherogenesis such as oxidative modification of lipoproteins and phospholipids, endothelial cell activation, and macrophage infiltration/activation are facilitated by vascular oxidative stress and inhibited by endothelial NO. Therefore, prevention of vascular oxidative stress and improvement of endothelial NO production represent reasonable therapeutic strategies in addition to the treatment of established risk factors (hypercholesterolemia, hypertension, and diabetes mellitus) [Förstermann *et al.*, 2017].

Above results show that in many environments, elevated levels of ROS are an important trigger for insulin resistance. ROS is involved in the interaction between macrophages and endothelial cells in diabetes complications. In type 1 diabetes and type 2 diabetes, especially type 2 diabetes, ROS is closely related to diabetes caused by over-nutrition and obesity.

11.3.2   *Nitric oxide, oxidative stress, and diabetes*

It is found that nitric oxide (NO) is closely related to diabetes. Increased oxidative stress and reduced NO bioavailability play a causal role in endothelial cell dysfunction occurring in the vasculature of diabetic patients. In addition to ROS causing oxidative stress in diabetes, reactive nitrogen (RNS) also plays an important role in oxidative stress and diabetes. Endogenous NO is a small signal molecule that stimulates the production of mitochondria. NO has an important role in the regulation of vascular tone and impaired NO activity could be implicated in the development of diabetic vasculopathy. Diabetes complications are characterized by endothelial (vascular) dysfunction. A number of studies have suggested that RNS play an important role in the pathogenesis of diabetic vascular dysfunction as the pathways of diabetic complications have a close relationship to oxidative stress. In particular, increased glucose leads to increased mitochondrial formation of RNS. Superoxide is a ROS that produces peroxynitrite with reaction of NO, a substance with stronger oxygen activity, which can lead to oxidative stress. It has also been documented in human studies that endothelial cells in diabetes fail to produce sufficient amount of NO and fail to relax in response to endothelium-dependent vasorelaxants (e.g., acetylcholine, bradykinin, shear stress, etc.) [Avogaro *et al.*, 2006]. There is a systematic review and meta-analysis about NO levels in patients with diabetes mellitus. These papers describe a meta-analysis conducted to evaluate if there is a relationship between NO levels and type 1 or type 2 diabetes. A literature search was done to identify all studies that investigated NO levels between type 1 or type 2 patients and non-diabetic subjects (controls). NO levels were increased in European type 1 diabetes patients compared with controls. No other ethnicity was evaluated in type 1 studies. NO levels were also increased in both European and Asian type 2 patients, but not in Latin American patients compared with controls. These data of meta-analysis detected a significant increase in NO levels in European type 1 diabetes patients as well as European and Asian type 2 patients. Further studies in other ethnicities are necessary to confirm these data [Assmann *et al.*, 2016].

Animal models of diabetes are associated with reduced bioavailability of NO and impaired endothelium-dependent relaxation [Durante *et al.*, 1988]. These results suggest a major role for NO in destruction by O_2^- in diabetes-associated vascular dysfunction. Interestingly, endothelial NO synthase (eNOS) knockout mice exhibited accelerated diabetic nephropathy [Zhao *et al.*, 2006]. Inducible NO synthase (iNOS) is very relevant to diabetic pathophysiology. Recent reports reveal that decreased expression of eNOS accompanies increased expression of iNOS and nitro-tyrosine during the progression of diabetes in rats [Nagareddy *et al.*, 2005]. The decrease of bioavailability of NO is related to diabetic macro-vascular and microvascular diseases. Diabetes is related to the decrease of nitric oxide produced by vascular endothelial dysfunction. Nitric oxide synthase (NOS) catalyzes the formation of NO. The role of NOS in patho-logical conditions such as obesity, diabetes, and heart disease is consid-ered. The uncoupling of eNOS and increased generation of free radicals in hyperglycemic conditions can also lead to the formation of highly RNS such as peroxynitrite, which can lead to DNA damage, carcinogenic muta-tions, and activation of critical pathways involved in cell proliferation and apoptosis [Milsom *et al.*, 2002].

RNS is related with endothelial function in diabetes. An increasing body of evidence suggests that oxidant stress is involved in the pathogen-esis of many cardiovascular diseases, including hypercholesterolemia, atherosclerosis, hypertension, heart failure related with diabetes. Recent studies have also provided important new insights into potential mecha-nisms underlying the pathogenesis of vascular disease induced by diabe-tes. Glycosylation of proteins and lipids, which can interfere with their normal function, activation of protein kinase C with subsequent alteration in growth factor expression, promotion of inflammation through the induction of cytokine secretion, and hyperglycemia-induced oxidative stress, are some of these mechanisms. Hyperglycemia-induced RNS con-tribute to cell and tissue dysfunction in diabetes. A variety of enzymatic and non-enzymatic sources of RNS exist in the blood vessels. These include NADPH oxidase, mitochondrial electron transport chain, xanthine oxidase, and nitric oxide synthase [Fatehi-Hassanabad *et al.*, 2010].

An important reason of oxidative stress injury in diabetes caused by RNS is that NO is produced together with superoxide anion free radical. NO reacts with superoxide anion to produce peroxynitrite, which is a substance with stronger oxygen activity and can lead to oxidative stress in diabetes.

11.3.3 *Mitochondrial oxidative damage and insulin resistance*

The role of altered mitochondria function has recently emerged as an important mechanism for the development of diabetic complications. Also, the mitochondria are a substantial source of superoxide production, preferentially during states of elevated intracellular glucose concentrations. The mitochondria function is regulated by several factors including nitric oxide, oxidative stress, mammalian target of rapamycin, ADP, and P(i) availability, which result in a complex regulation of ATP production and oxygen consumption, but also superoxide generation. Mitochondrial generated ROS in type 1 diabetes. The complex etiology of type 1 diabetes (T1D) is the outcome of failures in regulating immunity in combination with β cell perturbations. Mitochondrial dysfunction in β cells and immune cells may be involved in type 1 diabetes pathogenesis. Mitochondrial energy production is essential for the major task of β cells, which is the secretion of insulin in response to glucose. Mitochondria are a major site of ROS production. Under immune attack, mitochondrial ROS (mtROS) participate in β cell damage. Similarly, T cell fate during immune responses is tightly regulated by mitochondrial physiology, morphology, and metabolism. Production of mtROS is essential for signaling in antigen-specific T cell activation. Mitochondrial dysfunction in T cells has been noted as a feature of some human autoimmune diseases. Sensitivity of β cells to mtROS is associated with genetic type 1 diabetes risk loci in human and the type 1 diabetes-prone non-obese diabetic (NOD) mouse. Mitochondrial dysfunction and altered metabolism have also been observed in immune cells of NOD mice and patients with type 1 diabetes. This immune cell mitochondrial dysfunction has been linked to deleterious functional changes [Chen *et al.*, 2018].

Mitochondrial dysfunction plays a role in the pathogenesis of insulin resistance. Endoplasmic reticulum stress, through mainly increased oxidative stress, also plays important role in the etiology of insulin resistance, especially seen in non-alcoholic fatty liver disease. Lipotoxicity plays a role in insulin resistance and pancreatic beta cell dysfunction. Increased circulating levels of lipids and the metabolic alterations in fatty acid utilization and intracellular signaling, have been related to insulin resistance in muscle and liver. Different pathways, like novel protein kinase c pathways and the JNK-1 pathway are involved as the mechanisms of how lipotoxicity leads to insulin resistance in non-adipose tissue organs, such as liver and muscle. Endoplasmic reticulum stress, through mainly increased oxidative stress, also plays important role in the etiology of insulin resistance, especially seen in non-alcoholic fatty liver disease. Visceral adiposity and insulin resistance both increase the cardiometabolic risk and lipotoxicity seems to play a crucial role in the pathophysiology of these associations [Yazıcı & Sezer *et al.*, 2017]. Studies have shown that insulin resistance is related to mitochondrial dysfunction, reduction of mitochondrial number and decrease of ATP production [Kelley *et al.*, 2002; Nunomura *et al.*, 2001]. In pre-diabetes and diabetic patients, skeletal muscle involved in oxidative phosphorylation (OXPHOS) gene expression decreased [Mootha *et al.*, 2004]. Mitochondria is an important component in the body that produces ROS. If the efficiency of oxidative phosphorylation decreases, the genes involved in energy metabolism in the mitochondrial genome are eliminated, and ATP synthesis is accompanied by the production of more superoxide free radicals (O_2^-). Therefore, improving the function of mitochondria by reducing oxidative damage is a reasonable way to treat/improve insulin resistance.

Nitric oxide was found to trigger mitochondrial biogenesis in cells as diverse as brown adipocytes and 3T3-L1, U937, and HeLa cells. Moreover, the mitochondrial biogenesis induced by exposure to cold was markedly reduced in brown adipose tissue of endothelial nitric oxide synthase null-mutant (eNOS-/-) mice, which had a reduced metabolic rate and accelerated weight gain as compared to wild-type mice. Thus, a nitric oxide-cGMP-dependent pathway controls mitochondrial biogenesis and body energy balance [Nisoli *et al.*, 2003]. Oxide can promote the action of insulin, and hydrogen peroxide as a second messenger can stimulate the activation of insulin downstream signaling pathway. In addition, sugar restriction can prolong the life span of *Caenorhabditis elegans*, which is mainly thought to induce mitochondrial respiration and increase oxidative stress. Overexpression of glutathione peroxidase 1 can increase the development of insulin resistance in mice. The reason may be that the increase of glutathione peroxidase 1 inhibits endogenous reactive oxygen species and interferes with the effect of insulin.

Visceral adiposity and insulin resistance both increase the cardio-metabolic risk and lipotoxicity seems to play a crucial role in the pathophysiology of these associations [Schriner *et al.*, 2005]. More and more studies show that type 2 diabetes is closely related to mitochondrial dysfunction. Mitochondria are the main site of producing ROS in cells and also the main target of ROS. Oxidative damage of mitochondrial DNA includes common deletion and gene deletion involving oxidative phosphorylation. Heterologous mutations in mitochondrial DNA8468 13446 loci were found in skeletal muscles of elderly patients with diabetes or impaired glucose tolerance, which is considered to be a "common" deletion [Liang *et al.*, 1997]. Their research confirmed that there is a susceptibility to deletion of mitochondrial DNA in insulin resistant rats, and the increase of ROS caused by high glucose can induce the variation of mitochondrial DNA *in vitro*. It is speculated that high glucose-related oxidative stress and possible hyperinsulinemia can change the integrity of mitochondrial gene [Fukagawa *et al.*, 1999]. Compared with normal subjects, prediabetes and diabetic patients showed that the genes involved in mitochondrial oxidative phosphorylation were downregulated by three genes at the transcriptional level.

A series of natural compounds and nutrients such as antioxidants can promote the formation and function of mitochondria. These mitochondrial nutrients and compositions can effectively upregulate the activity of peroxisome proliferators-activated receptor γ coactivator 1 alpha (PGC-1α) and promote the formation of mitochondria, so as to prevent insulin resistance. At the same time, mitochondria are rich in a variety of enzymes, structural proteins, membrane lipids, and nucleic acids, which are also the targets of direct attack by oxidants such as ROS. Mitochondrial structure damage, DNA mutation, and enzyme activity decrease, which in turn further aggravate mitochondrial damage, resulting in changes in mitochondrial protein expression, disorder of cell energy metabolism, decline of mitochondria loss of cell function, and leading to diseases such as diabetes and aging [Harman, 1956, 1981; Beckman & Ames, 1998].

More and more research findings and evidence suggest that many kinds of tissue and cell oxidative stress processes play a key role in the pathogenesis of type 2 diabetes. Obesity is accompanied by the resistance of human target tissue cells to adipokine and leptin secreted by adipose tissue, leading to insulin resistance. In addition, adiponectin, another adipokine with the function of improving tissue sensitivity to insulin, decreased significantly in blood. The decrease of insulin sensitivity leads to the chronic accumulation and abnormal increase of insulin, leptin, glycolipid, free fatty acid, and other metabolites in circulating blood, resulting in the impairment and decline of islet cell function in susceptible individuals. Recent studies have shown that nutritional imbalance leads to the destruction of the balance of energy metabolism, resulting in the excessive production of ROS in mitochondria, the cell energy plant, in the process of oxygen metabolism, leading to cellular oxidative stress. ROS excessive production can destroy the dynamic balance between oxidation and antioxidant systems in the cell system, leading to the decline of mitochondria itself (mitochondrial decay), lipid molecule peroxidation, abnormal DNA and protein modification, and cell function damage. Hyperglycemia and hyperlipidemia can enhance oxidative stress and endothelial dysfunction and the self-defense ability of diabetics is significantly reduced. This is manifested in the reduction of some special antioxidant levels or antioxidant enzyme activity. β-cell dysfunction leads to impaired regulation of glucose metabolism and a special pathological state of metabolic syndrome. In a study, it was observed that Goto-Kakizaki (GK) diabetic rats had disorder of immune function accompanied by oxidative damage and mitochondrial dysfunction.

Mitochondrial dysfunction is an important cause of insulin resistance. Mitochondrial dysfunction caused by ROS leads to a vicious circle of ROS and mitochondrial dysfunction. It plays an important role in the process of insulin resistance in type 1 and 2 diabetes. Therefore, promoting mitochondrial production and preventing mitochondrial dysfunction are the main strategies for the prevention and treatment of insulin resistance and diabetes.

11.3.4 *Inflammation directly causes oxidative stress in diabetes*

The occurrence of diabetes and its complications is related to oxidative stress and low chronic inflammation. Oxidative stress is an imbalance between cellular oxidants and antioxidant system, which is the result of excessive production of free radicals and related ROS and RNS. High glucose increases the markers of chronic inflammation, promotes ROS and RNS production and oxidative stress injury, leading to diabetic complications, including vascular dysfunction. In addition, the increase of ROS and RNS level will reduce insulin secretion and damage the signal transduction of insulin sensitivity and insulin responsive tissues. Therefore, the appropriate treatment of high glucose and inhibition of excessive production of ROS and RNS is crucial for delaying the occurrence and development of diabetes and preventing subsequent complications.

ROS and RNS and pro-inflammatory cytokines play a key role in type 1 diabetes pathogenesis. Type 1 diabetes (T1D) is a T cell-mediated autoimmune disease characterized by the destruction of insulin-secreting pancreatic β cells. In humans with T1D and in non-obese diabetic (NOD) mice (a murine model for human T1D), autoreactive T cells cause β-cell destruction, as the transfer or deletion of these cells induces or prevents disease, respectively. $CD4^+$ and $CD8^+$ T cells use distinct effector mechanisms and act at different stages throughout T1D to fuel pancreatic β-cell destruction and disease pathogenesis. While these adaptive immune cells employ distinct mechanisms for β-cell destruction, one central means for enhancing their auto-reactivity is by the secretion of pro-inflammatory cytokines, such as IFN-γ, TNF-α, and IL-1. In addition to their production by diabetogenic T cells, pro-inflammatory cytokines are induced by ROS and RNS via redox-dependent signaling pathways. Highly reactive molecules, pro-inflammatory cytokines, are produced upon lymphocyte infiltration into pancreatic islets and induce disease pathogenicity by directly killing β cells, which characteristically possess low levels of antioxidant defense enzymes. In addition to β-cell destruction, pro-inflammatory cytokines are necessary for efficient adaptive immune maturation, and in the context of T1D they exacerbate autoimmunity by intensifying adaptive immune responses [Padgett *et al.*, 2013].

CD^+ cell (T cell with CD^+ receptor) is an important immune cell in the human immune system. CD^+ is mainly expressed by helper T cells; it is the receptor of helper T cell engineered T-cell receptor (TCR) recognition antigen. It binds to the non-polypeptide region of major histocompatibility complex (MHC) class II molecules and participates in the process of helper T cell TCR (T cell receptor-engineered T cells) recognition antigen. CD4$^+$ T cells are closely related to inflammation. ROS implicates on CD4$^+$ T cells in type 1 diabetes. Lots of work indicated that type 1 diabetes pathology is highly driven by ROS and RNS. One way in which ROS and RNS shape the autoimmune response demonstrated in type 1 diabetes is by promoting CD4$^+$ T cell activation and differentiation. As CD4$^+$ T cells are a significant contributor to pancreatic β-cell destruction in T1D, understanding how ROS and RNS impact their development, activation, and differentiation is critical. CD4$^+$ T cells themselves generate ROS and RNS via nicotinamide adenine dinucleotide phosphate (NADPH) oxidase expression and electron transport chain activity. Moreover, T cells can also be exposed to exogenous ROS and RNS generated by other immune cells (e.g., macrophages and dendritic cells) and β cells. Genetically modified animals and ROS and RNS inhibitors have demonstrated that ROS and RNS blockade during activation results in CD4$^+$ T cell hypo-responsiveness and reduced diabetes incidence. ROS and redox have also been shown to play roles in CD4$^+$ T cell-related tolerogenic mechanisms, including thymic selection and regulatory T cell-mediated suppression [Previte & Piganelli, 2018]. Increased oxidative stress and inflammation can lead to insulin resistance and impaired insulin secretion. Proper treatment of hyperglycemia and inhibition of ROS and RNS over-production is crucial for delaying onset of diabetes and for prevention of cardiovascular complications. It is imperative to determine the mechanisms involved in the progression from prediabetes to diabetes including a clarification of how old and new medications affect oxidative and immune mechanisms of diabetes. The relationship between oxidative stress and hyperglycemia along with links between inflammation and prediabetes need to be studied. Additionally, the effects of hyperglycemic memory, micro-vesicles, micro-RNA, and epigenetic regulation on inflammation, oxidative state, and glycemic control also need to be high-lighted [Luc *et al.*, 2019].

Oxidative stress and inflammatory are the markers in pre-diabetes and insulin resistance and impaired β-cell function are often already present in pre-diabetes. Patients with pre-diabetes may upregulate markers of chronic inflammation and contribute to increased ROS and RNS genera-tion, which cause vascular dysfunction, pancreatic β-cell destruction, and ultimately diabetes.

11.4 Effects of tea polyphenols on diabetes

Diabetes mellitus is one of the major public health problems worldwide. Recent evidence suggests that the cellular reduction-oxidation (redox) imbalance leads to oxidative stress and subsequent occurrence and development of diabetes and related complications by regulating certain signaling pathways involved in β-cell dysfunction and insulin resistance. ROS and RNS can also directly oxidize certain proteins (defined as redox modification) involved in the diabetes process. There are a number of potential problems in the clinical application of antioxidant therapies. Novel antioxidant tea polyphenols may overcome pharmacokinetic and stability problem and improve the selectivity of scavenging ROS and RNS. As a worldwide drink, green tea has many beneficial effects, especially green tea polyphenols (GTPs), mainly EGCG, can reduce the incidence of diabetes, prevent oxidative stress induced by other diseases, and complications of diabetes. Epidemiological investigation shows that drinking green tea can significantly reduce the death rate caused by abnormal lipid metabolism. Tea polyphenols can significantly reduce the content of triglycerides and cholesterol in blood, and reduce the plasma glucose, oxidation (redox) imbalance, oxidative stress, and subsequent occurrence and development of diabetes. Tea polyphenols can significantly reduce diabetes caused by body weight and body fat content [Zhao 2002, 2008, 2020]. Therefore, next, we will focus on the role of oxidative stress and tea polyphenols therapies in the pathogenesis of diabetes mellitus. Precise therapeutic interventions against ROS and downstream targets are now possible and provide important new insights into the treatment of diabetes.

11.4.1 *Clinical results*

There is a multi-center, cross-sectional study about tea consumption and risk of diabetes in the Chinese population. It explored the influence of tea consumption on diabetes mellitus in the Chinese population. This multi-center, cross-sectional study was conducted in eight sites from south, east, north, west, and middle regions in China by enrolling 12,017 subjects aged 20–70 years. A standard procedure was used to measure anthropometric characteristics and to obtain blood samples. The diagnosis of diabetes was determined using a standard 75-g oral glucose tolerance test. In the final analysis, 10,825 participants were included, and multiple logistic models and interaction effect analysis were applied for assessing the association between tea-drinking with diabetes. Compared with non-tea drinkers, the multivariable-adjusted or for newly diagnosed diabetes were 0.80, 0.88, and 0.86 for daily tea drinkers, occasional tea drinkers, and seldom tea drinkers, respectively. Furthermore, drinking tea daily was related to decreased risk of diabetes in females by 32%, elderly (>45 years) by 24%, and obese (BMI > 30 kg/m^2) by 34%. Moreover, drinking dark tea was associated with reduced risk of diabetes by 45% ($p < 0.01$) [Chen *et al.*, 2020].

A study determined the effects of black tea on postprandial plasma glucose and insulin concentrations in healthy humans in response to an oral glucose load. A four-way randomized, crossover trial was designed in which 16 healthy fasted subjects would consume 75 g of glucose in either 250 ml of water (control), 250 ml of water plus 0.052 g of caffeine (positive control), or 250 ml of water plus 1.0 g or 3.0 g of instant black tea. Chemical analysis showed that the tea was rich in polyphenolic compounds (total, 350 mg/g). Plasma glucose concentrations < 60 minutes in response to the drinks were similar, but were significantly reduced at 120 minutes ($p < 0.01$), following ingestion of the 1.0 g tea drink, relative to the control and caffeine drinks. Tea consumption resulted in elevated insulin concentrations compared with the control and caffeine drinks at 90 minutes ($p < 0.01$) and compared with caffeine drink alone at 150 minutes ($p < 0.01$). In conclusion, the 1.0 g tea drink reduced the late phase plasma glucose response in healthy humans with a corresponding increase in insulin [Bryans *et al.*, 2007].

A single-blind, crossover study compared the effect of isomaltulose together with green tea on glycemic response and antioxidant capacity. A total of 15 healthy subjects (eight women and seven men of ages 23.5 ± 0.7 years, with body mass index of 22.6 ± 0.4 kg/m²) consumed five beverages: 1) 50 g sucrose in 400 mL water; 2) 50 g isomaltulose in 400 mL of water; 3) 400 mL of green tea; 4) 50 g sucrose in 400 mL of green tea; 5) 50 g isomaltulose in 400 mL of green tea. Incremental area under postprandial plasma glucose, insulin, ferric reducing ability of plasma, and malondialdehyde (MDA) concentration were determined during 120 minutes of administration. Following the consumption of isomaltulose, the incremental 2-hour area under the curve indicated a higher reduction of postprandial glucose (43.4%) and insulin concentration (42.0%) than the consumption of sucrose. The addition of green tea to isomaltulose produced a greater suppression of postprandial plasma glucose (20.9%) and insulin concentration (37.7%). In accordance with antioxidant capacity, consumption of sucrose (40.0%) and isomaltulose (28.7%) caused the reduction of green tea-induced postprandial increases in reducing ability of plasma. A reduction in postprandial MDA after drinking green tea was attenuated when consumed with sucrose (34.7%) and isomaltulose (17.2%) [Suraphad *et al.*, 2017]. There is a study regarding the relationship between green tea and total caffeine intake and risk for self-reported type 2 diabetes among Japanese adults (a total of 17,413 persons, 40–65 years of age). During the five-year follow-up, there were 444 self-reported new cases of diabetes, among which are 231 men and 213 women. Consumption of green tea and coffee was inversely associated with risk for diabetes after adjustment for age, sex, body mass index, and other risk factors. No association was found between consumption of black or oolong teas and the risk for diabetes [Iso *et al.*, 2006].

Not all the experimental results are consistent. Two studies have found that tea drinking and tea polyphenols are not negatively correlated with diabetes.

To study the effects of the intake of green tea and polyphenols on insulin resistance and systemic inflammation, a randomized controlled trial was conducted on 66 patients aged 32–73 years old (53 males and 13 females) with borderline diabetes or diabetes. Subjects in the intervention group were asked to take a packet of green tea extracts/powder containing 544 mg polyphenols (456 mg catechins) daily, which was a dose that could be taken without difficulty. They were asked to divide the green tea extracts/powder in a packet into three or four fractions dissolved in hot water every day and to take a fraction after every meal or snack for two months, in addition to daily food intake. To calculate the level of green tea polyphenol intake that the subject usually drank at home, the subject was asked to taste three teas of different strengths (1%, 2%, and 3%) and the tea that was closest to the one that the subject drank at home, was selected by each subject. After two months, the mean daily polyphenol intake in the intervention group was 747 mg, which was significantly higher than that of 469 mg in the control group. In the intervention group, the body weight, BMI, systolic and diastolic blood pressures, blood glucose level, glycosylated hemoglobin, insulin level and insulin resistance index after taking the supplementation for two months, were lower than the respective value before intervention. However, these parameters in the intervention group at two months did not significantly differ from those in the control group. Within the intervention group, changes in insulin level tended to be associated with changes in polyphenol intake. In addition, changes in BMI were associated with changes in blood glucose level and insulin level. In conclusion, the daily supplementary intake of 500 mg green tea polyphenols did not have clear effects on blood glucose level, glycosylated hemoglobin level, insulin resistance, or inflammation markers. The positive correlation between the level of polyphenol intake and insulin level warrants further studies about the effect of green tea on insulin resistance [Fukino *et al.*, 2005].

There is another double-blind randomized study about the effect of an extract of green and black tea on glucose control in adults with type 2 diabetes mellitus. They evaluated the ability of an extract of green and black tea to improve glucose control over a three-month period. A double-blind, placebo-controlled, randomized multiple-dose (0 mg, 375 mg, or 750 mg per day for three months) study in adults with type 2 diabetes mellitus not taking insulin was performed. The primary end point was changed in glycosylated hemoglobin at three months. The 49 subjects who completed this study were predominantly white with an average age of 65 years and a median duration of diabetes of six years, and 80% of them reported using hypoglycemic medication. After 3 months, the mean changes in glycosylated hemoglobin were +0.4 (95% confidence interval, 0.2-0.6), +0.3 (0.1-0.5), and +0.5 (0.1-0.9) in the placebo, 375-mg, and 750-mg arms, respectively. Although 375 mg decreased, the changes were not significantly different between study arms [Mackenzie *et al.*, 2007].

It can be seen from the above clinical results that after drinking green tea, the risk of diabetes is significantly reduced. Among women, the elderly and obese people, drinking tea every day is negatively correlated with the risk of diabetes. Green tea can promote the decrease of postprandial glucose and insulin concentration. In addition, drinking black tea is also associated with reducing the risk of diabetes. However, there are some exceptions and uncertain results, which may be mainly caused by the different dosage and time of diabetic patients and tea polyphenols. Therefore, further research is needed in future. In particular, the quantitative and timing study of tea polyphenols on diabetes should be strictly carried out.

11.4.2 *Animal and cellular experiment results*

A large number of experimental studies have found that green tea polyphenols (GTPs) significantly reduce the blood glucose level and increase glucose tolerance in animals and cellular system.

1. GTPs ameliorate diabetic phenotype of KK-ay

Serum glucose level is an important index for diabetes. We studied the effect of tea polyphenols GTPs on blood glucose in diabetic rats fed with high-fat diet. GTPs feeding for four weeks decreased the random blood glucose content by ~30.4% (low concentration) and 51.2% (high concentration), the fasting blood glucose content by ~31.6% (low concentration) and 43.3% (high concentration) and the two-hour blood glucose content by ~26.5% (low concentration) and 49.7% (high concentration), respectively. The standards of normal random blood glucose, fasting blood glucose, and two-hour blood glucose are 11 mM, 7 mM, and 11 mM respectively. The mean fasting blood glucose of high GTPs group was 6.7 mM, which was lower than normal standard. The mean random blood glucose and two-hour blood glucose of high GTPs group were 16.2 mM and 13.5 mM, respectively, which were all near to the normal standards [Yan *et al.*, 2012] (Fig. 11-1).

GTCs feeding decreased random blood glucose content (RBG)(a). GTCs feeding increased glucose tolerance of KK-ay mice(b). Values are mean ± SE of 10 animals per group for each measurement. *, $p < 0.05$ indicates a significant difference between GTCs feeding with vehicle feeding groups.

A study found that GTP intervention significantly reduced serum ALT and AST levels. Fasting serum glucose, insulin resistance, and hepatic

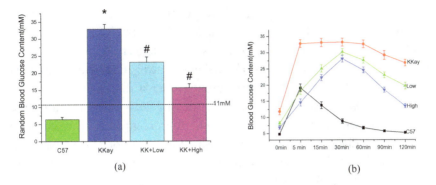

Figure 11-1. GTCs feeding decreased blood glucose content of KK-ay mice.

lipid levels were all decreased in the GTP-treated rats. GTP also significantly decreased the levels of TNF-α, IL-6, and malondialdehyde. In contrast, superoxide dismutase levels were increased in the liver. Furthermore, GTP also significantly increased phosphorylation of non-alcoholic fatty liver disease (NAFLD) and attenuated histopathological changes indicative of injury in liver tissue. GTP has a protective effect on high-fat diet (HFD)-induced hepatic steatosis, insulin resistance, and inflammation, and the underlying mechanism may involve the AMPK pathway [Xia *et al.*, 2019].

2. Effect of tea polyphenols on streptozotocin-induced diabetes

A work studied the protective effects of green tea epicatechin against the toxic effects of streptozotocin (STZ) and green tea (-)-epicatechin (EC) against the toxic effects of STZ, a selective β cell toxin, on pancreatic islets *in vivo* and *in vitro*. Compared with STZ treatment alone, insulin release was increased and nitrite production was decreased in EC+STZ-treated islets. From the above result, EC appears to be helpful in protecting pancreatic islets against exposure to STZ in both *in vivo* and *in vitro* systems [Kim *et al.*, 2003]. White tea is very similar to green tea but it is exceptionally prepared only from the buds and young tea leaves of *Camelia sinensis* plant while green tea is prepared from the matured tea leaves. The present study was investigated to examine the effects of a 0.5% aqueous extract of white tea in a STZ-induced diabetes model of rats. Six-week-old male Sprague-Dawley rats were divided into three groups of six animals in each group, namely: normal control, diabetic control, and diabetic white tea. Diabetes was induced by an intraperitoneal injection of STZ (65 mg/kg BW) in diabetic control and diabetic white tea groups, except the normal control group. After a four-week feeding of 0.5% aqueous extracts of white tea, the drink intake was significantly ($p <$ 0.05) increased in the diabetic white tea group compared to the diabetic control and diabetic control groups. Blood glucose concentrations were significantly decreased and glucose tolerance ability was significantly improved in the diabetic white tea group compared to the diabetic control group. Liver weight and liver glycogen were significantly increased and serum total cholesterol and LDL-cholesterol were significantly decreased in the diabetic white tea group compared to the diabetic control group. The food intake, body weight gain, serum insulin, and fructosamine concentrations were not influenced by the consumption of white tea. Data of this study suggest that the 0.5% aqueous extract of white tea is effective to reduce most of the diabetes associated abnormalities in a STZ-induced diabetes model of rats [Islam, 2011].

3. Green tea polyphenol prevents diabetic rats from acute kidney injury

Diabetes can cause kidney damage. Nephropathy in diabetes is one of the major microvascular complications of diabetes, and the incidence rate in diabetes is about 20–40%. If patients with diabetes have a medical history of 5–10 years, they may be associated with diabetes-related kidney disease, especially those with hypertension, fundus disease, or long-term poor blood sugar control, diabetes-related kidney damage will be more obvious.

EGCG possesses a reno-protective effect through its diverse biochemical properties and assessed the effect on renal function for diabetic rats. EGCG group exhibited milder tubular injury histologically ($p <$ 0.0001) and reduced expression of kidney injury molecule-1, a biomarker for renal tubular injury ($p < 0.0001$) and 8-hydroxy-2′-deoxyguanosine ($p < 0.01$), indicating attenuated oxidant stress. Administration of EGCG ameliorates acute kidney injury (AKI) in a common complication accompanying cardiopulmonary bypass (CPB) model of diabetic rats through anti-oxidative properties [Funamoto *et al.*, 2016]. The increased renal oxidative stress and the loss of renal function induced by the HFD were ameliorated by GTPs. Renal ketogenesis and SIRT3 expression and activity levels, which were reduced by the HFD, were restored by GTPs. *In vitro*, HEK293 cells were transfected with the eukaryotic expression plasmid pcDNA 3-hydroxy-3-methylglutaryl-CoA (HMG-CoA) synthase 2 (HMGCS2). GTP treatment could upregulate HMGCS2 and sirtuin 3(SIRT3) expression. Although SIRT3 expression was not affected by HMGCS2 transfection, the 4-hydroxy-2-nonenal (4-HNE) level and the acetyl-MnSOD (K122)/MnSOD ratio were reduced in HMGCS2-transfected cells in the context of H_2O_2 [Yi *et al.*, 2017].

4. Dark Pu'er tea regulates high blood sugar

Dark tea, comprising one of the six major teas, has many biological activities, which originate from their active substrates, such as polyphenols, polysaccharides, and so on. The hypoglycemic effect is one of its most prominent activities. During a study, it separately analyzed the phytochemical composition, glycosidase inhibition, and free radical scavenging activities, and hypoglycemic activity in type 2 diabetes mellitus mice, as well as the alleviation of insulin resistance in HepG2 cells of four dark tea aqueous extracts. The results showed that the phytochemical composition of dark tea aqueous extracts was significantly different, and they all had good glycosidase inhibition and free radical scavenging activities, *in vivo* hypoglycemic activity and alleviation of insulin resistance, and could also activate the phosphatidylinositol 3-kinase-Akt-perixisome proliferation-activated receptor cascade signaling pathway to regulate glucose and lipid metabolism, change the key enzyme activities related to glucose metabolism and antioxidant activity, and reduce oxidative stress and inflammatory factor levels. Among them, Liubao brick tea (LBT) and Pu-erh tea (PET) possessed better glycosidase inhibitory activity, *in vivo* hypoglycemic activity, and improved insulin resistance activity, whereas Qingzhuan brick tea and Fuzhuan brick tea had better free radical scavenging activity, which may be explained by their distinct phytochemical compositions, such as tea proteins, polysaccharides, polyphenols, catechins, tea pigments, and some elements [Zhu *et al.*, 2021].

Theabrownin is one of the most active and abundant pigments in Pu-erh tea, and it is a brown pigment with multiple aromatic rings and attached residues of polysaccharides and proteins. A study found that theabrownin (TB) isolated from Pu-erh tea regulates Bacteroidetes to improve metabolic syndrome of rats induced by high-fat, high-sugar, and high-salt diet. In high-fat, high-sugar, and high-salt diet (HFSSD) mode, prevotella_sp._CAG:1031 was one of the main dominant characteristic bacteria of TB targeting regulation, while roseburia_sp._1XD42-69, mainly inhibitory intestinal bacteria, helped to reduce body weight, TG, and blood sugar levels of HFSSD rats. Glycerophospholipid metabolism, arachidonic acid metabolism, glycolysis/gluconeogenesis, and insulin resistance were the critical pathway. TB has a high application potential in reducing the risk of metabolic diseases [Yue *et al.*, 2020]. TB has been shown to be hypolipidemic and displays fasting blood glucose (FBG)-lowering properties in rats fed a high-fat diet. This study aimed to determine the effect of TB in treating diabetes and explore the underlying mechanism of action of intestinal microbes by using Goto-Kakizaki (GK) rats. Diabetic GK rats were treated up to eight weeks with TB (GK-TB). Following treatment, the body weight, triglyceride (TG) content, fasting blood glucose (FBG) content, and Homeostatic Model Assessment for Insulin Resistance (HOMA-IR) were significantly lower in the GK-TB group than in the GK control group ($p < 0.05$). Meanwhile, the circulating adiponectin (ADPN), leptin, and glucokinase levels in the serum of the GK-TB group were significantly higher than those in the GK group, while there was little difference in hepatic lipase (HL) and hormone-sensitive triglyceride lipase (HSL) enzyme activities ($p > 0.05$). Furthermore, with the extension of treatment time, the number of unique intestinal microorganisms in GK rats greatly increased and an interaction among intestinal microorganisms was observed. The Firmicutes/Bacteroides ratio was decreased significantly, and the composition of Actinobacteria and Proteobacteria was increased. The use of multiple omics technologies show that TB is involved in the targeted regulation of the core characteristic intestinal flora including Bacteroides thetaiotaomicron (BT), Lactobacillus murinus (LM), Parabacteroides distasonis (PD), and Bacteroides_acidifaciens (BA), which improved the glucose and lipid metabolism of GK rats via the AMP-activated protein kinase signaling pathway, insulin signaling pathway, bile secretion, and glycerophospholipid metabolism. Intragastric administration of BT, LM, PD, or BA, led to a significantly reduced HOMA-IR in GK rats. Furthermore, BT significantly reduced serum lipid TG and total cholesterol (TC), and BA significantly reduced the serum lipid TC and low-density lipoprotein (LDL). PD significantly reduced serum LDL, while the effect of LM was not significant. However, LM and PD significantly increased the content of ADPN in serum [Yue *et al.*, 2022].

Dark tea and Pu-erh tea are highly attractive candidates for developing anti-diabetic food. Dark tea and Pu-erh tea may be good natural sources of agricultural products with anti-diabetic effects.

11.5 The mechanism of tea polyphenols in improving diabetes

Epidemiological and animal study data have suggested that drinking green tea and tea polyphenols are negatively associated with diabetes. Growing evidence from animal studies supports the anti-diabetic properties of some dietary polyphenols, suggesting that dietary polyphenols could be one dietary therapy for the prevention and management of type 2 diabetes. Moreover, adipose oxidative stress may have a central role in causing insulin resistance, according to recent findings. To study the new mechanism for green tea's anti-insulin resistance effect, we used obese KK-ay mouse, high-fat diet induced obese rat, and induced-insulin resistant 3T3-L1 adipocytes as models. Insulin sensitivity and adipose ROS level were detected in animal and adipocyte. The oxidative stress assay, glucose uptake ability assay, and effect of EGCG on insulin signals were detected. It was found that GTPs significantly decreased glucose level and increased glucose tolerance in animals. GTPs reduced ROS content in adipose tissues. EGCG attenuated Dexamethasone and TNF-α promoted ROS generation and increased glucose uptake ability. EGCG also decreased JNK phosphorylation and promoted GLUT-4 translocation. EGCG and GTPs could improve adipose insulin resistance and exact this effect on their ROS scavenging functions. The mechanism of tea polyphenols in improving diabetes mainly focuses on the following aspects, which will be covered in detail.

11.5.1 *GTPs attenuate oxidative stress in KK-ay adipose tissues*

Adipose oxidative stress induced by fat accumulation causes oxidative stress in the whole body. Our results showed ROS content in KK-ay adipose tissues was increased (Fig. 11-2a). Also, ROS was secreted by adipocyte increased liver, serum, and total blood MDA contents (Figs. 11-2b, 11-2c, 11-2d). GTPs feeding decreased ROS content of white adipose tissues (Fig. 11-2a) and decreased MDA content in liver, serum, and blood of KK-ay mice (Figs. 11-2b, 11-2c, 11-2d). These data suggested GTPs feeding reduced adipose ROS content and adipose ROS secretion [Yan *et al.*, 2012].

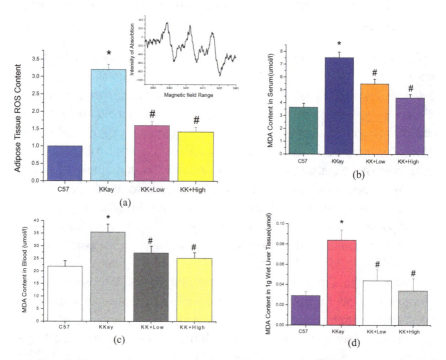

Figure 11-2. GTCs feeding ameliorated oxidative stress in KK-ay mice. GTCs feeding decreased (a) ROS content in serum, (b) MDA content in serum, (c) blood, and (d) liver. ROS was detected by the ERS spin trapping technique and MDA was detected by the TAB method, respectively, in the samples of mice after treatment by GTPs for 45 days. Values are mean ± SE of 10 animals per group for each measurement. *, $p < 0.05$ indicates a significant difference between GTPs feeding and vehicle feeding groups.

As liver, muscles, and adipose tissues store most of the glucose after eating and ROS in adipocyte has been reported as the common trigger for different types of insulin resistance [Houstis *et al.*, 2006], we detected ROS level in animal adipose tissues. GTPs feeding significantly decreased ROS level in adipose tissues. ROS produced by adipose tissues not only influence adipose redox, but also results in oxidative stress in other tissues [Furukawa *et al.*, 2004]. A study examined in the blood of overweight men aged from 62–83 years, the relationships between age and insulin resistance, selected parameters of the oxidative stress, and the antioxidant defense system. The results of the study did not show significant differences between groups investigated with respect to concentrations of TBARS, GSH, and GPx. However, significantly higher concentrations of glucose and antibodies against oxLDL ($p < 0.05$) were observed in the group of men >74 years old in comparison to the group of "young-old" men. It was indicated that the increased insulin resistance and hyperglycemia in elderly men are related to body mass and that they cause intensified oxidative modifications of LDL [Karolkiewicz *et al.*, 2006]. An aqueous solution of green tea polyphenols was found to inhibit lipid peroxidation, scavenge hydroxyl, and superoxide radicals *in vitro*. Sabu *et al.* studied the anti-diabetic activity of green tea polyphenols and their role in reducing oxidative stress in experimental diabetes. Concentration needed for 50% inhibition of superoxide, hydroxyl, and lipid peroxidation radicals were 10 micro g/ml, 52.5 micro g/ml, and 136 micro g/ml, respectively. Administration of green tea polyphenols (500 mg/kg b.wt.) to normal rats increased glucose tolerance significantly ($p < 0.005$) at 60 minutes. Green tea polyphenols was also found to reduce serum glucose level in alloxan diabetic rats significantly at a dose level of 100 mg/kg b.wt. Continued daily administration (15 days) of the extract 50 mg/kg b.wt. and 100 mg/kg b.wt. produced 29% and 44% reduction in the elevated serum glucose level produced by alloxan administration, respectively. Elevated hepatic and renal enzymes produced by alloxan were found to be reduced ($p < 0.001$) by green tea polyphenols. The serum lipid peroxidation level, which was increased by alloxan, was reduced by significantly ($p < 0.001$) by the administration of 100 mg/kg b.wt. of green tea polyphenols. Decreased liver glycogen after alloxan administration showed a significant ($p < 0.001$) increase after green tea polyphenols treatment. Green tea polyphenols-treated group showed increased antioxidant potential, as seen from improvements in superoxide dismutase and glutathione levels. However, catalase, lipid peroxidation, and glutathione peroxidase levels were unchanged [Sabu *et al.*, 2002]. These results indicate that alterations in the glucose utilizing system and oxidation status in rats increased by alloxan were partially reversed by the administration of the glutamate pyruvate transaminase.

Our results showed GTPs feeding ameliorated both blood and liver oxidative stress. Epidemical investigation results showed oxidative stress in many tissues were proved to be consistent with insulin resistance, and pathological research also found the important role of oxidative stress in insulin resistance. Total plasma antioxidative capacity, erythrocyte, and plasma-reduced glutathione level were significantly decreased in obese-diabetic patients, but also in obese-healthy subjects, compared to the values in controls. The plasma lipid peroxidation products and protein carbonyl groups were significantly higher in obese diabetics, more than in obese healthy subjects, compared to the control healthy subjects. The increase of erythrocyte lipid peroxidation at basal state was shown to be more pronounced in obese diabetics, but the apparent difference was obtained in both the obese healthy subjects and obese diabetics, compared to the control values, after exposing of erythrocytes to oxidative stress induced by H_2O_2. Positive correlation was found between the MDA level and index of insulin sensitivity. Decreased ROS level in adipose and other tissues might increase insulin sensitivity. That might be the mechanism of GTPs' insulin-sensitizing effects.

11.5.2 *GTPs suppress oxidative-related signal and increase glucose transporter 4 (GLUT-4) expression in KK-ay mice*

Obesity is closely associated with insulin resistance and establishes the leading risk factor for type 2 diabetes mellitus. Extraordinary phosphorylation of c-Jun amino-terminal kinases (JNK) in adipocyte and muscles induced insulin resistance. The JNK can interfere with insulin action in cultured cells and are activated by inflammatory cytokines and free fatty acids, molecules that have been implicated in the development of type 2 diabetes. JNK is a crucial mediator of obesity and insulin resistance and a potential target for therapeutics [Hirosumi *et al.*, 2002].

The c-Jun N-terminal kinase JNK pathway is known to be activated under diabetic conditions and to possibly be involved in the progression of insulin resistance. Suppression of the JNK pathway in liver exerts greatly beneficial effects on insulin resistance status and glucose tolerance in both genetic and dietary models of diabetes [Nakatani *et al.*, 2004]. Oxidative stress suppressed glucose transporter 4 (GLUT-4) expression and translocation to plasma membrane by activating JNK. GTPs feeding reduced JNK phosphorylation in adipose tissues (Fig. 11-3) and increased GLUT-4 expression as a result (Fig. 11-3).

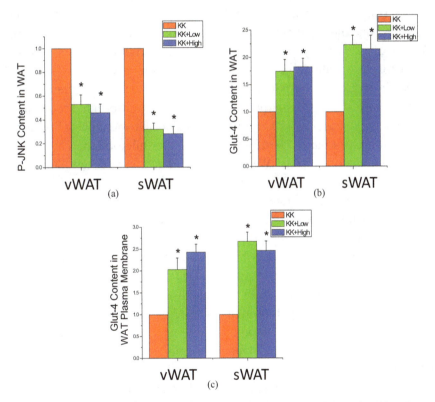

Figure 11-3. Effects of GTC on phospho-JNK (P-JNK) and GLUT-4 level and in subcutaneous visceral white adipose tissues and content of GLUT-4 in plasma membrane protein of KK-ay mice. GTCs feeding reduced the (a) phospho-JNK (P-JNK) and (b) GLUT-4 level in subcutaneous and visceral white adipose tissues (sWAT and vWAT). GTCs feeding reduced the content of (c) GLUT-4 in both total cellular protein and plasma membrane protein. The column graphs are the statistical results of band intensities. Values are mean ± SE of 10 animals per group for each measurement. *, $p < 0.05$ indicates a significant difference between GTCs feeding and vehicle feeding groups.

GTPs feeding also increased GLUT-4 content in plasma membrane which meant that GLUT-4 translocation had been enhanced (Fig. 11-3). These data suggest that GTPs increased insulin sensitivity by suppressing JNK pathway [Yan *et al.*, 2012].

11.5.3 *GTPs improve diet-induced obese rat glucose metabolism*

Growing evidence from animal studies supports the anti-diabetic properties of some dietary tea polyphenols, suggesting that dietary polyphenols could be one dietary therapy for the prevention and management of type 2 diabetes. Dietary tea polyphenols may inhibit α-amylase and α-glucosidase, inhibit glucose absorption in the intestine by sodium-dependent glucose transporter 1 (SGLT1), stimulate insulin secretion, and reduce hepatic glucose output. Polyphenols may also enhance insulin-dependent glucose uptake, activate 5′ adenosine monophosphate-activated protein kinase (AMPK), modify the microbiome, and have anti-inflammatory effects. We studied the effects of GTP on blood glucose metabolism and ROS in diet induced obese rats, and it was found that in diet-induced obese rats, serum glucose content was increased compared with chow diet control. High-fat food also caused an increase of ROS content in rat adipose tissues. GTPs feeding decreased glucose content in diet-induced obese rats and reduced the ROS production of adipose tissues. These data suggested GTPs feeding increased glucose absorption and decreased ROS content in diet induced obese animals [Yan *et al.*, 2012]. Tea polyphenols reversed diet-induced weight gain by reducing visceral fat, total blood cholesterol, and circulating FFA ($p \leq$ 0.05). Treatment with tea polyphenols improved adipokine regulation depicted by reduced leptin, adiponectin ratio (LAR) ($p < 0.05$). Treatment with AP improved parameters of glucose homeostasis as demonstrated by reduced fasting blood glucose (FBG) and homeostatic model assessment of insulin resistance (HOMA-IR) ($p \leq 0.05$) and increased GLUT 4 ($p <$ 0.05) [Mokwena *et al.*, 2021].

Tea infusion possess anti-obesity and anti-inflammatory properties, improved glucose uptake and reduce insulin resistance in diet-induced metabolic syndrome in rats which could be attributed to its richness in polyphenols. Therefore, tea polyphenols could have potential benefits against type 2 diabetes and obesity which are components of metabolic syndrome validating its pharmacologicaluse.

11.5.4 *EGCG recovers the impaired insulin-stimulated glucose uptake of 3T3-L1 adipocyte*

Drugs that reverse insulin resistance are important as insulin resistance is one of the most important characters of metabolic syndrome and frequently associated with type 2 diabetes [Mackenzie *et al.*, 2007]. Insulin resistance is a cardinal feature of type 2 diabetes and is characteristic of a wide range of other clinical and experimental settings. DEX and TNF-α caused insulin resistance by inducing oxidative stress in adipocyte. Gene expression analysis suggests that ROS levels are increased in both models ROS have previously been proposed to be involved in insulin resistance, although evidence for a causal role has been scant [Houstis *et al.*, 2006]. Apocynin (1-(4-hydroxy-3-methoxyphenyl) ethanone), is an inhibitor of NADPH-oxidase and could reduce intracellular ROS production. NADPH-oxidase was widely considered to be the major source of oxidative stress caused by fat accumulation. DEX and TNF-α can quickly increase blood sugar and cause blood sugar fluctuation. Our experiment found that DEX or TNF-α incubation reduced insulin-stimulated 2-deoxy-D-[^{3}H]-glucose uptake. As a positive control, apocynin prevent 2-deoxy-D-[^{3}H]-glucose uptake from DEX or TNF-α impairment (Fig. 11-4). EGCG treatment attenuated the effect of DEX or TNF-α and increased the impaired glucose uptake. These data suggested that EGCG could protect adipocyte glucose uptake from oxidative stress disturbance. In our laboratory, the effects of GTP on type 2 diabetes were studied and the mechanism of GTP on insulin sensitive effect was investigated by feeding of GTPs to obese spontaneous type 2 diabetes (KK-ay) mice, diet-induced obese rats, and 3T3-L1 adipocyte; all ameliorated insulin resistance to varying degrees. KK-ay mouse was genetically predisposed to obesity and type 2 diabetes, accompanied with a high blood glucose level and the glucose intolerance. GTPs feeding improved random blood glucose (RBG), fasting blood glucose (FBG), and two-hour postprandial blood glucose (2HBG). For the glucose tolerance assay, GTP feeding delayed the appearance of glucose peak

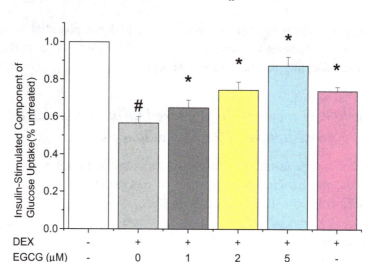

Figure 11-4. Effects of EGCG on insulin-stimulated glucose uptake in 3T3-L1 adipocytes. Insulin stimulated glucose uptake was assayed with 2-deoxy glucose and 2-deoxy-D-[^{3}H]-glucose. Fully differentiation 3T3-L1 adipocytes were incubated with EGCG and DEX or DEX and apocynin (20 μM) for eight days. Insulin stimulated glucose uptake was assayed with 2-deoxy glucose and 2-deoxy-D-[^{3}H]-glucose. Values are mean ± SE. *, $p <$ 0.05 indicates a significant difference between EGCG groups with DEX or TNF-α group; #, $p <$ 0.05 indicates a significant difference between DEX or TNF-α groups with control group.

value and decreased glucose level rapidly, compared with the control group. That meant that GTP feeding significantly increased glucose uptake ability of KK-ay and enhanced insulin sensitivity. In diet-induced obese rats, similar results were also got [Yan *et al.*, 2012].

11.5.5 *EGCG reduces DEX- or TNF-α-induced ultra ROS generation*

Nitroblue tetrazoli (NBT) can be reduced by ROS to a dark-blue, insoluble form of NBT called formazan. Formazan production was increased by DEX or TNF-α, and EGCG reduced formazan production. Formazan could be dissolved in 50% acetic acid, and the absorbance was determined at 560 nm. The absorbance of adipocyte homogenates after NBT dyeing was increased by DEX or TNF-α but decreased by EGCG. Detection of ROS content with the redox-sensitive dye 2′,7′-dichlorofluorescin diacetate also showed that DEX or TNF-α increased ROS contents in adipocyte and EGCG attenuate these increases. These data suggested that EGCG ameliorated oxidative stress induced by DEX or TNF-α. Oxidative stress in adipocyte plays a central role in whole-body insulin resistance and ROS secreted by adipocyte leads to insulin resistance in many tissues. We detected the effect of GTP on insulin signals in 3T3-L1 adipocyte. To detect the effect of GTPs in different types of insulin resistance, we induced insulin resistance in mature 3T3-L1 adipocyte by treatment with the cytokine TNF-α and the other with the glucocorticoid DEX. TNF-α induced oxidative stress and insulin resistance in adipocyte through its membrane receptor while DEX through inflammatory reactions [Houstis *et al.*, 2006]. In KK-ay adipose tissues, NADPH oxidase was proved as the major source of oxidative stress; an inhibitor of NADPH oxidase, apocynin, was used to treat adipocyte in this experiment as a positive control. After stimulated by insulin, adipocytes got a rapid increase in glucose uptake ability. Our results showed that apocynin partially ameliorated glucose-uptake ability by TNF-α and DEX. This means that both TNF-α- and DEX-induced insulin resistance relied on a ROS production process. EGCG increased the insulin stimulated glucose uptake ability of adipocyte. As a potent antioxidant, EGCG may ameliorate TNF-α- and DE-induced insulin resistance by scavenging ROS.

Glut-4 is the major adipose glucose inner transporter after being stimulated by insulin [Rea & Donnelly, 2004]. EGCG decreased DEX- and TNF-α-induced JNK phosphorylation and enhanced Glut-4 translocation. EGCG increased the glucose uptake ability of adipocyte. This explained why EGCG promoted adipocyte to uptake more 2-deoxy-D-[^{3}H]-glucose than DEX or TNF-α impairing groups, when added to medium together with DEX or TNF-α. We detected ROS in adipocyte after treatment. Both NBT dye and 2′,7′-dichlorofluorescein diacetate (DCF-DA) detection showed that TNF-α and DEX increased ROS level in adipocyte; this effect was suppressed by EGCG treatment. NBT was incubated with cells for three hours while DCF-DA was incubated with the cells for 30 minutes. As NBT was reduced by ROS to formazan constantly; DCF-DA results reflected the final levels of ROS in adipocytes while NBT results reflected the accumulative effect of treatment of DEX or TNF-α with or without EGCG [Furukawa *et al.*, 2004]. Over-loaded ROS in adipocyte induced insulin resistance partly by activating JNK phosphorylation [Kaneto *et al.*, 2005; Bennett *et al.*, 2003]. Phosphor-JNK interferes insulin signals and weakens downstream reactions. Clinical research showed that extraordinarily high levels of phosphor-JNK were found in the bodies of people with insulin resistance [Liu & Rondinone, 2005; Zhou *et al.*, 2005]. EGCG treatment decreased DEX- and TNF-α-induced JNK phosphorylation by scavenging ROS.

11.5.6 *EGCG reduces JNK phosphorylation in adipocyte*

Phosphor-JNK interferes insulin signals and delays glucose transporter 4 (Glut-4) translocation to plasma membrane [Derave *et al.*, 2003]. We found that DEX or TNF-α caused an increased level of phosphor-JNK (Fig. 11-5). When adipocytes were incubated with EGCG, increases of

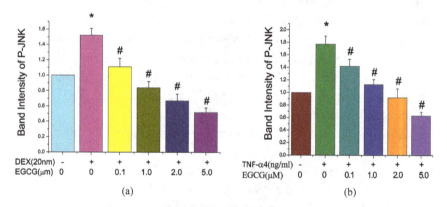

(a)   (b)

Figure 11-5.   Effect of EGCG on the phosphorylation level of JNK in 3T3-L1 adipocytes. Phosphorylation level of JNK in 3T3-L1 adipocytes incubated with (a) EGCG and DEX or with (b) EGCG and TNF-α was detected by Western blot at Day 16 after the cells were treated by EGCG and DEX or EGCG and TNF-α. Values are mean ± SE. #, $p < 0.05$ indicates a significant difference between EGCG groups and DEX or TNF-α group; *, $p < 0.05$ indicates a significant difference between DEX or TNF-α groups and control group; &, $p < 0.05$ indicates a significant difference between DEX or TNF-α groups and control group considering the effects of EGCG on the phosphorylation level of JNK in 3T3-L1 adipocytes.

JNK phosphorylation were attenuated. These data suggested that DEX or TNF-α activated JNK pathway, but EGCG suppressed this activation. Pre-treatment with EGCG significantly restored the activation of AKT and GSK in HepG2 cells and primary hepatocytes exposed to high glucose [Yan *et al.*, 2012]. In HepG2 cells and primary hepatocytes, glycogen synthesis was improved by EGCG treatment in a dose-dependent manner. High glucose significantly stimulated the production of ROS, while EGCG protected high glucose-induced ROS production. ROS is known to serve a major role in high glucose-induced insulin resistance by increasing JNK and IRS1 serine phosphorylation. EGCG was observed to enhance the insulin-signaling pathway. EGCG ameliorated high glucose-induced insulin resistance in the hepatocytes by potentially decreasing ROS-induced JNK/IRS1/AKT/GSK signaling [Ma *et al.*, 2017].

11.5.7 *EGCG enhances translocation of GLUT-4 to plasma membrane*

GLUT4 translocation using GLUT4-eGFP chimeric protein and dynamics of its intracellular distribution after addition of stimulants were detected with a confocal microscope. Our experiment found that treatment of adipocytes with DEX or TNF-α alone attenuated the fluorescence intensity of GLUT-4-eGFP (enhanced green fluorescent protein) in plasma membrane after insulin stimulating. However, EGCG elevated the translocation of confusion GLUT-4-eGFP to plasma membrane [Yan *et al.*, 2012]. A work studied the effect of green tea supplementation on insulin sensitivity in Sprague-Dawley rats. Results demonstrated that green tea increases insulin sensitivity in Sprague-Dawley rats and that green tea polyphenol is one of the active components [Wu *et al.*, 2004]. For mechanisms of the anti-obesity actions, green tea significantly reduced glucose uptake accompanied by a decrease in translocation of glucose transporter 4 (GLUT4) in adipose tissue, while it significantly stimulated the glucose uptake with GLUT4 translocation in skeletal muscle. Moreover, green tea suppressed the expression of peroxisome proliferator-activated receptor gamma and the activation of sterol regulatory element binding protein-1 in adipose tissue [Ashida *et al.*, 2004]. In this report, we studied the effects of teas and tea catechins on the small intestinal sugar transporters, SGLT1 and

GLUTs (GLUT1, GLUT2, and GLUT5). Green tea extract (GT), oolong tea extract (OT), and black tea extract (BT) inhibited glucose uptake into the intestinal Caco-2 cells with GT being the most potent inhibitor, followed by OT and BT. Catechins present in teas were the predominant inhibitor of glucose uptake into Caco-2 cells and gallated catechins were the most potent: CG > ECG > EGCG ≥ GCG, when compared to the non-gallated catechins (C, EC, GC, and EGC). In Caco-2 cells, individual tea catechins reduced the SGLT1 gene, but not protein expression levels. In contrast, GLUT2 gene and protein expression levels were reduced after two hours exposure to catechins but increased after 24 hours. These *in vitro* studies suggest that teas containing catechins may be useful dietary supplements capable of blunting postprandial glycaemia in humans, including those with or at risk to type 2 diabetes mellitus [Ni *et al.*, 2020].

Above results suggested that green tea modulates the glucose uptake system in adipose tissue and skeletal muscle and suppresses the expression and/or activation of adipogenesis-related transcription factors, as the possible mechanisms of its anti-obesity actions.

11.5.8 *EGCG alters synthesis of adipose factor*

Adiponectin is an endogenous bioactive polypeptide or protein secreted by adipocytes. Adiponectin is an insulin-sensitizing hormone, which can improve the insulin resistance of mice. Resistin is an adipose tissue hormone proposed to contribute to metabolic disease. Adiponectin and reisitin were synthesized in adipocytes. Our results found that the effects of DEX and EGCG on mRNA levels of adiponectin and reisitin were measured (Fig. 11-6). DEX increased the mRNA level of reisitin but decreased the mRNA level of adiponectin. EGCG reduced the effects of DEX on adipocyte, decreased adiponectin gene expression, and increased resistin gene expression [Yan *et al.*, 2012]. Serisier *et al.* [2008] studied effects of green tea on insulin sensitivity, lipid profile and expression of PPARα and PPARγ, and their target genes in obese dogs. This study was designed to examine the effects of green tea on insulin sensitivity and plasma lipid concentrations in an obese insulin-resistant dog model. Obese dogs were divided into two groups: a green tea group and a control group. Dogs in the green tea group were given green tea extract (80 mg/kg per day) orally,

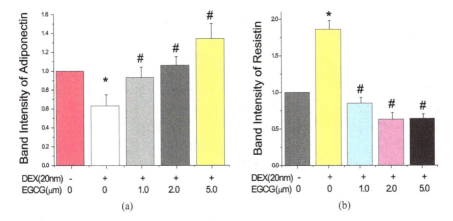

Figure 11-6. Effects of EGCG on expression of (a) adiponecin and (b) resistin in adipocytes. Expression of adiponectin and resisitin in 3T3-L1 adipocytes incubated with EGCG and DEX and detected by RT-PCR at day 16 after the cells were treated by EGCG and DEX. Values are mean ± SE. *, $p < 0.05$ indicates a significant difference with EGCG groups with DEX group; #, $p < 0.05$ indicates a significant difference with DEX groups with control group.

just before their single daily meal, for 12 weeks. At week 12 in the green tea group, mean insulin sensitivity index was 60% higher ($p < 0.05$) and total cholesterol concentration was 50% lower ($p < 0.001$), than baseline. PPARγ, GLUT4, lipoprotein lipase (LPL), and adiponectin expression were significantly higher in both adipose tissues, whilst PPARα and LPL expression were significantly higher in skeletal muscle compared with baseline [Serisier *et al.*, 2008]. A study found that the malondialdehyde (MDA) level in the HF diet group was significantly higher than that in the control (CON) group ($p < 0.05$). Decreased peroxisome proliferator-activated receptor (PPAR)-α and sirtuin 3 (SIRT3) expression and increased manganese superoxide dismutase (MnSOD) acetylation levels were also detected in the HF diet group ($p < 0.05$). GTP treatment upregulated SIRT3 and PPARα expression, increased the pparα mRNA level, reduced the MnSOD acetylation level, and decreased MDA production in rats fed a HF diet ($p < 0.05$). No significant differences in total renal MnSOD and PPAR-γ coactivator-1α (PGC1-α) expression were detected. The reduced oxidative stress detected in kidney tissues after GTP treatment was partly due to the higher SIRT3 expression, which was likely mediated by PPARα [Yang *et al.*, 2015].

These findings show that nutritional doses of green tea extract may improve insulin sensitivity and lipid profile and alter the expression of genes involved in glucose and lipid homeostasis. Diabetes could induce insulin resistance for oxidative stress in adipocyte, and ROS could play important roles in various types of insulin resistance. Based on results *in vivo* and *in vitro*, it was found that EGCG and GTPs could improve adipose insulin resistance and exert this effect on their ROS scavenging functions. Although there are still many debates on the clinical usage of GTPs on type 2 diabetes, these results have found and explained the anti-adipose insulin resistance effects of GTPs.

11.6 Conclusion

The above epidemiological investigation and clinical and animal experiments show that both green tea and black tea polyphenols are effective in inhibiting diabetes. They may inhibit the digestion and absorption of lipids and sugars and reduce calorie intake. Polyphenols play an active role in inhibiting diabetes, which may involve three main mechanisms: 1) Inhibiting the digestion, absorption and intake of starch and sugars, thereby reducing blood glucose; 2) Inhibiting insulin resistance; tea polyphenols can clear ROS and RNS, inhibit verification, and protect mitochondria; 3) Tea polyphenols can also pass through GLUT-4, TNF-α, JNK, and other signaling pathways that promote adipocytes to take in more glucose and block the pathological process and related complications of diabetes. Dietary tea polyphenols may also inhibit α-amylase and α-glucosidase, inhibit glucose absorption in the intestine by sodium-dependent glucose transporter 1 (SGLT1), stimulate insulin secretion, and reduce hepatic glucose output. Tea polyphenols may also enhance insulin-dependent glucose uptake, activate 5' adenosine monophosphate-activated protein kinase (AMPK), modify the microbiome, and have anti-inflammatory effects. However, human epidemiological and intervention studies have shown inconsistent results. Further intervention studies are essential to clarify the conflicting findings and confirm or refute the

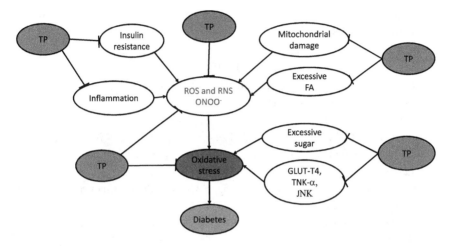

Figure 11-7. The pathway and the mechanism of tea polyphenols in prevention and treatment of diabetes. TP: tea polyphenols.

anti-diabetic effects of dietary tea polyphenols. To sum up, the ways and mechanisms of tea polyphenols in preventing and treating diabetes are shown in Figure 11-7.

References

Ashida H, Furuyashiki T, Nagayasu H, Bessho H, Sakakibara H, Hashimoto T, Kanazawa K. (2004) Anti-obesity actions of green tea: possible involvements in modulation of the glucose uptake system and suppression of the adipogenesis-related transcription factors. *Biofactors*, **22**, 135–140.

Assmann TS, Brondani LA, Bouças AP, Rheinheimer J, de Souza BM, Canani LH, Bauer AC, Crispim D. (2016) Nitric oxide levels in patients with diabetes mellitus: a systematic review and meta-analysis. *Nitric Oxide*, **61**, 1–9.

Avogaro A, Fadini GP, Gallo A, Pagnin E, de Kreutzenberg S. (2006) Endothelial dysfunction in type 2 diabetes mellitus. *Nutr Metab Cardiovasc Dis*, **16**(Suppl 1), S39–S45.

Babu PV, Sabitha KE, Shyamaladevi CS. (2006) Therapeutic effect of green tea extract on oxidative stress in aorta and heart of streptozotocin diabetic rats. *Chem Biol Interact*, **162**, 114–120.

Beckman KB, Ames BN. (1998) The free radical theory of aging matures. *Physiol Rev*, **78**, 547–581.

Bennett BL, Satoh Y, Lewis AJ. (2003) JNK: a new therapeutic target for diabetes. *Curr Opin Pharmacol*, **3**, 420–425.

Briaud I, Harmon JS, Kelpe CL, Segu VB, Poitout V. (2001) Lipotoxicity of the pancreatic β-cell is associated with glucose-dependent esterification of fatty acids into neutral lipids. *Diabetes*, **50**, 315–321.

Bryans JA, Judd PA, Ellis PR. (2007) The effect of consuming instant black tea on postprandial plasma glucose and insulin concentrations in healthy humans. *J Am Coll Nutr*, **26**(5), 471–477.

Camp HS, Ren D, Leff T. (2002) Adipogenesis and fat-cell function in obesity and diabetes. *Trends Mol Med*, **8**, 442–7.

Charles N. Rotimi, Guanjie Chen, *et al.* (2004) A genome-wide search for type 2 diabetes susceptibility genes in West Africans. *Diabetes*, **53**, 838–841.

Chen J, Stimpson SE, Fernandez-Bueno GA, Mathews CE. (2018) Mitochondrial reactive oxygen species and type 1 diabetes. *Antioxid Redox Signal*, **29**(14), 1361–1372.

Chen Y, Li W, Qiu S, *et al.* (2020) Tea consumption and risk of diabetes in the Chinese population: a multi-centre, cross-sectional study. *Br J Nutr*, **123**(4), 428–436.

Christ-Crain M, Gaisl O. (2021) Diabetes insipidus. *Presse Med*, **50**(4), 104093.

Czech MP, Lawrence JC Jr, Lynn WS. (1974) Evidence for the involvement of sulfhydryl oxidation in the regulation of fat cell hexose transport by insulin. *Proc Natl Acad Sci USA*, **71**, 4173–4177.

Davi G, Chiarelli F, Santilli F, Pomilio M, Vigneri S, Falco A, Basili S, Ciabattoni G, Patrono C. (2003) Enhanced lipid peroxidation and platelet activation in the early phase of type 1 diabetes mellitus: role of interleukin-6 and disease duration. *Circulation*, **107**, 3199–3203.

Derave W, Eijnde BO, Verbessem P, Ramaekers M, Van Leemputte M, Richter EA, Hespel P. (2003) Combined creatine and protein supplementation in conjunction with resistance training promotes muscle GLUT-4 content and glucose tolerance in humans. *J Appl Physiol*, **94**, 1910–1916.

Durante W, Sen AK, Sunahara FA. (1988) Impairment of endothelium-dependent relaxation in aortae from spontaneously diabetic rats. *Br J Pharmacol*, **94**(2), 463–468.

Fatehi-Hassanabad Z, Chan CB, Furman BL. (2010) Reactive oxygen species and endothelial function in diabetes. *Eur J Pharmacol*, **636**(1–3), 8–17.

Flier JS. (2004) Obesity wars: molecular progress confronts an expanding epidemic. *Cell*, **116**, 337–350.

Förstermann U, Xia N, Li H. (2017) roles of vascular oxidative stress and nitric oxide in the pathogenesis of atherosclerosis. *Circ Res*, **20**(4), 713–735.

Freidenberg GR, Reichart D, Olefsky JM, Henry RR. (1988) Reversibility of defective adipocyte insulin receptor kinase activity in non-insulin-dependent diabetes mellitus. Effect of weight loss. *J Clin Invest* **82**:1398–1406.

Fukagawa NK. *et al.* (1999) Aging and high concentrations of glucose potentiate injury to mitochondrial DNA. *Free Radic Biol Med*, **27**, 1437–1443.

Fukino Y, Shimbo M, Aoki N, Okubo T, Iso H. (2005) Randomized controlled trial for an effect of green tea consumption on insulin resistance and inflammation markers. *J Nutr Sci Vitaminol (Tokyo)*, **51**, 335–342.

Funamoto M, Masumoto H, Takaori K, Taki T, Setozaki S, Yamazaki K, Minakata K, Ikeda T, Hyon SH, Sakata R. (2016) Green tea polyphenol prevents diabetic rats from acute kidney injury after cardiopulmonary bypass. *Ann Thorac Surg*, **101**(4), 1507–1513.

Furukawa S, Fujita T, Shimabukuro M, Iwaki M, Yamada Y, Nakajima Y, Nakayama O, Makishima M, Matsuda M, Shimomura I. (2004) Increased oxidative stress in obesity and its impact on metabolic syndrome. *J Clin Invest*, **114**, 1752–1761.

Goldstein BJ, Mahadev K, Wu X. (2005) Redox paradox: insulin action is facilitated by insulin-stimulated reactive oxygen species with multiple potential signaling targets. *Diabetes*, **54**, 311–321.

Guilherme A, Virbasius JV, Puri V, Czech MP. (2008) Adipocyte dysfunctions linking obesity to insulin resistance and type 2 diabetes. *Nat Rev Mol Cell Biol*, **9**, 367–377.

Gomez-Cabrera MC, *et al.* (2008) Oral administration of vitamin C decreases muscle mitochondrial biogenesis and hampers training-induced adaptations in endurance performance. *Am J Clin Nutr*, **87**, 142–149.

Halliwell B, Gutteridge JMC. (1999) *Free Radicals in Biology and Medicine (3rd edition)*. Oxford University Press, Inc.

Harman D. (1956) Aging: a theory based on free radical and radiation chemistry. *J Gerontol*, **11**, 298–300.

Harman, D. (1981) The aging process. *Proc Natl Acad Sci USA*, **78**, 7124–7128.

Higaki Y, Mikami T, Fujii N, Hirshman MF, Koyama K, Seino T, Tanaka K, Goodyear LJ. (2008) Oxidative stress stimulates skeletal muscle glucose uptake through a phosphatidylinositol-3-kinasedependent pathway. *Am J Physiol Endocrinol Metab*. 294:E889–E897.

Hirosumi J, Tuncman G, Chang L, Gorgun CZ, Uysal KT, Maeda K, Karin M, Hotamisligil GS. (2002) A central role for JNK in obesity and insulin resistance. *Nature*, **420**, 333–336.

Houstis N, Rosen ED, Lander ES. (2006) Reactive oxygen species have a causal role in multiple forms of insulin resistance. *Nature*, **440**, 944–948.

Islam MS. (2011) Effects of the aqueous extract of white tea (Camellia sinensis) in a streptozotocin-induced diabetes model of rats. *Phytomedicine*, **19**(1), 25–31.

Iso H, Date C, Wakai K, Fukui M, Tamakoshi A. (2006) The relationship between green tea and total caffeine intake and risk for self-reported type 2 diabetes among Japanese adults. *Ann Intern Med*, **144**, 554–562.

Jacqueminet S, Briaud I, Rouault C, Reach G, Poitout V. (2000) Inhibition of insulin gene expression by long-term exposure of pancreatic β-cells to palmitate is dependent on the presence of a stimulatory glucose concentration. *Metabolism*, **49**, 532–536.

Janle EM, Portocarrero C, Zhu Y, Zhou Q. (2005) Effect of long-term oral administration of green tea extract on weight gain and glucose tolerance in Zucker diabetic (ZDF) rats. *J Herb Pharmacother*, **5**, 55–65.

Kaneto H, Matsuoka TA, Nakatani Y, Kawamori D, Matsuhisa M, Yamasaki Y. (2005) Oxidative stress and the JNK pathway in diabetes. *Curr Diabetes Rev*, **1**, 65–72.

Karolkiewicz J, Pilaczynska-Szczesniak L, Maciaszek J, Osinski W. (2006) Insulin resistance, oxidative stress markers and the blood antioxidant system in overweight elderly men. *Aging Male*, **9**, 159–163.

Kelley DE, He J, Menshikova EV, Ritov VB. (2002) Dysfunction of mitochondria in human skeletal muscle in type 2 diabetes. *Diabetes*, **51**, 2944–2950.

Kim MJ, Ryu GR, Chung JS, Sim SS, Min DS, Rhie DJ, Yoon SH, Hahn SJ, Kim MS, Jo YH. (2003) Protective effects of epicatechin against the toxic effects of streptozotocin on rat pancreatic islets: in vivo and in vitro. *Pancreas*, **26**, 292–299.

Kolb H, Martin S. (2017) Environmental/lifestyle factors in the pathogenesis and prevention of type 2 diabetes. *BMC Med*, **15**(1), 131.

Liang P, Hughes V, Fukagawa NK. (1997) Increased prevalence of mitochondrial DNA deletions in skeletal muscle of older individuals with impaired glucose tolerance: possible marker of glycemic stress. *Diabetes*, **46**, 920–923.

Luc K, Schramm-Luc A, Guzik TJ, Mikolajczyk TP. (2019) Oxidative stress and inflammatory markers in prediabetes and diabetes. *J Physiol Pharmacol*, **70**(6).

Liu G, Rondinone CM. (2005) JNK bridging the insulin signaling and inflammatory pathway. *Curr Opin Investig Drugs*, **6**, 979–987.

Liu Y, Geng T, Wan Z, Lu Q, Zhang X, Qiu Z, Li L, Zhu K, Liu L, Pan A, Liu G. (2022) Associations of Serum folate and vitamin b12 levels with cardiovascular disease mortality among patients with type 2 diabetes. *JAMA Netw Open*, **5**(1), e2146124.

Ma SB, Zhang R, Miao S, Gao B, Lu Y, Hui S, Li L, Shi XP, Wen AD. (2017) Epigallocatechin-3-gallate ameliorates insulin resistance in hepatocytes. *Mol Med Rep*, **15**(6), 3803–3809.

Mackenzie T, Leary L, Brooks WB. (2007) The effect of an extract of green and black tea on glucose control in adults with type 2 diabetes mellitus: double-blind randomized study. *Metabolism*, **56**, 1340–1344.

McClung JP, *et al.* (2004) Development of insulin resistance and obesity in mice overexpressing cellular glutathione peroxidase. *Proc Natl Acad Sci USA*, **101**, 8852–8857.

Mason TM, Goh T, Tchipashvili V, Sandhu H, Gupta N, Lewis GF, Giacca A. (1999) Prolonged elevation of plasma free fatty acids desensitizes the insulin secretory response to glucose in vivo in rats. *Diabetes*, **48**, 524–530.

Maximo V, Soares P, Lima J, Cameselle-Teijeiro J, Sobrinho-Simoes M. (2002) Mitochondrial DNA somatic mutations (point mutations and large deletions) and mitochondrial DNA variants in human thyroid pathology — a study with emphasis on Hurthle cell tumors. *American Journal of Pathology*, **160**, 1857–1865.

Milsom AB, Jones CJ, Goodfellow J, Frenneaux MP, Peters JR, James PE. (2002) Abnormal metabolic fate of nitric oxide in type I diabetes mellitus. *Diabetologia*, **45**(11), 1515–1522.

Montague CT, O'Rahilly S. (2000) The perils of portliness: causes and consequences of visceral adiposity. *Diabetes*, **49**, 883–888.

Mokwena MAM, Engwa GA, Nkeh-Chungag BN, Sewani-Rusike CR. (2021) Athrixia phylicoides tea infusion (bushman tea) improves adipokine balance, glucose homeostasis and lipid parameters in a diet-induced metabolic syndrome rat model. *BMC Complement Med Ther*, **21**(1), 292.

Mootha VK, *et al.* (2004) Erralpha and Gabpa/b specify PGC-1alpha-dependent oxidative phosphorylation gene expression that is altered in diabetic muscle. *Proc Natl Acad Sci USA*, **101**, 6570–6575.

Nagareddy PR, Xia Z, McNeill JH, MacLeod KM. (2005) Increased expression of iNOS is associated with endothelial dysfunction and impaired pressor responsiveness in streptozotocin-induced diabetes. *Am J Physiol Heart Circ Physiol*, **289**(5), H2144–H2152.

Nakatani Y, Kaneto H, Kawamori D, Hatazaki M, Miyatsuka T, Matsuoka TA, Kajimoto Y, Matsuhisa M, Yamasaki Y, Hori M. (2004) Modulation of the JNK pathway in liver affects insulin resistance status. *J Biol Chem*, **279**, 45803–45809.

Ni D, Ai Z, Munoz-Sandoval D, Suresh R, Ellis PR, Yuqiong C, Sharp PA, Butterworth PJ, Yu Z, Corpe CP. (2020) Inhibition of the facilitative sugar transporters (GLUTs) by tea extracts and catechins. *FASEB J*, **34**(8), 9995–10010.

Nisoli E. *et al.* (2003) Mitochondrial biogenesis in mammals: the role of endogenous nitric oxide. *Science*, **299**, 896–899.

Nunomura A, *et al.* (2001) Oxidative damage is the earliest event in Alzheimer disease. *J Neuropathol Exp Neurol*, **60**, 759–767.

Padgett LE, Broniowska KA, Hansen PA, Corbett JA, Tse HM. (2013) The role of reactive oxygen species and proinflammatory cytokines in type 1 diabetes pathogenesis. *Ann N Y Acad Sci*. **1281**(1), 16–35.

Panigrahy SK, Bhatt R, Kumar A. (2017) Reactive oxygen species: sources, consequences and targeted therapy in type 2 diabetes. *J Drug Target*, **25**(2), 93–101.

Permutt MA, Wasson JC, *et al.* (2001) A genome scan for type 2 diabetes susceptibility loci in a genetically isolated population. *Diabetes*, **50**, 681–685.

Previte DM, Piganelli JD. (2018) Reactive oxygen species and their implications on CD4(+) T cells in Type 1 diabetes. *Antioxid Redox Signal*, **29**(14), 1399–1414.

Rea R, Donnelly R. (2004) Resistin an adipocyte-derived hormone. Has it a role in diabetes and obesity? *Diabetes Obes Metab*, **6**, 163–170.

Rebolledo OR, Marra CA, Raschia A, Rodriguez S, Gagliardino JJ. (2008) Abdominal adipose tissue: early metabolic dysfunction associated to insulin resistance and oxidative stress induced by an unbalanced diet. *Horm Metab Res*.

Rendra E, Riabov V, Mossel DM, Sevastyanova T, Harmsen MC, Kzhyshkowska J. (2019) Reactive oxygen species (ROS) in macrophage activation and function in diabetes. *Immunobiology*, **224**(2), 242–253.

Ryu OH, Lee J, Lee KW, Kim HY, Seo JA, Kim SG, Kim NH, Baik SH, Choi DS, Choi KM. (2006) Effects of green tea consumption on inflammation, insulin resistance and pulse wave velocity in type 2 diabetes patients. *Diabetes Res Clin Pract*, **71**, 356–358.

Sabu MC, Smitha K, Kuttan R. (2002) Anti-diabetic activity of green tea polyphenols and their role in reducing oxidative stress in experimental diabetes. *J Ethnopharmacol*, **83**, 109–116.

Schriner SE, *et al.* (2005) Extension of murine lifespan by overexpression of catalase targeted to mitochondria. *Science*.

Schulz TJ, *et al.* (2007) Glucose restriction extends Caenorhabditis elegans life span by inducing mitochondrial respiration and increasing oxidative stress. *Cell Metab*, **6**, 280–293.

Serisier S, Leray V, Poudroux W, Magot T, Ouguerram K, Nguyen P. (2008) Effects of green tea on insulin sensitivity, lipid profile and expression of PPARalpha and PPARgamma and their target genes in obese dogs. *Br J Nutr*, **99**, 1208–1216.

Shi D, Motamed M, Mejía-Benítez A, Li L, Lin E, Budhram D, Kaur Y, Meyre D. (2021) Genetic syndromes with diabetes: a systematic review. *Obes Rev*, **22**(9), e13303.

Spiegelman BM, Flier JS. (2001) Obesity and the regulation of energy balance. *Cell*, **104**, 531–543.

Suraphad P, Suklaew PO, Ngamukote S, Adisakwattana S, Mäkynen K. (2017) The effect of isomaltulose together with green tea on glycemic response and antioxidant capacity: a single-blind, crossover study in healthy subjects. *Nutrient*, **9**(5), 464.

Svegliati-Baroni G, Candelaresi C, Saccomanno S, Ferretti G, Bachetti T, Marzioni M, De Minicis S, Nobili L, Salzano R, Omenetti A, *et al.* (2006) A model of insulin resistance and nonalcoholic steatohepatitis in rats: role of peroxisome proliferator-activated receptor-alpha and n-3 polyunsaturated fatty acid treatment on liver injury. *Am J Pathol*, **169**, 846–860.

Wu LY, Juan CC, Ho LT, Hsu YP, Hwang LS. (2004) Effect of green tea supplementation on insulin sensitivity in Sprague-Dawley rats. *J Agric Food Chem*, **52**, 643–648.

Wang P, Mariman E, Renes J, Keijer J. (2008) The secretory function of adipocytes in the physiology of white adipose tissue. *J Cell Physiol*, **216**(1), 3–13.

Xia HM, Wang J, Xie XJ, Xu LJ, Tang SQ. (2019) Green tea polyphenols attenuate hepatic steatosis, and reduce insulin resistance and inflammation in high-fat diet-induced rats. *Int J Mol Med*, **44**(4), 1523–1530.

Yan J, Zhao Y, Suo S, Liu Y, Zhao B-L. (2012) Green tea catechins ameliorate adipose insulin resistance by improving oxidative stress. *Free Radical Biology & Medicine*, **52**, 1648–1657.

Yang H, Zuo XZ, Tian C, He DL, Yi WJ, Chen Z, Zhang PW, Ding SB, Ying CJ. (2015) Green tea polyphenols attenuate high-fat diet-induced renal oxidative stress through SIRT3-dependent deacetylation. *Biomed Environ Sci*, **28**(6), 455–459.

Yazıcı D, Sezer H. (2017) Insulin resistance, obesity and lipotoxicity. *Adv Exp Med Biol*, **960**, 277–304.

Yi W, Xie X, Du M, Bu Y, Wu N, Yang H, Tian C, Xu F, Xiang S, Zhang P, Chen Z, Zuo X, Ying C. (2017) Green tea polyphenols ameliorate the early renal damage induced by a high-fat diet via ketogenesis/SIRT3 pathway. *Oxid Med Cell Longev*, **2017**, 9032792.

Yue S, Peng C, Zhao D, Xia X, Tan C, Wang Q, Gong J. (2020) Theabrownin isolated from Pu-erh tea regulates Bacteroidetes to improve metabolic syndrome of rats induced by high-fat, high-sugar and high-salt diet. *J Sci Food Agric*, **102**(10), 4250-4265.

Yue S, Shan B, Peng C, Tan C, Wang Q, Gong J. (2022) Theabrownin-targeted regulation of intestinal microorganisms to improve glucose and lipid metabolism in Goto-Kakizaki rats. *Food Funct*, **13**(4), 1921–1940.

Zhao B-L. (2002) Oxygen free radicals and natural antioxidants (Revised Edition). Science Press, Beijing.

Zhao B-L. (2008) Nitric oxide free radical. Science Press, Beijing.

Zhao B-L. (2020) The pros and cons of drinking tea. *Tradit Med Mod Med*, **3**(3), 1–12.

Zhao HJ, Wang S, Cheng H, Zhang MZ, Takahashi T, Fogo AB, Breyer MD, Harris RC. (2006) Endothelial nitric oxide synthase deficiency produces accelerated nephropathy in diabetic mice. *J Am Soc Nephrol*, **17**(10), 2664–2669.

Zhou J, Deo BK, Hosoya K, Terasaki T, Obrosova IG, Brosius FC, Kumagai AK. (2005) Increased JNK phosphorylation and oxidative stress in response to increased glucose flux through increased GLUT1 expression in rat retinal endothelial cells. *Invest Ophthalmol Vis Sci*, **46**, 3403–3410.

Zhou YP, Grill VE. (1994) Long-term exposure of rat pancreatic islets to fatty acids inhibits glucose-induced insulin secretion and biosynthesis through a glucose fatty acid cycle. *J Clin Invest*, **93**, 870–876.

Zhou YP, Grill V. (1995) Long term exposure to fatty acids and ketones inhibits B-cell functions in human pancreatic islets of Langerhans. *J Clin Endocrinol Metab*, **80**, 1584–1590.

Zhu J, Yu C, Zhou H, Wei X, Wang Y. (2021) Comparative evaluation for phytochemical composition and regulation of blood glucose, hepatic oxidative stress and insulin resistance in mice and HepG2 models of four typical Chinese dark teas. *J Sci Food Agric*, **101**(15), 6563– 6577.

Chapter 12

Bactericidal and Antiviral Effects of Tea Polyphenols

Baolu Zhao

Institute of Biophysics, Chinese Academy of Sciences, Beijing, China

12.1 Introduction

Bacterial and virus infection is an acute systemic infection caused by pathogenic bacteria or conditional pathogenic bacteria invading the blood circulation, growing, and reproducing, producing toxins and other metabolites, causing symptoms such as fever and high fever. Many pathogenic bacteria and virus secrete powerful internal and external toxins, enzymes, and pathogenic factors, which are highly invasive. If the treatment is not timely and complications arise, it can develop into sepsis or sepsis leading to death. Due to the wide use of antibiotics, antivirus, and immunosuppressive drugs, most infectious diseases can be treated. However, the bacterial species have also changed, resulting in drug resistance and thereby, reducing the efficacy of antibiotics. Viral infection refers to the process that viruses invade the body through various ways and proliferate in susceptible host cells. Human virus refers to the virus that can infect or cause disease to human beings. The essence of virus infection is the process of interaction between virus and organism, or virus and susceptible cells. Viral infection often produces different degrees of damage or viral diseases due to different types of viruses and body states.

Studies have found that both bacterial and viral infections are related to oxidative stress. Nitric oxide (NO) and reactive oxygen species (ROS) produced during infection are involved critically in host defense mechanisms. Infection and inflammation-mediated ROS are markers underlying pathology, clinical significance. The consequences of persistent bacterial and viral infections potentially include increased morbidity and mortality from the infection itself as well as an increased risk of dissemination of disease.

It has been found that antioxidant tea polyphenols treatment can inhibit oxidative stress caused by bacterial and viral infections and reduce disease severity during secondary bacterial infection. In this chapter, we will discuss the oxidative stress caused by bacterial and viral infection and the inhibitory effect of tea polyphenols on oxidative stress caused by bacterial and viral infection.

12.2   Hazards of bacterial and viral infections

The infectious diseases caused by bacterial and viral infections will lead to major disasters and plagues. Bacteria are very different from viruses because while there are good and bad bacteria, viruses are not good.

12.2.1   *Bacteria*

Bacteria is one of the main groups of organisms in the world. It is also the most abundant of all organisms, with an estimated total of about 5×10^{30} types. The shape of bacteria is very diverse, mainly spherical, rod, and spiral. Bacteria also have a great impact on human activities. On the one hand, there are beneficial bacteria, for example, there are a lot of beneficial bacteria in our intestines that help us with digestion. Humans often use bacteria in food, such as cheese, yogurt, and wine, as well as antibiotics and wastewater treatment. On the other hand, bacteria are the pathogens of many diseases. They can spread many diseases in normal people through various ways, such as contact, digestive tract, respiratory tract, insect bite, and so on. They are highly contagious and do great harm to society. The characteristics of bacteria parasitic, proliferating, and pathogenic in human body are called bacterial pathogenicity or pathogenicity. The pathogenicity is that there are many kinds of bacteria, such as pestis caused by *Yersinia pestis* and tuberculosis caused by *Mycobacterium tuberculosis*. The degree of pathogenicity varies with the virulence of various bacteria and vary according to different host species and environmental conditions. The pathogenicity of pathogenic bacteria is closely related to its virulence, the number of invading organisms, the mode of invasion, and the immune status of organisms. Invasiveness refers to the ability of bacteria to invade and spread in the body. Toxins released by bacteria can be divided into exotoxins and endotoxins.

12.2.2 *Virus*

Biological virus is a small, simple, non-cellular organism that contains only one kind of nucleic acid (DNA or RNA) and must parasitize in living cells and proliferate by replication. Virus is a non-cellular life form, which is composed of a long nucleic acid chain and a protein shell. The virus has no own metabolic mechanism and no enzyme system. Therefore, when the virus leaves the host cell, it becomes a chemical substance without any life activities and cannot reproduce independently. Its ability to replicate, transcribe, and translate is carried out in the host cell. When it enters the host cell, it can use the materials and energy in the cell to complete life activities and produce a new generation of virus like it according to the genetic information contained in its own nucleic acid. Recently, human immunodeficiency virus (HIV) and coronavirus-19 (COVID-19) diseases, which cause great harm to human beings, do the greatest harm to human beings. HIV and COVID-19 are caused by a typical virus.

12.2.3 *Hazards of bacteria and virus infection*

Plague is an infectious disease caused by some highly pathogenic substances, such as bacteria and viruses. Plague is a malignant infectious disease, also known as a pandemic, which is mainly caused by bacteria and viruses. Plague spreads in many regions and even the world, but not all the diseases that cause a pandemic may cause death. It is generally caused by poor environmental sanitation after natural disasters. There are both mild and severe plagues. Severe acute respiratory syndrome (SARS), smallpox, influenza, and pestis have a serious impact on future generations. In human history, there have been 10 plagues that are unforgettable to mankind.

From 1629–1631, a series of pestis broke out in Italy, which was called the Milan plague. The plague killed about 280,000 people. The third pestis pandemic occurred in 1885–1950. A major pestis began in 1855 in Yunnan Province, China. The pestis spread to Europe and Africa and spread to more than 60 countries in 77 ports in 10 years. In India and China alone, more than 12 million people died from the pestis. In 164–180 AD, the plague, smallpox, and measles of Antony in ancient Rome spread to the people of Antony. At that time, 2,000 people died of diseases in a day in Rome, equivalent to one fourth of the infected people. The total estimated death toll is as high as 5 million. In some places, the plague killed one third of the total population. London plague is a large-scale plague that occurred in England from 1665–1666. In this plague, 75,000–100,000 people died, more than one fifth of the total population of London at that time. From 1918–1920, the Spanish influenza caused about 1 billion people to be infected and 25 million to 40 million people died in the world; its global average mortality rate was about 2.5–5%. At that time, about 8 million people in Spain were infected with the disease, even the king of Spain was infected with the disease, so it was called Spanish influenza. In 1347–1351, the black death caused 75 million deaths worldwide, including 25 million to 50 million deaths in Europe. In 427 BC, pestis plague almost destroyed the whole city of Athens, and even dogs died from the disease. In 1720–1722, the Marseille plague affected the whole city and surrounding cities, causing 100,000 deaths. In 541–542, the Justinian plague, the first large-scale pestis outbreak in the Mediterranean world, caused extremely serious losses. A total of 5,000–7,000 people, even tens of thousands of people, died in one day. The American Plague occurred in the 6th century, during which many died in the decades of the 16th century, some historians even called it the largest genocide in human history.

In recent years, Acquired Immune Deficiency Syndrome (AIDS) and COVID-19, two terrible viral infectious diseases, have brought great and heavy disasters to mankind. AIDS is caused by human immunodeficiency virus (HIV), also known as acquired immunodeficiency. It is the last stage of human immunodeficiency virus infection. AIDS is the last stage of human immunodeficiency virus infection. According to statistics, the number of AIDS-related deaths reached a peak in 2006, with the number of deaths falling from 1.5 million in 2006 to 950,000 in 2017. Global antiretroviral treatment coverage increased from 2.98 million in 2006 to 21 million in 2017. New HIV infections peaked in 1999, reaching 3.16 million people.

The prevalence of COVID-19 was declared as a pandemic coronavirus-19 pneumonia by the World Health Organization (WHO) in early March 2020. According to the WHO report, since the outbreak occurred in 2019, the virus has spread to almost all countries in the world, causing hundreds of millions of confirmed cases of COVID-19 and millions of deaths. It also imposes a huge economic and social burden on the world.

Just the above-mentioned disasters of infectious diseases caused by bacteria and viruses can be seen that we must carefully study the infectious diseases caused by bacteria and viruses to find out countermeasures. To prevent, control, and treat infectious diseases caused by bacteria and viruses and protect human health, studies have found that infectious diseases caused by bacteria and viruses are related to oxidative stress injury. Additionally, natural antioxidants tea polyphenols can prevent and treat infectious diseases caused by bacteria and viruses. This chapter discusses this aspect in detail.

12.3 Bacteria and viruses and oxidative stress

The consequences of persistent bacterial and virus infections potentially include increased morbidity and mortality from the infection itself as well as an increased risk of dissemination of disease. Studies have shown that both bacteria and viruses can cause fever and severe inflammation and oxidative stress. Much evidence have shown that bacterial and viral infections are closely related to oxidative stress injury. Oxidative stress defines a condition in which the pro-oxidant-antioxidant balance in the cell is disturbed, resulting in DNA hydroxylation, protein denaturation, lipid peroxidation, and apoptosis, ultimately compromising cells' viability.

Oxidative stress caused by bacteria and viruses seriously damages human health, thereby requiring prevention and serious treatment.

12.3.1 *Bacteria causes oxidative stress*

Bacterial pathogens are highly adaptable organisms, a quality that enables them to overcome changing hostile environments. Once it enters the body, it will cause infection under appropriate conditions. It is known that there are many kinds of pathogenic bacteria. Different bacteria cause different symptoms, studies have shown that various bacteria can cause fever, severe inflammation and oxidative stress, and production of ROS and RNS free radicals. These free radicals are powerful mutagens that can lead to DNA breakage, point mutation, and protein DNA crosslinking, all of which can lead to genomic instability of colorectal cancer [Cooke *et al.*, 2003]. The consequences of persistent bacterial infections may include increased incidence rate and mortality due to the infection itself, as well as an increased risk of disease transmission. The relationship between bacterial infection and oxidative stress is illustrated with the examples of *Vibrio cholerae, Escherichia coli,* and *Helicobacter pylori.*

1. Oxidative stress caused by *Vibrio cholerae*

Cholera is an acute diarrhoeal infectious disease caused by the contamination of food or water caused by *Vibrio cholerae* (*V. cholerae*). Every year, it is estimated that there are 3–5 million cholera cases and 100,000–120,000 deaths. The peak of the disease is in summer, which can cause diarrhea, dehydration, and even death within a few hours. *V. cholerae* can cause high fever and diarrhea, but it will not cause a disease epidemic. Recently, new mutant strains have been found in some regions of Asia and Africa. It is observed that these strains can cause more serious cholera disease and higher mortality. Since *V. cholerae* exists in water, the most common cause of infection is drinking water contaminated by patients' feces, where it produces cholera toxin and causes secretory diarrhea. Rice-washing feces are the characteristics of cholera [Yoon & Waters, 2019]. *V. cholerae*, the causative agent of cholera, is able to colonize host small intestines and combat host-produced ROS during infection. In this study, another OxyR homolog in *V. cholerae*, which was named OxyR2, was identified and renamed the previous OxyR OxyR1. OxyR2 is required to activate its divergently-transcribed gene ahpC, encoding an alkylhydroperoxide reductase independently of H_2O_2. A conserved cysteine residue in OxyR2 is critical for this function. Mutation of either oxyR2 or AhpC rendered *V. cholerae* more resistant to H_2O_2 RNA sequencing analyses, which indicated that OxyR1-activated oxidative stress-resistant genes were highly expressed in OxyR2 mutants even in the absence of H_2O_2. Further genetic analyses suggest that OxyR2-activated AhpC modulates OxyR1 activity by maintaining low intracellular concentrations of H_2O_2. Furthermore, it showed that ΔOxyR2 and ΔAhpC mutants were less fit when anaerobically grown bacteria were exposed to low levels of H_2O_2 or incubated in seawater [Wang *et al.*, 2018a]. Many types of extra-cytoplasmic stresses, caused either by genetic alterations of outer membrane constituents or by chemical or physical damage to the cell envelope, induce common signaling pathways that ultimately lead to internal oxidative stress and mis-regulation of iron homeostasis [Sikora *et al.*, 2009]. These results suggest that the formation of radical oxygen species, induction of oxidative stress, and changes in iron physiology are likely general responses to cell envelope damage. OxyR2 and AhpC play important roles in the *V. cholerae* oxidative stress response. Oxidative stress represents natural conditions provoking induction of Shiga toxin-converting prophages as a consequence of H_2O_2 excretion by either neutrophils in infected humans or protist predators outside the human body.

2. Oxidative stress caused by *Escherichia coli*

The gut microbiota acts as a real organ. The symbiotic interactions between resident micro-organisms and the digestive tract highly contribute to maintaining the gut homeostasis and health of the human body. However, alterations to the microbiome caused by environmental changes (e.g., infection, diet and/or lifestyle) can disturb this symbiotic relationship and promote disease, such as inflammatory bowel diseases and cancer. *Escherichia coli* (*E. coli*) is one of the most characteristic bacteria associated with inflammatory bowel disease. It is reported that in patients with inflammatory bowel disease, adhesive and invasive *E. coli* colonize the intestinal mucosa abnormally [Darfeuille-Michaud *et al.*, 2004]. In addition, it was showed that *E. coli* associated with colorectal cancer can induce the expression of pro-inflammatory gene COX-2 in macrophages, supporting the regulation of inflammation by bacteria during the occurrence of colorectal cancer [Raisch *et al.*, 2015]. The results revealed that the *E. coli*-infected mice exhibited increased cardiac index, contents of IL-1β, IL-6, IL-8, TNF-α, leptin, and resistin, levels of apoptotic proteins (caspase-3 and caspase-9, and bax/bcl-2 ratio), cardiac pathological changes, and oxidative stress [Guo *et al.*, 2021]. The higher intracellular ROS in *E. coli* reacted with membrane unsaturated fatty acids by peroxidation, and then reduced cell membrane fatty acid saturation, accumulated MDA in cells, and further caused damage to cell membrane, reduced the ATPase activity, and eventually resulted in inactivation or apoptosis of *E. coli* [Yang *et al.*, 2017]. The concentration of endogenous H_2O_2 was submicromolar in a mutant strain *E. coli*, which was enough to cause damage to DNA and proteins as well as concomitant cell growth and metabolism.

3. Oxidative stress caused by *Helicobacter pylori*

Helicobacter pylori (*H. pylori*) is a microaerophilic, gram-negative pathogen in human stomach. *H. pylori* has been identified as an etiologic agent in the development of gastric ulcer, peptic ulcer, gastritis, and many other stomach-related diseases even gastric cancer. In corpus mucosa, 8-OHdG, and 8-NG production were significantly associated with the degree of glandular atrophy, infiltration of chronic inflammatory cells, and intestinal metaplasia in the glandular epithelial cells. *H. pylori* infection can induce inflammatory cells infiltration, which evokes DNA damage of gastric epithelial cells through ROS and RNS production [Katsurahara *et al.*, 2009]. Despite the chronic active gastritis that develops following colonization, *H. pylori* is able to persist unharmed in the stomach for decades. Much of the damage caused by gastric inflammation results from the accumulation of ROS and RNS within the stomach environment, which can induce oxidative damage in a wide range of biological molecules. *H. pylori*-induced oxidative stress according to *in vitro* and *in vivo* studies, some enterococci, especially *Enterococcus faecalis*, produce hydroxyl free radicals [Huycke *et al.*, 2001]. In addition, the role of *H. pylori* in gastric carcinogenesis by inducing oxidative stress has been clearly demonstrated, and this species can produce and induce ROS through immune cells. In this way, *H. pylori* affects many signal transduction pathways in gastric cells, thus promoting the occurrence of gastric cancer. Therefore, it can infer the fact that the presence of chemical, bacterial and/or immune-induced inflammation in the large intestine also induces the recruitment of neutrophils and macrophages, which is the main source of ROS, leading to genetic and epigenetic changes associated with colorectal cancer [Bartsch & Nair, 2002]. The intestinal microbiota also promotes the production of NO and its secondary nitric oxide synthase by the host, especially by activating macrophages in the inflammatory response, resulting in DNA damage. Some bacterial species can directly produce NOS [Lundberg *et al.*, 2004]. It has been reported that *Lactobacillus* and *Bifidobacterium* produced significant levels of NOS in sterile and single related mice, and a nitrate rich diet increased NOS production [Sobko *et al.*, 2006]. ROS and RNS can play an important role in cellular injury and carcinogenesis of gastric epithelial cells infected with *H. pylori*.

4. Host defense and oxidative stress signaling pathways in bacterial infection

It was reported that in APCMin/+ and IL-10-/- mice model susceptible to colorectal cancer, *Bacteroides fragilis* (*B. fragilis*) toxin (BFT) induced Th17 pro-inflammatory response, which is related to the occurrence of early tumors. However, both enterotoxigenic *B. fragilis* (ETBF) and non-toxigenic *B. fragilis* (NTBF) chronically colonize mice, only ETBF triggers colitis and strongly induces colonic tumors in multiple intestinal neoplasia (Min) mice. ETBF induces robust, selective colonic signal transducer, and activator of transcription-3 (Stat3) activation with colitis characterized by a selective T helper type 17 (T(H)17) response distributed between CD4+ T cell receptor-alphabeta (TCRalphabeta)+ and CD4-8-TCRgammadelta+ T cells. Antibody-mediated blockade of interleukin-17 (IL-17) as well as the receptor for IL-23, a key cytokine amplifying T(H)17 responses, inhibits ETBF-induced colitis, colonic hyperplasia, and tumor formation [Wu *et al.*, 2009]. In addition, the contribution of *Streptococcus bovis* to colorectal cancer is related to the increased expression of pro-inflammatory genes, such as IL-1, IL-8, and COX-2. In addition, APCMin/+ and IL-10-/- mice infected with *Clostridium nucleatum* and *Enterococcus faecalis*, respectively, increased immune cell infiltration and pro-inflammatory cytokines such as TNF-α, IL-1β, IL-6, and IL-8. A study discovered a unique second messenger, 8-nitroguanosine 3′,5′-cyclic monophosphate (8-nitro-cGMP) that mediates electrophilic signal transduction during oxidative stress and other cellular redox signaling in general. 8-Nitro-cGMP is formed via guanine nitration with NO and ROS, and in fact, NO-dependent 8-nitro-cGMP formation and heme oxygenase (HO-1) induction were identified in Salmonella-infected mice. HO-1 induction was regulated solely by 8-nitro-cGMP formed in cells, and more important, its potent anti-apoptotic function was evident in such a Salmonella infection. 8-Nitro-cGMP has a potent cyto-protective function, of which the signaling appears to be mediated via protein sulfhydryls to generate a post-translational modification called protein S-guanylation. 8-Nitro-cGMP, specifically S-guanylates Keap1, is a negative regulator of transcription factor Nrf2, which in turn upregulates transcription of HO-1. Recent study revealed that the autophagy might be involved in the 8-nitro-cGMP-dependent antimicrobial effect [Akaike, 2015]. These results indicate that 8-nitro-cGMP signaling was regulated by reactive sulfur species and that bacterial infection is associated with ROS, RNS free radicals, and increased expression of pro-inflammatory genes, such as IL-1, IL-8, and COX-2. Autophagy may be involved in the 8-nitro-cGMP-dependent antibacterial effect. 8-nitro-cGMP signaling was also found to be regulated by active sulfur species with excellent antioxidant activity and unique signaling function. It functions during bacterial infection by forming a unique cellular signaling molecule, 8-nitro-cGMP.

12.3.2 *Viruses causes oxidative stress*

Studies have shown that HIV, COVID-19, and other viruses can cause fever, severe inflammation, and oxidative stress, and produce ROS and RNS free radicals. These free radicals are powerful mutagens, which can lead to DNA breakage, point mutation, and protein DNA crosslinking, all of which can lead to a variety of disease symptoms. The consequences of persistent viral infection may include increased incidence rate and mortality due to the infection itself, as well as an increased risk of disease transmission. The relationship between viral infection and oxidative stress is illustrated with HIV and COVID-19 as examples.

1. Oxidative stress caused by HIV infection

Human immunodeficiency virus (HIV), also known as acquired immune deficiency (AIDS) is the last stage of human immunodeficiency virus infection. Studies on HIV virology and pathogenesis address the complex mechanisms that result in the HIV infection of the cell and destruction of the immune system. These studies are focused on both the structure and the replication characteristics of HIV and on the interaction of the virus with the host, to provide continuous updating of knowledge on structure, variability, and replication of HIV, as well as the characteristics of the host immune response. HIV mainly invades the body's immune system, including CD4+T lymphocytes and mononuclear macrophages. The AIDS virus is broken into the human blood through the skin or the mucous membrane. The target cells that attack and destroy mainly are T4 lympho-cytes (T4 lymphocytes play a central regulatory role in the cellular immune system, which can promote B cells to produce antibodies), which causes T4 cells to lose their normal immunity. When the T4 cells activated by immunization are almost completely eliminated by HIV, the number of T4 cell suppressor cells will increase dramatically. On the contrary, the number of T4 cells in the patients will suddenly decrease, resulting in complete immune system failure of the patients. HIV infection increases the risk of colds, flu, and pneumonia, and increases the risk of pulmonary arterial hypertension (PAH). PAH refers to a hemodynamic and patho-physiological state in which the pulmonary artery pressure increases beyond a certain threshold, which can lead to right heart failure. It can be an independent disease, a complication or a syndrome. Over time, PAH, which is caused by HIV, can make the heart tensed [Fanales-Belasio *et al.*, 2010]. HIV infection will lead to opportunistic infections and tumors. These complications will have an impact on people's health and may even threaten lives. Hence, AIDS virus can cause people to panic. The virus attacks and gradually destroys the human immune system, leaving the host unprotected when infected. People who are infected with human immunodeficiency virus, often die of secondary infection or cancer. Acute infection often occurs in the first 2–4 weeks after HIV infection. Close to 15–20% of the infected people may have fever, sweating, fatigue, myal-gia, joint pain, anorexia, rash, lymphadenopathy, and other symptoms within 2–6 weeks after infection.

HIV-1 infection has a pervasive impact on brain function. It was found that the increase in energy supply requirement by HIV-1-infected neuro-immune cells as well as the deterrence of nutrient uptake across the blood-brain barrier significantly depletes the energy source and neuro-environment homeostasis in the central nervous system (CNS) [Agas *et al.*, 2021]. Astrocytes are the major cellular component of the CNS, and they play multiple roles in brain development, normal brain function, and CNS responses to pathogens and injury. The functional versatility of astrocytes is linked to their ability to respond to a wide array of biological stimuli through finely-orchestrated changes in cellular gene expression. Dysregulation of gene expression programs, generally by chronic exposure to pathogenic stimuli, may lead to dysfunction of astrocytes and contribute to neuro-pathogenesis [Borjabad *et al.*, 2010]. Neurons themselves are rarely infected by HIV-1, but HIV-1 infects resident microglia, periventricular macrophages, leading to increased production of cytokines and to release of HIV-1 proteins, the most likely neurotoxins, among which are the envelope glycoprotein gp120 and HIV-1 trans-acting protein Tat (Tetanus antitoxin). Human immunodeficiency virus with proteins gp120 and Tat induce oxidative stress in the brain, leading to neuronal apoptosis/death [Louboutin & Strayer, 2014].

HIV infection is associated with oxidative stress. In addition to HIV itself, some antiretroviral drugs can also increase oxidative stress and reduce virus replication. To study the changes of oxidative stress parameters and thiol disulfide homeostasis in HIV, infected patients were treated with antiretroviral therapy based on integrase inhibitors. A total of 30 untreated HIV-infected adult were prospectively included in the study. The oxidative stress index and disulfide level were calculated by a mathematical equation. The results showed that the levels of total oxidative status, oxidative stress index), malondialdehyde (MDA), and protein carbonyl in untreated HIV-infected patients were significantly higher than those in art-treated patients ($p < 0.001$). The total and natural mercaptans in the two HIV infected groups were significantly lower than those in the control group ($p < 0.001$). The levels of protein carbonyl and MDA in the two HIV infected groups were significantly higher than those in the control group ($p < 0.001$). In a correlation analysis, in untreated HIV-infected people, MDA was negatively correlated with age, while total oxidative status was positively correlated with CD4+T cell count. In the healthy control group, age was positively correlated with total oxidative status. Therefore, it is concluded that antiretroviral therapy based on integrase inhibitors can reduce the oxidative stress caused by HIV infection, which may be a good treatment choice for HIV infected people [De Bruyne *et al.*, 1999; Akkoyunlu *et al.*, 2020].

The above evidence show that HIV mainly invades the human immune system, including CD4+T lymphocytes and monocyte macrophages, resulting in a series of disease symptoms such as fever, sweating, fatigue, myalgia, joint pain, anorexia, rash, lymph node enlargement, increasing the risk of cold, influenza and pneumonia, and increasing the risk of pulmonary hypertension and tumor. Oxidative stress, total oxidative state, serum MDA, and protein carbonyl levels increased as well as damaged tissue, cell and neuron energy homeostasis, and autophagy.

2. Oxidative stress caused by COVID-19 infection

SARS-CoV-2 causes severe pneumonia (COVID-19) that affects all people, especially elderly people. The prevalence of COVID-19 was declared as pandemic coronavirus-19 pneumonia by the World Health Organization (WHO) in early March 2020. According to WHO, novel coronavirus pneumonia cases have been spreading over almost all countries since the outbreak in 2019, resulting in hundreds of millions of confirmed cases of new crown pneumonia and millions of deaths. It has also created a huge economic and social burden on the world. Moreover, the symptoms of COVID-19 vary greatly within individuals. Now, there have been a variety of variants, and some new variants have more concealment and stronger infectivity. The most common symptoms of COVID-19 are fever, dry cough, fatigue, mild dyspnea, sore throat, headache, diarrhea, nausea, and vomiting. COVID-19 causes the most serious death by inflammation of lung diseases.

The high neutrophil to lymphocyte ratio observed in critically ill patients with COVID-19 is associated with excessive levels of ROS, which promotes a cascade of biological events that drive pathological host responses. ROS induce tissue damage, thrombosis, and red blood cell dysfunction, which contribute to COVID-19 disease severity [Kostic *et al.*, 2020]. COVID-19 viruses cause respiratory inflammation augment

ROS production in host cells. Factors leading to inflammation and immune response include nitric oxide (NO/ROS) ratio, activation of classically activated (M1) macrophages, and red blood cell (RBC) injury. Normal NO/ROS balance is essential for normal vascular function. NO is an endothelium-derived relaxing factor, which plays a key role in vascular signal transduction, blood flow regulation, and host defense. ROS, such as superoxide species, also act as host defense and are induced during stress, such as viral infection. Excessive ROS (such as virus overload) activate M1 macrophages, recruit neutrophils, and promote the production of peroxynitrite together with NO, so as to effectively respond to the invading virus, accompanied by endothelial dysfunction, osmotic vascular, and lipid membrane peroxidation. Macrophages produce a large number of highly reactive nitrogen- and oxygen-derived molecular species and pro-inflammatory cytokines, such as interleukin (IL)-2, IL-6, IL-8, interferon IFN-α/β, and tumor necrosis factor, TNF-α. They neutralize invading organisms but also exacerbate vascular damage. In addition, inflammation-induced platelet activation (which can be alleviated by NO) can lead to increased coagulation, which is an important consequence of the disease. Recently, evidence that COVID-19 destroys erythrocyte hemoglobin has been reported, which may lead to significant oxidative stress. Subsequent hemolysis can lead to anemia, but more importantly, it can exacerbate the inflammatory process. Acellular hemoglobin clears coagulation regulators such as endothelial NO, while the release of pro-inflammatory heme and iron activates platelets. Enhanced coagulation and slow blood flow lead to systemic hypoxia in oxygen-sensitive organs such as kidney. Under the promotion of NO, it stops the pro-inflammatory injury to the tissue and starts the repair process and debris removal. Unchecked macrophage response produces inflammatory cascade and cytokine storm, resulting in the formation of cell debris and edema due to vascular leakage, which is manifested as respiratory distress syndrome in the lung [Banu *et al.*, 2020].

The high neutrophil to lymphocyte ratio observed in critically ill patients with COVID-19 is associated with excessive levels of ROS, which promote a cascade of biological events that drive pathological host responses. ROS induce tissue damage, thrombosis, and red blood cell dysfunction, which contribute to COVID-19 disease severity. Decreased expression of SOD3 in the lungs of elderly COVID-19 patients correlates with disease severity [Laforge *et al.*, 2020]. Blood selenoprotein P, which is secreted from the liver for selenium delivery, is dramatically lowered in severe COVID-19 patients, probably due to pronounced hypoxia and/or marked IL-6 elevation that suppress hepatic selenoprotein P [Becker *et al.*, 2014].

Angiotensin-converting enzyme inhibits NO production and promotes ROS and inflammation. The binding of spinous-process protein of COVID-19 to angiotensin-converting enzyme 2 may downregulate it and lead to unchecked downstream effects of angiotensin-converting enzyme, including increased vascular permeability and decreased anti-inflammatory mediators (such as NO). This imbalance of NO/ROS is transformed into a potential intervention area of the current pandemic. Both hypoxia/reoxygenation and ischemia/reperfusion initiate the pro-inflammatory cycle through ROS- and heparinase-mediated degradation of glycocalyx and endothelial lining. In the case of blood flow stagnation and microvascular embolism, ventilation and rapid recovery of blood flow at hypoxic and ischemic sites may contribute to the pathogenesis of respiratory distress syndrome. Anti-inflammatory molecules such as N-acetylcysteine can potentially reduce the damage of ROS to cells and restoring the balance of NO/ROS can reduce the damage of reoxygenation and reperfusion to cells [Alvarez *et al.*, 2020].

In severe cases of COVID-19, viral pneumonia progresses to respiratory failure. Peripheral blood analysis showed that viral infection usually led to a significant increase in pro-inflammatory cytokines and chemokines and developed into a powerful cytokine storm. When high inflammation lasts for a long time, it will cause damage to multiple tissues and organs. In addition, high inflammation leads to a serious imbalance of NO/ROS in the body, which leads to oxidative stress. A large number of activated pro-inflammatory cytokines and chemokines, including IL-2, IL-6, IL-10, and TNF-α were found in the serum of patients with severe COVID-19. Pro-inflammatory cytokines promote the production of excessive ROS in mitochondria by blocking mitochondrial oxidative phosphorylation and adenosine triphosphate production, leading to increased membrane permeability and kinetic changes, mitochondrial dysfunction, and apoptosis. Mitochondria are the main source of ROS. When there is excessive ROS, apoptosis, activation of transcription factors (NFκB, AP-1), and overexpression of inflammatory cytokines and adhesion molecules (ICAM-1, VCAM-1, E-selectin) will aggravate endothelial injury and significantly reduce the production of NO. In addition, ROS changes vascular tension by increasing intracellular calcium concentration and reducing the bioavailability of NO. How NO level and bioavailability affect COVID-19 patients are closely related to their unique characteristics [Ricciardolo *et al.*, 2020]. A study reported that sera from patients with COVID-19 (n = 50 patients, n = 84 samples) had elevated levels of cell-free DNA, myelo-peroxidase (MPO)-DNA, and citrullinated histone H3 (Cit-H3); the latter two are highly specific markers of neutrophil extracellular traps (NETs). Highlighting the potential clinical relevance of these findings, cell-free DNA strongly correlated with acute phase reactants, including C-reactive protein, D-dimer, and lactate dehydrogenase, as well as absolute neutrophil count. MPO-DNA was associated with both cell-free DNA and absolute neutrophil count, while Cit-H3 correlated with platelet levels. Importantly, both cell-free DNA and MPO-DNA were higher in hospitalized patients receiving mechanical ventilation as compared to hospitalized patients breathing room air. Finally, sera from individuals with COVID-19 triggered NET release from control neutrophils *in vitro* [Zuo *et al.*, 2020].

Respiratory virus induced oxidative damage and ROS facilitated SARS-CoV-2 replication. The lung is responsible for maintaining adequate oxygenation in the organism. Exposure to this condition favors the increase of ROS from mitochondria, as from NADPH oxidase, xanthine oxidase/reductase, and nitric oxide synthase enzymes, as well as establishing an inflammatory process. Hypoxia enhances the generation of superoxide anions and increases release of ROS from the inner mitochondrial membrane to the intermembrane space in the lung, resulting in disruption of redox balance. Imbalance of redox state in lungs induced by hypoxia has been suggested as a participant in the changes observed in lung function in the hypoxic context, such as hypoxic vasoconstriction and pulmonary edema, in addition to vascular remodeling and chronic pulmonary hypertension. Hypoxia also increases pulmonary ROS production via inducing NOX4, xanthine oxidase/reductase, or endothelial/inducible nitric oxide synthases [Araneda & Tuesta, 2012].

Hypoxia-induced pulmonary oxidative stress promotes vasoconstriction, edema, inflammation, vascular remodeling, and pulmonary hypertension [Fresquet *et al.*, 2006]. In COVID-19, macrophage infiltration into the lung causes a rapid and intense cytokine storm, finally leading to multi-organ failure and death. Age-related mitochondrial dysfunction is proposed as an enhancing factor in COVID-19 disease. Mitochondrial dysfunction is one of the hallmarks of aging and COVID-19 risk factors. Dysfunctional mitochondria is associated with defective immunological response to viral infections and chronic inflammation. Chronic inflammation caused by mitochondrial dysfunction is responsible for the explosive release of inflammatory cytokines causing severe pneumonia, multi-organ failure, and finally death in COVID-19 patients [Moreno Fernandez-Ayala & Navas, 2020]. In patients infected by SARS-CoV-2 who experience an exaggerated inflammation leading to pneumonia, monocytes likely play a major role. Monocytes from COVID-19 patients exhibited mitochondrial dysfunction. Basal extracellular acidification rate was also diminished, suggesting reduced capability to perform aerobic glycolysis. Although COVID-19 monocytes had a reduced ability to perform oxidative burst, they were still capable of producing TNF and IFN-γ *in vitro*. A significantly high number of monocytes had depolarized mitochondria and abnormal mitochondrial ultrastructure. A redistribution of monocyte subsets with a significant expansion of intermediate/pro-inflammatory cells and high amounts of immature monocytes were found, along with a concomitant compression of classical monocytes and an increased expression of inhibitory checkpoints like PD-1/PD-L1. High plasma levels of several inflammatory cytokines and chemokines, including GM-CSF, IL-18, CCL2, CXCL10, and osteopontin, finally confirm the importance of monocytes in COVID-19 immuno-pathogenesis. [Gibellini *et al.*, 2020].

The above evidence shows that the infection of COVID-19 mainly invades the human respiratory system, leading to lung infection, causing macrophages to infiltrate the lungs, which will lead to a rapid and strong cytokine storm. The pro-inflammatory cytokines promote the production of excess ROS and RNS in mitochondria by blocking mitochondrial oxidative phosphorylation and adenosine triphosphate production. The activated pro-inflammatory cytokines and chemokines include IL-2, IL-6, IL-10, and TNF-α. These lead to the increase of membrane permeability and dynamic changes, mitochondrial dysfunction, and apoptosis, and finally leading to multiple organ failure and death. Thereafter causing a series of disease symptoms, such as imbalance of pulmonary redox state, pulmonary edema, vascular remodeling, and chronic pulmonary hypertension. The blood vessels leak and form cell debris and edema, resulting in pulmonary respiratory distress syndrome and death.

3. Oxidative stress caused by influenza virus

Influenza, or flu for short, is an acute respiratory disease caused by Influenza A, B, and C viruses, belonging to Class C infectious disease. Influenza is common in winter and spring. Its clinical manifestations are mainly high fever, fatigue, headache, cough, muscle soreness, and other systemic poisoning symptoms, while respiratory symptoms are mild. Influenza virus mutates easily, is highly infectious, and has a high incidence rate. It has caused many outbreaks in the world in history and is an important public health issue of global concern. Influenza is a disease caused by a highly infectious virus. Influenza virus is mainly transmitted through droplets in the air, contact between susceptible and infected people, or contact with contaminated objects. Generally, autumn and winter are the high incidence periods. The risk of dying from heart disease after catching the flu has increased by one-third. Researchers have found that acute pneumonia caused by influenza infection can destroy the stability of atherosclerotic plates in arteries. Electron Spin Resonance (ESR) spectroscopy was used to study the oxidative stress level of mice infected with Influenza A virus FM1 subpopulation when inhaled intra-nasally in a laboratory. The magnitude of ESR spectrum (28.65 ± 10.71 AU) in mice infected with influenza virus was significantly higher than those of healthy control mice (19.10 ± 3.61 AU). Serum in mice treated with Ribavirin, ascorbic acid, SOD, and Kegan Liyan oral prescription (KGLY), a proprietary Chinese medicine for influenza and common cold, declined to 19.70 ± 6.05 AU, 18.50 ± 2.93 AU, 16.25 ± 3.59 AU, and 18.40 ± 2.14 AU, respectively. The aforementioned data indicates a close resemblance to the signal height observed in healthy controls. This resemblance is achieved through the downregulation of oxidative stress levels caused by the influenza virus, resulting in a decline in the lung index of pneumonia compared to untreated healthy mice and mice with influenza virus-induced pneumonia. The antiviral effect of Ribavirin, ascorbic acid, SOD, and KGLY, are to treat influenza and common cold symptoms through their redox reactions [Duan *et al.*, 2009].

12.4 Antibacterial and antiviral effects of tea polyphenols

Tea is one of the most popular drinks in the world, and research have shown that tea is rich in polyphenols and is beneficial to our health because it contributes to the prevention of many diseases. Tea polyphenols have many important biological functions, including antibacterial, antiviral, antioxidant, and anti-inflammatory effects. Many studies have shown that tea polyphenols have antibacterial and antiviral properties. Both bacteria and viruses can cause fever and severe inflammation and oxidative stress. Studies have shown that tea polyphenols can inhibit or even kill bacteria and viruses, inhibit oxidative stress, and prevent and treat the diseases caused by bacteria and viruses. In this part, the antibacterial and antiviral effects of tea polyphenols on various bacteria and viruses, such as HIV and COVID-19 were discussed.

12.4.1 *Tea polyphenols have antibacterial properties*

Tea polyphenolics such as catechins are known to have the potential to inhibit many bacterial pathogens. The antibacterial effects of tea polyphenol epigallocatechin gallate (EGCG), a common phytochemical with a number of potential health benefits, are well known. A large number of studies have shown that tea polyphenols have antibacterial properties, such as the microbiota in gut and oral microorganism, *Haemophilus parasuis*.

1. Tea polyphenols have anti-microbiota effect in gut
Green tea, oolong tea, and black tea all can have effects on human intestinal microbiota and each sample can induce the proliferation of certain beneficial bacteria and inhibited *Bacteroides*, *Prevotella*, and *Clostridium histolyticum* [Sun *et al.*, 2018]. Green tea can affect the intestinal microbiota by stimulating the growth of specific species or preventing the

development of harmful species. At the same time, intestinal bacteria can metabolize green tea compounds and produce smaller bioactive molecules. Therefore, the benefits of green tea may come from bioactive metabolites of beneficial bacteria or microorganisms. The intestinal microbiota may act as an intermediary, at least in terms of some of the health benefits of green tea. Many of the health promoting effects of green tea seem to be related to the interaction between green tea and the intestinal microbiota. On one hand, facts have proved that green tea can correct microbial disorders in many cases, such as obesity or cancer. On the other hand, tea compounds can affect the growth of bacterial species involved in the inflammatory process, such as the release of LPS or the regulation of interleukin (IL) production, thus, affecting the development of different chronic diseases. Many studies have attempted to link green tea or green tea polyphenols to health benefits through the intestinal microbiota. Enterotoxigenic *Escherichia coli* (ETEC) increased the expression of pro-inflammatory cytokines, activated oxidative stress, inhibited the activities of antioxidant enzymes, induced phosphorylation of p38 MAPK gene and protein expression, and destroyed intestinal barrier function, resulting in diarrhea caused by ETEC, resulting in high incidence rate and mortality. Due to the abuse of antibiotics and the emergence of drug resistance, antibiotics are no longer considered to be the only beneficial but also possibly harmful drug. Supplements that inhibit bacterial growth are expected to replace antibiotics. A study used a mouse model to study the role of tea polyphenols in ETEC K88 infection. Tea polyphenol pretreatment can reduce the symptoms caused by ETEC K88. In addition, in the cell adhesion test, tea polyphenols inhibited the adhesion between ETEC K88 and ipec-j2 cells. When the cells were infected with ETEC K88, the mRNA and protein levels of claudin-1 decreased significantly compared with the control cells. However, when cells were pretreated with tea polyphenols and infected with ETEC K88, claudin-1 mRNA and protein levels were higher than those of untreated cells. After the cells pretreated with tea polyphenols were infected with ETEC K88, the level of toll-like receptor 2 (TLR2) mRNA was also higher. These data suggest that tea polyphenols can upregulate the expression of claudin-1 by activating TLR2, thereby increasing the barrier integrity of ipec-j2 cells. Tea polyphenols have beneficial effects on epithelial barrier function. Therefore, tea polyphenols can be used as a new strategy to control and treat infection caused by ETEC K88 [Ma *et al.*, 2021].

For the low bioavailability, most tea polyphenols are thought to remain in the gut and metabolized by intestinal bacteria. In the gut, the unabsorbed tea polyphenols are metabolized to a variety of derivative products by intestinal flora, which may accumulate to exert beneficial effects. Numerous studies have shown that tea polyphenols can inhibit obesity and its related metabolism disorders effectively. Meanwhile, it has demonstrated that tea polyphenols and their derivatives may modulate intestinal micro-ecology [Zhao & Zhang, 2020]. Green tea polyphenols (GTP) decreased the relative abundance of *Bacteroidetes* and *Fusobacteria* and increased the relative abundance of *Firmicutes* as revealed by 16S rRNA gene sequencing analysis. The relative proportion of *Acidaminococcus, Anaerobiospirillum, Anaerovibrio, Bacteroides, Blautia, Catenibactetium, Citrobacter, Clostridium, Collinsella,* and *Escherichia* were significantly associated with GTPs-induced weight loss. GTP significantly ($p < 0.01$) decreased expression levels of inflammatory cytokines, including TNF-α, IL-6, and IL-1β, and inhibited induction of the TLR4 signaling pathway compared with high-fat diet [Li *et al.*, 2020].

It was found that the mechanism of its bactericidal action was oxidative stress. Using *E. coli* as a model organism, it is argued here that H_2O_2 synthesis by EGCG is not attributed to its inhibitory effects. In contrast, the bactericidal action of EGCG was a result of increased intracellular ROS and blunted adaptive oxidative stress response in *E. coli* due to the co-administration of antioxidant N-acetylcysteine, and not on account of exogenous catalase. Furthermore, it noted a synergistic bactericidal effect for EGCG when combined with paraquat. However, under anaerobic conditions, the inhibitory effect of EGCG was prevented [Xiong *et al.*, 2017].

The above results show that tea polyphenols not only inhibit and kill harmful intestinal bacteria, but also benefit probiotics, reduce the relative abundance of intestinal *Bacteroides* and *Clostridium*, and increase the relative abundance of *chlamydosporium*. EGCG can increase the endogenous oxidative stress of *E. coli* and inhibit its growth. Green tea can correct microbial disorders in many cases.

2. Tea polyphenols inhibit *Helicobacter pylori*

H. pylori induces gastric damage and may be involved in the pathogenesis of gastric cancer. *H. pylori*-vacuolating cytotoxin, Vacuolar cytotoxin (VacA), is one of the important virulence factors and is responsible for *H. pylori*-induced gastritis and ulceration. Infection by the stomach-dwelling bacterium *H. pylori* constitutes a major risk factor for gastritis, peptic ulcer, gastric cancer, and mucosa-associated lymphoid tissue lymphoma. A study investigated the effect of nine tea extracts — three different brands representing four different processed types (white, green, oolong, and black) — on the inhibition of *H. pylori*. Most five-minute extracts showed inhibitive effects on *H. pylori*. Extracts that showed inhibition were further evaluated for their effect on beneficial lactic acid bacteria. None of the samples showed inhibition, suggesting that tea might be able to inhibit *H. pylori* without affecting the beneficial lactic acid bacteria [Ankolekar *et al.*, 2011]. VacA is a major virulence factor of the widespread stomach-dwelling bacterium, *H. pylori*. It causes cell vacuolation and tissue damage by forming anion-selective, urea-permeable channels in plasma and endosomal membranes. Green tea, which contain many of the compounds such as polyphenols, can also potently inhibit the toxin [Tombola *et al.*, 2003]. Polyphenols had the strongest inhibitory activity in a concentration-dependent manner for VacA binding to its receptors, RPTP (alpha) and RPTP (beta), VacA uptake, and VacA-induced vacuolation in susceptible cells. In addition, oral administration of polyphenols with VacA to mice reduced VacA-induced gastric damage at 48 hours [Yahiro *et al.*, 2005]. *H. pylori* infection resulted in increased BrdU-labeled cells in both the antrum and the bodies. Administration of polyphenols suppressed this increased proliferation. *H. pylori* infection increased apoptotic cells in both the antrum and the corpus in comparison with controls. This increase was not seen in *H. pylori*-infected mice given polyphenols. We conclude that the administration with polyphenols might suppress gastric carcinogenesis that is in part related to *H. pylori* infection [Akai *et al.*, 2007].

These studies indicate that tea can be potentially used as a low-cost dietary support to combat *H. pylori*-linked gastric diseases without affecting the beneficial intestinal bacteria.

3. Tea polyphenols inhibit oral microorganism

In the human oral cavity, there are more than 750 different species of bacteria living together within dental plaque. Some of the bacteria are pathogens that contribute to the development of oral diseases such as dental caries, periodontitis, pulpitis, mucosal disease, or halitosis through their virulence factors and their metabolites. Until now, many studies have reported that tea polyphenols have evident inhibitory effects on some oral pathogenic microorganisms by suppressing pivotal steps of their pathogenic processes. The effectiveness and mechanisms of tea polyphenols in inhibiting microorganisms provide new ideas for the prevention and treatment of oral diseases and to contribute to the global dental public health [Li *et al.*, 2019].

It has been proven that tea polyphenols can inhibit the sucrose-dependent adherence of oral cariogenic bacterium to the tooth surface. Changing the concentrations of tea polyphenols solution from 1.0 mg/ml to 4.0 mg/ml, could decrease the number of *Streptococcus mutans* 3a3 and *Actinomyces viscosus* ATCC 19246 and the inhibition rates rose with the increase of the concentrations of tea polyphenols solution [Xiao *et al.*, 2000]. Gingival epithelium is a kind of stratified squamous tissue. As the interface between the external environment and the underlying connective tissue, it plays an active role in maintaining periodontal health. This paper was to study the ability of green tea catechin to enhance the function of gingival epithelial barrier and prevent the destruction of epithelial integrity caused by *Porphyromonas gingivalis*. Both green tea extract and EGCG increased the cross epithelial resistance (TER) of gingival keratinocyte model in a dose- and time-dependent manner and decreased the permeability of cell monolayer to fluorescein iso-thiocyanate conjugated 4.4-kda dextran. This is related to the increased expression of two tight junction proteins, occlusive Band-1 (ZO-1) and occludin. Treatment of gingival keratinocyte monolayer with *Pseudomonas gingivalis* can reduce TER, affect the distribution of ZO-1 and occludin, and make *Pseudomonas gingivalis* shift through the cell monolayer. Green tea extract and EGCG eliminate these harmful effects mediated by *Pseudomonas gingivalis*. This protective effect may be partly related to the ability of tea catechins to inhibit the protease activity of *Pseudomonas gingivalis* [Lagha *et al.*, 2018].

Halitosis is caused mainly by volatile sulfur compounds (VSCs) such as H_2S and CH_2SH produced in the oral cavity. Oral microorganisms degrade proteinaceous substrates into cysteine and methionine, which are then converted to VSCs. Most treatments for halitosis focus on controlling the number of microorganisms in the oral cavity. Since tea polyphenols have been shown to have antimicrobial and deodorant effects, green tea powder reduces VSCs such as H_2S and CH_3SH produced in the oral cavity. Green tea also demonstrated strong deodorant activities *in vitro* [Lochhead *et al.*, 2013]. The administration of the oolong tea extract and the isolated polyphenol compound into drinking water resulted in significant reductions in caries development and plaque accumulation in the rats infected with mutans streptococci. The active components in the oolong tea extract were presumptively identified as polymeric polyphenols which were specific for oolong tea leaves. These results indicate that the oolong tea polyphenolic compounds could be useful for controlling dental caries [Ooshima *et al.*, 1993].

From the above characteristics, green tea and oolong tea polyphenols may represent promising molecules for the prevention and treatment of periodontal disease. Green tea and oolong tea are very effective in reducing oral malodor temporarily because of its disinfectant and deodorant activities. Moreover, green tea and oolong tea may prevent enamel from caries by decreasing the adherence of main cariogenic bacterium to salivary acquired pellicle.

4. Tea polyphenols suppress *Vibrio cholerae*

Vibrio cholerae (*V. cholerae*) is the pathogen of human cholera. Cholera is one of the ancient and widespread infectious diseases. It has caused many pandemics in the world, mainly manifested as severe vomiting, diarrhea, dehydration, and high mortality. It is an international quarantine infectious disease. A study investigated the anti-virulence activity of a polyphenolic fraction previously isolated from Kombucha, a 14-day fermented beverage of sugared black tea, against *V. cholerae*. The isolated fraction was mainly composed of the polyphenols catechin and isorhamnetin. The fraction, the individual polyphenols and the combination of the individual polyphenols significantly inhibited bacterial swarming motility and expression of flagellar regulatory genes motY and flaC, even at sub-inhibitory concentrations [Bhattacharya *et al.*, 2020]. Catechin, one of the primary antibacterial polyphenols in tea was also found to be present in Kombucha. The overall study suggests that Kombucha can be used as a potent antibacterial agent against entero-pathogenic bacterial infections, which mainly is attributed to its polyphenolic content [Bhattacharya *et al.*, 2016].

The above results implied that tea polyphenols isolated from red tea might be considered as a potential alternative source of anti-virulence polyphenols against *V. cholerae.*

5. Tea polyphenols suppress growth and infections of *Haemophilus parasuis* and other bacteria

Haemophilus parasuis (*H. parasuis*) colonises healthy pigs and is the aetiological agent of Glässer's disease. The pathogenicity of *H. parasuis* is poorly characterized, while prevention and control of Glässer's disease continues to be challenging. Infection by *H. parasuis* requires adhesion to and invasion of host cells, resistance to phagocytosis by macrophages, resistance to serum complement, and induction of inflammation. Identification of virulence factors involved in these mechanisms has been limited by difficulties in producing mutants in *H. parasuis* [Costa-Hurtado & Aragon, 2013]. Currently, there are no effective vaccines that can confer protection against all *H. parasuis* serovars. Therefore, the present study aimed to investigate the effect of tea polyphenols on growth, expression of virulence-related factors, and biofilm formation of *H. parasuis*, as well as to evaluate their protective effects against *H. parasuis*. It findings demonstrated that tea polyphenols can inhibit *H. parasuis* growth in a dose-dependent manner and attenuate the biofilm formation of *H. parasuis*. In addition, tea polyphenols exerted inhibitory effects on the expression of *H. parasuis* virulence-related factors. Moreover, tea polyphenols could confer protection against a lethal dose of *H. parasuis* and can reduce pathological tissue damage induced by *H. parasuis* [Guo *et al.*, 2018]. A study reported that EGCG has marked suppressive activity on murine macrophages infected with the intracellular bacterium *Legionella pneumophila* (Lp), an effect mediated by enhanced production of both tumor necrosis factor-alpha (TNF-alpha) and gamma-interferon (IFN-gamma). This study also show that the EGCG catechin has a marked effect in modulating production of these immune-regulatory cytokines in stimulated primary murine bone marrow (BM)-derived dendritic cells, which are important for antimicrobial immunity, especially innate immunity [Rogers *et al.*, 2005].

Alpha-hemolysin (Hla), the virulence factor secreted by *Staphylococcus aureus* (*S. aureus*), plays a critical role in infection and inflammation, which is a severe health burden worldwide. EGCG have inhibitory effect on Hla-induced NLRP3 inflammasome activation. Moreover, Hla-induced expression of NLRP3, ASC, and caspase-1 protein and the generation of IL-1β and IL-18 in damaged liver tissue of mice were also significantly suppressed by EGCG in a dose-dependent manner [Liu *et al.*, 2021]. Oolong tea extract (OTE) was found to inhibit the water-insoluble glucan-synthesizing enzyme, glucosyltransferase I (GTase-I). Catechins and all other low-molecular-weight polyphenols, except theaflavin derived from black tea, did not show significant GTase-inhibitory activities [Nakahara *et al.*, 1993].

In summary, these results demonstrated the promising use of tea polyphenols as a novel treatment for infections caused by *H. parasuis*, Lp, and other bacteria.

6. Mechanism of tea polyphenols inhibiting bacterial infection

Tea polyphenols EGCG on Hla-induced NLRP3 inflammasome activation *in vitro* and *in vivo* and elucidated the potential molecular mechanism. It was found that EGCG attenuated the hemolysis of Hla by inhibiting its secretion. Besides, EGCG significantly decreased overproduction of ROS and activation of MAPK signaling pathway induced by Hla, thereby markedly attenuating the expression of NLRP3 inflammasome-related proteins in THP-1 cells. Notably, EGCG could spontaneously bind to Hla with affinity constant of 1.71×10^{-4} M, thus blocking the formation of the Hla heptamer. Moreover, Hla-induced expression of NLRP3, ASC, and caspase-1 protein and generation of IL-1β and IL-18 in the damaged liver tissue of mice were also significantly suppressed by EGCG in a dose-dependent manner. Collectively, EGCG could be a promising candidate for alleviating Hla-induced the activation of NLRP3 inflammasome, depending on ROS-mediated MAPK signaling pathway and inhibition of Hla secretion and heptamer formation. The co-action of a high-fat diet and tea polyphenol on gut microbiota and lipid metabolism using a human flora-associated high-fat diet C57BL/6J mice model is studied. Tea polyphenols reduced serum total cholesterol, triglyceride, low density lipoprotein, glucose, and insulin levels of high-fat diet mice in a dose-dependent manner ($p < 0.01$). Tea polyphenols also significantly increased acetic acid and butyric acid levels in HFA mice [Wang *et al.*, 2018b].

As above result indicated, tea polyphenols may play a role in inhibiting bacterial infection through the following ways: 1) Tea polyphenols act as antioxidants to inhibit the formation of active oxygen and free radicals; 2) Specific induction of detoxification enzymes; 3) Its molecular regulatory function on cell growth, development, and apoptosis; 4) Selectively improve intestinal flora function; 5) Suppressed bacterial infection induced expression of NLRP3, ASC, and caspase-1 protein and generation of IL-1β and IL-18 in the damaged tissue.

12.4.2 *Tea polyphenols have antivirus properties*

EGCG has been shown to have many different biological effects, including antiviral activities. EGCG is one of the most abundant polyphenolic catechin found in green tea. EGCG has been tested for its antiviral activity against several viruses and found to be a potential treatment option over synthetic chemical drugs. EGCG has been shown to possess a broad spectrum of antiviral activities against RNA viruses such as hepatitis C virus, human immunodeficiency virus. It is recognized as a multi-functional bioactive molecule exhibiting anti-tumorigenic, anti-inflammatory, anti-bacterial, anti-oxidative, and anti-proliferative properties in addition to its antiviral effects. As evident from the mechanisms of action of EGCG in various viruses, it is a wide spectrum antiviral agent with its mechanism differing from infection to infection. A large number of studies have shown that tea polyphenols can inhibit HIV and COVID-19 virus, as well as prevent and treat pneumonia caused by HIV and COVID-19. Studies found that tea polyphenols EGCG and theaflavins have highlight their potential to prevent transmitted infections caused by HIV and COVID-19 and other virus. HIV and COVID-19 pandemic are currently ongoing worldwide and causes a lot of deaths in many countries. Pre-clinical studies highlighted the antiviral activities of epigallocatechin gallate (EGCG), a catechin primarily found in green tea, and theaflavin found in black tea against various viruses, including HIV and COVID-19.

1. Antiviral activity of green tea polyphenols against HIV

HIV is mainly transmitted through semen during sexual intercourse. A study found the main green tea polyphenol EGCG counteracts semen-mediated enhancement of HIV infection. Green-tea polyphenols can be expected to prolong the efficacy of drug therapy in subjects infected with the human immunodeficiency virus. To determine the effects of EGCG on HIV infection, peripheral blood lymphocytes were incubated with either LAI/IIIB or Bal HIV strains and EGCG. It was found that EGCG strongly inhibited the replication of both virus strains [Fassina *et al.*, 2002]. Several reports also have shown that catechin has a protective effect against HIV infection, part of which is mediated by inhibiting virions to bind to the target cell surface. Peptide fragments, derived from prostatic acidic phosphatase, are secreted in large amounts into human semen and form amyloid fibrils. These fibrillar structures, termed semen-derived enhancer of virus infection (SEVI), capture HIV virions and direct them to target cells. Thus, SEVI appears to be an important infectivity factor of HIV during sexual transmission. EGCG, the major active constituent of green tea, targets semen-derived enhancer of virus infection for degradation. Furthermore, studies have shown that EGCG inhibits semen-derived enhancer of virus infection activity and abrogates semen-mediated enhancement of HIV-1 infection in the absence of cellular toxicity [Hauber *et al.*, 2009]. Semen harbors amyloid fibrils formed by proteolytic fragments of prostatic acid phosphatase (PAP248-286 and PAP85-120) and semenogelins (SEM1 and SEM2) that potently enhance HIV infectivity. Amyloid, but not soluble forms of these peptides, enhance HIV infection. Thus, agents that remodel these amyloid fibrils could prevent HIV transmission. It confirmed that the green tea polyphenol EGCG slowly remodels fibrils formed by prostatic acid phosphatase PAP248-286, termed semen-derived enhancer of virus infection, and exerts a direct anti-viral effect. EGCG remodels PAP85-120, SEM1(45-107), and SEM2(49-107) fibrils more rapidly than semen-derived enhancer of virus infection fibrils. It established EGCG as the first small molecule that can remodel all four classes of seminal amyloid. The combined anti-amyloid and antiviral properties of EGCG could have utility in preventing HIV transmission [Castellano *et al.*, 2015]. The green tea flavonoid, epigallo-catechin gallate (EGCG), has an anti-HIV-1 effect by preventing the binding of HIV-1 gp120 to the CD4 molecule on T cells. EGCG is one of the most abundant polyphenolic catechin found in green tea.

A study investigated the effect of EGCG on the expression of CD4 molecules and on its ability to bind gp120, an envelope protein of HIV-1. It was found that EGCG efficiently inhibited binding of anti-CD4 antibody to its corresponding antigen. This effect was mediated by the direct binding of EGCG to the CD4 molecule, with consequent inhibition of antibody binding, as well as gp120 binding [Kawai *et al.*, 2003]. A report suggested that EGCG inhibits the production of p24 antigen on isolated CD4 receptor cells, macrophages, and CD4±T cells depending on its dose. Another report suggested that even at concentrations obtained from the consumption of green tea, EGCG is seen effective in inhibiting gp120-CD4 attachment [Williamson *et al.*, 2006]. It also found that EGCG binds to the CD4 molecule at the gp120 attachment site and inhibits gp120 binding at physiologically relevant levels, thus establishing EGCG as a potential therapeutic treatment for HIV-1 infection. EGCG has also been found effective against HIV-1 and inhibits viral replication by acting at various stages. It blocks interaction of gp120 with CD4 by interfering with reverse transcriptase. Addition of CD4 to EGCG produced a linear decrease from EGCG but not from the control, (-)-catechin. Addition of 5.8 micromol/L CD4 to 310 micromol/L EGCG produced strong saturation at the aromatic rings of EGCG, but identical concentrations of (-)-catechin produced much smaller effects, implying that EGCG/CD4 binding is strong enough to reduce gp120/CD4 binding substantially. A binding site for EGCG is found in the D1 domain of CD4, the pocket that binds gp120. Physiologically relevant concentrations of EGCG (0.2 micromol/L) inhibited binding of gp120 to isolated human CD4+T cells. These results demonstrated clear evidence of high-affinity binding of EGCG to the CD4 molecule and inhibition of gp120 binding to human CD4+T cells [Sodagari *et al.*, 2016].

EGCG is a strong antioxidant that has previously been shown to reduce the number of plaques in HIV-infected cultured cells. The structure-activity relationship of different catechins has been found that the difference in the antiviral activity can be attributed to the number of hydroxyl groups present on the benzene ring and galloyl group, together with the pyrogallol group which is responsible for exhibiting diverse mechanisms. EGCG has the highest activity, and the 3-galloyl and 5'-OH groups of EGCG appear crucial for the antiviral activity [Kaihatsu *et al.*, 2018]. Mechanistic insights from computational modeling found that the favorable binding of EGCG with CD4 can effectively block gp120-CD4 binding. Green tea polyphenols EGCG can prevent HIV-1 infection [Hamza & Zhan, 2006].

EGCG, the major component of tea polyphenol, has been reported to have various physiologic modulatory activities. EGCG binds to CD4 molecules to block the inflammatory response and inhibit HIV-1 infection. Therefore, EGCG appears to be a promising supplement to antiretroviral microbicides to reduce the sexual transmission of HIV-1. The multiple mechanisms, whereby catechins contains a galloyl moiety, can target key proteins to inhibit sexual transmission of HIV-1, as well as HIV-1 fusion, HIV-1 reverse transcriptase, HIV-1 integrase, and HIV-1 protease. Furthermore, catechins with a galloyl moiety can mediate host cell factors such as nitric oxide synthase, nuclear factor-κB, and casein kinase II to inhibit HIV-1 infection. The most significant inhibitory effect is blocking gp120 binding to isolated human CD4+T cells. Therefore, EGCG seems to be a promising supplement of antiretroviral microbicides to reduce the sexual transmission of HIV-1. EGCG binds with CD4 molecule to block inflammatory reaction and inhibit HIV-1 infection. Green tea polyphenols are expected to prolong the efficacy of drug therapy for patients infected with human immunodeficiency virus.

2. Antiviral activity of tea polyphenols theaflavin against HIV

Black tea polyphenols have been reported to exhibit antiviral activities against various viruses, especially positive-sense single-stranded RNA viruses. The antiviral activities of several polyphenols are derived from black tea, such as theaflavin polyphenols, in particular theaflavin (TF1), theaflavin-3-monogallate (TF2A), theaflavin-3'-monogallate (TF2B), and theaflavin-3,3'-digallate (TF3). The polyphenols, especially those with galloyl moiety, can inhibit HIV-1 replication with multiple mechanisms of action. These tea polyphenols could inhibit HIV-1 entry into target cells by blocking HIV-1 envelope glycoprotein-mediated membrane fusion. The fusion inhibitory activity of the tea polyphenols was correlated with their ability to block the formation of the gp41 six-helix bundle, a fusion-active core conformation. TF2B was found to be the most potent among all the other tea polyphenols in the inhibition of HIV at a concentration of 1 µM and a selectivity index greater. The number of galloyl groups on the TF had a direct relation on its activity and it is estimated that these molecules can interact with the gp41 six-helix bundle to prevent viral entry into the host cell [Liu *et al.*, 2005]. To develop theaflavins as affordable anti-HIV-1 microbicide for preventing HIV sexual transmission, a study used an economic natural preparation containing 90% of theaflavins. Its antiviral activity against HIV-1 strains and the mechanism by which theaflavins inhibits HIV-1 infection was evaluated. The results suggested theaflavins exhibited potent anti-HIV-1 activity on lab-adapted and primary HIV-1 strains with IC(50) less than 1.20 µM. It also effectively inhibited infection by T-20 resistant HIV-1 strains. The mechanism studies suggest that TFmix mainly inhibit the HIV-1 entry by targeting gp41 since it is effective in inhibiting gp41 six-helix bundle (6-HB) formation and HIV-1 envelope protein-mediated cell-cell fusion. Theaflavins could also inhibit HIV-1 reverse transcriptase (RT) activity, but the IC(50) is about 8-fold higher than that for inhibiting gp41 6-HB formation, suggesting reverse transcriptase is not a major target for theaflavins. In conclusion, theaflavins is an economic natural product preparation containing high content of potent anti-HIV-1 activity by targeting the viral entry step through the disruption of gp41 6-HB core structure [Yang *et al.*, 2012]. Theaflavins have effect on the amyloid fibril formation of semen-derived enhancer of virus infection peptide. Theaflavins gel could degrade semen-derived enhancer of virus infection-specific amyloid fibrils and showed low cytotoxicity to epithelial cells of the female reproductive tract. Its potent anti-HIV-1 activity is marked stability at acidic condition, low mucosal toxicity, and lack of systemic absorption; theaflavins gel can be considered as an inexpensive and safe microbicide candidate for the prevention of HIV sexual transmission [Yang *et al.*, 2012].

Like EGCG, antiviral properties of theaflavin polyphenols and their derivatives have also been explored in several viral diseases. It was showed that the theaflavin derivatives had more potent anti-COVID-19 activity than catechin derivatives. It has a potential to be developed as a safe and affordable topical microbicide for preventing transmission of COVID-19.

3. Antiviral activity of tea polyphenols on COVID-19

COVID-19 is an infectious viral respiratory disease caused by a novel positive-sense single-stranded RNA coronavirus called as SARS-CoV-2. This viral disease is known to infect the respiratory system, eventually leading to pneumonia. Studies of the viral structure reveal its mechanism of infection as well as active binding sites and the drug-gable targets as scope for treatment of COVID-19. The possible drug targets (Chymotrypsin-like protease, RNA-dependent RNA polymerase, Papain-like protease, Spike RBD, and ACE2 receptor with spike RBD) are vital proteins. These receptors were docked against tea polyphenols, EGCG from green tea. A comparative study of docking scores and the type of interactions of EGCG, with the possible targets of COVID-19 showed that the tea polyphenols had good docking scores with significant *in silico* activity. [Mhatre *et al.*, 2021]. In order to find potent main protease inhibitors, a study selected eight polyphenols from green tea, as these are already known to exert antiviral activity against many RNA viruses. It elucidated the binding affinities and binding modes between these polyphenols including a well-known main protease inhibitor N3 and main protease using molecular docking studies. All eight polyphenols exhibit good binding affinity toward main protease. However, only three polyphenols (epigallocatechin gallate, epicatechingallate, and gallocatechin-3-gallate) interact strongly with one or both catalytic residues of main protease. These complexes are highly stable, experience less conformational fluctuations, and share similar degree of compactness. Estimation of total number of intermolecular H-bond and MM-GBSA analysis affirm the stability of these three main protease-polyphenol complexes. These polyphenols possess favorable drug-likeness characteristics [Ghosh *et al.*, 2021]. In a study evaluating potential medicinal herbs for main protease inhibition, green tea extract is highly effective in inhibiting main protease of SARS-COV-2 [Upadhyay *et al.*, 2020].

Genes associated with Nrf2-dependent antioxidant response are highly suppressed in lung biopsies from COVID-19 patients, and Nrf2 inducers (4-octyl-itaconate and dimethyl fumarate) inhibit SARS-CoV-2 replication and inflammatory response [Cuadrado *et al.*, 2020]. These lines of evidence suggest that Nrf2 activation is a promising strategy to prevent the infection of SARS-CoV-2 and reduce the severity of COVID-19. The main protease of SARS CoV-2, a key component of this viral replication, is considered as a prime target for anti-COVID-19 drug development. In order to find potent main protease inhibitors, eight polyphenols from green tea were selected, as these are already known to exert antiviral activity against many RNA viruses. It elucidated the binding affinities and binding modes between these polyphenols including a well-known main protease inhibitor N3 and main protease using molecular docking studies. All eight polyphenols exhibit good binding affinity toward main protease. However, only three polyphenols (EGCG and gallocatechin-3-gallate) interact strongly with one or both catalytic residues of main protease. Pharmacokinetic analysis additionally suggested that these polyphenols possess favorable drug-likeness characteristics [Ghosh *et al.*, 2021]. EGCG interferes with SARS-CoV-2 spike-receptor interaction and blocks the entry of SARS-CoV-2 pseudo-typed lentiviral vectors [Henss *et al.*, 2021]. Based on these studies, it suggested that EGCG may serve as a broad spectrum therapeutic in asymptomatic and symptomatic COVID-19 patients [Chourasia *et al.*, 2021].

EGCG can enhance key enzymatic activities of hepatic thioredoxin and glutathione systems in selenium-optimal mice but activates hepatic Nrf2 responses in selenium-deficient mice. EGCG can directly scavenge multiple types of ROS and induce antioxidant and detoxifying enzymes, such as heme oxygenase 1, quinone reductase, glutamate cysteine ligase, Glutathione transferase (GST), thioredoxin reductase, glutaredoxin, glutathione reductase, SOD, catalase, and glutathione peroxidase (GPX) [Dong *et al.*, 2016]. The replicase gene of SARS-CoV-2 encodes two overlapping polyproteins for viral replication and transcription. EGCG may suppress SARS-CoV-2 replication via inhibiting main protease. Green tea extract or EGCG shows a dose-dependent inhibitory activity against main protease of SARS-CoV-2 *in vitro* [Zhu & Xie, 2020]. A recent study found that EGCG from 1–20 μg/mL inhibited main protease activity and replication of HCoV-OC43 (a type of beta coronavirus, similar to SARS-CoV-2) in a dose-dependent manner, and even 1 μg/mL

EGCG was able to significantly reduce levels of HCoV-OC43 proteins in the infected cells [Jang *et al.*, 2021]. EGCG auto-oxidation leads to the formation of EGCG quinone, which can react with protein cysteinyl thiol to form quinone proteins [Ishii *et al.*, 2008]. It is possible that EGCG can inhibit main protease of SARS-CoV-2 by covalent bonding to Cys145 and this possibility remains.

EGCG, with eight phenolic groups, provides multiple electron acceptors and donors for hydrogen bonding to a variety of molecules, especially to proteins. That is one of the reasons EGCG can bind to many different proteins with high affinity and inhibit their activities [Guo *et al.*, 1996]. With these chemical re-activities, EGCG has been shown to have many different biological effects, including anti-viral activities. EGCG mainly inhibits the early stages of viral infection, such as attachment, entry, and membrane fusion, by interfering with either viral membrane proteins or host cellular proteins or both. EGCG-fatty acid monoesters bind more effectively to viral and cellular membranes. A large number of studies have shown that EGCG induces Nrf2-mediated antioxidant enzyme expression [Dong *et al.*, 2016; Na *et al.*, 2008]. Genes associated with Nrf2-dependent antioxidant response are highly suppressed in lung biopsies from COVID-19 patients, and Nrf2 inducers inhibit SARS-CoV-2 replication and inflammatory response [Olagnier *et al.*, 2020]. Nrf2, the cyto-protective transcription factor, regulates expression of a wide array of genes involved in anti-oxidation, detoxification, inflammation, immunity, and antiviral responses. EGCG may reduce SARS-CoV-2 infection via activation of Nrf2. EGCG as an Nrf2 activator can inhibit the entry of SARS-CoV-2 into host cells and prime host cells against SARS-CoV-2 infection [McCord *et al.*, 2020]. A study demonstrated that EGCG, the main active ingredients of green tea is potentially effective to inhibit SARS-CoV-2 activity. Coronaviruses require the 3CL-protease for the cleavage of its polyprotein to make individual proteins functional. EGCG showed inhibitory activity against the SARS-CoV-2 3CL-protease in a dose-dependent manner for EGCG [Jang *et al.*, 2020].

EGCG, via activating Nrf2, can suppress ACE2 (a cellular receptor for SARS-CoV-2) and TMPRSS2, which mediate cell entry of the virus. Through inhibition of SARS-CoV-2 main protease, EGCG may inhibit viral reproduction. EGCG via its broad antioxidant activity may protect against SARS-CoV-2 evoked mitochondrial ROS (which promote SARS-CoV-2 replication) and against ROS burst inflicted by neutrophil extracellular traps. By suppressing ER-resident GRP78 activity and expression, EGCG can potentially inhibit SARS-CoV-2 life cycle. EGCG also shows protective effects against cytokine storm-associated acute lung injury/ acute respiratory distress syndrome, thrombosis via suppressing tissue factors and activating platelets, sepsis by inactivating redox-sensitive HMGB1, and lung fibrosis through augmenting Nrf2 and suppressing NF-κB. These activities remain to be further substantiated in animals and humans. The possible concerted actions of EGCG suggest the importance of further studies on the prevention and treatment of COVID-19 in humans [Zhang *et al.*, 2021]. The rapid spread of novel coronavirus called SARS-CoV-2 or nCoV has caused countries all over the world to impose lockdowns and undertake stringent preventive measures. This new positive-sense single-stranded RNA strain of coronavirus spreads through droplets of saliva and nasal discharge. Many studies have shown that tea polyphenols can inhibit COVID-19 virus, prevent, and treat pneumonia caused by COVID-19. The antiviral activities of two polyphenols are derived from green tea, EGCG. Green tea polyphenols have been reported to exhibit antiviral activities against various viruses, especially positive-sense single-stranded RNA viruses.

Coronaviruses require the 3CL-protease for the cleavage of its polyprotein to make individual proteins functional. Theaflavin showed inhibitory activity against the SARS-CoV-2 3CL-protease in a dose-dependent manner for theaflavin [Jang *et al.*, 2020]. A study analyzed the inhibitory effect of theaflavin-3′-o-gallate on SARS-CoV-2 RdRp. It was found that theaflavin-3′-o-gallate black tea component blocked the active site of SARS-CoV-2 RNA-dependent RNA polymerase [Banerjee *et al.*, 2021]. Compared with green tea polyphenols, theaflavin and theaflavin gallate (including 3,3′-di-o-theaflavin gallate) have more significant antiviral properties [Ohgitani *et al.*, 2021].

The above research show that EGCG is a broad-spectrum antiviral drug and its ways of action varies from infection to infection. The difference in antiviral activity can be attributed to the number of hydroxyl groups on the benzene ring and gallic acid group, as well as the pyrogallol group responsible for showing different mechanisms. Experimental evidence also highlights the potential use of EGCG as an alternative treatment option for SARS CoV-2 infection. EGCG is a kind of nutritional immunity, which is used to regulate the effective innate immune response to fight against COVID-19 and other infectious diseases.

4. EGCG can "anti" other virus

In addition to anti-HIV and COVID-19, several studies have reported that EGCG has anti-infective properties by virus, such as herpes simplex virus (HSV-1). Antiviral activities of EGCG with different modes of action have been demonstrated on diverse families of viruses, such as *Retroviridae*, *Orthomyxoviridae*, and *Flaviviridae* and include important human pathogens like human immunodeficiency virus, influenza A virus, and the hepatitis C virus [Steinmann *et al.*, 2013]. Modified EGCG, palmitoyl-EGCG (p-EGCG), is of interest as a topical antiviral agent for HSV-1 infections. This study evaluated the effect of p-EGCG on HSV-infected Vero cells. Results of cell viability and cell proliferation assays indicate that p-EGCG is not toxic to cultured Vero cells and show that modification of the green tea polyphenol EGCG with palmitate increases the effectiveness of EGCG as an antiviral agent. Furthermore, p-EGCG is a more potent inhibitor of herpes simplex virus 1 (HSV-1) than EGCG and can be topically applied to skin, one of the primary tissues infected by HSV. Viral binding assay, plaque forming assay, PCR, real-time PCR, and fluorescence microscopy were used to demonstrate that p-EGCG concentrations of 50 μM and higher block the production of infectious HSV-1 particles. EGCG was found to inhibit HSV-1 adsorption to Vero cells. Thus, EGCG may provide a novel treatment for HSV-1 infections [de Oliveira *et al.*, 2013]. Cytotoxic effects of HSV-1 on the viability of oral epithelial cells were evidently reduced in the presence of EGCG (25 μg/ml). Viral yields were also significantly reduced by treatment of cells with EGCG. Expression of viral immediate early protein, infected cell protein 0 (ICP0), was greatly inhibited when cells were treated with EGCG. Effects of EGCG were more evident for the expression of viral thymidine kinase, ICP5, and glycoprotein D. EGCG significantly reduced the levels of viral particles and viral DNA during viral entry phase [Wu *et al.*, 2021].

Above studies have shown that EGCG and theaflavin has remarkable anti-infective properties by virus activity. According to *in vitro* experiments, animal models, and human experiments, tea polyphenols, micronutrients, and vitamins have the potential to regulate and enhance the innate immune response. Theaflavin is rich in black tea and is of great significance in controlling viral infection. Theaflavin is a non-toxic, non-invasive, antioxidant, anticancer, and antiviral molecule. Tea polyphenols, including green tea polyphenols and theaflavins, can be used as a natural immune modulator against viruses.

12.5 Antiviral mechanism of tea polyphenols on COVID-19

Similar to the COVID-19, severe acute respiratory syndrome (SARS) is an acute respiratory infectious disease caused by SARS coronavirus, which is named as severe acute respiratory syndrome by the WHO. COVID-19 causes immune cells to overproduce cytokines, resulting in mucosal inflammation, lung damage, and multiple organ failure in patients. The cytokine storm is characterized as a sudden acute increase in circulating levels of different pro-inflammatory cytokines. SARS-CoV-2 infection in some individuals induces a hyperactive and uncontrolled immune response. Several types of immune cells including T-lymphocytes, macrophages, dendritic cells, and neutrophils secret immense amounts of pro-inflammatory cytokines (IFNα, IFNγ, IL-6, IL-1β, IL-12, etc.) and chemokines, leading to acute respiratory distress syndrome (ARDS) and systemic inflammatory responses [Zhao *et al.*, 1989]. In addition, through the activation of Nrf-2-regulated heme oxygenase 1, EGCG can mediate antiviral responses by increasing the expression of type 1 interferons and alleviating SARS-CoV-2-initiated inflammatory responses through crosstalk of Nrf2 and NF-κB in inflamed tissues, where innate immune cells are recruited [Cuadrado *et al.*, 2020]. Several studies have found that epigallocatechin-3-gallate and melatonin upregulate sirtuins proteins, which leads to the downregulation of pro-inflammatory gene transcription and NF-κB, protecting organisms from oxidative stress in autoimmune, respiratory, and cardiovascular illnesses. EGCG, a polyphenol belonging to the flavonoid family in tea, has potent anti-inflammatory and anti-oxidative properties that helps to counter the inflammation and oxidative stress associated with many neurodegenerative diseases. EGCG perhaps modulate sirtuin-signaling pathways that counteract cytokine storm and oxidative stress, the root causes of severe inflammation and symptoms in these patients [Chattree *et al.*, 2022].

Many studies have shown that EGCG is highly effective in attenuating acute lung injury (ALI)/ARDS induced by virus, lipopolysaccharides (LPS), and other factors. Following the infection of H9N2 swine influenza virus in mice, administration of EGCG daily for five consecutive days significantly prolongs mouse survival and reduces death rate from 65% to 35%, attenuates lung histological lesions (inflammatory cell margination and infiltration, alveolar and interstitial edema, and bronchiolitis), decreases lung wet/dry ratio, suppresses total white blood cell count and leukocyte differential counts in bronchoalveolar lavage fluid, and lowers cytokines levels in the lung by downregulating TLR4 and NF-κB [Xu, *et al.*, 2017]. EGCG is also effective in other rodent models of ALI/ARDS; EGCG, immediately after the infliction of thermal injury in rats, significantly reduces plasma concentrations of inflammatory mediator and severity of acute respiratory distress syndrome (ARDS) [Liu *et al.*, 2017]. EGCG, 10 minutes before inflicting hip fracture in rats, significantly reduces the elevated lung injury and infiltration of inflammatory cells in the bronchoalveolar lavage fluid [Zhao *et al.*, 2007]. EGCG, one hour prior to pleural injection of carrageenan in mice, reduces acute inflammatory responses (interstitial haemorrhage, polymorphonuclear leukocyte infiltration, increased TNFα, and activated STAT1) in lung tissues [Di Paola *et al.*, 2005].

Sepsis, an infection-triggered systemic inflammatory response syndrome, is a common complication of severe and critical COVID-19 patients. Studies have shown that EGCG can prevent sepsis. In mouse model of LPS-induced lethal endotoxemia, EGCG or green tea polyphenols significantly increase survival rate [Li *et al.*, 2007]. High mobility group box 1 (HMGB1), released by activated monocytes, macrophages, or neutrophils, acts as a late mediator of sepsis and is a crucial therapeutic target for preventing sepsis-triggered lethality [Yang *et al.*, 2004]. EGCG also reduces cytoplasmic HMGB1 levels in endotoxin-stimulated macrophages. In addition to the redox modification of cysteines, the inhibitory effect of EGCG on STAT1 activation [Menegazzi *et al.*, 2020] may also play a role in reducing cytoplasmic HMGB1. JAK/STAT1 pathway regulates HMGB1 cytoplasmic accumulation. JAK/STAT1 inhibition decreases HMGB1 release and enhances survival in animal models of lethal sepsis and endotoxemia [Liu *et al.*, 2014].

Evidence show that many patients have persistent respiratory symptoms months after their initial illness [Fraser, 2020], and suggests that SARS-CoV-2 infection may have pulmonary fibrosis sequelae [George *et al.*, 2020]. EGCG and green tea extract may be useful for the prevention and treatment of pulmonary fibrosis in COVID-19 patients. Studies have shown that EGCG can prevent lung fibrosis. COVID-19 shows three main histological patterns in the lung, namely epithelial, vascular, and fibrotic with interstitial fibrosis. Oral administration of green tea extract in drinking water (EGCG doses of 300–400 mg/kg) to mice significantly reduces pulmonary fibrosis induced by intratracheal challenge of fluorescein isothiocyanate; interstitial and peribronchial fibrosis is nearly completely prevented [Donà *et al.*, 2003]. EGCG, to patients with idiopathic pulmonary fibrosis, reverses pro-fibrotic biomarkers in their diagnostic biopsies and serum samples [Chapman *et al.*, 2020]. Overall, with demonstrated activities in alleviating pulmonary fibrosis in rodent models and humans.

COVID-19 patients with diabetes comorbidity are at great risk of lengthy hospitalization and dire consequences, including ICU admission and mortality. EGCG and green tea extract may be useful for the prevention of diabetes comorbidity risk. The main feature of diabetes, hyperglycemia, promotes the replication of SARS-CoV-2 and the development of NETs. Mounting evidence demonstrate that EGCG is helpful in controlling blood glucose in diabetic subjects [Zhao *et al.*, 2020]. EGCG at doses from 300–900 mg/day dose-dependently increases plasma levels of soluble receptor for advanced glycation end products (RAGE), a RAGE variant acting as a ligand decoy that competes with RAGE [Huang *et al.*, 2013]. These inhibitory activities of EGCG, in rodent models and humans, suggest that EGCG could reduce the risk of ICU admission and mortality of COVID-19 patients with diabetes comorbidity.

The above research show that EGCG has protective effects on cytokine storm-related acute lung injury/acute respiratory distress syndrome, preventing thrombosis by inhibiting tissue factor and activating platelets, preventing sepsis by inactivating redox sensitizers, and increasing Nrf2 and inhibiting NF-κB to prevent pulmonary fibrosis. Above experimental evidence highlight the potential use of EGCG as an alternative therapeutic choice for the treatment of SARS-CoV-2 infection. Innate immunity impairment led to disruption in cascade of signaling pathways upregulating pro-inflammatory cytokines, diminish interferons, depleted natural killer cells, and activate reactive oxygen species production. These conditions severely affect the body's ability to fight against infectious diseases and play a pivotal role in disease progression. EGCG is on nutritional immunity for regulating effective innate immune response for combating against infectious diseases like novel COVID-19. Drawing from discoveries on *in vitro* experiments, animal models, and human trials, tea polyphenols micronutrients have the potential to modulate and enhance innate immune response. Tea infusion is rich in polyphenol EGCG, and major green tea polyphenols are the innate immunity modulator. Theaflavin is rich in black tea and is of great significance in controlling viral infection. Theaflavin is a non-toxic, non-invasive, antioxidant, anticancer and antiviral molecule [Zhang *et al.*, 2021].

The difference in antiviral activity can be attributed to the number of hydroxyl groups on the benzene ring and gallic acid group, as well as the pyrogallol group responsible for showing different mechanisms. EGCG has shown protective effects on 1) cytokine storm related acute lung injury/acute respiratory distress syndrome, 2) preventing thrombosis by inhibiting tissue factor and activating platelets, 3) preventing sepsis by inactivating redox sensitizers, and 4) increasing Nrf2 and inhibiting NF-κB to prevent pulmonary fibrosis. As discussed above, EGCG has shown many redox and specific inhibitory activities, which may be applicable for the prevention and treatment of COVID-19. EGCG has direct and indirect antioxidant activities to possibly protect against SARS-CoV-2 evoked oxidative stress. EGCG has shown activities with the potential to prevent against SARS-CoV-2 infection, suppress SARS-CoV-2 life cycle, and curb SARS-CoV-2 triggered cytokine storm, oxidative stress, endoplasmic reticulum stress, thrombosis, sepsis, and lung fibrosis. The antiviral mechanism of tea polyphenols on COVID-19 is represented in Figure 12-1.

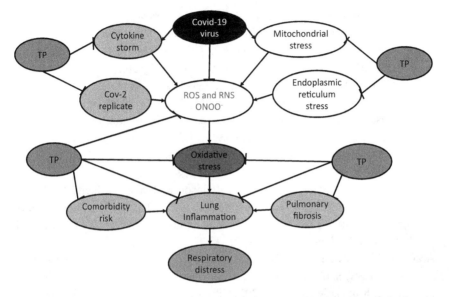

Figure 12-1. The schematic diagram about the antiviral mechanism of tea polyphenols on COVID-19. TP: tea polyphenols.

12.6 Conclusion

This chapter discusses the relationship between bacteria and virus and oxidative stress damage, the inhibition of tea polyphenols on bacteria and virus and oxidative stress, and the prevention of bacteria and virus and oxidative stress damage. Through this discussion, we found that both bacteria and viruses can cause body damage through oxidative stress. Tea polyphenols can not only inhibit the reproduction of bacteria and viruses, but also inhibit the damage to the body caused by oxidative stress due to bacteria and viruses. Tea polyphenols are an auxiliary means of resisting bacteria and viruses.

References

Agas A, Kalluru J, Leiser B, Garcia R, Kataru H, Haorah J. (2021) Possible mechanisms of HIV neuro-infection in alcohol use: Interplay of oxidative stress, inflammation, and energy interruption. *Alcohol*, **94**, 25–41.

Akai Y, Nakajima N, Ito Y, Matsui T, Iwasaki A, Arakawa Y. (2007) Green tea polyphenols reduce gastric epithelial cell proliferation and apoptosis stimulated by Helicobacter pylori infection. *J Clin Biochem Nutr*, **40**(2), 108–115.

Akaike T. (2015) Host defense and oxidative stress signaling in bacterial infection. *Nihon Saikingaku Zasshi*, **70**(3), 339–349.

Akkoyunlu Y, Kocyigit A, Okay G, Guler EM, Aslan T. (2020) Integrase inhibitor-based antiretroviral treatments decrease oxidative stress caused by HIV infection. *Eur Rev Med Pharmacol Sci*, **24**(23), 12389–12394.

Alvarez RA, Berra L, Gladwin MT. (2020) Home nitric oxide therapy for COVID-19. *Am J Respir Crit Care Med*, **202**, 16–20.

Ankolekar C, Johnson D, Pinto Mda S, Johnson K, Labbe R, Shetty K. (2011) Inhibitory potential of tea polyphenolics and influence of extraction time against Helicobacter pylori and lack of inhibition of beneficial lactic acid bacteria. *J Med Food*, **14**(11), 1321–1329.

Araneda OF, Tuesta M. (2012) Lung oxidative damage by hypoxia. *Oxid Med Cell Longev*, 856918.

Banerjee A, Kanwar M, Maiti S. (2021) Theaflavin-3′-O-gallate a black-tea constituent blocked SARS CoV-2 RNA dependant RNA polymerase active-site with better docking results than remdesivir. *Drug Res (Stuttg)*, **71**(8), 462–472.

Banu N, Panikar SS, Leal LR, Leal AR. (2020) Protective role of ACE2 and its downregulation in SARS-CoV-2 infection leading to Macrophage Activation Syndrome: therapeutic implications. *Life Sci*, **256**, 117905.

Bartsch H, Nair J. (2002) Potential role of lipid peroxidation derived DNA damage in human colon carcinogenesis: studies on exocyclic base adducts as stable oxidative stress markers. *Cancer Detect Prev*, **26**, 308–312.

Becker NP, Martitz J, Renko K, Stoedter M, Hybsier S, Cramer T. (2014) Hypoxia reduces and redirects selenoprotein biosynthesis. *Metall*, **6**(5), 1079–1086.

Bhattacharya D, Bhattacharya S, Patra MM, Chakravorty S, Sarkar S, Chakraborty W, Koley H, Gachhui R. (2016) Antibacterial activity of polyphenolic fraction of kombucha against enteric bacterial pathogens. *Curr Microbiol*, **73**(6), 885–896.

Bhattacharya D, Sinha R, Mukherjee P, Howlader DR, Nag D, Sarkar S, Koley H, Withey JH, Gachhui R. (2020) Anti-virulence activity of polyphenolic fraction isolated from Kombucha against Vibrio cholerae. *Microb Pathog*, **140**, 103927.

Borjabad A, Brooks AI, Volsky DJ. (2010) Gene expression profiles of HIV-1-infected glia and brain: toward better understanding of the role of astrocytes in HIV-1-associated neurocognitive disorders. *J Neuroimmune Pharmacol*, **5**(1), 44–62.

Castellano LM, Hammond RM, Holmes VM, Weissman D, Shorter J. (2015) Epigallocatechin-3-gallate rapidly remodels PAP85-120, SEM1(45-107), and SEM2(49-107) seminal amyloid fibrils. *Biol Open*, **4**(9), 1206–1212.

Chapman HA, Wei Y, Montas G, *et al.* (2020). Reversal of TGFbeta1-driven profibrotic state in patients with pulmonary fibrosis. *N Engl J Med*, **382**(11), 1068–1070.

Chattree V, Singh K, *et al.* (2022) A comprehensive review on modulation of SIRT1 signaling pathways in the immune system of COVID-19 patients by phytotherapeutic melatonin and epigallocatechin-3-gallate. *J Food Biochem*, e14259.

Chourasia M, Koppula PR, Battu A, Ouseph MM, Singh AK. (2021) EGCG, a green tea catechin, as a potential therapeutic agent for symptomatic and asymptomatic SARS-CoV-2 infection. *Molecules*, **26**(5).

Cooke MS, Evans MD, Dizdaroglu M, Lunec J. (2003) Oxidative DNA damage: mechanisms, mutation, and disease. *FASEB J*, **17**, 1195–1214.

Costa-Hurtado M, Aragon V. (2013) Advances in the quest for virulence factors of Haemophilus parasuis. *Vet J*, **198**(3), 571–576.

Cuadrado A, Pajares M, Benito C, *et al.* (2020) Can activation of Nrf2 be a strategy COVID-19? *Trends Pharmacol Sci*, **41**(9), 598–610.

Fanales-Belasio E, Raimondo M, Suligoi B, Buttò S. (2010) HIV virology and pathogenetic mechanisms of infection: a brief overview. *Ann Ist Super Sanita*, **46**(1), 5–14.

Darfeuille-Michaud A, Boudeau J, Bulois P, *et al.* (2004) High prevalence of adherent-invasive Escherichia coli associated with ileal mucosa in Crohn's disease. *Gastroenterology*, **127**, 412–421.

De Bruyne T, Pieters L, Witvrouw M, De Clercq E, Vanden Berghe D, Vlietinck AJ. (1999) Biological evaluation of proanthocyanidin dimmers and related polyphenols. *J Nat Prod*, **62**, 954–958.

de Oliveira A, Adams SD, Lee LH, Murray SR, Hsu SD, Hammond JR, Dickinson D, Chen P, Chu TC. (2013) Inhibition of herpes simplex virus type 1 with the modified green tea polyphenol palmitoyl-epigallocatechin gallate. *Food Chem Toxicol*, **52**, 207–215.

Di Paola R, Mazzon E, Muià C, et al. (2005) Green tea polyphenol extract attenuates lung injury in experimental model of carrageenan-induced pleurisy in mice. *Respiratory Research*, **6**(1), 66.

Dong R, Wang D, Wang X, Zhang K, Chen P, Yang CS. (2016) Epigallocatechin-3-gallate enhances key enzymatic activities of hepatic thioredoxin and glutathione systems in selenium-optimal mice but activates hepatic Nrf2 responses in selenium-deficient mice. *Redox Biology*, **10**, 221–232.

Donà M, Dell'Aica I, Calabrese F, Benelli R, Morini M, Albini A. (2003) Neutrophil restraint by green tea: Inhibition of inflammation, associated angiogenesis, and pulmonary fibrosis. *J Immunol*, **170**(8), 4335–4341.

Duan S, GuL, Wang Y, Zheng R, Lu j, Yin J, Guli L, Ball M. (2009) Regulation of influenza virus-caused oxidative stress by Kegan Liyan oral prescription, as monitored by ascorbyl radical ESR signals. *Am J Chinese Med*, **37**(6), 1167–1177.

Fassina G, Buffa A, Benelli R, Varnier OE, Noonan DM, Albini A. (2002) Polyphenolic antioxidant (-)-epigallocatechin-3-gallate from green tea as a candidate anti-HIV agent. *AIDS*, **16**(6), 939–941.

Fresquet F, Pourageaud F, Leblais V, Brandes RP, Savineau JP, Marthan R. (2006) Role of reactive oxygen species and gp91phox in endothelial dysfunction of pulmonary arteries induced by chronic hypoxia. *Br J Pharmacol*, **148**(5), 714–723.

Fraser E. (2020) Long term respiratory complications of COVID-19. *Br Med J*, **70**, 3001.

George PM, Wells AU, Jenkins RG. (2020) Pulmonary fibrosis and COVID-19: the potential role for antifibrotic therapy. *Lancet Respir Med*, **8**(8), 07–815.

Ghosh R, Chakraborty A, Biswas A, Chowdhuri S. (2021) Evaluation of green tea polyphenols as novel corona virus (SARS CoV-2) main protease (Mpro) inhibitors — an in silico docking and molecular dynamics simulation study. *J Biomol Struct Dyn*, **39**(12), 4362–4374.

Gibellini L, De Biasi S, Paolini A, *et al.* (2020) Altered bioenergetics and mitochondrial dysfunction of monocytes in patients with COVID-19 pneumonia. *EMBO Molecular Medicine*, **12**(12).

Guo H, Zuo Z, Wang F, Gao C, Chen K, Fang J, Cui H, Ouyang P, Geng Y, Chen Z, Huang C, Zhu Y, Deng H. (2021) Attenuated Cardiac oxidative stress, inflammation and apoptosis in Obese Mice with nonfatal infection of Escherichia coli. *Ecotoxicol Environ Saf* **225**, 112760.

Guo L, Guo J, Liu H, Zhang J, Chen X, Qiu Y, Fu S. (2018) Tea polyphenols suppress growth and virulence-related factors of Haemophilus parasuis. *J Vet Med Sci*, **80**(7), 1047–1053.

Guo Q, Zhao B-L, Li M-F, Shen S-R, Xin W-J. (1996) Studies on protective mechanisms of four components of green tea polyphenols (GTP) against lipid peroxidation in synaptosomes. *Biochem Biopphys Acta*, **1304**, 210–222.

Hamza A, Zhan CG. (2006) How can (-)-epigallocatechin gallate from green tea prevent HIV-1 infection? Mechanistic insights from computational modeling and the implication for rational design of anti-HIV-1 entry inhibitors. *J Phys Chem B*, **110**(6), 2910–2917.

Hauber I, Hohenberg H, Holstermann B, Hunstein W, Hauber J. (2009) The main green tea polyphenol epigallocatechin-3-gallate counteracts semen-mediated enhancement of HIV infection. *Proc Natl Acad Sci U S A*, **106**(22), 9033–9038.

Henss L, Auste A, Schurmann C, Schmidt C, von Rhein C, Muhlebach MD. (2021) The green tea catechin epigallocatechin gallate inhibits SARS-CoV-2 infection. *J Gen Virol*, **102**(4).

Huang SM, Chang YH, Chao YC, *et al.* (2013) EGCG-rich green tea extract stimulates sRAGE secretion to inhibit S100A12-RAGE axis through ADAM10-mediated ectodomain shedding of extracellular RAGE in type 2 diabetes. *Mol Nutr Food Res*, **57**(12), 2264–2268.

Huycke MM, Moore D, Joyce W, Wise P, Shepard L, Kotake Y, Gilmore MS. (2001) Extracellular superoxide production by Enterococcus faecalis requires demethylmenaquinone and is attenuated by functional terminal quinol oxidases. *Mol Microbiol*, **42**, 729–740.

Ishii T, Mori T, Tanaka T, Mizuno D, Yamaji R, Kumazawa S, Akagawa M. (2008) Covalent modification of proteins by green tea polyphenol (-)-epigallocatechin-3-gallate through autoxidation. *Free Radic Biol Med*, **45**(10), 1384–1394.

Jang M, Park YI, Cha YE, Park R, Namkoong S, Lee JI, Park J. (2020) Tea polyphenols EGCG and theaflavin inhibit the activity of SARS-CoV-2 3CL-protease in vitro. *Evid Based Complement Alternat Med*, **16**. 5630838.

Jang M, Park R, Park YI, Cha YE, Yamamoto A, Lee JI. (2021) EGCG, a green tea polyphenol, inhibits human coronavirus replication in vitro. *Biochem Biophys Res Commun*, **547**, 23–28.

Kaihatsu K., Yamabe M., Ebara Y. (2018) Antiviral mechanism of action of epigallocatechin-3-O-gallate and its fatty acid esters. *Molecules*, **23**(10), 2475.

Katsurahara M, Kobayashi Y, Iwasa M, Ma N, Inoue H, Fujita N, Tanaka K, Horiki N, Gabazza EC, Takei Y. (2009) Reactive nitrogen species mediate DNA damage in Helicobacter pylori-infected gastric mucosa. *Helicobacter*, **14**(6), 552–558.

Kawai K, Tsuno NH, Kitayama J, Okaji Y, Yazawa K, Asakage M, Hori N, Watanabe T, Takahashi K, Nagawa H. (2003) Epigallocatechin gallate, the main component of tea polyphenol, binds to CD4 and interferes with gp120 binding. *J Allergy Clin Immunol*, **112**(5), 951–957.

Kostic AD, Chun E, Robertson L, Glickman JN, Gallini CA, Michaud M, Clancy TE, Chung DC, Laforge M, Elbim C, Frère C, *et al*. (2020) Tissue damage from neutrophil-induced oxidative stress in COVID-19. *Nat Rev Immunol*, **20**(9), 515–516.

Laforge M, Elbim C, Frère C, *et al*. (2020) Tissue damage from neutrophil-induced oxidative stress in COVID-19. *Nat Rev Immunol*, **20**(9), 515–516.

Lagha AB, Groeger S, Meyle J, Grenier D. (2018) Green tea polyphenols enhance gingival keratinocyte integrity and protect against invasion by Porphyromonas gingivalis. *Pathog Dis*, **76**(4).

Lochhead P, Hold GL, *et al*. (2013) Fusobacterium nucleatum potentiates intestinal tumorigenesis and modulates the tumor-immune microenvironment. *Cell Host Microbe*, **14**, 207–215.

Li W, Ashok M, Li J, Yang H, Sama AE, Wang H. (2007) A major ingredient of green tea rescues mice from lethal sepsis partly by inhibiting HMGB1. *PLoS One*, **2**(11), e1153.

Li Y, Jiang X, Hao J, Zhang Y, Huang R. (2019) Tea polyphenols: application in the control of oral microorganism infectious diseases. *Arch Oral Biol*, **102**, 74–82.

Li Y, Rahman SU, Huang Y, Zhang Y, Ming P, Zhu L, Chu X, Li J, Feng S, Wang X, Wu J. (2020) Green tea polyphenols decrease weight gain, ameliorate alteration of gut microbiota, and mitigate intestinal inflammation in canines with high-fat-diet-induced obesity. *J Nutr Biochem*, **78**, 108324.

Liu C, Hao K, Liu Z, Liu Z, Guo N. (2021) Epigallocatechin gallate (EGCG) attenuates staphylococcal alpha-hemolysin (Hla)-induced NLRP3 inflammasome activation via ROS-MAPK pathways and EGCG-Hla interactions. *Int Immunopharmacol*, **100**, 108170.

Liu S, Lu H, Zhao Q, He Y, Niu J, Debnath AK, Wu S, Jiang S. (2005) Theaflavin derivatives in black tea and catechin derivatives in green tea inhibit HIV-1 entry by targeting gp41. *Biochimica et Biophysica Acta (BBA) — General Subjects*, **1723**, 270–281.

Liu R, Xu F, Si S, Zhao X, Bi S, Cen Y. (2017) Mitochondrial DNA-induced inflammatory responses and lung injury in thermal injury rat model: protective effect of epigallocatechin gallate. *J Burn Care Res*, **38**(5), 304–311.

Liu W, Dong M, Bo L, *et al*. (2014). Epigallocatechin-3-gallate ameliorates seawater aspiration-induced acute lung injury via regulating inflammatory cytokines and inhibiting JAK/STAT1 pathway in rats. *Mediators Inflamm*, **2014**, 612593.

Louboutin JP, Strayer D. (2014) Role of oxidative stress in HIV-1-associated neurocognitive disorder and protection by gene delivery of antioxidant enzymes. *Antioxidants (Basel)*, **3**(4), 770–797.

Lundberg JO, Weitzberg E, Cole JA, Benjamin N. (2004) Nitrate, bacteria and human health. *Nat Rev Microbiol*, **2**, 593–602.

Ma T, Peng W, Liu Z, Gao T, Liu W, Zhou D, Yang K, Guo R, Duan Z, Liang W, Bei W, Yuan F, Tian Y. (2021) Tea polyphenols inhibit the growth and virulence of ETEC K88. *Microb Pathog*, **152**, 104640.

McCord JM, Hybertson BM, Cota-Gomez A, Geraci KP, Gao B. (2020) Nrf2 activator PB125® as a potential therapeutic agent against COVID-19. *Antioxidants*, **9**(6), 518.

Menegazzi M, Campagnari R, Bertoldi M, Crupi R, Di Paola R, Cuzzocrea S. (2020) Protective effect of epigallocatechin-3-gallate (EGCG) in diseases with uncontrolled immune activation: Could such a scenario be helpful to counteract COVID-19? *Int J Mol Sci*, **21**(14), 5171.

Mhatre S, Naik S, Patravale V. (2021) A molecular docking study of EGCG and theaflavin digallate with the druggable targets of SARS-CoV-2. *Comput Biol Med*, **129**, 104137.

Mhatre S, Srivastava T, Naik S, Patravale V. (2021) Antiviral activity of green tea and black tea polyphenols in prophylaxis and treatment of COVID-19: a review. *Phytomedicine*, **85**, 153286.

Moreno Fernandez-Ayala DJ, Navas P. (2020) Age-related mitochondrial dysfunction as a key factor in COVID-19 disease. *Exp Gerontol*, **142**, 111147.

Na HK, Kim EH, Jung JH, Lee HH, Hyun JW, Surh YJ. (2008) (-)-Epigallocatechin gallate induces Nrf2-mediated antioxidant enzyme expression via activation of PI3K and ERK in human mammary epithelial cells. *Arch Biochem Biophys*, **476**(2), 171–177.

Nakahara K, Kawabata S, Ono H, Ogura K, Tanaka T, Ooshima T, Hamada S. (1993) Inhibitory effect of oolong tea polyphenols on glycosyltransferases of mutans Streptococci. *Appl Environ Microbiol*, **59**(4), 968–973.

Ohgitani E, Shin-Ya M, Ichitani M, Kobayashi M, Takihara T, Kawamoto M, Kinugasa H, Mazda O. (2021) Significant inactivation of SARS-CoV-2 in vitro by a green tea catechin, a catechin-derivative, and black tea galloylated theaflavins. *Molecules*, **26**(12), 3572.

Olagnier D, Farahani E, Thyrsted J, *et al.* (2020) SARS-CoV2-mediated suppression of NRF2-signaling reveals potent antiviral and anti-inflammatory activity of 4-octyl-itaconate and dimethyl fumarate. *Nat Commun*, **11**(1), 4938.

Ooshima T, Minami T, Aono W, Izumitani A, Sobue S, Fujiwara T, Kawabata S, Hamada S. (1993) Oolong tea polyphenols inhibit experimental dental caries in SPF rats infected with mutans streptococci. *Caries Res*, **27**(2), 124–129.

Raisch J, Rolhion N, Dubois A, Darfeuille-Michaud A, Bringer MA. (2015) Intracellular colon cancer-associated Escherichia coli promote protumoral activities of human macrophages by inducing sustained COX-2 expression. *Lab Invest*, **95**, 296–307.

Ricciardolo FL, Bertolini F, Carriero V, Högman M. (2020) Nitric oxide's physiologic effects and potential as a therapeutic agent against COVID-19. *J Breath Res*, **15**(1), 014001.

Rogers J, Perkins I, van Olphen A, Burdash N, Klein TW, Friedman H. (2005) Epigallocatechin gallate modulates cytokine production by bone marrow-derived dendritic cells stimulated with lipopolysaccharide or muramyldipeptide, or infected with Legionella pneumophila. *Exp Biol Med (Maywood)*, **230**(9), 645–651.

Sikora AE, Beyhan S, Bagdasarian M, Yildiz FH, Sandkvist M. (2009) Cell envelope perturbation induces oxidative stress and changes in iron homeostasis in Vibrio cholerae. *J Bacteriol*, **191**(17), 5398–5408.

Sodagari H, Bahramsoltani R, Farzaei MH, Abdolghaffari AH, Rezaei N, Taylor-Robinson AW. (2016) *J Natl Remedies*, **16**, 60–72.

Steinmann J, Buer J, Pietschmann T, Steinmann E. (2013) Anti-infective properties of epigallocatechin-3-gallate (EGCG), a component of green tea. *Br J Pharmacol*, **168**(5), 1059–1073.

Sobko T, Huang L, Midtvedt T, Norin E, Gustafsson LE, Norman M, Jansson EA, Lundberg JO. (2006) Generation of NO by probiotic bacteria in the gastrointestinal tract. *Free Radic Biol Med*, **41**, 985–991.

Sun H, Chen Y, Cheng M, Zhang X, Zheng X, Zhang Z. (2018) The modulatory effect of polyphenols from green tea, oolong tea and black tea on human intestinal microbiota in vitro. *J Food Sci Technol*, **55**(1), 399–407.

Tombola F, Campello S, De Luca L, Ruggiero P, Del Giudice G, Papini E, Zoratti M. (2003) Plant polyphenols inhibit VacA, a toxin secreted by the gastric pathogen Helicobacter pylori. *FEBS Lett*, **543**(1-3), 184–189.

Upadhyay S, Tripathi PK, Singh M, Raghavendhar S, Bhardwaj M, Patel AK. (2020) Evaluation of medicinal herbs as a potential therapeutic option against SARS-CoV-2 targeting its main protease. *Phytother Res*, **34**(12), 3411–3419.

Wang H, Xing X, Wang J, Pang B, Liu M, Larios-Valencia J, Liu T, Liu G, Xie S, Hao G, Liu Z, Kan B, Zhu J. (2018a) Hypermutation-induced in vivo oxidative stress resistance enhances Vibrio cholerae host adaptation. *PLoS Pathog*, **14**(10), e1007413.

Wang L, Zeng B, Liu Z, Liao Z, Zhong Q, Gu L, Wei H, Fang X. (2018b) Green tea polyphenols modulate colonic microbiota diversity and lipid metabolism in high-fat diet treated HFA mice. *J Food Sci*, **83**(3), 864–873.

Williamson MP, McCormick TG, Nance CL, Shearer WT. (2006) Epigallocatechin gallate, the main polyphenol in green tea, binds to the T-cTea polyphenols as natural products for potential future management of HIV infection — an overview.ell receptor, CD4: potential for HIV-1 therapy. *J Allergy Clin Immunol*, **118**, 1369–1374.

Wu CY, Yu ZY, Chen YC, Hung SL. (2021) Effects of epigallocatechin-3-gallate and acyclovir on herpes simplex virus type 1 infection in oral epithelial cells. *J Formos Med Assoc*, **120**(12), 2136–2143.

Wu S, Rhee KJ, Albesiano E, Rabizadeh S, Wu X, Yen HR, Huso DL, Brancati FL, Wick E, McAllister F, *et al.* (2009) A human colonic commensal promotes colon tumorigenesis via activation of T helper type 17 T cell responses. *Nat Med*, **15**, 1016–1022.

Xiao Y, Liu T, Zhan L, Zhou X. (2000) The effects of tea polyphenols on the adherence of cariogenic bacterium to the salivary acquired pellicle in vitro. *Hua Xi Kou Qiang Yi Xue Za Zhi*, **18**(5), 336–339.

Xiong LG, Chen YJ, Tong JW, Huang JA, Li J, Gong YS, Liu ZH. (2017) Tea polyphenol epigallocatechin gallate inhibits Escherichia coli by increasing endogenous oxidative stress. *Food Chem*, 217, 196–204.

Xu J, Xu Z, Zheng WM. (2017) A review of the antiviral role of green tea catechins. *Molecules*, 22(8):1337.

Yahiro K, Shirasaka D, Tagashira M, Wada A, Morinaga N, Kuroda F, Choi O, Inoue M, Aoyama N, Ikeda M, Hirayama T, Moss J, Noda M. (2005) Inhibitory effects of polyphenols on gastric injury by Helicobacter pylori VacA toxin. *Helicobacter*, 10(3), 231–239.

Yang J, Li L, Tan S, Jin H, Qiu J, Mao Q, Li R, Xia C, Jiang ZH, Jiang S, Liu S. (2012) A natural theaflavins preparation inhibits HIV-1 infection by targeting the entry step: potential applications for preventing HIV-1 infection. *Fitoterapia*, **83**(2), 348–355.

Yang J, Li L, Jin H, Tan S, Qiu J, Yang L, Ding Y, Jiang ZH, Jiang S, Liu S. (2012) Vaginal gel formulation based on theaflavin derivatives as a microbicide to prevent HIV sexual transmission. *AIDS Res Hum Retroviruses*, **28**(11), 1498–1508.

Yang M, Li Y, Ye JS, Long Y, Qin HM. (2017) Effect of PFOA on oxidative stress and membrane damage of *Escherichia coli*. *Huan Jing Ke Xue*, **38**(3):1167–1172.

Yoon SH, Waters CM. (2019) Vibrio cholerae. *Trends Microbiol*, **27**(9), 806–807.

Zhang Z, Zhang X, Bi K, He Y, Yan W, Yang CS, Zhang J. (2021) Potential protective mechanisms of green tea polyphenol EGCG against COVID-19. *Trends Food Sci Technol*, **114**, 11–24.

Zhao B-L. (2002) Oxygen free radicals and natural antioxidants (Revised Edition). Science Press, Beijing.

Zhao B-L. (2007) Free radicals and natural antioxidants and health. China Science and Culture Press, Hong Kong.

Zhao B-L, Li XJ, He RG, Cheng SJ, Xin WJ. (1989) Scavenging effect of extracts of green tea and natural antioxidants on active oxygen radicals. *Cell Biophys*, **14**, 175–184.

Zhao G, Wu X, Wang W, Yang CS, Zhang J. (2020) Tea drinking alleviates diabetic symptoms via upregulating renal water reabsorption proteins and downregulating renal gluconeogenic enzymes in db/db mice. *Mol Nutr Food Res*, e2000505.

Zhao Y, Zhang X. (2020) Interactions of tea polyphenols with intestinal microbiota and their implication for anti-obesity. *J Sci Food Agric*, **100**(3), 897–903.

Zhu Y, Xie DY. (2020) Docking characterization and in vitro inhibitory activity of flavan-3-ols and dimeric proanthocyanidins against the main protease activity of SARS-CoV-2. *Front Plant Sci*, **11**, 601316.

Zuo Y, Yalavarthi S, Shi H, Gockman K, Zuo M, Madison JA, *et al.* (2020) Neutrophil extracellular traps (NETs) as markers of disease severity in COVID-19. Neutrophil extracellular traps in COVID-19. *JCI Insight*, **5**(11).

https://doi.org/10.1142/9789811274213_0013

Chapter 13

Tea Polyphenols and Immune and Inflammatory

Baolu Zhao

Institute of Biophysics, Chinese Academy of Sciences, Beijing, China

13.1 Introduction

Immunity is a physiological function of the human body. The human body relies on this function to identify "self" and "non-self" components to destroy and reject the antigenic substances (such as bacteria) entering the human body or the damaged cells and tumor cells produced by the human body itself, in order to maintain the health of the human body. The immune system is an important system for the body to perform immune response and immune function. It has the function of identifying and eliminating antigenic foreign bodies and coordinating with other systems of the body to jointly maintain the stability of the environment and physiological balance in the body. It can find and remove foreign bodies, foreign pathogenic microorganisms, and other factors that cause fluctuations in the internal environment. Studies have found that oxidative stress plays an important role in the immune response process, and peroxynitrite produced by the reaction of oxygen free radicals (ROS) and nitric oxide (NO) free radicals plays a particularly important role in the immune response.

Inflammation is a defensive response of the body to stimulation, manifested as redness, swelling, heat, pain, and dysfunction. Inflammation caused by infection can be either an infectious or non-infectious inflammation. Usually, inflammation is a beneficial, automatic defense response of the human body, but sometimes, inflammation can also be harmful, such as attack on the human body's own tissue, for example, the inflammation in transparent tissue. However, the oxidative stress injury and fever caused by inflammation exceed a certain degree or long-term fever, which can affect the metabolic process of the body and cause the functional disorder of multiple systems, especially the central nervous system. Chronic inflammatory diseases affect millions of people globally and the incidence rate is on the rise. While inflammation contributes to the tissue healing process, chronic inflammation can lead to life-long debilitation and loss of tissue function and organ failure. Chronic inflammatory diseases include hepatic, gastrointestinal, and neurodegenerative complications which can lead to malignancy.

The rise of body temperature to a certain extent can enhance the metabolism of the body, promote the formation of antibodies, and enhance the phagocytic function of phagocytes and the barrier detoxification function of the liver, thus improving the defense function of the body. Monocyte phagocyte system cell proliferation is a manifestation of the body's defense response. In the process of inflammation, especially the inflammation caused by pathogenic microorganisms, the cells of mononuclear phagocyte system often proliferate to varying degrees. It often shows local lymphadenopathy, hepatomegaly, and splenomegaly. Macrophages in bone marrow, liver, spleen, and lymph nodes proliferate, and their phagocytic and digestive abilities are enhanced. B and T lymphocytes in lymphoid tissue also proliferate, and their functions of releasing lymphokines and secreting antibodies are enhanced. Therefore, immunity and inflammation are inseparable twin brothers.

A large number of studies have shown that antioxidants such as tea polyphenols can improve the body's immunity, remove reactive oxygen species (ROS) and reactive nitrogen species (RNS) free radicals, reduce the damage caused by oxidative stress in inflammation, promote blood circulation, remove blood stasis, and prevent and reduce the damage of inflammation to the body. This chapter discusses the role of oxidative stress in the process of inflammation and immunity, and the role of tea polyphenols in eliminating inflammation and improving immunity.

13.2 Immunity and inflammation

Although immunity and inflammation are inseparable twin brothers, both can be beneficial or harmful to the body. Both immunity and inflammation are closely related to oxidative stress, but they still have many differences. Immunity is essential to ensure and maintain human health. While on one hand we may seek to improve our immunity, on the other hand, the cause for immune diseases lies in over-immunization. Although inflammation is an immune response, which is beneficial to the body in a certain temperature range, inflammation is the main damage to the human body, and we need to treat and control inflammation.

13.2.1 *Immunization*

Immunity refers to the function of the immune system recognizing self and alien substances, then removing antigenic foreign bodies, so as to maintain the physiological balance of the body. Immunity is a physiological function of the human body. The human body relies on this function to destroy and repel the antigenic substances entering the human body or tumor cells generated by the human body itself, to maintain the health of the human body. Immunity can also be defined as the state of resisting or preventing infection by microorganisms or parasites or other unwanted biological invasion. Immunity involves specific and nonspecific components. Nonspecific components do not need to be exposed in advance, can respond immediately, and effectively prevent the invasion of various pathogens. Specific immunity is developed in the life span of the subject, which is specifically targeted at a pathogen.

Human immunity can be divided into natural immunity and acquired immunity: natural immunity is innate to individuals, generally nonspecific immunity, such as the role of phagocytes. Acquired immunity includes automatic acquired immunity and passive acquired immunity. Automatic acquired immunity generally means that, the immunization lasts a long time and can last for a lifetime, such as acquired immunity against measles and smallpox; the time of passive acquired immunity was short. The human body has three lines of defense. The first line of defense is composed of skin and mucous membrane, which can not only prevent pathogens from invading the human body, but also their secretions have the effect of sterilization. There are cilia on the respiratory mucosa, which can remove foreign bodies. The second line of defense is through the use of bactericidal substances and phagocytes in body fluids. These first two lines of defense are the natural defense functions gradually established by human beings in the process of evolution; the characteristic is that everyone is born with them. They do not target a specific pathogen but have a defense effect against a variety of pathogens, so they are called non-specific immunity. The third line of defense is specific immunity. It is mainly composed of immune organs (thymus, lymph nodes, spleen, etc.) and immune cells (lymphocytes). Among them, lymphoid B cells are responsible for humoral immunity, while lymphoid T cells are "responsible" for cellular immunity.

The immune system is defined as the ability of the human body to resist invasion of pathogens and their toxic products, so that people can avoid infectious diseases and prevent pathogenic microorganisms from invading the body. When the function is too hyperactive, hypersensitivity reaction occur. When the function is too low, immune-deficiency disease occurs. Immunity is usually beneficial to the body, but it can also cause damage to the body under certain conditions. Immunity is essential to ensure and maintain human health. Therefore, it is crucial for humans to improve their immunity. Studies have found that both immune killing of foreign microorganisms and excessive immunity are related to oxidative stress. Many studies have shown that natural antioxidant tea polyphenols can not only improve immunity, but also inhibit excessive immune oxidative stress injury.

13.2.2 *Inflammation*

Inflammation is a defensive response of the body to stimulation. It is divided into acute and chronic inflammation. The main symptoms of acute inflammation are redness, swelling, and pain, that is, inflammation mainly composed of vascular system response. In chronic inflammation, local blood vessels dilate, blood is slow, and blood components such as plasma and neutrophils exude into tissues. Any factor that can cause tissue damage, can become the cause of inflammation, which is also known as an inflammatory agent. Bacteria, viruses, rickettsia, mycoplasma, fungi, spirochetes, and parasites are the most common causes of inflammation. High temperature, low temperature, radioactive substances, ultraviolet radiation and mechanical damage, exogenous chemicals such as strong acid, strong base, turpentine, mustard gas, and foreign bodies entering the human body can cause inflammatory reactions. When the body's immune response is abnormal, it can cause inappropriate or excessive immune response, resulting in damage and inflammation in the tissue and cell.

Inflammatory lesions are mainly local, but local lesions and the whole affect each other. In severe inflammatory diseases, especially when pathogenic microorganisms spread in the body, there are often obvious systemic reactions, such as fever in acute inflammation, especially acute inflammation caused by bacterial infection, which may increase the risk of leukemia. During serious inflammation, the action of pathogenic microorganisms and their toxins, as well as the influence of local blood circulation disorders, fever, and other factors, the parenchymal cells of heart, liver, kidney, and other organs, may undergo varying degrees of degeneration, necrosis, and organ dysfunction. In the process of inflammation, a series of local reactions centered on the vascular system limit and eliminate injury factors and promote the healing of damaged tissues. Therefore, inflammation is a natural local reaction based on defense, which may be beneficial to the body.

Many studies have shown that inflammation leads to oxidative stress, and tea polyphenols can eliminate the free radicals produced in the process of inflammation and inhibit the damage of oxidative stress to the body.

13.3 Oxidative stress in inflammation and immunity

Oxidative stress is not only an important tool in immunity, but also an important injury factor of inflammatory response. Oxidative stress and inflammation play an important role in fighting infectious pathogens. The process of inflammatory response contributes to the normal immune defense mechanism. The regulation of leukocyte signal transduction is an important mechanism of immune and inflammatory response [Lauridsen, 2019]. A large number of studies have shown that inflammation and immunity cause oxidative stress, and oxidative stress plays important functions for body health. Immune activation, inflammation, oxidative stress, mitochondrial bio-energetic changes, and autophagy are considered to be involved in some important pathophysiological events. Therefore, the oxidative stress of immunity is closely related to the oxidative stress of inflammatory response. In the following sections, we will discuss the oxidative stress of immunity and inflammation.

13.3.1. *Phagocytes and immunity, inflammation, and oxidative stress*

Leukemia, especially phagocytes, plays an important role in immune and inflammatory responses. Inflammation is a protective response after the body is invaded by external microorganisms. Phagocytes play an important role in the immune and inflammatory reaction.

1. The role of phagocytes in immunity, inflammation, and oxidative stress

Immune and inflammation is a complex and potentially life-threatening condition that involves the participation of a variety of chemical mediators, signaling pathways, and cell types. During the inflammatory reaction, they phagocytize bacteria, activate by stimulation, produce respiratory burst, and release a large number of reactive ROS and RNS and various enzymes. These products will not only kill invaders, but also generate oxidative stress and damage normal body tissues. In 1960, it was proposed that phagocytic lysosomal particles can transfer substances into lysosomes and release lysosomes. In 1961, it was found that phagocytes can convert formate into carbon dioxide. In 1966, it was found that the system composed of hydrogen peroxide, myeloperoxidase, and halogen had the function of killing microorganisms. In 1967, it was found that granulocytes of patients with acute granuloma (CGD) had normal phagocytosis, but there was no respiratory outbreak and could not kill bacteria. In 1973, it was found that respiratory burst during phagocytosis not only included oxygen consumption, increased phosphorylation of peroxide and hexose, but also produced peroxide. In 1979, respiratory burst products were found to have the function of killing tumor cells. In recent years, people have paid special attention to the study of oxygen and NO free radicals produced by stimulated or sensitized phagocytes. The research in this field will have important applications in anti-microbial and anti-tumor biology and medicine [Henson, 1984; Hewitt, 1987].

As early as the 19th century, scientists in the former Soviet Union observed the phagocytosis of bacteria by animal blood cells, that is, the phenomenon that phagocytes tend to invade around particles and surround and dissolve them into the cytoplasm with cell membrane. In the late stage of inflammation, the activity of monocytes is less than neutrophils in phagocytes. Once they leave the circulation and enter the inflammatory area, they differentiate into macrophages, and their lysosomal enzyme content, metabolic activity, motility, phagocytosis, and ability to kill microorganisms are greatly enhanced. Macrophages are larger than neutrophils and have no scaly nuclei. They can also be found in the lymphatic system and spleen in addition to forming in the inflammatory phase. Pulmonary macrophages are located in the alveolar wall and are the main defenders against inhaled bacteria and other particulate matter. Astrocytes formed in the hepatic sinuses are also macrophages. Macrophages can devour bacteria and kill cells, decompose red blood cells, and digest a large amount of insoluble substances. Phagocytes produce respiratory burst during phagocytosis or when stimulated, consume oxygen, activate hexose phosphorylation branch, and release superoxide anion free radical, hydrogen peroxide, and singlet oxygen. Using ESR spin trapping and chemiluminescence techniques, we directly captured ROS and NO free radicals produced by respiratory burst of human polymer macrophage leukocytes (PMN) stimulated by cancer promoter phorbol-1,2-myristate-1,3-acetate (phorbol alcohol, PMA) [Zhao *et al.*, 1989a, 1989b, 1996a, 1996b, 1996c, 1996d; Li *et al.*, 1990; Yang *et al.*, 1991].

The most studied biochemical mechanism of phagocytosis is pulmonary macrophages and neutrophils in blood because they can be easily separated from bronchial perfusion fluid and peripheral blood. At the beginning of phagocytosis, they all significantly increased oxygen consumption, 10–20 times higher than the rest state. At the same time, starting the hexose phosphorylation branch, glucose consumption increased sharply, which is called respiratory burst. Cyanide cannot inhibit this oxygen consumption, indicating that it has nothing to do with the respiratory chain. The initiation of respiratory burst depends on the disturbance to the membrane, because not only conditioned bacteria, but also small latex particles, conditioned yeast polysaccharide and some chemical reagents can induce respiratory burst, including PMA, fluoride ion, small molecular peptide N-formyl-thiocarbamylphenylalanine (fmet Leu) PNE and concanavalin A, therefore, phagocytosis is not required for respiratory burst. It was found that the entry of phagocytes into the wound site is key to the development of inflammatory response. The entry of these phagocytes (neutrophils, monocytes, and macrophages) constitutes the primary response of inflammatory response, which can phagocytize and decompose invading bacterial particles. In the inflammatory reaction, phagocytes exude from the circulatory system through the blood vessel wall through a series of steps, and finally produce phagocytosis to kill invasives. It has long been known that phagocytes, especially neutrophils, have two sides of protecting and damaging the body. In acute inflammatory injury, the inflow of phagocytes corresponds to the severity of inflammatory reaction and tissue injury. The depletion of neutrophils technique shows that neutrophils are actually mediators of vascular injury [Yoshino, 1985].

In inflammatory reaction, phagocytes can produce various bioactive products. At first, it was thought that the local vascular tissue injury might be caused by the accumulation of lactic acid, but soon it was believed to be caused by the protease of leukocytes. During phagocytosis, various proteases are released into extracellular media to attack various target sites, such as intercellular substances, hyaluronic acid, insoluble elastin, and collagen. *In vitro* studies have shown that proteases secreted by activated neutrophils can decompose these substances, Collagen and elastin breakdown products are found in the urine of inflammatory patients. However, some later studies found that leukocyte protease was not directly included in the inflammatory reaction, and the purified protease only caused little tissue damage. The vascular damage of mice lacking neutrophil protease was similar to that of normal mice in the inflammatory reaction — protease inhibitors cannot protect the tissue damage of animals with acute inflammatory reaction, which shows that protease may not play a direct and important role in the tissue damage of inflammatory reaction [Hoffsten *et al.*, 1988].

The depolymerization of hyaluronic acid by superoxide anion radical was observed in 1974. Oxygen free radicals can also denature elastin, inactivate various enzymes, and oxidize polyunsaturated fatty acids. Studies have shown that oxygen free radicals produced by neutrophils and macrophages can directly poison eukaryotic cells and damage endothelial cells, erythrocytes, fibroblasts, platelets, and sperm. Leukocytes themselves can also be damaged by oxygen free radicals produced by themselves. The damaging effect of superoxide anion radical *in vivo* has also been reported. The treatment of carrageenan-induced pleurisy with SOD can reduce the accumulation of pleural effusion and leukocytes. SOD can also inhibit arthritis, reduce the precipitation of IgG in glomerulus, and interfere with the accumulation of leukocytes in the inflammatory site of glomerulus. SOD can inhibit the changes of organ permeability caused by acute tracheitis caused by xanthine/xanthine oxidase and reduce the accumulation of polymorphonuclear leukocytes in the lungs. These studies show that superoxide anion free radicals also seem to produce some trend factors in tissue injury, resulting in the accumulation of polymorphonuclear leukocytes in tissue [Li *et al.*, 1991].

2. ROS and NO free radicals produced by phagocytes

ROS and NO free radicals produced by phagocytes play an important role in immunity and inflammation. We studied the effects of superoxide anion radical, hydroxyl radical, and hydrogen peroxide. We directly captured the oxygen free radicals produced by PMA-stimulated polymorphonuclear leukocyte respiratory burst by ESR spin trapping technology with spin trapping agent DMPO. It is a mixed spectrum superimposed by two signals, namely DMPO-OOH and DMPO-OH. The confirmation of this spectrum is done by computer simulation. In order to determine the source of superoxide anion radical and hydroxyl radical in the spectrum, SOD and catalase were added into the reaction system, respectively. The results showed that after the addition of catalase, the ESR line decreased, but it was still composed of these two lines. After the addition of SOD, these two lines almost disappeared (Figs. 13-1, 13-2). It shows that the spectral lines produced in the system mainly come from superoxide anion free radicals produced by PMA-stimulated PMN respiratory burst and

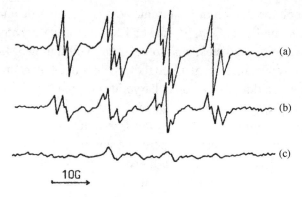

Figure 13-1. The ESR spectrum obtained from (a) 1.5×10^7 polymorphonuclear leukocytes/ml, 0.1 mmol/l, 100 ng/ml PMA, and adding 0.1mol/l DMPO after holding for two minutes (37°C); (b) adding 300 u/ml catalase to (a); (c) adding 300 u/ml SOD to (a).

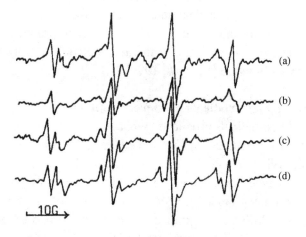

Figure 13-2. The ESR spectrum obtained from (a) 1.5×10^7 polymorphonuclear leukocytes/ml, 100 ng/ml PMA, and adding 0.1 mol/l DMPO after holding for two minutes (37°C); (b) adding 300 u/ml SOD to (a); (c) added 300 u/ml catalase to (a); (d) ESR spectra obtained by adding DMPO to a at the beginning of incubation for two minutes.

hydrogen peroxide and hydroxyl free radicals produced by disproportionation reaction of superoxide anion free radicals.

Phagocytes produce respiratory burst during phagocytosis or when stimulated, consume oxygen, and activate hexose phosphorylation branch, which not only releases superoxide anion free radical, hydrogen peroxide,

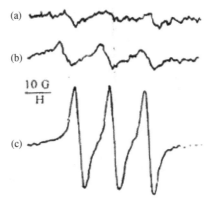

Figure 13-3. ESR spectra of nitric oxide free radicals produced by (a) PMA-stimulated PMN with spin trapping complex (MGD) $2Fe^{2+}$; (b) added NOS inhibitor into the system; (c) add the NOS substrate L-arginine into the system.

and singlet oxygen, but also produces nitric oxide (NO) free radical. ESR spin trapping and chemiluminescence technique was used to directly capture ROS and NO radicals produced by PMN respiratory burst of human PMN stimulated by cancer promoter PMA. We also captured the ESR spectrum of nitric oxide free radicals produced by PMA-stimulated PMN using the self-trapping complex (MGD) $2Fe^{2+}$ (Fig. 13-3). [Zhao *et al.*, 1989b, 1996b; Wang *et al.*, 1997; Li *et al.*, 1990, 1996; Yang *et al.*, 1991]. Lymphocytes can also produce respiratory burst and oxygen radicals as polymorphonuclear leukocytes [Zhao *et al.*, 1990].

By using ultra weak chemiluminescence technology, we found that PMA can induce rat macrophage to emit chemiluminescence. The kinetic curve of the chemiluminescence clearly showed two peaks when a proper ratio of PMA and macrophage were mixed (Fig. 13-4). It was proved that the first one mainly came from the oxygen free radicals and the second one from NO generated in the PMA-stimulated cells by using the substitute and inhibitor of NO synthase [Li *et al.*, 1996].

Leukocytes play a key role in immune response. Previous studies have shown that cells release many reactive oxygen free radicals in the process of immune killing as a weapon to kill foreign invading microorganisms. Additionally, studies have found that leukocytes, especially macrophages, release not only reactive oxygen species but also a large number of NO

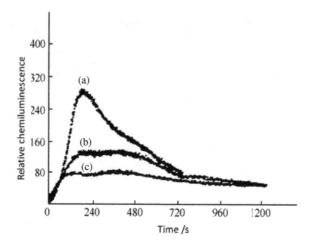

Figure 13-4. Nitric oxide free radicals produced from macrophages stimulated by PMA detected by ultra-weak chemiluminescence technique from (a) 1×10^7 polymorphonuclear leukocytes/ml; (b) 3×10^6 polymorphonuclear leukocytes/ml; (c) 1×10^6 polymorphonuclear leukocytes/ml.

free radicals in this process. These two radicals can react quickly to form peroxynitrite anion (k = 3.7×10^{-9} [mol $\cdot$ L^{-1} $\cdot$ s^{-1}]. Under alkaline conditions, peroxynitrite is relatively stable, but under slightly lower than neutral pH, it decomposes immediately to form hydroxyl like substances and NO_2 with stronger oxidation.

$$\bullet O_2^- + NO \rightarrow ONOO^- + H+ \rightarrow ONOOH \rightarrow \bullet OH + NO_2$$

From the perspective of free radical molecular toxicology, this reaction mechanism is very meaningful. Both $\bullet O_2^-$ and NO are free radicals, but their oxidizability is not very strong. They both have certain biological functions *in vivo*, and combine to form peroxynitrite anion. Under the condition of higher than physiological pH, peroxynitrite is quite stable, allowing it to diffuse from the generation position to a long distance, once the surrounding pH is slightly lower than the physiological conditions, it will immediately decompose into hydroxyl free radical and NO_2 free radicals. These two free radicals have strong oxidation and cytotoxicity, which is of great advantage in killing invading microorganisms and tumor cells.

PMN is stimulated by PMA and produces respiratory burst during phagocytosis, and releases $\bullet O_2^-$ and NO free radicals at the same time. NO and $\bullet O_2^-$ free radical can react to form $ONOO^-$, which has important physiological significance in immunity, prevention, and killing of foreign invading microorganisms and cancer cells. The half-life of $ONOO^-$ is about one second at pH 7.4, which makes it stable enough under physiological conditions and diffuses from the position where it is generated to the target molecules of the reaction, such as the sulfhydryl groups of membrane lipids and proteins. Once acidified, it is immediately transformed into peroxynitrite with strong oxidizing vibration excited state and produces NO_2 and $\bullet OH$ like substances, which are similar to oxygen free radicals, but they are a protective way and mechanism for the body with stronger lethality than oxygen free radicals.

NO and ROS exert multiple modulating effects on inflammation and play a key role in the regulation of immune responses. They affect virtually every step of the development of inflammation. Low concentrations of NO produced by constitutive and neuronal nitric oxide synthases inhibit adhesion molecule expression, cytokine and chemokine synthesis, and leukocyte adhesion and transmigration. Large amounts of NO, generated primarily by iNOS can be toxic and pro-inflammatory. However, the actions of NO are not dependent primarily on the enzymatic source, but rather on the cellular context, NO concentration, and initial priming of immune cells. These observations may explain difficulties in determining the exact role of NO in Th1 and Th2 lymphocyte balance in normal immune responses and in allergic disease. Similarly $\bullet O_2^-$ produced by NAD(P)H oxidases present in all cell types participating in inflammation may lead to toxic effects when produced at high levels during oxidative burst, but may also modulate inflammation in a far more discrete way when continuously produced at low levels by NOXs. The effects of both NO and $\bullet O_2^-$ in immune regulation are exerted through multiple mechanisms, which include interaction with cell signaling systems like cGMP, cAMP, G-protein, JAK/STAT, or MAPK-dependent signal transduction pathways. They may also lead to modification of transcription factors activity and in this way, modulate the expression of multiple other mediators of inflammation [Guzik *et al.*, 2003].

The microcirculation, which is critical for the initiation and perpetuation of an inflammatory response, exhibits several characteristic functional and structural changes in response to inflammation. These include vasomotor dysfunction, the adhesion and transendothelial migration of leukocytes, endothelial barrier dysfunction, blood vessel proliferation, and enhanced thrombus formation. These diverse responses of the microvasculature largely reflect the endothelial cell dysfunction that accompanies inflammation and the central role of these cells in modulating processes as varied as blood flow regulation, angiogenesis, and thrombogenesis. The importance of endothelial cells in inflammation-induced vascular dysfunction is also predicated on the ability of these cells to produce and respond to ROS and NOS. Inflammation seems to upset the balance between NO and superoxide within endothelial cells, which is necessary for normal vessel function. ROS and RNS contribute significantly to the diverse vascular responses in inflammation and support efforts that are directed at targeting these highly reactive species to maintain normal vascular health in pathological conditions that are associated with acute or chronic inflammation [Kvietys & Granger, 2012].

From above discussion, we can clearly see that phagocytes and their produced ROS and NOS free radicals play an important role in immunity and inflammation.

13.3.2 *Oxidative stress and inflammation caused by diseases*

Several studies have shown that almost most diseases are related to inflammation, which leads to the occurrence and development of many diseases. As we studied in the previous chapters, we discussed cardiovascular and cerebrovascular diseases, neurodegenerative diseases, metabolic diseases, diabetes, and so on [Wan *et al.*, 2011; Zhao *et al.*, 1990, 1996a; Shen *et al.*, 1998; Yan *et al.*, 2012b, 2013]. Acute inflammatory reaction to kill harmful foreign bacteria is beneficial to the body but any abnormal activation of phagocytes will cause destructive reaction, for example, the main cause of gout, which is caused by the increase of uric acid concentration in blood. Arthritis is caused by the precipitation of sodium urate crystals in the joints. These crystals exacerbate inflammation through different mechanisms, such as stimulating neutrophils to produce respiratory burst and leukotrienes to attract more neutrophils. Allopurinol, an inhibitor of xanthine oxidase, can be used to treat gout. The most prominent consequence of abnormal activation of phagocytes is autoimmune diseases.

1. Inflammation, oxidative stress, and cardiovascular diseases

Multiple candidates are proposed, especially those involved in oxidative and inflammatory processes in cardiovascular disease, such as $\bullet O_2^-$, NO, and peroxynitrite. We found that ROS and RNS free radicals are produced during myocardial ischemia-reperfusion, which react to produce peroxynitrite ions, and further lead to myocardial oxidative stress injury. ROS dysregulation is a hallmark of cardiovascular disease, characterized by an imbalance in the synthesis and removal of ROS. ROS such as superoxide ($\bullet O_2^-$), hydrogen peroxide (H_2O_2), hydroxyl ($\bullet OH$), and peroxynitrite ($ONOO^-$) have a marked impact on cardiovascular function, contributing to the vascular impairment and cardiac dysfunction associated with diseases such as angina, hypertension, diabetes, and heart failure. Central to the vascular dysfunction is a reduction in bioavailability and/or physiological effects of cardiovascular protective NO, leading to vasoconstriction, inflammation, and vascular remodeling. In a cardiac context, increased ROS generation can also lead to modification of key proteins involved in cardiac contractility. Whilst playing a key role in the pathogenesis of cardiovascular disease, ROS dysregulation also limits the clinical efficacy of current therapies, such as Nitro dilators [Zhao *et al.*, 1996b, 1996c; Shen *et al.*, 1998; Kemp-Harpe *et al.*, 2021].

2. Inflammation, oxidative stress, and obesity

Obesity is a chronic disease of multifactorial origin and can be defined as an increase in the accumulation of body fat. Among the adipose factors, some were found with inflammatory functions, for example, Interleukin-6 (IL-6). These adipose factors induce the production of ROS, generating a process known as oxidative stress. There are several mechanisms by which obesity produces oxidative stress. The first of this is the mitochondrial and peroxisomal oxidation of fatty acids, which can produce ROS in oxidation reactions, while another mechanism is the over-consumption of oxygen, which generates free radicals in the mitochondrial respiratory chain that is found coupled with oxidative phosphorylation in mitochondria. Lipid-rich diets are also capable of generating ROS because they can alter oxygen metabolism. Upon the increase of adipose tissue, the activity of antioxidant enzymes such as superoxide dismutase (SOD), catalase (CAT), and glutathione peroxidase (GPx), was found to be significantly diminished. Finally, high ROS production and the decrease in antioxidant capacity can lead to various abnormalities, including endothelial dysfunction, which is characterized by a reduction in the bioavailability of vasodilators, particularly NO, and an increase in endothelium-derived contractile factors, favoring atherosclerotic disease [Fernández-Sánchez *et al.*, 2011; Yan *et al.*, 2013].

3. Inflammation, oxidative stress, and COVID-19

Recent studies have shown a correlation between COVID-19, caused by severe acute respiratory syndrome coronavirus 2 (SARS-CoV-2) infection, and the distinct, exaggerated immune response titled "cytokine storm". This immune response leads to excessive production and accumulation of ROS that cause clinical signs characteristic of COVID-19 such as decreased oxygen saturation, alteration of hemoglobin properties, decreased NO bioavailability, vasoconstriction, elevated cytokines, cardiac and/or renal injury, enhanced D-dimer, leukocytosis, and an increased neutrophil to lymphocyte ratio. Particularly, neutrophil myeloperoxidase (MPO) is thought to be especially abundant and, as a result, contributes substantially to oxidative stress and the pathophysiology of COVID-19 [Camp *et al.*, 2021]. Inflammation, oxidative stress is seen also in other chronic viral infections, such as AIDS and viral hepatitis. Influenza and paramyxo-viruses activate in phagocytes in the generation of superoxide by a mechanism involving the interaction between the viral surface glycoproteins and the phagocyte's plasma membrane. Interestingly, viruses that activate this host defense mechanism are toxic when injected in the bloodstream of animals. Mice infected with influenza virus undergo oxidative stress. In addition, a wide array of cytokines are formed in the lung, contributing to the systemic effects of influenza [Duan *et al.*, 2009; Fassina *et al.*, 2002]. Oxidant production in viral hepatitis may contribute to the emergence of hepatocellular carcinoma, a tumor seen in patients after years of chronic inflammation of the liver.

4. Inflammation, oxidative stress, and arthritis

Rheumatoid arthritis (RA) is a chronic, systemic, autoimmune disorder, predominantly symmetric, which causes joint inflammation, cartilage degeneration, and bone erosion, resulting in deformity and the loss of physical function. RA can regulate intracellular signaling pathways and can generate different immune responses through some key factors (i.e., MAPK, IL-1, and IL-6), tumor necrosis factor (TNF), nuclear factor light k chain promoter of activated receptor (NF-κB), and c-Jun N-terminal kinases (JNK). The critical function of the Toll-like receptor (TLR)-dependent mitogen-activating protein kinase (MAPK) signaling pathway in mediating the pathogenic characteristics of RA has been briefly discussed. Oxidative stress can trigger a change in transcription factors, which leads to the different expression of some genes involved in the inflammatory process [Behl *et al.*, 2021]. A vast amount of circumstantial evidence implicates oxygen-derived free radicals, especially ROS and NO as mediators of inflammation and/or tissue destruction in inflammatory and arthritic disorders, as it relates to the roles of NO and ROS in the pathogenesis of this condition. ROS can initiate a wide range of toxic oxidative reactions. These include initiation of lipid peroxidation, direct inhibition of mitochondrial respiratory chain enzymes, inactivation of glyceraldehyde-3 phosphate dehydrogenase, inhibition of membrane sodium/potassium ATP-ase activity, inactivation of membrane sodium channels, and other oxidative modifications of proteins. All these toxicities are likely to play a role in the pathophysiology of inflammation. ROS are all potential reactants capable of initiating DNA single-strand breakage, with subsequent activation of the nuclear enzyme poly (ADP ribose) synthetase (PARS), leading to eventual severe energy depletion of the cells and necrotic-type cell death. Furthermore, there is an evolving picture of a pathway that contributes to tissue damage both directly via the formation of reactive oxygen species with them associated toxicities and indirectly through the amplification of the inflammatory response [Cuzzocrea, 2006].

5. Inflammation, oxidative stress, and acute kidney injury

Ischemia-reperfusion injury (IRI)-induced acute kidney injury (IRI-AKI) is characterized by elevated levels of reactive oxygen species (ROS), mitochondrial dysfunction, and inflammation.

Under normal, steady-state conditions, the oxygen supply to the renal tissues is well regulated and utilized not only for mitochondrial production of ATP (mainly for Na reabsorption), but also for the production of NO and the ROS needed for physiological control of renal function. However, under pathological conditions, such as inflammation, shock, or sepsis, the renal microcirculation becomes compromised, which results in a disruption of the homeostasis of NO, ROS, and oxygen supply and utilization. This imbalance results in these compounds exerting pathogenic effects, such as hypoxemia and oxidative stress, resulting in further deterioration of renal microcirculatory function [Aksu *et al.*, 2011]. Uncoupling protein 2 (UCP2) is involved in the maintenance of mitochondrial function, immune response, and regulation of oxidative stress under physiological or pathological conditions. Compared to the control group, LPS treatment increased UCP2 expression *in vitro* and *in vivo*. *In vitro*, UCP2 overexpression protected HK-2 cells from LPS-induced injury by suppression of apoptosis, inflammation, oxidative stress, MMP loss, and ROS production, increase of ATP production and mtDNA content, and amelioration of damage to the mitochondrial ultrastructure. Additionally, inhibition of UCP2 expression by si-UCP2 resulted in decreased HK-2 cell resistance to LPS toxicity, as shown by increased apoptosis, inflammation, mitochondrial dysfunction, and oxidative stress. *In vivo*, UCP2 downregulation aggravated the LPS-induced renal injury, inflammation, macrophages infiltration, mitochondrial dysfunction, and oxidative stress [Ding *et al.*, 2019].

6. Immune system, inflammation, and hypertension

The immune system, inflammation, and hypertension are related to each other. Innate and adaptive immunity system triggers an inflammatory process, in which may increase blood pressure, stimulating organ damage. Cells in innate immune system produce ROS, such as superoxide and hydrogen peroxide, which are aimed at killing pathogens. Long-term inflammation process increases ROS production, causing oxidative stress which leads to endothelial dysfunction. The endothelial function is to regulate blood vessel tone and structure. When inflammation lasts, NO bioavailability decreases, disrupting its main function as vasodilator, so that blood vessels relaxation and vasodilatation are absent. Effector

T cells and regulatory lymphocytes, part of the adaptive immune system, play role in blood vessels constriction in hypertension [Agita & Alsagaff, 2017; Wójcik *et al.*, 2021].

13.3.3 *Oxidative stress and autoimmune diseases*

Autoimmune diseases are a kind of diseases that seriously endanger human health, such as lupus erythematosus and rheumatoid arthritis. It is also a kind of diseases related to inflammation. Autoimmune diseases, including psoriasis, systemic lupus erythematosus (SLE), and rheumatic arthritis (RA), are caused by a combination of environmental and genetic factors that lead to over-activation of immune cells and chronic inflammation. ROS and RNS free radicals produced by fever and inflammation react to produce peroxynitrite ions, which further leads to myocardial oxidative stress injury. However, in some immune system diseases, ROS and RNS free radicals and oxidative stress play the opposite role in damaging the body. Naturally, it is closely related to oxygen free radicals and oxidative stress. Oxidative stress occurs in many autoimmune diseases, along with the excess production of ROS and RNS. The sources of such reactive species include NADPH oxidases (NOXs), the mitochondrial electron transport chain, nitric oxide synthases, nitrite reductases, and the hydrogen sulfide, producing enzymes cystathionine-β synthase and cystathionine-γ lyase. Superoxide undergoes a dismutation reaction to generate hydrogen peroxide which, in the presence of transition metal ions (e.g., ferrous ions), forms the hydroxyl radical.

1. Oxidative stress and systemic lupus erythematosus autoimmune diseases

Systemic lupus erythematosus (SLE) is an autoimmune inflammatory disease whose etiology remains largely unknown. Lupus erythematosus mainly infects young women and causes extensive damage to skin, kidney, muscles, joints, heart, and blood vessels. There are various antibodies, including circulating antibodies against DNA and RNA, antibodies against erythrocytes, antibodies against subcellular organelles, and plasma proteins. Rheumatoid arthritis is a disease characterized by acute arthritis, especially hand and leg arthritis. Since oxidative stress is a common feature of autoimmune diseases, which activates leukocytes to intensify inflammation, antioxidants could reduce the severity of these diseases. The uncontrolled oxidative stress in SLE contributes to functional oxidative modifications of cellular protein, lipid, and DNA and consequences of oxidative modification play a crucial role in immunomodulation and trigger autoimmunity. Measurements of oxidative modified protein, lipid, and DNA in biological samples from SLE patients may assist in the elucidation of the pathophysiological mechanisms of the oxidative stress-related damage, the prediction of disease prognosis, and the selection of adequate treatment in the early stage of disease [Shah *et al.*, 2014]. In a study, 30 patients with SLE and 30 healthy controls were selected to donate their saliva samples. After centrifugation of unstimulated saliva, biological activity of peroxidase (POD), superoxide dismutase (SOD), and catalase (CAT) were evaluated on their appropriate substrates using spectrophotometric methods and the results were statistically analyzed. The results showed that activities of antioxidant enzymes SOD and CAT were significantly reduced in saliva of SLE patients as compared to controls [Zaieni *et al.*, 2015].

Above results suggested that autoimmune diseases systemic lupus erythematosus, are oxidative attacks on normal biological molecules that may produce new antigens, which may be one of the causes of autoimmune diseases. More serious than this are autoimmune diseases with extensive damage and self-antibodies that attack multiple tissues. Oxygen free radicals produced by activated macrophages can change the antigenic behavior of immunoglobulins, further activate macrophages, and produce inflammation and oxidative stress.

2. Oxidative stress and rheumatoid diseases

Rheumatoid arthritis is a disease characterized by acute arthritis, especially hand and foot arthritis. There are often self-antibodies attacking immunoglobulin IgG in serum and joint synovial fluid in rheumatoid arthritis. Obviously, IgG is regarded as an antigen here, which may be the result of oxidative modification of IgG in the process of inflammation. In addition to activating leukocytes, oxidative stress increases the production of lipid mediators, notably of endocannabinoids and eicosanoids, which are products of enzymatic lipid metabolism that act through specific receptors. Generally, there are active and static periods, which makes drug treatment more difficult. For example, the injection of SOD should last for a long time. The knee synovial fluid of rheumatoid patients contains high concentrations of conjugated dienes and total bile acid reactants, indicating that there are more lipid peroxides in the body. Moreover, the concentration of total bile acid reactants in synovial fluid has a good correlation with the severity of clinical diseases and biochemical indexes, such as acute phase protein concentration. The synovial fluid of the patient hardly contained catalase, glutathione peroxidase, and glutathione, and only contained a small amount of SOD activity. In this way, the ROS produced by phagocytes in the joints of patients with rheumatism cannot be effectively eliminated. In the presence of iron ion, hydroxyl radical will be produced. Ascorbic acid in synovial fluid and plasma of rheumatic patients is significantly lower than that of normal people, and mostly exists in the form of oxidative dehydroascorbic acid, which may be caused by the oxidation of oxygen free radicals and hypochlorous acid produced by activated neutrophils and oxidative stress injury [Axford, 1987].

It is known that rheumatoid arthritis is often accompanied by abnormal iron metabolism. At the beginning of inflammation, the total iron content in plasma decreases rapidly, followed by the decrease of hemoglobin concentration and the increase of iron precipitation on synovial membrane. The decrease of plasma iron content is closely related to the inflammatory process. Iron in synovial fluid of rheumatoid arthritis mainly exists in ferritin. If there is ferritin or hemosiderin in early rheumatism, its recovery will be very poor. The treatment of anemia in patients with rheumatism with oral iron salt is often ineffective and even aggravates their condition. Subcutaneous injection of iron and glucose also often has problems. Some people blame it on glucose. In fact, iron does not play a good role. When injecting iron glucose into rheumatic patients, when transferrin in plasma is saturated, unbound iron will appear in body fluid, aggravating the inflammatory reaction. Patients with spontaneous hemochromatosis and iron excess are often accompanied by arthritis. Peroxidation injury can cause ferritin and hemosiderin to release iron, leading to free radical reaction and lipid peroxidation, and resulting in oxidative stress injury [Giordano, 1984]. However, SOD injection into the knee joint of rheumatoid arthritis has achieved good results. Deferoxamine can treat experimental rats with autoimmune disease allergic encephalo-myelitis, which may be due to the ability of deferoxamine to block free radical reaction dependent on iron [Theofilopoulous & Dixon, 1982]. High dose of deferoxamine has certain anti-inflammatory effect on acute arthritis, and low dose of deferoxamine can aggravate inflammation, but repeated medication can shorten the acute phase. Before the experiment, rats fed iron-deficient food caused mild anemia, which can reduce inflam-mation, which may be due to the reduction of iron-catalyzed hydroxyl radical formation. In order to effectively control inflammation, it may be beneficial to use hydroxyl radical scavengers, especially since inflamma-tion is damage caused by oxidative stress [Biemond, 1986; Bowern, 1984].

From the above discussion, it can be seen that the oxidative damage of oxygen free radicals produced by phagocyte activation plays an impor-tant role in the synovial fluid of rheumatoid arthritis, although phagocyte activation products also include proteolytic enzymes, prostaglandins, and other arachidonic acid derivatives.

3. Nitric oxide and oxygen free radicals in asthma

In early onset asthma, there is an association between increased eosino-philic inflammation and ROS and RNS, whereas in late onset, there is a correlation with lower levels of nitric oxide (NO) and predominantly non-T2 inflammation. There are probably multiple pathways by which oxida-tive stress impacts asthma; airway and systemic oxidative stress has been proposed as a mechanism that could potentially explain the oxidative stress mediated increased comorbidity and poor response to treatment. More likely than not, oxidative stress is an epiphenomenon of a very diverse set of processes driven by complex changes in airway and sys-temic metabolism [Grasemann & Holguin, 2021]. Asthma is a complex inflammatory disease characterized by airway inflammation and hyper-responsiveness. Inflammation in asthma, produces a large amount of NO, because NO produced in the immune process is mainly produced by inducible iNOS. Although the main regulatory substances produced in the inflammatory process, such as IFN-γ, TNF-α, leukotrienes, and most prostaglandins, are pro-inflammatory, cyclopentane prostaglandins are anti-inflammatory. Therefore, any regulatory factors produced in the inflammatory process cannot be absolutely said to be good or bad. Of course, in the inflammatory process, not only NO, but also ROS free radicals are produced, because NO is a complex combination of toxicity, regulation, apoptosis induction, and anti-apoptosis to different cells at different stages of the inflammatory process. The role of NO and reactive oxygen free radicals in inflammation induced by contact allergy is also complex. Low concentration promotes inflammation by relaxing and adsorbing neutrophils. At high concentration, it regulates adhesion factors, inhibits activation, and induces apoptosis of inflammatory cells [Kharitonov *et al.*, 1994]. Significant contributions of Toll-like receptors (TLRs) and transcription factors such as NF-κB, have been reported as major contributors to inflammatory pathways [Mishra *et al.*, 2018].

These results revealed the involvement of NO and ROS and various cell types and activation of intracellular signaling pathways that result in activation of inflammatory genes in asthma.

4. Oxidative stress in vitiligo and thyroid diseases

Vitiligo is a multifactorial polygenic disorder with a complex pathogenesis, linked with both genetic and non-genetic factors. Vitiligo is an acquired dermatological disease frequently associated with autoimmune thyroid disorders. So far, there have been several theories proposed to unravel the complex vitiligo pathogenesis. Theories regarding loss of melanocytes are based on autoimmune, cytotoxic, oxidant-antioxidant, and neural mechanisms. ROS in excess have been documented in active vitiligo skin. Currently, the auto-cytotoxic and the autoimmune theories are the most accredited hypothesis, since they are sustained by several important clinical and experimental evidence. A growing body of evidence show that autoimmunity and oxidative stress strictly interact to determine melanocyte loss. In this scenario, associated thyroid autoimmunity might play an active and important role in triggering and maintaining the depigmentation process of vitiligo [Colucci *et al.*, 2015]. Various factors lead to ROS overproduction in the melanocytes of vitiligo. The exogenous and endogenous stimuli that cause ROS production, include low levels of enzymatic and non-enzymatic antioxidants, disturbed antioxidant pathways, and polymorphisms of ROS-associated genes. These factors synergistically contribute to the accumulation of ROS in melanocytes, finally leading to melanocyte damage and the production of autoantigens through apoptosis, accumulation of misfolded peptides and cytokines induced by endoplasmic reticulum stress as well as the sustained unfolded protein response, and an "eat me" signal for phagocytic cells triggered by calreticulin. Subsequently, autoantigens presentation and dendritic cells maturation occurred, mediated by the release of antigen-containing exosomes, adenosine triphosphate, and melanosomal autophagy. With the involvement of inducible heat shock protein 70, cellular immunity-targeting autoantigens takes the essential place in the destruction of melanocytes, which eventually results in vitiligo [Xie *et al.*, 2016].

The above results show that various factors lead to the excessive production and accumulation of ROS in melanocytes of vitiligo, and finally lead to the damage and loss of melanocytes through a variety of ways.

5. Oxidative stress and apoptosis in immune diseases

Oxidative stress is involved in immune diseases as discussed above; ROS excess and deficiency of antioxidants lead to apoptosis and virus activation. Antigenic stimuli increase ROS that influence T-cell activation by interfering with the oxidant-antioxidant balance. Oxidative stress takes place when excess of ROS production is not counterbalanced by antioxidant mechanisms and bcl-2 gene product that inhibits apoptosis by interacting with mitochondrial superoxide dismutase. Excess ROS induces apoptosis both by the activation of NF-kB-dependent genes and DNA damage. The latter has been shown to elicit the activation of poly-ADP-ribose transferase and the accumulation of p53, thus determining apoptosis. Additionally, oxidative stress may induce the formation of cell membrane oxidized lipids, potent inducers of apoptosis. ROS produced at sites of chronic inflammation, have genotoxic effects. As a consequence, abnormalities of the p53 genes might explain the conversion from an inflammatory phase into autonomous progression of rheumatoid arthritis or other chronic inflammatory disorders [Agostini *et al.*, 2002]. GSH-4-hydroxy-t-2,3-nonenal is present in the serum of a significant percentage of patients with various diseases characterized by immune-mediated endothelial dysfunction, including systemic lupus erythematosus and carotid atherosclerosis. These autoantibodies increased intracellular levels of 4-hydroxy-t-2,3-nonenal, decreased levels of GSH, and activated C-Jun

NH2 Kinase signaling (JNK), thus inducing oxidative stress-mediated endothelial cell apoptosis. The dietary antioxidant alpha-tocopherol counteracted endothelial cell demise [Margutti *et al.*, 2008].

These findings show that autoantibodies play a pathogenic role in immune-mediated diseases and are important biological indicators of immune-mediated diseases. Various factors lead to excessive production and accumulation of ROS in cells, and ultimately lead to cell damage and apoptosis through a variety of ways.

13.4 Tea polyphenols limit inflammation

Inflammatory processes can occur over short periods in response to pathogens or can be activated by intracellular ROS. The use of antioxidants has been shown to be beneficial in stimulating immune cells to increase phagocytosis and upregulate the cellular processes to limit acute inflammation. Tea polyphenols have many functions, such as anti-mutation, anti-cancer, regulating immunity, anti-allergy, and anti-aging. In recent years, the research on the mechanism of the biological effect of tea polyphenols mainly focuses on its antioxidant activity. It is generally believed that it can promote cell survival by regulating the redox state *in vivo*/cell, scavenging free radicals, upregulating antioxidant enzyme system and antioxidant damage, and anti-tumor. Studies shown that oral tea polyphenols can induce the antioxidant enzyme system of small intestine, liver, and lung in mice. Tea polyphenols may upregulate antioxidant enzyme system. Transcription factors directly related to these enzymes act on or affect the signal pathways of these enzymes' transcription. The intermediates of tea polyphenols' metabolism/reaction *in vivo* may induce oxidative stress and upregulate antioxidant enzymes. A large number of studies have shown that drinking tea and tea polyphenols can reduce inflammation and improve immunity [Khan *et al.*, 1992]. It is well known that green tea polyphenols are potent antioxidants with important roles in regulating vital signaling pathways. These comprise transcription nuclear factor-kappa B-mediated I kappa B kinase complex pathways, programmed cell death pathways like caspases, and B-cell lymphoma-2 and intervention with the surge of inflammatory markers like cytokines and production of cyclooxygenase-2.

13.4.1 *Anti-inflammatory action of green tea polyphenols*

In the previous chapters, we discussed the benefits of tea polyphenols on some acute inflammation. The benefits of tea polyphenols on some chronic inflammation will be discussed in this part. Clinical trials and meta-analyses reveal that constant consumption of diets rich in polyphenols help to protect against chronic inflammatory diseases, including cardiovascular and neurodegenerative diseases in humans. Chronic inflammatory diseases affect millions of people globally and the incidence rate is on the rise. While inflammation contributes to tissue healing process, chronic inflammation can lead to life-long debilitation and loss of tissue function and organ failure. Chronic inflammatory diseases include hepatic, gastrointestinal, and neurodegenerative complications which can lead to malignancy. Despite the millennial advancements in diagnostic and therapeutic modalities, there remains no effective cure for patients who suffer from inflammatory diseases. It is well known that green tea polyphenols are potent antioxidants with important roles in regulating vital signaling pathways. Green tea, rich in antioxidant and anti-inflammatory catechins, especially EGCG, has been shown to reduce surrogate markers of atherosclerosis and lipid peroxidation, particularly LDL oxidation and malondialdehyde concentrations, in several *in vitro*, animal, and limited clinical studies. Through cellular, animal, and human experiments, green tea and its major component, EGCG, have been demonstrated to have anti-inflammatory effects and suppress the gene and/or protein expression of inflammatory cytokines and inflammation-related enzymes. These comprise transcription nuclear factor-kappa B-mediated I kappa B kinase complex pathways, programmed cell death pathways like caspases and B-cell lymphoma-2, and intervention with the surge of inflammatory markers like cytokines and production of cyclooxygenase-2.

1. Scavenging effect of tea polyphenols on oxygen free radicals produced by respiratory burst of polymorphonuclear leukocytes

Oxygen free radicals produced by PMNs play an important role in the process of immunity and inflammation, but they can also damage normal cells. Oxygen free radicals produced by this system are often used to test the scavenging effect of antioxidants on oxygen free radicals produced by cell system. From our studies, we found that the oxygen free radicals produced by the respiratory burst of human PMN stimulated by cancer promoter PMA captured by spin trapping technology, are composed of superoxide anion and hydroxyl radical. The addition of two kinds of tea polyphenols almost completely eliminated the oxygen free radicals produced by this system, which is much stronger than vitamin C, curcumin, rose flavor, and vitamin E. The measurement of oxygen consumption by spin oximetry proves that tea polyphenols are capable of scavenging oxygen free radicals produced by PMNs, rather than inhibiting the production of oxygen free radicals by PMNs. Quantitatively determining the scavenging effect of tea polyphenols on superoxide anion radical has shown that the superoxide anion radical was produced by riboflavin / EDTA system under light. When tea polyphenols and other antioxidants were added to the system, superoxide anion free radicals were removed in varying degrees. Tea polyphenols and other antioxidants can remove superoxide anion free radicals. Quantitatively determining indicated the scavenging effect of tea polyphenols on hydroxyl radical was produced by Fenton reaction. When tea polyphenols and other antioxidants were added to the system, hydroxyl radicals were eliminated in varying degrees. Tea polyphenols and these antioxidants can scavenge hydroxyl radicals (Fig. 13-5) [Zhao *et al.*, 1989b].

A study showed that EGCG significantly inhibited the expression of extracellular matrix metalloproteinase inducer (EMMPRIN) and matrix metalloproteinases (MMP-9) and activation of extracellular signal-regulated kinase 1/2 (ERK1/2), p38, and c-Jun N-terminal kinase (JNK) in PMA-induced macrophages. Downregulation of EMMPRIN by gene silencing hindered PMA-induced MMP-9 secretion and expression, indicating an important role of EMMPRIN in the inhibition of MMP-9 by EGCG. It is well documented that the overexpression of EMMPRIN and MMPs by monocytes/macrophages plays an important role in atherosclerotic plaque rupture. Green tea polyphenol, EGCG, has a variety of pharmacological properties and exerts cardiovascular protective effects. Recently, the 67-kD laminin receptor (67LR) has been identified as a cell surface receptor of EGCG [Wang *et al.*, 2016].

The above results show that tea polyphenols can scavenge oxygen free radicals produced by neutrophils in inflammation. The effect of EGCG on PMA-induced macrophages has the potential mechanism for the expression of EMMPRIN and MMP-9.

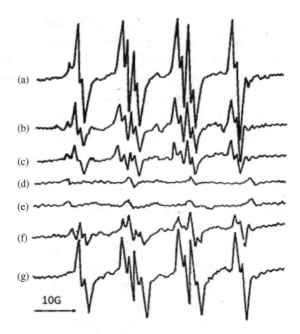

Figure 13-5.   Scavenging effects of GTP on oxygen radicals generated from PMA stimulated PMN. (a) ESR spectrum produced from 1.5×10^7 mg/ml PMN stimulated by 100 nM PMA in the presence of 0.1 M DMPO and 0.1 mM DETAPAC at 37°C; (b) Conditions were same as (a), except that 200 mg/ml curity was present; (c) Conditions were same as (a), except that 200 mg/ml rosmary was present; (d) Conditions were same as (a), except that 200 mg/ml green tea polyphenols was present; (e) Conditions were same as (a), except that 200 mg/ml F6 was present; (f) Conditions were same as (a), except that 200 mg/ml Vitamin C was present; (g) Conditions were same as (a), except that 200 mg/ml Vitamin E was present.

2. Inhibitory effect of tea polyphenols on peroxynitrite oxidation activity

ROS and NO free radicals are often produced at the same time in the inflammatory process. NO has important biological functions *in vivo*. For example, NO can prevent platelet aggregation. It is an endothelial cell relaxation factor (EDRF) but it is still a free radical. Therefore, it has active chemical properties, it can react with ROS free radicals to form strong reactivity, cytotoxicity, and plays an important role in the process of immune killing and inflammatory process. Our studies have found that when cells produce NO, they also produce superoxide anion free radicals.

NO and superoxide anion radicals have a very high reaction rate and react to form peroxynitrite (ONOO⁻). ONOO⁻ is a highly oxidizing substance, which can oxidize the thiohydrogen groups of cell membrane lipids and proteins, leading to cell damage and disease. Experiments show that such reactions occur in many pathological states, so the research on peroxynitrite is very interesting. Many components of ONOO⁻ damaged cells are formed through the free radical mechanism. We directly captured the free radicals produced had used ONOO⁻ by spin trapping technology. Using this free radical generation system, we studied the scavenging effect of tea polyphenols and some natural antioxidants on methyl free radicals and the inhibition of ONOO⁻ oxidation activity. The results showed that tea polyphenols could effectively inhibit peroxynitrite and its inflammatory reaction [Zhao *et al.*, 1996a].

3. Signal pathways of anti-inflammation effects of tea polyphenols

Regular consumption of tea polyphenols are known for various health-promoting functions such as antimicrobial and anti-inflammatory effects. Different relevant investigations have revealed green tea polyphenols as potent antioxidants to play important roles in several signaling pathways involved in inflammation. These include transcription NF-κB-mediated I kappa B kinase complex (IKK) pathways [Yang *et al.*, 2001], TNFα [Oz *et al.*, 2013], downregulation of cyclooxygenase (Cox)-2 and B-cell lymphoma-2 (Bcl-2) activities, and upregulation of protective programmed cell death pathways [Beckman, 2000]. Tea polyphenols have shown a variety of possible applications, including increasing antioxidants (e.g., GSH, cysteine) depots in vital organs [Oz & Ebersole, 2008]. Culture supernatants from GTP-treated mice showed decreased spontaneous interferon-gamma and tumor necrosis factor-alpha secretion compared with that of controls. At six weeks, the GTP group had less severe colitis as demonstrated by lower histologic scores and wet colon weights. This was associated with lower plasma levels of serum amyloid A, increased weight gain, and improved hematocrits [Varilek *et al.*, 2001]. Studies revealed GTPs attenuate inflammatory responses in signaling pathways, by downregulating IKK, NF-κB (0.4 mg/mL), cytokines like TNFα, inflammatory markers, Cox-2, and Bcl-2, to protect against hepatic [Yang *et al.*, 2001], colonic [Najafzadeh *et al.*, 2009], respiratory inflammatory

[Wheeler *et al.*, 2004], neurodegenerative complications [Renaud *et al.*, 2015], and various anti-malignancy effects [Ju *et al.*, 2003].

These results show that GTP attenuated inflammation through transcription NF-κB-mediated I kappa B kinase complex (IKK) and TNFα, Cox-2, and Bcl-2 pathways, thereby suggesting a role for GTP in treating chronic inflammatory diseases such as inflammatory bowel disease.

13.4.2 *Benefits of green tea polyphenols on inflammatory diseases*

Studies have shown that tea polyphenols have many benefits for some diseases caused by chronic inflammation, which can be used to prevent and treat the diseases, such as intestinal diseases, respiratory diseases, liver diseases, neurological diseases caused by inflammation, and inflammatory diseases caused by ultraviolet radiation and heavy metal poisoning.

1. Green tea polyphenols have anti-inflammatory properties in Colitis

Inflammatory bowel disease (IBD) is a collection of inflammatory conditions of colon and small intestine which affect millions of individuals worldwide and the prevalence amount is on the rise. The organ failure as well as loss of tissue function is because of the inflammatory reaction, which is the major contributor of tissue healing leading to lifelong debilitation. Unfortunately, there is no cure for autoimmune chronic inflammatory bowel disease. In a bold animal study, green tea polyphenols (GTP), EGCG, were compared to sulfasalazine for their anti-inflammatory properties. Wild-type mice were given dextran sodium sulfate (DSS) for a chemically-induced ulcerative colitis model [Oz *et al.*, 2013]. Interleukin-10 (IL-10) deficient mice spontaneously develop IBD when exposed to the normal gut microbiota from their control wild-type background to provoke enterocolitis similar to Crohn's disease. Colitis and enterocolitis animals tolerated treatments with GTP, EGCG, which was added into the diets. Treated animals similarly developed less severe symptoms compared to the sham-treated animals. The inflammatory markers (TNFα, IL-6, serum amyloid A) were significantly upregulated along with pathological symptoms but drastically decreased with GTP, EGCG treatment.

While hepatic and colonic antioxidants (glutathione, cysteine) are depleted in IBD patients and colitic models [Oz & Ebersole, 2008], GTP and EGCG significantly restored antioxidant concentrations and attenuated colitis symptoms similar to sulfasalazine administration. In addition, GTP decreased disease activity and inhibited inflammatory responses in interleukin-2-deficient (IL-2–/–) mouse models for chronic inflammatory disease. Colonic explants and lipoprotein lipase cultures from GTP-treated mice had decreased spontaneous interferon-gamma and TNFα secretions [Varilek *et al.*, 2001]. In another study, lymphocytes from IBD patients and healthy subjects were chemically damaged (hydrogen peroxide) *in vitro* and then treated with epicatechin. A significant reduction in induced-DNA damage was discovered in lymphocytes from patients (48.6%) and normal controls (35.2%) when compared with lymphocytes from untreated subjects (both $p < 0.001$) [Najafzadeh *et al.*, 2009]. Therefore, epicatechin significantly decreased oxidative stress in lymphocytes and supported beneficial effects of epicatechin inclusion in diets for IBD patients.

In a study, the BALB/c mice received dextran sodium sulfate (DSS) to induce colitis (ulcerative colitis model). Exposure of IL-10 deficient mice (BALB/c-background) to normal microbiota provoked enterocolitis (mimics Crohn's disease). DSS-treated animals developed severe bloody diarrhea and colitis (score 0-4, 3.2 ± 0.27). IL-10 deficient mice developed severe enterocolitis as manifested by diarrhea, rectal prolapse, and colonic lesions. Animals tolerated green tea polyphenols, EGCG, with no major side effects, and further developed less severe colitis. IL-10-deficient animals became moribund on high dose, while tolerated low and medium doses with significant improved symptoms of enterocolitis. The inflammatory markers TNFα (3-fold), IL-6 (14-fold), and serum amyloid A (40-fold) increased in colitic animals and significantly decreased with treatment regiments. In contrast, circulatory leptin levels decreased in colitic animals (2-fold). Additionally, EGCG reduced leptin levels ($p < 0.01$), while green tea polyphenols and sulfasalazine had no effect on leptin levels ($p < 0.05$). Hepatic and colonic antioxidants were significantly depleted in colitic animals and treatment regiments significantly restored antioxidants levels [Oz *et al.*, 2013]. In another study, it was shown that GTPs play significant roles in downregulating signaling pathways because

GTPs exert effective antioxidant properties and regulate TLR4 expression via certain receptor, inhibited endotoxin-mediated tumor necrosis factor alpha (TNF-α) production by blocking transcription NF-κB activation, and upstream of mediated I kappa B kinase complex pathway activities, as well as intrusion with the flow of cytokines and synthesis of cyclooxygenase-2 (COX-2) [Rahman *et al.*, 2018].

Alteration of the gut microbiota may contribute to the development of inflammatory bowel disease. The above study showed that EGCG is beneficial in IBD alleviation. A study revealed that anti-inflammatory effect and colonic barrier integrity were enhanced by oral, but not rectal, EGCG. It observed a distinct EGCG-mediated alteration in the gut microbiome by increasing *Akkermansia* abundance and butyrate production. It also demonstrated that the EGCG pre-supplementation induced similar beneficial outcomes to oral EGCG administration. Prophylactic EGCG attenuated colitis and significantly enriched short-chain fatty acids (SCFAs)-producing bacteria such as *Akkermansia* and SCFAs production in DSS-induced mice. To validate these discoveries, we performed fecal microbiota transplantation (FMT) and sterile fecal filtrate (SFF) to inoculate DSS-treated mice. Microbiota from EGCG-dosed mice alleviated the colitis over microbiota from control mice and SFF shown by superiorly effect and colonic barrier integrity, and also enriched bacteria such as *Akkermansia* and SCFAs. Collectively, the attenuation of colitis by oral EGCG suggests an intimate involvement of SCFAs-producing bacteria *Akkermansia* and SCFAs, which was further demonstrated by prophylaxis and FMT. This study provides the first data indicating that oral EGCG ameliorated the colonic inflammation in a gut microbiota-dependent manner [Wu *et al.*, 2021]. EGCG improved the intestinal mucosal injury in rats, inhibited production of inflammatory factors, maintained the balance of Th1/Th2, and reduced the expression of TLR4, MyD88, and NF-κB. After TLR4 antagonism, the protective effect of EGCG on intestinal mucosal injury was weakened in rats with ulcerative colitis and the expressions of inflammatory factors were upregulated [Bing *et al.*, 2017]. EGCG has been shown to inhibit signaling pathways involved in inflammation, including NF-κB and activator protein-1 (AP-1), which are important inducers of pro-inflammatory mediators.

A study evaluated the therapeutic efficacy of EGCG in experimental colitis, which was induced by rectal administration of trinitrobenzenesulfonic acid (TNBS) in C57/BL6 mice. Mice were treated twice daily with vehicle or with EGCG (10 mg/kg) intraperitoneally, and sacrificed on days one, three, and seven after TNBS administration. After induction of colitis, vehicle-treated mice experienced bloody diarrhea and loss of body weight. A remarkable colonic damage with hemorrhage, ulcers, and edema was observed and was associated with neutrophil infiltration as evaluated by myeloperoxidase (MPO) activity. Elevated plasma levels of tumor necrosis factor alpha, IL-6, IL-10, and keratinocyte-derived chemokine were also found. These events were paralleled by increased DNA binding of NF-kB and AP-1 in the colon of the vehicle-treated group. In contrast, the EGCG-treated mice experienced very mild diarrhea and no weight loss. Damage of the colon was characterized by edema and hyperemia only. Tissue levels of MPO were also significantly reduced when compared to vehicle-treated mice. These beneficial effects of EGCG were associated with a significant reduction of NF-kB and AP-1 activation [Abboud *et al.*, 2008].

These results indicate that GTP can reduce chronic inflammatory bowel disease such as colitis by inhibition of TNFα, IL-6, interleukin, NFκB, IL-6, and other signals.

2. EGCG inhibits inflammation in respiratory system

Respiratory tract inflammation is a common disease, such as influenza, and it is also a very dangerous disease. Some examples include Jiangxi respiratory distress and COVID-19, discussed in Chapter 12. Studies have found that tea polyphenols have many preventive and therapeutic effects on these respiratory diseases. It reduced wet/dry weight ratio, histological severities, and neutrophil accumulation in the lungs in mice given LPS.

A study showed that EGCG inhibits IL-1 beta-dependent pro-inflammatory signal transduction in cultured respiratory epithelial cells and EGCG inhibits tumor necrosis factor-alpha (TNF-alpha)-mediated activation of the nuclear factor-kappa B (NF-κB) pathway, partly through inhibition of I kappa B kinase (IKK). The NF-κB pathway may also be activated in response to interleukin-1 beta (IL-1 beta) stimulation through a distinct signal transduction pathway. EGCG may inhibit IL-1 beta-mediated activation of the NF-κB pathway. Because the gene expression of interleukin-8 (IL-8), the major human neutrophil chemoattractant, is dependent on activation of NF-κB, IL-8 gene expression in human lung epithelial (A549) cells treated with human IL-1 beta was used as a model of IL-1 beta signal transduction. The EGCG markedly inhibited IL-1 beta-mediated IL-1 beta receptor-associated kinase (IRAK) degradation and the signaling events downstream from IRAK degradation: IKK activation, I kappa B alpha degradation, and NF-κB activation. In addition, EGCG inhibited phosphorylation of the p65 subunit of NF-κB. The functional consequence of this inhibition was evident by inhibition of IL-8 gene expression [Wheeler *et al.*, 2004].

The above results show that tea polyphenols have preventive and therapeutic effects on these respiratory inflammatory diseases. Moreover, tea polyphenol EGCG can inhibit IL-1 in respiratory epithelial cells β-dependent pro-inflammatory signal transduction, inhibit TNF-α NF-κB activation of pathways partly through inhibition of I-κB kinase, and play the role inhibition for inflammatory.

3. EGCG inhibits UV radiation exposure-induced skin inflammation

Epidemiological, clinical, and laboratory studies have implicated solar ultraviolet (UV) radiation in various skin diseases, including premature aging of the skin and melanoma and non-melanoma skin cancers. Chronic UV radiation exposure-induced skin diseases or skin disorders are caused by the excessive induction of inflammation, oxidative stress, and DNA damage, etc. A study conducted on human skin has demonstrated that GTP prevents UV-B-induced cyclobutane pyrimidine dimers (CPD), which are considered to be mediators of UVB-induced immune suppression and skin cancer induction. GTP-treated human skin prevented the penetration of UV radiation, which was demonstrated by the absence of immunostaining for CPD in the reticular dermis. The topical application of GTP or its most potent chemo-preventive constituent EGCG prior to exposure to UVB protects against UVB-induced local as well as systemic immune suppression in laboratory animals. Additionally, studies have shown that EGCG treatment of mouse skin inhibits UVB-induced infiltration of CD11b+ cells. CD11b is a cell surface marker for activated macrophages and neutrophils, which are associated with the induction of UVB-induced suppression of contact hypersensitivity responses. EGCG treatment also results in the reduction of UVB-induced immune-regulatory cytokine IL-10 in skin as well as in draining lymph nodes, where exist an elevated amount of IL-12 [Katiyar *et al.*, 2001].

These *in vivo* observations suggest that GTPs are photoprotective and can be used as pharmacological agents for the prevention of solar UVB light-induced skin disorders associated with immune suppression and DNA damage. In Chapter 15, we will discuss in more detail about the effects of tea polyphenols on various diseases and inflammation caused by ultraviolet radiation.

4. Benefits of tea polyphenols on gastric ulcer-related diseases

Gastric cancer is one of the most prevalent causes of morbidity and mortality. Additionally, gastric cancer patients are at risk of developing severe complications including diarrhea, nausea, and abdominal pain during chemotherapy, mainly due to cytotoxic effects of anticancer drugs. Gastrointestinal ulcer and inflammation not only cause gastrointestinal pain, but also lead to gastrointestinal tumors. Studies have found that tea polyphenols can prevent and treat gastrointestinal ulcers and inflammation. A study showed that tea polyphenols have preventive and treatment effects of on gastric injury in mice induced by HCl/ethanol. Tea polyphenols inhibited the formation and further deterioration of gastric mucosal lesions, reduced the secretion of gastric juice, and raised gastric juice pH compared to the control. The tea polyphenols-treated group had lower serum levels of motilin, substance P, and endothelin than the control group, but they had higher serum levels of vasoactive intestinal peptide and somatostatin. Mice treated with tea polyphenols had lower serum levels of cytokines IL-6, IL-12, tumor necrosis factor-α (TNF-α), and interferon-γ, than the control group. The activities of SOD, nitric oxide, and glutathione peroxidase (GSH-Px) were higher in the gastric tissues of tea polyphenols-treated mice, but the malondialdehyde content was lower. The mRNA expression of occludin, epidermal growth factor (EGF), EGF receptor (EGFR), vascular EGF (VEGF), inhibitor kappaB-α, cuprozinc-superoxide dismutase, manganese-superoxide dismutase, GSH-Px, neuronal nitric oxide synthase, and endothelial NOS increased significantly in the gastric tissues of tea polyphenols-treated mice. However, the activated B cell, inducible NOS, cyclooxygenase-2, TNF-α, IL-1 beta, and IL-6 mRNA expression levels in the tea polyphenols group were lower than those in the control group [Wang *et al.*, 2021]. When epithelial cells were treated with GTP (0.4–0.8 mg/mL), they induced DNA fragmentation in a dose-responsive fashion. In higher concentrations (>0.8 mg/mL), GTP caused a mixture of cytolysis and apoptosis. In addition, epithelial cells exposed to GTP and EGCG, but not other polyphenols (i.e., EC, EGC), had increased caspase-3, caspase-8, and caspase-9 activities, but caspase inhibitors could rescue cells from imminent apoptosis. Furthermore, GTP caused activation of Fas-associated proteins with (Fas)-associated protein with death domain (FADD) recruitment to Fas/CD95 domains. Indeed, GTP blocked NF-κB activation, yet NF-κB inhibitor (MG132) only promoted cytolysis and not apoptosis [Oz & Ebersole, 2010].

These results indicate that the tea polyphenols played a role in the prevention of gastric damage. GTP- and EGCG-induced apoptosis in intestinal epithelia and the activation of death pathways mediated by the caspase-8 through FADD-dependent pathways present promising results as possible anti-cancer agents. In Chapter 14, we will discuss in more detail about the effects of tea polyphenols on various diseases and inflammation caused by gastroenteritis and tumor.

5. Benefits of green tea polyphenols on hepatitis and its complications

Hepatitis is a general term for liver inflammation. It usually refers to the destruction of liver cells and the impairment of liver function caused by a variety of pathogenic factors, such as viruses, bacteria, parasites, chemical poisons, drugs, alcohol, autoimmune factors, which can cause a series of discomfort symptoms and abnormal liver function indicators, Hepatitis, mostly referred to viral hepatitis caused by hepatitis A, B, C and other hepatitis viruses. Hepatitis C is a viral hepatitis caused by hepatitis C virus (HCV), which is mainly transmitted through blood transfusion, acupuncture, drug abuse, etc. According to the statistics of the World Health Organization, the global HCV infection rate is about 3%, and it is estimated that about 180 million people are infected with HCV, with about 35,000 new cases of hepatitis C every year. Hepatitis C is a global epidemic, which can lead to chronic inflammation, necrosis, and fibrosis of the liver. Some patients can develop cirrhosis and even hepatocellular carcinoma (HCC). In the next 20 years, the mortality related to HCV infection (death caused by liver failure and hepatocellular carcinoma) will continue to increase, which is very harmful to the health and life of patients and has become a serious social and public health problem. Many studies have shown that tea polyphenols, especially catechins and their derivatives, have antiviral properties against viral hepatitis [Song, 2018]. A study found that polyphenols extracted from green tea have already shown anti-HCV activity as entry inhibitors. This study identified the polyphenol, EGCG, as an inhibitor of HCV entry. Green tea catechins, such as EGCG and its derivatives, epigallocatechin (EGC), epicatechin gallate (ECG), and epicatechin (EC), have been previously found to exert antiviral and antioncogenic properties. EGCG had no effect on HCV RNA replication, assembly, or release of progeny virions. However, it potently inhibited cell-culture-derived HCV (HCVcc) entry into hepatoma cell lines as well as primary human hepatocytes. The effect was independent of the HCV genotype, and both infection of cells by extracellular virions and cell-to-cell spread were blocked. Pretreatment of cells with EGCG before HCV inoculation did not reduce HCV infection, whereas the application of EGCG during inoculation strongly inhibited HCV infectivity. EGCG inhibits viral attachment to the cell, thus disrupting the initial step of HCV cell entry [Ciesek *et al.*, 2011]. Three different theaflavins, theaflavin (TF1), theaflavin-3'-monogallate (TF2), and theaflavin-3-3'-digallate (TF3), which are major polyphenols from black tea, were tested against HCV in cell culture. The results showed that all theaflavins inhibit HCV infection in a dose-dependent manner in an early step of infection. Results obtained with HCV pseudo-typed virions confirmed their activity on HCV entry and demonstrated their pan-genotypic action. No effect on HCV replication was observed by using HCV replicon. Investigation on the mechanism of action of black tea theaflavins showed that they can act directly on the virus particle and are able to inhibit cell-to-cell spread. Combination study with inhibitors most widely used in anti-HCV treatment regimen demonstrated that TF3 exerts additive effect [Chowdhury *et al.*, 2018].

Acetaminophen [N-acetyl-p-aminophenol (APAP)] has been widely used as an over-the-counter anti-pyretic and analgesic drug since 1955. APAP overdose is a common cause of acute hepatic failure and mortality. In an investigation, mice were given a toxic dose of APAP (0.75 mg/g) by oral gavage. Animals developed profound upregulation of inflammatory markers, TNFα and Serum Amyloid A (SAA) release, as well as Cox-2 activities and Bcl-2 production. The inflammatory markers caused extensive centrilobular apoptosis, necrosis, and severe infiltration of leukocytes accompanied with generation of ROS and depletion of hepatic GSH concentration. GTP supplementation in the diet prior to APAP injection significantly improved the concentration of hepatic GSH, attenuated inflammatory markers and liver lesions, and downregulated Cox-2 and Bcl-2 expression. In addition, GTP normalized pathologically elevated hepatic enzyme activity of alanine aminotransferase (ALT) released by damaged hepatocytes and protected against liver injury. Therefore, GTP-attenuated hepatotoxicity through normalizing antioxidants, inflammatory markers, and Cox-2 and Bcl-2 activation [Oz & Ebersole, 2008], suggesting a potential for GrTP additives protecting against APAP toxicity.

Non-alcoholic fatty liver disease (NASH) is manifested with obesity and other complications with severe life-threatening consequences. A double-blind, randomized clinical trial was reported in NASH patients with diagnostic ultrasonography symptoms and elevated hepatic enzymes, glutamic pyruvic transaminase (GPT or ALT) > 31 mg/dL and glutamic oxaloacetic transaminase (GOT or AST) > 41 mg/dL. Subjects were given green tea extract (500 mg tablet/day) or placebo for 90 days. Green tea significantly decreased hepatic enzymes, ALT and AST, compared to placebo ($p < 0.001$) [Pezeshki *et al.*, 2016]. EGCG (0.05 mg/g/day) oral gavage regulated hepatic mitochondrial respiratory cascades and improved lipid metabolism and insulin sensitivity in obese mice. EGCG increased energy expenditure and prevented oxidation of lipid substrates stimulated by mitochondria and hepatic steatosis in this obesity model. EGCG is reported to specifically inhibit activated hepatic stellate cells by upregulating *de novo* biosynthesis of GSH [Fu *et al.*, 2008]. Further, GTP is reported to protect against NASH by decreasing hepatic steatosis and NF-κB activation in a model on a high fat diet given for eight weeks.

The above research results show the benefits of GTPs in liver complications and its application in the treatment of hepatitis. These results also indicate theaflavins in black tea are new inhibitors of HCV entry and hold promise for developing in therapeutic arsenal for HCV infection.

6. Neurodegenerative disorders and green tea polyphenols

In the previous Chapters 5, 6, and 7, we discussed a lot about the therapeutic effect of tea polyphenols on neurodegenerative diseases. Here, we will briefly discuss the inhibitory effect of tea polyphenols on inflammation of neurodegenerative diseases. Neurodegeneration is related to neuro-inflammation of the central nervous system (CNS). Microglia are primary immune cells which release pro-inflammatory cytokines (e.g., TNFα) and neurotoxins in the brain and contribute to neuro-inflammation. EGCG was reported to protect neuronal cells from microglia-induced cytotoxicity and to suppress amyloid β-induced TNFα release. In a rat model for Parkinson's, green tea extract or EGCG reversed pathological and behavioral modifications, demonstrating neuroprotection by decreasing rotational and increased locomotor activities. Additionally, green tea extracts and EGCG improved cognitive dysfunction by antioxidant and anti-inflammatory properties [Bitu *et al.*, 2015]. In a double blinded, randomized trial, daily consumption of 2000 mg green tea powder (containing 220 mg of catechins) for 12 months did not significantly improve cognitive function in elderly Japanese (nursing home) participants. However, levels of markers for oxidative stress, malondialdehyde-modified low-density lipoprotein, were significantly lower in the green tea group ($p = 0.04$) compared to those in the placebo arm [Ide *et al.*, 2016]. Stroke is a major cerebrovascular disease which results in disability and mortality. EGCG has anti-angiogenic properties and a possible preventive effect against ischemic stroke is via the nuclear factor erythroid 2-related factor 2 (Nrf2) signaling pathway. EGCG therapy for the acute phase of ischemic stroke has been reported to promote angiogenesis in a mouse model of transient middle cerebral artery occlusion (MCAO), conceivably by upregulating the Nrf2 signaling pathway [Pang *et al.*, 2016]. Additionally, EGCG was shown to augment proliferation and differentiation of neural progenitor cells (NPCs) isolated from the ipsilateral subventricular zone with subsequent spontaneous recovery after ischemic stroke [Zhang *et al.*, 2017]. Taken together, green tea and EGCG may exert a beneficial effect on neurogenesis, stroke recovery, and prevention. Diabetic retinopathy is a recurrent complication of diabetes (type 1, type 2) which results in increased inflammation, oxidative stress, and vascular dysfunction. The inflammation and neurodegeneration may occur even before the development of clinical signs of diabetes. During the process of diabetes, the retina triggering pro-inflammatory signaling pathways becomes chronically activated, leading to retinal neurodegeneration and the loss of vision [Arroba & Valverde, 2017]. In a case-control clinical trial, 100 patients with diabetic retinopathy were recruited along with 100 age- and sex-matched diabetic controls without retinopathy in China. Diabetic retinopathy was confirmed from retinal photographs and the pattern of green tea consumption was collected using a face-to-face interview. The odds ratio for green tea consumption for diabetic retinopathy patients was 0.49 (95% CI: 0.26–0.90).

The above results show that tea polyphenols have an inhibitory effect on nervous system (CNS) neuritis. Green tea and EGCG have preventive and therapeutic effects on neurological diseases caused by neuro-inflammation.

13.4.3 *EGCG inhibits heavy metal-induced inflammation and apoptosis through suppression of oxidative stress*

Studies have shown that heavy metal poisoning can cause oxidative stress injury, inflammation, and cell apoptosis, and tea polyphenols can inhibit these three problems. Green tea polyphenol was demonstrated to reduce heavy metal toxicity in such cells and tissues as testis, liver, kidney, and neural cells. Several protective mechanisms that seem to play a pivotal role in EGCG-induced effects, including reactive oxygen species scavenging, heavy metals chelation, activation of nuclear factor erythroid 2-related factor 2 (Nrf2), anti-inflammatory effects, and protection of mitochondria, are described. However, some studies, especially *in vitro* experiments, reported potentiation of harmful heavy metals actions in the presence of EGCG. The adverse impact of EGCG on heavy metals toxicity may be explained by events such as auto-oxidation of EGCG, EGCG-mediated iron (Fe^{3+}) reduction, depletion of intracellular glutathione (GSH) levels, and disruption of mitochondrial functions [Zwolak, 2021].

1. EGCG inhibits arsenic-induced inflammation and apoptosis through suppression of oxidative stress

Arsenic is the most common environmental toxins and toxicants to humans. Arsenic toxins and toxicants may impact human health at the molecular (DNA, RNA, or protein), organelle (mitochondria, lysosome, or membranes), cellular (growth inhibition or cell death), tissue, organ, and systemic levels. Formation of reactive radicals, lipid peroxidation, inflammation, genotoxicity, hepatotoxicity, embryo-toxicity, neurological alterations, apoptosis, and carcinogenic events are some of the mechanisms mediating the toxic effects of the environmental toxins and toxicants. Green tea, the non-oxidized and non-fermented form of tea that contains several polyphenols, including green tea catechins, exhibits protective effects against arsenic toxins and toxicants in preclinical studies and, to a much-limited extent, in clinical trials. The protective effects are collectively mediated by antioxidant, anti-inflammatory, anti-mutagenic, hepatoprotective and neuroprotective, and anti-carcinogenic activities. In addition, green tea modulates signaling pathway including NF-κB and ERK pathways, preserves mitochondrial membrane potential, inhibits caspase-3 activity, downregulates pro-apoptotic proteins, and induces the phase II detoxifying pathway. EGCG, the main and active polyphenolic catechin present in green tea, has shown potent antioxidant, anti-apoptotic, and anti-inflammatory activity *in vivo* and *in vitro* [Chen *et al.*, 2017]. This study investigated the protective effects of EGCG against arsenic-induced inflammation and immune-toxicity in mice. The results showed that arsenic treatment significantly increased oxidative stress levels (as indicated by catalase, malonyldialdehyde, superoxide dismutase, glutathione, and reactive oxygen species), increased levels of inflammatory cytokines, and promoted apoptosis. Arsenic exposure increased the relative frequency of the $CD8^+$ (Tc) cell subpopulation (from 2.8% to 18.9%) and decreased the frequency of $CD4^+$ (Th) cells (from 5.2% to 2.7%). Arsenic exposure also significantly decreased the frequency of T (CD3) (from 32.5% to 19.2%) and B(CD19) cells (from 55.1% to 32.5%). All these effects induced by $NaAsO_2$ were attenuated by EGCG. These results indicate that EGCG attenuates not only $NaAsO_2$-induced immunosuppression but also inflammation and apoptosis [Yu *et al.*, 2017].

2. EGCG attenuates cadmium-induced chronic renal injury and fibrosis

Cadmium (Cd) pollution is a serious environmental problem. Kidney is a main target organ of Cd toxicity. This study was undertaken to investigate the potential protective effects of EGCG against chronic renal injury and fibrosis induced by $CdCl_2$. Rat model was induced by exposing to 250 mg/L $CdCl_2$ through drinking water. It was found that EGCG ameliorated the $CdCl_2$-induced renal injury, inhibited the level of oxidative stress, normalized renal enzymatic antioxidant status and E-cadherin level, as well as attenuated the over generation of TGF-β1, pp-Smad3, vimentin, and α-SMA. EGCG also decreased the production of miR-21 and miR-192, and enhanced the levels of miR-29a/b/c. These results showed that EGCG could attenuate Cd-induced chronic renal injury [Chen *et al.*, 2016]. Owing to the ability of tea polyphenols to bind ions of Cd and the strong anti-oxidative potential of these compounds, as well as their abundance in dietary products, it seems to be of high importance to consider the possibility of using tea polyphenols as potential preventive and therapeutic agents against Cd renal and hepatotoxicity, determined by its strong pro-oxidative properties.

The above research shows that heavy metals such as arsenic and cadmium poisoning can cause oxidative stress damage, inflammation, and apoptosis. Tea polyphenols can inhibit oxidative stress injury, inflammation, and apoptosis through NF-κ Signal pathways, including B and ERK pathways, protect mitochondrial membrane potential, inhibit caspase-3 activity, downregulate apoptosis promoting proteins, and induce phase II detoxification pathway.

13.5 Tea polyphenols can improve immunity

A large number of studies have shown that tea polyphenols can enhance the body's immunity and improve the body's immune system. Health-promoting multi-functionality has been well illustrated for tea polyphenols, especially anti-inflammatory, antioxidant, and anticancer effects. As natural immune modulators, tea polyphenols could control and defeat disorders affecting immune system through up- or downregulating immune responses without undesired adverse effects.

13.5.1 *Immune potentiating effects of tea polyphenols*

A large number of studies have shown that tea polyphenols can potentiate cellular-immune, humoral immunity, human microbiota, and corresponding immunological implications, immune response, and reduce immune system damage by several signal pathways.

1. Effects of green tea polyphenols on cellular immunity

Cellular immunity is also called cell-mediated immunity. Cellular immunity refers to the immune response mediated by T cells, that is, after stimulated by antigens, T cells differentiate, proliferate, and transform into sensitized T cells, which directly kill antigens and synergistically kill cytokines released by sensitized T cells. T cell-mediated immune response is characterized by inflammatory response dominated by monocyte infiltration and/or specific cytotoxicity. Cellular immunity should also include primitive phagocytosis and natural killer (NK) cell-mediated cytotoxicity. Cellular immunity is the most effective defense response to eliminate intracellular parasitic microorganisms, and it is also an effective means to reject allografts or tumor antigens. T cells could release various cytokines, stimulate B cell activation and proliferation, and improve immune response. The CD4$^+$ T cells participate in recognizing antigens presented by MHC type II (exogenous Ag) and CD8$^+$ T cells participate in the recognition of antigens presented by MHC type I (endogenous Ag). The ratio of CD4$^+$/CD8$^+$ (<1) refers to a sign of immunodeficiency [Lee & Wan, 2000]. Dietary GTPs enhance defense ability of the host against exogenous infection through promoting immune cell proliferation, activating T lymphocytes, elevating the percentage of CD4$^+$ T cell, elevating the ratio of CD4$^+$/CD8$^+$, improving T lymphocyte transformation (LTT), and recovering cells from immune damage caused by oxidative stress in piglets [Deng *et al.*, 2010]. Much of the cancer chemopreventive properties of green tea are mediated by EGCG that induces apoptosis and promotes cell growth arrest, by altering the expression of cell cycle regulatory proteins, activating killer caspases, and suppressing nuclear factor kappa-B activation. Besides, it regulates and promotes IL-23 dependent DNA repair and stimulates cytotoxic T cells activities in a tumor microenvironment. It also blocks carcinogenesis by modulating the signal transduction pathways involved in cell proliferation, transformation, inflammation, and metastasis [Butt & Sultan, 2009].

The above results show that tea polyphenols can promote the proliferation of immune cells, activate T lymphocytes, improve T lymphocyte transformation, make cells recover from the immune damage caused by oxidative stress, enhance the host's defense against exogenous infection, and improve the killing ability for foreign invading microorganisms and tumor cells.

2. Effects of green tea polyphenols on humoral immunity

Humoral immunity is an immune mechanism in which plasma cells produce antibodies to achieve the purpose of protection. B cells are responsible for humoral immunity. The relative molecular weight of antigens of humoral immunity is more than 10,000, consisting mostly of proteins and polysaccharides. GTPs and their derived substances could stimulate B-cell proliferation and antibody production effectively. In BALB/c mice, tea polyphenols exerted stimulatory effects on humoral immune response through increasing the number of antibody-secreted cells in spleen and significantly increasing the immunoglobulin M (IgM)-mediated and IgG-mediated immune response to non-particulate antigen bovine serum albumin (BSA) and particulate antigen sheep erythrocyte (SRBCs) in a dose-dependent manner. In particular, splenocytes reached the highest ($344 \pm 10/106$) on day 14 post-immunization ($p < 0.001$) [Khna *et al.*, 2016]. When tea polyphenols were used, the mice showed a sharp increase in IgM and IgG antibodies on day 14 and day 21, respectively. In addition, GTPs significantly reduced total IgG and type II collagen-specific IgG levels in serum and arthritic joints, as well as the neutral endopeptidase activity. *In vitro*, EGCG could strongly enhance the direct plaque-forming cell (PFC) response to sheep red blood cells (SRBC) and show strong mitogenic activity towards mouse splenic B-cells. In particular, the galloyl group on EGCG was responsible for enhancement. ECG, EGCG, and theaflavin digallate (TFDG) displayed significant enhancement of the spontaneous proliferation of B-cells, though with quite different potencies. In healthy Wuchang bream juveniles, dietary GTPs could elevate the content and mRNA levels of splenic IL-1β, TNF-α, and IgM [Hu *et al.*, 1992].

The above results show that tea polyphenols can stimulate the humoral immune response by increasing the number of antibody-secreting cells in the spleen and increase the immune response to non-granular antigen and granular antigen mediated by IgM and IgG.

3. Benefits of green tea polyphenols on human microbiota and corresponding immunology

The microbiota has been widely considered to affect a variety of physiological processes, such as metabolism, neurotransmission, circulation, and immunity. Specifically, the microbiota shapes the host's immune system by regulating local and systemic immune responses. For example, beneficial intestinal microorganisms promote the antitumor activity of immunotherapeutic drugs through immune regulation, including immune checkpoint inhibitors. In view of the close relationship between microbiota and cancer, microbiological manipulation is being widely used as a promising strategy for the treatment and prevention of cancer. Green tea polyphenols could strongly reduce the body fat content, as well as hepatic triacylglycerol and cholesterol accumulation. In particular, the reduction was negatively correlated to the amount of *Akkermansia* and the total amount of intestinal bacteria. A dysfunctional gut microbiota might participate in the pathogenesis of type 2 diabetes. In C57BL/6J mice, green tea powder in combination with *Lactobacillus plantarum* could promote growth of *Lactobacillus* in the intestine, improve the diversity of intestinal bacteria, and attenuate high fat diet-induced inflammation [Axling *et al.*, 2010]. In volunteers who did not usually consume green tea, green tea consumption acted as a prebiotic and improved the colon environment by increasing the proportion of the *Bifidobacterium* species. Fermented green tea extract restored the changes in gut microbiota composition, including the ratios of *Firmicutes/Bacteroidetes* and *Bacteroides/Prevotella*, which is closely related with the development of obesity and insulin resistance. In has been speculated that the fermented green tea extract improved obesity and other associated symptoms through modulating composition of gut microbiota, as mRNA expression levels of lipogenic and inflammatory genes were significantly downregulated in the white adipose tissue of mice [Dae-Bang *et al.*, 2015]. Based on the high-throughput MiSeq sequencing and multivariate statistical analysis, green tea infusion consumption substantially increased diversity and altered the structure of gut microbiota in high-fat diet-induced obese C57BL/6J mice, which further exerted anti-obesity and anti-inflammatory activities. Diversity of the total bacterial community reached the maximum after GTP treatment for almost three weeks and then decreased when the mice were fed without GTP. In particular, the relative abundance of *Bacteroidetes* increased from 0.56 ± 0.06 (first week) to 0.60 ± 0.05 (third week), but *Firmicutes* decreased from 0.42 ± 0.06 (first week) to 0.37 ± 0.02 (third week). Interestingly, *Bacteroidetes* and *Proteobacteria* still increased, but *Firmicutes* decreased even when the mice were fed without GTP (fourth week) [Guo *et al.*, 2017]. GTP could benefit the stability of certain gut microbiota in an environment-triggered microbial imbalance situation, providing prebiotic-like activity contributing to anti-obesity and anti-inflammatory effects. GTPs could boost mammal energy conversion by modulating gut-microbial community structure, gene orthologs, and metabolic pathways. Following the increase of beneficial microbials in families *Clostridia*, *Ruminococcaceae*, *Lachnospiraceae*, and *Bacteroidaceae*, metabolic modulation could also have been achieved through enriching many gene orthologs. In rats, GTPs could enhance energy conversion through boosting mitochondrial tricarboxylic acid cycle and urea cycle of gut-microbiota. Remarkable changes of 39 metabolites in the mitochondrial tricarboxylic acid cycle and urea cycle were observed, which showed significant dose- and time-dependencies on the GTPs treatment (0–1.5% wt/vol) [Zhou *et al.*, 2020].

The above research results show that tea polyphenols can increase the number of beneficial microorganisms in a variety of beneficial human microbiota, such as *Bifidobacteria*, *Clostridaceae*, *Ruminal coccaceae*, *Lacrimaceae*, and *Bacteroideae*, restore the changes in the composition of intestinal microbiota, and thus increase human health.

4. Benefits of green tea polyphenols EGCG on immune response

Over-immunity will cause immune damage to the body. A study investigated the toxic effects of dietary copper and EGCG on bioaccumulation, antioxidant enzyme, and immune response of Korean bullhead, *Pseudobagrus fulvidraco* (*P. fulvidraco*). In this study, *P. fulvidraco* were exposed for four weeks to dietary Cu concentration of 0 mg Cu kg-1 (control), 700 mg Cu kg-1, 900 mg Cu kg-1, and 1100 mg Cu kg-1 dry feed to establish maximum tolerable levels of dietary Cu. All fish were then fed the dietary EGCG concentration of 100 mg EGCG kg/l and 500 mg

EGCG kg/l dry feed for a further two weeks to assess recovery. The Cu exposure induced a significant accumulation in the intestine, liver, and gill tissues, and the highest accumulation was observed in intestinal tissues (17–34 fold), but dietary EGCG exposure decreased (about 0.8-fold) Cu concentration in each tissue ($p < 0.05$). In antioxidant enzymes, SOD and CAT significantly increased by approximately 1.6-fold by dietary Cu exposure in the liver and gill tissue, respectively, but dietary EGCG exposure decreased SOD and CAT by about 1.1-fold, respectively ($p < 0.05$). For immune responses, lysozyme and phagocytosis in the blood significantly were decreased by approximately 1.5-fold, respectively, by dietary Cu exposure, but dietary EGCG exposure increased lysozyme and phagocytosis by about 1.1-fold, respectively ($p < 0.05$). During recovery period, bioaccumulation, antioxidant enzymes (SOD and CAT activity), and immune response (lysozyme and phagocytosis activity) tended to alleviate the significant changes by Cu exposure, and the tendency to return normal state was observed in high level of EGCG [Lee *et al.*, 2021]. Another study investigated that EGCG affects the antioxidant and immune defense of the rainbow trout, *Oncorhynchus mykiss*. EGCG was compared with vitamin E in terms of its effects on antioxidant defense and immune response of rainbow trout, by means of a feeding trial of eight weeks. Observation of tissue levels indicated that the high amount of EGCG helped to increase the availability of the lipid-soluble antioxidant vitamin E. The lower levels of lipid hydroperoxide in the liver of fish that were fed the higher amount of EGCG, suggested that it was an effective antioxidant. Considering the immune indices, EGCG and vitamin E at 100 mg had identical capabilities in improving phagocytic activity and controlling hydrogen peroxide production by leucocytes. However, EGCG could possibly be more effective at enhancing serum lysozyme activity and the alternative complement activity [Thawonsuwan *et al.*, 2010].

The above results show that tea polyphenol, EGCG, can reduce the stress-related effects immune dysfunction and tissue damage caused by copper toxicity and *Acinetobacter* infection.

5. Protective effect of EGCG on the immune function

The immune function will also be damaged by different factors, especially with the increase of age, the immune function will gradually decline and be vulnerable to various diseases, so it needs good protection. Studies have found that tea polyphenols can protect the immune function.

Dendritic cells (DCs) are professional antigen-presenting cells, capable of priming naive T cells, and play the key roles in the activation of T-cell-mediated immune responses. A paper studied the protective effect of EGCG on the immune function of DCs after ultraviolet B irradiation (UVB) and its underlying mechanisms. The monocytes were isolated from peripheral blood and cultivated into DCs with cytokines, such as granulocyte-macrophage colony-stimulating factor (GM-CSF) and interleukin (IL)-4. DCs were harvested after cultivation for seven days and subjected to irradiation with different dosages of UVB. Then, 200 mg/mL of EGCG was added in certain groups 24 hours before irradiation. DCs treated with only UVB or treated with both UVB and EGCG were co-cultured with lymphocytes, and mono-nuclear cell direct cytotoxicity was used to detect the ability of DCs to stimulate proliferation of lymphocytes. UVB irradiation was able to inhibit the ability of DCs to stimulate the proliferation of lymphocytes and surface expressions of CD80, CD86, HLA-DR, and CD40 on DCs in a dose-dependent manner. The inhibition rate of DCs was improved to some extent after treatment with 200 mg/mL of EGCG. UVB showed no significant influence on the secretion of IL-10 and IL-12 from DCs, while EGCG was able to downregulate the secretion level of IL-12 and upregulate that of IL-10 [Jin *et al.*, 2009]. However, another paper found that EGCG induced apoptosis and affected the phenotype of the developing DCs. The expressions of CD83, CD80, CD11c, and MHC class II, which are molecules essential for antigen presentation by DCs, were downregulated by EGCG. EGCG also suppressed the endocytotic ability of immature DCs, whereas dexamethasone-treated DCs had higher endocytotic ability than control DCs. Most importantly, mature DCs treated with EGCG inhibited stimulatory activity toward allogeneic T cells while secreting high amounts of IL-10. EGCG induces immunosuppressive alterations on human MODCs, both by induction of apoptosis and suppression of cell surface molecules and antigen presentation [Yoneyama *et al.*, 2008].

This study was performed to evaluate the effects of EGCG on lipopolysaccharide (LPS)-induced acute lung injury in a murine model. In the present study, production of TNF-alpha and MIP-2 and activation of extracellular signal-regulated kinases (ERK) 1/2, c-Jun amino terminal kinases (JNK), and p38 in RAW264.7 cells were measured. EGCG inhibited the production of TNF-alpha and MIP-2, and attenuated phosphorylation levels of ERK1/2 and JNK, but not p38 in RAW264.7 cells stimulated with LPS. Additionally, EGCG attenuated the production of TNF-alpha and MIP-2, and the phosphorylation of ERK1/2 and JNK in mice administered with LPS intratracheally. The results showed that EGCG attenuated LPS-induced lung injury by suppression of the MIP-2 and TNF-alpha production, and ERK1/2 and JNK activation in macrophage stimulated with LPS [Bae *et al.*, 2010]. EGCG has exhibited potential activity against *A. baumannii in vitro*. The aim of this study was to determine if EGCG could be used for pretreating stress-related effects, liver damage, and immune dysfunction caused by *A. baumannii* infection *in vivo*. Pretreatment with EGCG in the murine pneumonia model markedly reduced stress hormones, oxidative metabolites, and pro-inflammatory cytokine production. EGCG also increased the immune function by increasing the levels of secretory immunoglobulin A, T cells, and neutrophils after infection. Secretory immunoglobulin A forms an important immune protective layer on the surface of intestinal mucosa, which is the main effector of intestinal immune barrier. Moreover, pretreatment with EGCG significantly decreased the liver damage by inhibiting the levels of transaminases, oxidative stress metabolites, and cytokines, while maintaining the normal activity of CYP450 enzymes in the liver [Yan *et al.*, 2012b].

The above results show that tea polyphenol EGCG not only reduce the stress-related effects immune dysfunction, but also can protect immune functions. However, there are also reports that EGCG impairs immune function, so further research is needed.

6. Signal pathways for immune potentiating effects of tea polyphenols

GTPs and their derivatives act through stimulating multiple Toll-like receptor (TLR) signaling pathways in human and could effectively inhibit proliferation of murine lymphocytes [Wilasrusmee *et al.*, 2002]. Furthermore, EGCG also inhibited NF-κB activation induced by TLR4 and TLR2 agonists. EGCG could prevent the production of interferon gamma (IFNγ) by SEB (Staphylococcus aureus enterotoxin B)-stimulated peripheral blood mononuclear cells [Watson *et al.*, 2005]. EGCG pretreatment significantly delayed and reduced mortality through regulating expression of innate immune-related genes, such as immune deficiency (IMD), Phenol oxidase (proPO), QM, myosin, rhodopsin (Rho), Rab7, p53, TNF-α, MAPK, and NOS [Wang *et al.*, 2018]. EGCG could serve as an enhancer of immune parameters (total hemocyte count, phenoloxidase, and superoxide dismutase activities) and an inhibitor of apoptosis, finally delaying or even reducing mortality upon pathogen challenge. EGCG (2.5–15 mM) could inhibit T cell division and cycle progression in a dose-dependent manner, and the inhibitory effect was more pronounced in CD4$^+$ T cells than CD8$^+$ T cells. Compared with CD8$^+$ T cells, CD4$^+$ T cells were more responsive to EGCG [Bayer *et al.*, 2008]. At physiologically relevant concentrations (2.5–10 mM/L), EGCG inhibited splenocyte proliferation stimulated by T cell mitogen concanavalin A (Con A) in a dose-dependent manner. Moreover, T cell division and cell cycle progression could also be inhibited by EGCG [Wu *et al.*, 2009]. EGCG can directly suppress T cell proliferation through impaired IL-2 utilization and cell cycle progression. EGCG could also suppress the LPS-induced phenotypic and functional maturation of murine DCs through inhibiting the mitogen-activated protein kinases (MAPK) and NF-κB. In mice, EGCG (0.3%) directly inhibited T cell proliferative response to polyclonal and antigen-specific stimulation. EGCG can dose-dependently inhibit cell division and cell cycle progression and this effect of EGCG was more pronounced in CD4$^+$ than in CD8$^+$ T cells. EGCG directly inhibits T cell proliferative response to both polyclonal and antigen-specific stimulation. CD4$^+$ cells are more responsive to EGCG than CD8$^+$ cells [Pae *et al.*, 2010]. T cells secrete several inflammatory cytokines that play a critical role in the progression of atherosclerosis. In human primary T cells, EGCG (10 μM or 20 μM) tended to inactivate AP-1 DNA-binding activity and decreased INF-γ levels by 31.3% and 34.7%, IL-2 levels by 26.0% and 38.8%, IL-4 levels by 41.5% and 55.9%, as well as TNF-α levels by 23.0% and 37.6%, respectively [Huang *et al.*, 2021].

The above results show that tea polyphenol EGCG can pass through multiple signal pathways, such as TLR, IFNγ, IMD, propo, QM, myosin, rho, Rab7, p53, TNF-α, and MAPK and NOS have the protective effect on the immune system.

13.5.2 *Benefits of tea polyphenols toward immune-related disease*

Immune-related diseases are great harm to the human body and studies have found that tea polyphenols are beneficial to these immune-related diseases. Results show that tea polyphenols can enhance immune function through multiple signal pathways. It has been found that GTPs can prevent many types of chronic diseases if regularly ingested in diet due to antioxidant, anti-cancer, anti-inflammatory, anti-atherogenic, antidiabetic, antibacterial, and antiviral activities. Moreover, GTPs exert a protective effect on autoimmune diseases [Molina *et al.*, 2015]. The immunosuppressive effects of EGCG were tested on anti-CD3 plus anti-CD28-activated primary human T-lymphocytes in culture. EGCG significantly inhibited T-cell proliferation, and this effect was not due to toxicity. IL-2 production was also reduced by these agents, implicating this important T-cell cytokine in proliferation suppression. These results suggest the potential use of EGCG for treating autoimmune and transplant patients [Hushmendy *et al.*, 2009]. Besides genetic or acquired defects in immune tolerance or immune regulatory pathways, molecular mimicry to viral or bacterial proteins and impaired clearance of apoptotic cell materials are all the mechanisms of autoimmunity. Therefore, autoimmunity might be mostly due to copious production of autoantibodies and autoreactive cells. The autoimmune diseases could be classified into systemic and organ specific. Autoimmune disorders may be caused by multiple interactions predisposition, and environmental triggers also contribute a lot to the autoimmune diseases [Hayashi *et al.*, 2004]. GTPs have showed significantly therapeutic potential in a variety of autoimmune diseases.

1. Benefits of tea polyphenols toward Sjogren's syndrome

Sjogren's syndrome (SS), a relatively common auto-immune disease, is characterized by inflammatory cell infiltration and loss of function of the lacrimal and salivary glands, which will result in ocular and oral health problems. Selective inhibition of gland-cell apoptosis, autoantigen expression, and production of pro-inflammatory cytokines are the three potential strategies for ameliorating SS. EGCG (0.2% consumption) protected non-obese diabetes (NOD) mouse submandibular glands from autoimmune-induced inflammation through reducing lymphocyte infiltration in the salivary glands during disease advancing stages, inhibiting apoptotic activity within the lymphocytic infiltrates, decreasing levels of serum total anti-nuclear antibody, and suppressing the expression of cell nuclear proliferation markers, such as proliferating cell nuclear antigen (PCNA) and Ki-67 [Gillespie *et al.*, 2008]. Glandular cells may be important in initiating and sustaining Sjogren's syndrome, and the T-cell-mediated cytotoxicity and autoantibodies are also critical in the loss of gland function. Moreover, glandular epithelial cells contribute a lot to autoimmune process through secreting pro-inflammatory cytokines [Hayashi *et al.*, 2004]. Aberrant expression and translocation of nuclear auto-antigens onto acinar-cell membrane during apoptosis occurred, where they will be exposed to APCs (macrophages and dendritic cells) [Van Woerkom *et al.*, 2004]. Furthermore, the auto-antigens redistributed in glandular cells and formed apoptotic bodies and blebs, around which autoantigen proteins (SS-A/Ro, SS-B/La, Ku, PARP, fodrin, golgins, and NuMA) were clustered as subcellular structures. The structural changes in auto-antigens may also contribute to the altered configuration of autoantigen-cluster, resulting in autoimmune response. Naturally occurring phytochemicals in plant could be used in treating SS-associated disorders, such as GTPs, which could reduce the lymphocytic infiltration of submandibular gland in non-obese diabetic (NOD) mouse model for SS [Hsu *et al.*, 2007]. Sjogren's syndrome can also cause uveitis, increase the thickness of the retina and choroid, leading to dry eyes. Compared to a water-treated experimental autoimmune uveoretinitis (EAU) murine model, GTP attenuated clinical manifestations of uveitis, increased retinal-choroidal thicknesses (RCT) (1.100 ± 0.013 times versus 1.005 ± 0.012 times, $p < 0.001$) and retinal vessel dilation (308.9 ± 6.189 units versus 240.8 units, $p < 0.001$) in a dose-dependent manner [Li *et al.*, 2019].

Above results shown that GTP could partially alleviate Sjogren's syndrome phenotypes and recover visual function in the murine model of EAU. The treatment of GTP and its major component EGCG upregulated Th-17 associated pro-inflammatory genes, such as IL-1β, IL-6, IL-17A, and TNF-α. GTE consumption serves as a potent therapeutic agent and a food supplement to develop alternative treatments against autoimmune Sjogren's syndrome and uveitis.

2. Benefits of tea polyphenols in cutaneous immunity

Human skin could be divided into epidermal, dermal, and hypodermal layers. In particular, the epidermis consists of five layers, including stratum basale, stratum spinosum, stratum granulosum, stratum lucidum, and stratum corneum. Stratum basale, the innermost layer of epidermis, contains rapidly proliferating and differentiating keratinocytes, Merkel cells, and melanocytes. The stratum spinosum layer contains langerhans cells, which serve as part of the immune system. UV light could affect skin biology and immune system, and lead to immunosuppression, premature aging, oxidative stress, and even carcinogenesis. Immunosuppression induced by solar UV radiation (UVR) is a risk factor for melanoma and non-melanoma skin cancers, which could potentially allow undetected dysplastic cells to develop into neoplasms. Notably, UV-induced DNA damage, particularly in terms of the formation of cyclobutane pyrimidine dimers (CPD), serves as critical molecular triggers for UV-induced immunosuppression. As a 70-kDa heterodimer, IL-12 could induce Th1 responses and repair UV-induced DNA damage. The prevention of UVR-induced immunosuppression by IL-12 depended on DNA repair and the induction of nucleotide excision repair enzymes. Topical application of EGCG could effectively prevent UV-induced immunosuppression in rice through IL-12-dependent DNA repair, showing chemo-preventive activity in prevention of photo-carcinogenesis. In particular, EGCG reduced the number of CPDs+ cells and the migration of CPD+ APCs from skin to draining lymph nodes [Meeran *et al.*, 2006]. UV radiation could mediate inflammatory and immunological reactions through activation of receptors, induction of DNA/RNA damage, and production of ROS. UVR could activate multiple signaling cascades, including p38 MAPK, Jun N-terminal kinase, extracellular signal-regulated kinase 1/2, and NF-κB pathways in skin cells. After UV exposure, an early inflammatory event soon takes place, which is characteristic by erythema and redness due to vasodilation of cutaneous blood vessels. UV-induced TNF-α could diminish antigen presentation, reduce immune-surveillance, and initiate immunosuppression [Muthusamy and Piva, 2010]. In mice fed with purified GTP, dose-dependent decrease in UVR-induced immunosuppression was observed, which were performed through contact hypersensitivity response (CHR) to 2,4-dinitrofluorobenzene. Furthermore, the decrease in immunosuppression lasted four weeks, even after the resumption of a normal liquid diet. In UV-irradiated mice, GTP could reduce the migration of CPD positive cells to lymph nodes and improve nucleotide-excision repair mechanisms [Katiyar *et al.*, 2010].

In human and animal studies, GTP show significant protective effects against UV-induced skin damage and immunosuppression. In keratinocytes cells exposed to UV radiation, IL-12 could enhance the nucleotidase activity. DNA repair by topically-applied EGCG might be through an IL-12-dependent mechanism. EGCG could prevent UV-induced immunosuppression by enhancing the levels of IL-12. Moreover, Oral GTP significantly delayed pathogenic leukocyte entry into skin following UV-B treatment.

3. Benefits of tea polyphenols towards obesity-related immune disease

Obesity is a major public health problem and global epidemic associating with comorbidities, such as diabetes, dyslipidemia, hypertension, cardiovascular diseases, and some cancers. Strong relationship existed between adipose tissue and immune cells, as pro-inflammatory cytokines not only increase free fatty acid (FFA) levels, but also induce ROS production. In particular, adipose tissue inflammation play an important role in pathogenesis of diabetes. The energy-rich environment in obesity could damage immune cells present in blood stream and peripheral tissues [Chen *et al.*, 2015]. Due to over-activation of immune cells, obese individuals are more subject to chronic inflammatory diseases, which would further destroy the functionality of immune cells and cause certain inflammatory response [Grant *et al.*, 2013]. The increase in pro-inflammatory cytokines (IL-6, IL-1β, and TNF-α) and leptin released by adipocytes would drive lymphocytes to a Th1 phenotype in WAT. Obesity often provides danger signals mimicking bacterial infection, which will drive a shift in M1 macrophages, CD^{8+}, and CD^{4+} to Th1, Th2, and Th17, respectively. In contrast, the decrease in regulatory T cell (Treg) anti-inflammatory lymphocyte numbers could prevent WAT inflammation and insulin resistance [Chatzigeorgiou *et al.*, 2012].

Tea polyphenols are proposed to function via various mechanisms, the most important of which is related to ROS. These polyphenols exert conflicting dual actions as anti- and pro-oxidants. Their anti-oxidative actions help scavenge ROS and downregulate nuclear factor-κB to produce favorable anti-inflammatory effects. Meanwhile, pro-oxidant actions appear to promote ROS generation leading to the activation of 5'-AMP-activated protein kinase, which modulates different enzymes and factors with health beneficial roles [Ohishi *et al.*, 2021]. Cellular studies demonstrated that these dietary polyphenols reduce viability of adipocytes and proliferation of pre-adipocytes, suppress adipocyte differentiation and triglyceride accumulation, stimulate lipolysis and fatty acid β-oxidation, and reduce inflammation. Concomitantly, the polyphenols modulate signaling pathways including the adenosine-monophosphate-activated protein kinase, peroxisome proliferator activated receptor γ, CCAAT/enhancer binding protein α, peroxisome proliferator activator receptor gamma activator 1-alpha, sirtuin 1, sterol regulatory element binding protein-1c, uncoupling proteins 1 and 2, and NF-κB that regulate adipogenesis, antioxidant, and anti-inflammatory responses [Wang *et al.*, 2014]. Our research found that tea polyphenols can inhibit the inflammation promoting effect of obesity, scavenge the ROS produced, and reduce the activation of immune cells and infiltration into adipose tissue [Yan *et al.*, 2013].

The above results show that tea polyphenols can inhibit the production of ROS and pro-inflammatory cytokines released by adipocytes induced by obesity and avoid damaging the function of immune cells through a series of signal molecules.

4. Green tea EGCG mediates autoimmune encephalomyelitis

Autoimmune diseases are common, disabling immune disorders affecting millions of people. EGCGs have been shown to modulate immune cell functions and improve some autoimmune diseases in animal models. A study determined EGCG's effect on T-cell functions and its application in autoimmune diseases in a series of studies. They first observed that EGCG inhibited CD4$^+$ T-cell expansion induced by polyclonal (mitogens or anti-CD3/CD28) or antigen-specific stimulation. EGCG suppressed expansion and cell cycle progression of naïve CD4$^+$ T by modulating cell cycle-related proteins. EGCG also inhibited naive CD4$^+$ T-cell differentiation into Th1 and Th17 effector subsets by impacting their respective signaling transducers and transcription factors. Using the experimental autoimmune encephalomyelitis (EAE) mice, an animal model for human multiple sclerosis, they found that dietary supplementation with EGCG attenuated the disease's symptoms and pathology. These EGCG-induced changes are associated with findings in the immune and inflammation profiles in lymphoid tissues and the central nervous system: a reduction in proliferation of autoreactive T cells, production of pro-inflammatory cytokines, and Th1 and Th17 subpopulations, and an increase in regulatory T-cell populations. These results suggest that green tea or its active components may have a preventive and therapeutic potential in dealing with T-cell-mediated autoimmune diseases [Wu *et al.*, 2012; Pae & Wu, 2013; Wu, 2016]. Another paper found that EGCG treatment reduced EAE severity and macrophage inflammation in the CNS. Moreover, EAE severity was well correlated with the ratio of M1 to M2 macrophages, and EGCG treatment suppressed M1 macrophage-mediated inflammation in spleen. *In vitro* experiments showed that EGCG inhibited M1 macrophage polarization but promoted M2 macrophage polarization. These effects were likely related to the inhibition of nuclear factor-κB signaling and glycolysis in macrophages by EGCG [Cai *et al.*, 2021].

These results suggest that EGCG may improve T cell-mediated autoimmune diseases and may inhibit macrophage nuclear factor through $CD4^+$ T cell differentiation into Th1 and Th17 effector.

13.6 Conclusion and prospects

From the above discussion, we can clearly know that oxidative stress plays an important role in immune and inflammatory processes. Oxidative stress is not only an effective tool for immune and inflammatory invasion of microorganisms, but also can cause side effects of inflammation and abnormal immune cell and tissue damage. ROS and RNS produced by the respiratory burst of leukemia, especially macrophages, play a very important role in causing oxidative stress injury. Macrophages also play a key role in promoting T cell activation and subsequent proliferation.

Green tea is one of the most popular beverages around the world. Epidemiological studies have suggested that the consumption of green tea is associated with reduced risk of infections, and immune-regulation. Green tea polyphenols are useful in preventing onset and severity of arthritis. In particular, GTP possess significantly anti-inflammatory properties and is effective in inhibiting autoimmune diseases. Dietary GTP could enhance immune responses to alleviate oxidative stress and damage caused by ammonia, showing potentials as a preventive or therapeutic measure. The increased levels of glutathione is a critical mechanism underlying the beneficial effects of GTP against some inflammatory disorders, and other mechanisms may also have a significant contribution. GTP could not only divert immune types, but also reduce the burden of T lymphocytes under oxidative stress. The great potential of GTP to stimulate T lymphocytes immune response against exogenous antigens makes them ideal as vaccines against pathogens cells. Oral therapy with GTP could effectively prolong transplant survival and inhibit transplant-reactive T cell function *in vivo*, but the mechanisms need to be further discussed.

EGCG possess effectively hepatic- and immune-protective properties against restraint stress through the combination of anti-oxidant, anti-inflammatory, and immunomodulatory activities. EGCG could significantly reduce the release of stress hormones, which weakens the restraint stress response. Accessory cells (mainly macrophages and DC) play critical roles in facilitating T cell activation and consequent proliferation, EGCG could affect the function of accessory cells present in total splenocytes. EGCG could significantly weaken restraint stress response via preventing the release of H_2O_2, NOS, and 8-isoprostane, reducing the levels of IL-1β, IL-2, and IL-6, normalizing the level of cytochrome P450, 1A2, 2D22, 2E1, and 3A11, relieving the inhibition status of T cells subsets in serum and IgA in BALF, as well as improving hepatic damage through decreasing the serum levels of alanine amino-transaminase and aspartate transaminase. EGCG could significantly delay the onset of autoimmune diabetes and effectively protect salivary gland cells from autoimmune-induced damage at multiple levels, inferring that EGCG could be used to delay and manage autoimmune disorders. In addition, EGCG could inhibit NF-κB activation, activator of transcription 1-dependent cellular events, T cell proliferation, and cytokine production. Moreover, EGCG could interfere with maturation and functions of DCs, as well as induce regulatory T cell differentiation. In particular, the phenol rings of EGCG, comprising phenyl and hydroxyl group structures, possess significant anti-inflammatory, immunomodulatory, and antioxidant properties. Oral consumption of EGCG could protect secretory cells against autoimmune-induced damage and sustain cellular function in secretory glands via preventing cells from signals for proliferation or apoptosis. Compared with single compounds, green tea extract shows greater health promotion potential due to the synergistic effect of different compounds.

Although, a large number of studies have been carried out *in vitro* and *in vivo*, the exact mechanism is still unclear. Therefore, more careful design research is needed to further clarify the immune enhancement characteristics of cellular and humoral immune responses. In addition, it is also necessary to clarify the exact mechanism of action, the optimal therapeutic dose, the duration of treatment and the role of green tea polyphenols *in vitro* and *in vivo* model systems, especially in the treatment of inflammatory diseases. Due to its safety and immune regulation, green tea may provide a new source of chemo-preventive or therapeutic drugs for various chronic diseases. As an effective antioxidant to eliminate cytotoxic ROS, GTP can protect immune cells and regulate the immune response of animals to oxidative stress. The effect of GTPs on transplantation reactive T cell immunity shows that oral green tea can be used as an effective adjuvant therapy to prevent human transplantation rejection. However, further research to evaluate the immunosuppressive effect of GTP in depth is conducive to clinical trials aimed at inhibiting T-cell immune response to human transplantation antigens.

References

Abboud PA, Hake PW, Burroughs TJ, Odoms K, O'Connor M, Mangeshkar P, Wong HR, Zingarelli B. (2008) Therapeutic effect of epigallocatechin-3-gallate in a mouse model of colitis. *Eur J Pharmacol*, **579**(1–3), 411–417.

Agita A, Alsagaff MT. (2017) Inflammation, immunity, and hypertension. *Acta Med Indones*, **49**(2), 158–165.

Agostini M, Marco BD, Nocentini G, *et al.* (2002) Oxidative stress and apoptosis in immune diseases. *Int J Immunopathol Pharmacol*, **15**(3), 157–164.

Aksu U, Demirci C, Ince C. (2011) The pathogenesis of acute kidney injury and the toxic triangle of oxygen, reactive oxygen species and nitric oxide. *Contrib Nephrol*, **174**, 119–128.

Arroba AI, Valverde AM. (2017) Modulation of microglia in the retina: new insights into diabetic retinopathy. *Acta Diabetol*, **54**, 527–533.

Axford JS. (1987) Reduced B-cell galactosyltransferase activity in rhematoid arthrits. *Lanecet II*, 1486–1492.

Axling U, Olsson C, Xu J, Fernandez C, Larsson S, Ström K, Ahrnė S, Holm C, Molin G, Bae HB, Li M, Kim JP, Kim SJ, Jeong CW, Lee HG, Kim WM, Kim HS, Kwak SH. (2010) The effect of epigallocatechin gallate on lipopolysaccharide-induced acute lung injury in a murine model. *Inflammation*, **233**(2), 82–91.

Behl T, Upadhyay T, Singh S, Chigurupati S, Alsubayiel AM, Mani V, Vargas-De-La-Cruz C, Uivarosan D, Bustea C, Sava C, Stoicescu M, Radu AF, Bungau SG. (2021) Polyphenols targeting MAPK mediated oxidative stress and inflammation in rheumatoid arthritis. *Molecules*, **26**(21), 6570.

Bae HB, Li M, Kim JP, Kim SJ, Jeong CW, Lee HG, Kim WM, Kim HS, Kwak SH. (2010) The effect of epigallocatechin gallate on lipopolysaccharide-induced acute lung injury in a murine model. *Inflammation*, **33**(2), 82–91.

Bayer J, Gomer A, Demir Y, Amano H, Kish DD, Fairchild R, Heeger PS. (200) Effects of green tea polyphenols on murine transplant-reactive T cell immunity. *Clin Immunol*, **110**, 100–108.

Beckman CH. (2000) Phenolic-storing cells: keys to programmed cell death and periderm formation in wilt disease resistance and in general defense responses in plants? *Physiol Mol Plant Pathol*, **57**, 101–110.

Biemond P. (1986) Intraarticular ferritin-bound iron in rheumatoid arthritis. A factor that increases oxygen free radical-induced tissue destruction. *Arthritis Rheum*, **29**, 1187–1192.

Bing X, Xuelei L, Wanwei D, Linlang L, Keyan C. (2017) EGCG maintains Th1/Th2 balance and mitigates ulcerative colitis induced by dextran sulfate sodium through TLR4/MyD88/NF-kappaB signaling pathway in rats. *Can J Gastroenterol Hepatol*, **2017**, 3057268.

Bitu Pinto N, da Silva Alexandre B, Neves KR, Silva AH, Leal LK, Viana GS. (2015) Neuroprotective properties of the standardized extract from Camellia sinensis (Green Tea) and Its main bioactive components, epicatechin and epigallocatechin gallate, in the 6-OHDA model of Parkinson's Disease. *Evid Based Complement Altern Med*, **2015**, 161092.

Bowern N. (1984) Inhibition of autoimmune neuropathological process by treatment with an iron-chelating agent. *J Exp Med*, **160**, 1532–1542.

Butt MS, Sultan MT. (2009) Green tea: nature's defense against malignancies. *Crit Rev Food Sci Nutr*, **49**(5), 463–473.

Camp OG, Bai D, Gonullu DC, Nayak N, Abu-Soud HM. (2021) Melatonin interferes with COVID-19 at several distinct ROS-related steps. *J Inorg Biochem*, **223**, 111546.

Cai F, Liu S, Lei Y, Jin S, Guo Z, Zhu D, Guo X, Zhao H, Niu X, Xi Y, Wang Z, Chen G. (2021) Epigallocatechin-3 gallate regulates macrophage subtypes and immunometabolism to ameliorate experimental autoimmune encephalomyelitis. *Cell Immunol*, **368**, 104421.

Chatzigeorgiou A, Karalis KP, Bornstein SR, Chavakis T. (2012) Lymphocytes in obesityrelated adipose tissue inflammation. *Diabetologia*, **55**, 2583–2592.

Chen J, Du L, Li J, Song H. (2016) Epigallocatechin-3-gallate attenuates cadmium-induced chronic renal injury and fibrosis. *Food Chem Toxicol*, **96**, 70–78.

Cuzzocrea S. (2006) Role of nitric oxide and reactive oxygen species in arthritis. *Curr Pharm Des*, **12**(27), 3551–3570.

Chen L, Mo H, Zhao L, Gao W, Wang S, Cromie MM, Lu C, Wang JS, Shen CL. (2017) Therapeutic protecties of green tea against environmental insults. *J Nutr Biochem*, **40**, 1–13.

Chen S, Akbar SMF, Miyake T, Abe M, Al-Mahtab M, Furukawa S, Bunzo M, Hiasa Y, Onji M. (2015) Diminished immune response to vaccinations in obesity: role of myeloid-derived suppressor and other myeloid cells. *Obes Res Clin Pract*, **9**, 35–44.

Chowdhury P, Sahuc ME, Rouillé Y, Rivière C, Bonneau N, Vandeputte A, Brodin P, Goswami M, Bandyopadhyay T, Dubuisson J, Séron K. (2018) Theaflavins, polyphenols of black tea, inhibit entry of hepatitis C virus in cell culture. *PLoS One*, **13**(11), e0198226.

Ciesek S, von Hahn T, Colpitts CC, Schang LM, Friesland M, Steinmann J, Manns MP, Ott M, Wedemeyer H, Meuleman P, Pietschmann T, Steinmann E. (2011) The green tea polyphenol, epigallocatechin-3-gallate, inhibits hepatitis C virus entry. *Hepatology*, **54**(6), 1947–1955.

Colucci R, Dragoni F, Moretti S. (2015) Oxidative stress and immune system in vitiligo and thyroid diseases. *Oxid Med Cell Longev*, **2015**, 631927.

Dae-Bang S, Jeong HW, Cho D, Lee BJ, Lee JH, Choi JY, Il-Hong B, Sung-Joon L. (2015) Fermented green tea extract alleviates obesity and related complications and alters gut microbiota composition in diet-induced obese mice. *J Med Food*, **18**, 549–556.

Deng Q, Xu J, Yu B, He J, Zhang K, Ding X, Chen D. (2010) Effect of dietary tea polyphenols on growth performance and cell mediated immune response of post-weaning piglets under oxidative stress. *Arch Anim Nutr*, **64**, 12–21.

Ding Y, Zheng Y, Huang J, Peng W, Chen X, Kang X, Zeng Q. (2019) UCP2 ameliorates mitochondrial dysfunction, inflammation, and oxidative stress in lipopolysaccharide-induced acute kidney injury. *Int Immunopharmacol*, **71**, 336–349.

Duan S, Gu L, Wang Y, Zheng R, Lu j, Yin J, Guli L, Ball M. (2009) Regulation of influenza virus-caused oxidative stress by Kegan Liyan oral prescription, as monitored by ascorbyl radical ESR signals. *Am J Chin Med*, **37**(6), 1167–1177.

Fassina G, Buffa A, Benelli R, Varnier OE, Noonan DM, Albini A. (2002) Polyphenolic antioxidant (-)-epigallocatechin-3-gallate from green tea as a candidate anti-HIV agent. *AIDS*, **16**(6), 939–41.

Fernández-Sánchez A, Madrigal-Santillan E, Bautista M, *et al.* (2011) Inflammation, oxidative stress, and obesity. *Int J Mol Sci*, **12**(5), 3117–3132.

Fu Y, Zheng S, Lu SC, Chen A. (2008) Epigallocatechin-3-gallate inhibits growth of activated hepatic stellate cells by enhancing the capacity of glutathione synthesis. *Mol Pharmacol*, **73**, 1465–1473.

Gillespie K, Kodani I, Dickinson DP, Ogbureke K, Camba AM, Wu M, Looney S, Chu TC, Qin H, Bisch F, *et al.* (2008) Effects of oral consumption of the green tea polyphenol EGCG in a murine model for human Sjogren's syndrome, an autoimmune disease. *Life Sci*, **83**, 581–588.

Giordano N. (1984) Increased storage of iron and anaemia in rheumatoid arthritis: usefulness of desferrioxamine. *Br Med J*, **289**, 961–970.

Grant R, Youm YH, Ravussin A, Dixit VD. (2013) Quantification of adipose tissue leukocytosis in obesity. *Methods Mol Biol*, **1040**, 195–209.

Grasemann H, Holguin F. (2021) Oxidative stress and obesity-related asthma. *Paediatr Respir Rev*, **37**, 18–21.

Guo X, Cheng M, Zhang X, Cao J, Wu Z, Weng P. (2017) Green tea polyphenols reduce obesity in high-fat diet-induced mice by modulating intestinal microbiota composition. *Int J Food Sci Tech*, **52**, 1723–1730.

Guzik TJ, Korbut R, Adamek-Guzik T. (2003) Nitric oxide and superoxide in inflammation and immune regulation. *J Physiol Pharmacol*, **54**(4), 469–487.

Henson PM. (1984) Resolution of pulmonary inflammation. *Fed Proc*, **43**, 2799–2805.

Hayashi Y, Arakaki R, Ishimaru N. (2004) Apoptosis and estrogen deficiency in primary Sjögren's syndrome. *Curr Opin Rheumatol*, **16**, 522–526.

Hewitt SD. (1987) Effect of free radical altered IgG on allergic inflammation. *Ann Rheum Dis*, **46**, 866–872.

Hoffsten ST, Gennaro DE, and Meunier PC. (1988) Cytochromical demonstration of constitutive H2O2 production by macrophages in synovial tissue from rats with adjuvant arthritis. *Am J Pathol*, **130**, 120–127.

Hsu S, Dickinson DP, Qin H, Borke J, Ogbureke K, Winger JN, Walsh DS, Bollag WB, Stoppler H, Sharawy M, *et al.* (2007) Green tea polyphenols reduce autoimmune symptoms in a murine model for human Sjogren's syndrome and protect human salivary acinar cells from TNF-alpha-induced cytotoxicity. *Autoimmunity*, **40**, 138–147.

Hu ZQ, Toda M, Okubo S, Hara Y, Shimamura T. (1992) Mitogenic activity of (-)epigallocatechin gallate on B-cells and investigation of its structure function relationship. *Int J Immunopharmacol*, **14**, 1399–1407.

Huang SC, Kao YH, Shih SF, Tsai MC, Lin CS, Chen LW, Chuang YP, Tsui PF, Ho LJ, Lai JH, Chen SJ. (2021) Epigallocatechin-3-gallate exhibits immunomodulatory effects in human primary T cells. *Biochem Biophys Res Commun*, **550**, 70–76.

Hushmendy S, Jayakumar L, Hahn AB, Bhoiwala D, Bhoiwala DL, Crawford DR. (2009) Select phytochemicals suppress human t-lymphocytes and mouse splenocytes suggesting their use in autoimmunity and transplantation. *Nutr Res*, **29**, 568–578.

Ide K, Yamada H, Takuma N, Kawasaki Y, Harada S, Nakase J, Ukawa Y, Sagesaka YM. (2016) Effects of green tea consumption on cognitive dysfunction in an elderly population: a randomized placebo-controlled study. *Nutr J*, **15**, 49.

Jin SL, Zhou BR, Luo D. (2009) Protective effect of epigallocatechin gallate on the immune function of dendritic cells after ultraviolet b irradiation. *J Cosmet Dermatol*, **8**, 174–180.

Ju J, Liu Y, Hong J, Huang MT, Conney AH, Yang CS. (2003) Effects of green tea and high-fat diet on arachidonic acid metabolism and aberrant crypt foci formation in an azoxymethane-induced colon carcinogenesis mouse model. *Nutr Cancer*, 46, 172–178.

Katiyar SK, Bergamo BM, Vyalil PK, Elmets CA. (2001) Green tea polyphenols: DNA photodamage and photoimmunology. *J Photochem Photobiol B*, **65**(2-3), 109–114.

Katiyar SK, Vaid M, Van Steeg H, Meeran SM. (2010) Green tea polyphenols prevent UV-induced immunosuppression by rapid repair of DNA damage and enhancement of nucleotide excision repair genes. *Cancer Prev Res*, **3**, 179–189.

Kemp-Harper BK, Velagic A, Paolocci N, Horowitz JD, Ritchie RH. (2021) Cardiovascular therapeutic potential of the redox siblings, nitric oxide (NO) and nitroxyl (HNO), in the setting of reactive oxygen species dysregulation. *Handb Exp Pharmacol*, **264**, 311–337.

Kharitonov SA, Yates D, Robbins RA, *et al.* (1994) Increased nitric oxide in exhaled air of asthmatics. *Lancet*, **343**, 133–135.

Khan A, Ali NH, Santercole V, Paglietti B, Rubino S, Kazmi SU, Farooqui A. (2016) Camellia sinensis mediated enhancement of humoral immunity to particulate and non-particulate Antigens. *Phytother Res*, **30**, 41–48.

Khan SG, Katiyar SK, Agarwal R, Mukhtar, H. (1992) Enhancement of antioxidant and phase II enzymes by oral feeding of green tea polyphenols in drinking water to SKH-1 hairless mice: possible role in cancer chemoprevention. *Cancer Res*, **52**(14), 4050–4052.

Kvietys PR, Granger DN. (2012) Role of reactive oxygen and nitrogen species in the vascular responses to inflammation. *Free Radic Biol Med*, **52**(3), 556–592.

Lauridsen C. (2019) From oxidative stress to inflammation: redox balance and immune system. *Poult Sci*, **98**(10), 4240–4246.

Lee CYJ, Wan JMF. (2000) Vitamin E supplementation improves cell-mediated immunity and oxidative stress of Asian men and women. *J Nutr*, **130**, 2932–2937.

Lee H, Kim JH, Park HJ, Kang JC. (2021) Toxic effects of dietary copper and EGCG on bioaccumulation, antioxidant enzyme and immune response of Korean bullhead, Pseudobagrus fulvidraco. *Fish Shellfish Immunol*, **111**, 119–126.

Li H-T, Zhao B-L, Hou J-W, Xin W-J. (1996) Two peak kinetic curve of chemiluninencence in phorbol stimulated macrophage. *Biochem Biophys Res Commn*, **311**, 314–320.

Li X-J, Zhao B-L, Hou J-W, Xin W-J. (1990) Active oxygen radicals produced by leukocytes of malignant lymphoma. *Chinese Medical J*, **103**, 899–905.

Li X-J, Yan L-J, Zhao B-L, Xin W-J. (1991) Effects of oxygen radicals on the conformation of sulfhydryl groups on human polymorphonuclear leukocyte membrane. *Cell Biol Intern Report*, **12**, 667–675.

Margutti P, Matarrese P, Conti F, *et al.* (2008) Autoantibodies to the C-terminal subunit of RLIP76 induce oxidative stress and endothelial cell apoptosis in immune-mediated vascular diseases and atherosclerosis. *Blood*, **111**(9), 4559–4570.

Meeran SM, Mantena SK, Katiyar SK. (2006) Prevention of ultraviolet radiation — induced immunosuppression by (-)-epigallocatechin-3-gallate in mice is mediated through interleukin 12-dependent DNA repair. *Clin Cancer Res*, **12**, 2272–2280.

Mishra V, Banga J, Silveyra P. (2018) Oxidative stress and cellular pathways of asthma and inflammation: therapeutic strategies and pharmacological targets. *Pharmacol Ther*, **181**, 169–182.

Molina N, Bolin AP, Otton R. (2015) Green tea polyphenols change the profile of inflammatory cytokine release from lymphocytes of obese and lean rats and protect against oxidative damage. *Int Immunopharmacol*, **28**, 985–996.

Muthusamy V, Piva TJ. (2010) The UV response of the skin: a review of the MAPK, NFκB and TNFα signal transduction pathways. *Arch Dermatol Res*, **302**, 5–17.

Najafzadeh M, Reynolds PD, Baumgartner A, Anderson D. (2009) Flavonoids inhibit the genotoxicity of hydrogen peroxide (H_2O_2) and of the food mutagen 2-amino-3-methylimadazo[4,5-f]-quinoline (IQ) in lymphocytes from patients with inflammatory bowel disease (IBD). *Mutagenesis*, **24**, 405–411.

Ohishi T, Fukutomi R, Shoji Y, Goto S, Isemura M. (2021) The beneficial effects of principal polyphenols from green tea, coffee, wine, and curry on obesity. *Molecules*, **26**(2), 453.

Oz HS, Ebersole JL. (2008) Application of prodrugs to inflammatory diseases of the gut. *Molecules*, **13**(2), 452–474.

Oz HS, Chen T, de Villiers WJ. (2013) Green tea polyphenols and sulfasalazine have parallel anti-inflammatory properties in colitis models. *Front Immunol*, PMID: 23761791.

Oz HS, Ebersole J. (2010) Green tea polyphenols mediate apoptosis in Intestinal Epithelial Cells. *J Cancer Ther*, **1**, 105–113.

Pae M, Ren Z, Meydani M, Shang F, Meydani SN, Wu D. (2010) Epigallocatechin-3-gallate directly suppresses T cell proliferation through impaired IL-2 utilization and cell cycle progression. *J Nutr*, **140**(8), 1509–1515.

Pae M, Wu D. (2013) Immunomodulating effects of epigallocatechin-3-gallate from green tea: mechanisms and applications.

Pang J, Zhang Z, Zheng TZ, Bassig BA, Mao C, Liu X, Zhu Y, Shi K, Ge J, Yang YJ, *et al.* (2016) Green tea consumption and risk of cardiovascular and ischemic related diseases: a meta-analysis. *Int J Cardiol*, **202**, 967–974.

Pezeshki A, Safi S, Feizi A, Askari G, Karami F. (2016) The effect of green tea extract supplementation on liver enzymes in patients with nonalcoholic fatty liver disease. *Int J Prev Med*, **7**, 28.

Rahman SU, *et al.* (2018) Treatment of inflammatory bowel disease via green tea polyphenols: possible application and protective approaches. *Inflammopharmacology*, **26**(2), 319–330.

Renaud J, Nabavi SF, Daglia M, Nabavi SM, Martinoli MG. (2015) Epigallocatechin-3-gallate, a promising molecule for Parkinson's disease? *Rejuvenation Res*. 18:257–269.

Shah D, Mahajan N, Sah S, Nath SK, Paudyal B. (2014) Oxidative stress and its biomarkers in systemic lupus erythematosus. *J Biomed Sci*, **21**(1), 23

Shen J-G, Wang J, Zhao B-L, Hou J-W, Gao T-L, Xin W-J. (1998) Effects of EGb-761 on nitric oxide, oxygen free radicals, myocardial damage and arrhythmias in ischemia-reperfusion injury in vivo. *Biochim Biophys Acta*, **1406**, 228–236.

Song JM. (2018) Anti-infective potential of catechins and their derivatives against viral hepatitis. *Clin Exp Vaccine Res*, **7**(1), 37–42.

Thawonsuwan J, Kiron V, Satoh S, Panigrahi A, Verlhac V. (2010) Epigallocatechin-3-gallate (EGCG) affects the antioxidant and immune defense of the rainbow trout, Oncorhynchus mykiss. *Fish Physiol Biochem*, **36**(3), 687–697.

Theofilopoulous AN, Dixon FB. (1982) Autoimmune diseases: immunology and etiopathogenesis. *Am J Pathol*, **108**, 321–330.

Varilek GW, Yang F, Lee EY, deVilliers WJ, Zhong J, Oz HS, Westberry KF, McClain CJ. (2001) Green tea polyphenol extract attenuates inflammation in interleukin-2-deficient mice, a model of autoimmunity. *J Nutr*, **131**, 2034–2039.

Wan L, Nie G, Zhang J, Yunfeng Luo, Peng Zhang, Zhiyong Zhang, Baolu Zhao. (2011) β-amyloid peptide increases levels of iron content and oxidative stress in human cell and C. elegans models of Alzheimer's disease. *Free Rad Biol Med*, **50**, 122–129.

Wang QM, Wang H, Li YF, Xie ZY, Ma Y, Yan JJ, Gao YF, Wang ZM, Wang LS. (2016) Inhibition of emmprin and mmp-9 expression by epigallocatechin-3-gallate through 67-kda laminin receptor in pma-induced macrophages. *Cell Physiol Biochem*, **39**, 2308–2319.

Wang R, Sun F, Ren C, Zhai L, Xiong R, Yang Y, Yang W, Yi R, Li C, Zhao X. (2021) Hunan insect tea polyphenols provide protection against gastric injury induced by HCl/ethanol through an antioxidant mechanism in mice. *Food Funct*, **12**(2), 747–760.

Wang YY, Zhao BL, Li XJ, Su Z, Xin WJ. (1997) Spin-trapping technique studies on active oxygen radicals from human polymorphonuclear leukocytes during fluoride-stimulated respiratory burst *FLUORIDE*, **30**, 5–15.

Wang S, Moustaid-Moussa N, Chen L, Mo H, Shastri A, Su R, Bapat P, Kwun I, Shen CL. (2014) Novel insights of dietary polyphenols and obesity. *J Nutr Biochem*, **25**(1), 1–18.

Wang Z, Sun B, Zhu F. (2018) Epigallocatechin-3-gallate protects Kuruma shrimp *Marsupeneaus japonicus* from white spot syndrome virus and Vibrio alginolyticus. *Fish Shellfish Immun*, **78**, 1–9.

Watson JL, Vicario M, Wang A, Moreto M, McKay DM. (2005) Immune cell activation and subsequent epithelial dysfunction by Staphylococcus enterotoxin B is attenuated by the green tea polyphenol (-)-epigallocatechin gallate. *Cell Immunol*, **237**, 7–16.

Wheeler DS, Catravas JD, Odoms K, Denenberg A, Malhotra V, Wong HR. (2004) Epigallocatechin-3-gallate, a green tea-derived polyphenol, inhibits IL-1 beta-dependent proinflammatory signal transduction in cultured respiratory epithelial cells. *J Nutr*, **134**(5), 1039–44.

Wilasrusmee C, Kittur S, Siddiqui J, Bruch D, Wilasrusmee S, Kittur DS. (2002) In vitro immunomodulatory effects of ten commonly used herbs on murine lymphocytes. *J Altern Complem Med*, **8**, 467–475.

Wójcik P, Gęgotek A, Žarković N, Skrzydlewska E. (2021) Oxidative stress and lipd mediators modulate cell function in autoimmune diseases. *Int J Mol Sci*, **22**(2), 723.

Wu D, Guo Z, Ren Z, Guo W, Meydani SN. (2009) Green tea EGCG suppresses T cell proliferation through impairment of IL-2/IL-2 receptor signaling. *Free Radic Biol Med*, **47**, 636–643.

Wu D, Wang J, Pae M, Meydani SN. (2012) Green tea EGCG, T cells, and T cell-mediated autoimmune diseases. *Mol Aspects Med*, **33**(1), 107–118.

Wu D. (2016) Green tea EGCG, T-cell function, and T-cell-mediated autoimmune encephalomyelitis. *J Investig Med*, **64**(8), 1213–1219.

Wu Z, Huang S, Li T, Li N, Han D, Zhang B, Xu ZZ, Zhang S, Pang J, Wang S, Zhang G, Zhao J, Wang J. (2021) Gut microbiota from green tea polyphenol-dosed mice improves intestinal epithelial homeostasis and ameliorates experimental colitis. *Microbiome*, **9**(1), 184.

Van Woerkom JM, Geertzema JG, Nikkels PG, Kruize AA, Smeenk RJ, Vroom TM. (2004) Expression of Ro/SS-A and La/SS-B determined by immunohistochemistry in healthy, inflamed and autoimmune diseased human tissues: a generalized phenomenon. *Clin Exp Rheumatol* **22**, 285–292.

Xie H, Zhou F, Liu L, Zhu G, Li Q, Li C, Gao T. (2016) Vitiligo: how do oxidative stress-induced autoantigens trigger autoimmunity? *J Dermatol Sci*, **81**(1), 3–9.

Yan J, Zhao Y, Zhao B. (2013) Tea polyphenols prevent and treat obesity by regulating peroxisome proliferator activated receptor. *Sci China Life Sci*, **43**, 533–540.

Yan J, Zhao Y, Suo S, Liu Y, Zhao B. (2012) Green tea catechins ameliorate adipose insulin resistance by improving oxidative stress. *Free Radic Biol Med*, **52**, 1648–1657.

Yan Q, Hao S, Shi F, Zou Y, Song X, Li L, Li Y, Guo H, He R, Zhao L, Ye G, Tang H. (2012) Epigallocatechin-3-gallate reduces liver and immune system damage in Acinetobacter baumannii-loaded mice with restraint stress. *Int Immunopharmacol*, **92**, 107346.

Yang F, Oz HS, Barve S, de Villiers WJ, McClain CJ, Varilek GW. (2001) The green tea polyphenol (–)-epigallocatechin-3-gallate blocks nuclear factor-kappa B activation by inhibiting I kappa B kinase activity in the intestinal epithelial cell line IEC-6. *Mol Pharmacol*, **60**, 528–533.

Yang FJ, Zhao BL, Xin WJ. (1991) The activity of NADPH oxidase to produce O2 — was studied by chemiluminescence. *J Biophys*, **7**, 530–538.

Yoneyama S, Kawai K, Tsuno NH, Okaji Y, Asakage M, Tsuchiya T, Yamada J, Sunami E, Osada T, Kitayama J, Takahashi K, Nagawa H. (2008) Epigallocatechin gallate affects human dendritic cell differentiation and maturation. *J Allergy Clin Immunol*, **121**(1), 209–214.

Yoshino S. (1985) Effect of blood on the activity and persistence of antigen induced inflammation in the rat air pouch. *Ann Rheum Dis*, **44**, 485–490.

Yu NH, Pei H, Huang YP, Li YF. (2017) (-)-Epigallocatechin-3-gallate inhibits arsenic-induced inflammation and apoptosis through suppression of oxidative stress in mice. *Cell Physiol Biochem*, **41**(5), 1788–1800.

Zaieni SH, Derakhshan Z, Sariri R. (2015) Alternations of salivary antioxidant enzymes in systemic lupus erythematosus. *Lupus*, **24**(13), 1400–1405.

Zhang JC, Xu H, Yuan Y, Chen JY, Zhang YJ, Lin Y, Yuan SY. (2017) Delayed treatment with green tea polyphenol EGCG promotes neurogenesis after ischemic stroke in adult mice. *Mol Neurobiol*, **54**, 3652–3664.

Zhao B-L. Oxygen free radicals and natural antioxidants, Science Press, 2002 (Revised Edition), Beijing.

Zhao BL, Wang JC, Hou J, Xin W-J (1996d) NO and superoxide anion radicals produced by polymorphonuclear leukocytes mainly form ONOO⁻. *Chinese Science*, **26**, 406–413.

Zhao BL, Li XJ, Xin WJ. (1989a) ESR study on oxygen consumption during the respiratory burst of human polymophonuclear leukocytes. *Cell Biol Intern Report*, **13**, 317–326.

Zhao B-L, Li X-J, He R-G, Cheng S-J, Xin W-J. (1989b) Scavenging effect of extracts of green tea and natural antioxidants on active oxygen radicals. *Cell Biophys*, **14**, 175.

Zhao B-L, Duan S-J, Xin W-J. (1990) Lymphocytes can produce respiratory burst and oxygen radicals as polymorphonuclear leukocytes. *Cell Biophys*, **17**, 205.

Zhao B-L, Wang J-C, Hou J-W, Xin W-J. (1996a) Studied the nitric oxide free radicals generated from polymorphonuclear leukocytes (PMN) stimulated by phobol myristate (PMA). *Cell Biol Intern*, **20**, 343–350.

Zhao B-L, Shen J-G, Li M, Xin W-J. (1996b) Scavenging effect of Chinonin on NO and oxygen free radicals generated from ischemia reperfusion myocadium. *Biachem Biophys Acta*, **1317**, 131–137.

Zhao B, Wang J, Hou J, Xin W. (1996c) Scavenging effect of tea polyphenols on methyl radical produced by peroxynitrite oxidation dimethyl sulfoxide. *Sci Bull*, **41**, 925–927.

Zhou J, Tang L, Shen CL, Wang JS. (2020) Green tea polyphenols boost gut-microbiota- dependent mitochondrial TCA and urea cycles in Sprague-Dawley rats. *J Nutr Biochem*, **81**, 108395.

Zwolak I. (2021) Epigallocatechin gallate for management of heavy metal-induced oxidative stress: mechanisms of action, efficacy, and concerns. *Int J Mol Sci*, **22**(8), 4027.

Chapter 14

Tea Polyphenols and Cancer

Baolu Zhao

Institute of Biophysics, Chinese Academy of Sciences, Beijing, China

14.1 Introduction

Tumor is an organism formed by local tissue and cell proliferation under the action of various tumorigenic factors, which mostly presents space occupying massive processes. According to the cellular characteristics of new organisms and the degree of harm to the body, tumors can be classified as either benign tumors or malignant tumors. Benign tumors are not fatal while malignant tumors are cancerous. Malignant tumors first infiltrate the surrounding tissues locally and can also transfer to other tissues through the blood and lymphatic system. The occurrence and development of cancer can be divided into three stages: the initiation, the promotion and the formation, and development of cancer. Free radicals are produced and involved in each stage.

According to a research report released by the World Health Organization, the global cancer situation will become increasingly serious. In the next 20 years, the number of new patients will increase by 10 million to 15 million every year, and the number of cancer deaths will increase from 6 million to 10 million every year. At present, cancer mortality is second only to cardiovascular disease, which seriously threatens human life and health. Countries all over the world have invested a lot of human and material resources in cancer treatment and research.

Although great progress has been made in some aspects, there has been no fundamental progress on the pathogenesis of cancer. The incidence rate of breast cancer has been on the rise since the 1970s. The proportion of women with breast cancer in the United States is as high as 12.5%. In recent years, the incidence rate of breast cancer in China is increasing faster than that in developed countries. According to the latest statistics, the annual cancer mortality rate is 12.84%, of which 1 in 8 deaths is due to malignant tumors. The incidence rate of lung cancer is the highest in male cancer incidence, and breast cancer accounts for the first place in female cancer incidence. Furthermore, the incidence rate of colorectal cancer is higher than before. Moreover, the survey shows that China has been increasing in incidence rate and mortality rate of cancer in recent years.

The change of free radical content in the process of tumor formation and growth is complex, not a simple relationship of increase or decrease, as the content of free radical increases rapidly in the early stage of tumor formation. When the tumor grows to a certain size, the content of free radical reaches a maximum. Then with the growth of the tumor, the content of free radicals not only no longer increases, but also decreases. This shows that free radicals play an important role in tumor growth and development. The decrease of free radical content in the later stage of tumor growth may be caused by necrosis and hypoxia [Zhao & Zhang, 1984]. Studies have found that the antioxidant tea polyphenols have preventive and therapeutic effects on tumors. Studies have shown that the role of oxidative stress and antioxidants in tumor biology has always been the concern by scientists, medical circles, and the general public. This chapter will discuss this issue, especially the preventive and therapeutic effects of oxidative stress and antioxidant tea polyphenols on tumors.

14.2 Types, pathogenic factors, and treatment of tumors

Tumors can be divided into benign tumors and malignant tumors. Malignant tumors can be divided into cancer and sarcoma. Cancer refers to malignant tumors derived from epithelial tissue. Malignant tumors, such as colorectal cancer, skin cancer, gastric cancer, lung cancer, and malignant lymphoma, are collectively referred to as cancer. Leukemia is a malignant tumor of the blood system, commonly known as blood cancer. It is formed by the diffuse malignant growth of immature leukocytes in bone marrow, replacing normal bone marrow tissue and entering the blood. Benign tumor refers to mesenchymal tissue, including fibrous connective tissue, fat, muscle, vascular, bone and cartilage. In addition, doctors divide tumors into malignant and benign according to their pathological morphology, growth mode, and harm to patients.

Cancer has biological characteristics such as abnormal cell differentiation and proliferation, uncontrolled growth, invasion, and metastasis. Its occurrence is a multifactor and multi-step complex process, which is divided into three processes: carcinogenesis, cancer promotion, and evolution. It is closely related to smoking, infection, occupational exposure, environmental pollution, unreasonable diet, and genetic factors. At present, the pathogenesis of tumor has not been fully studied, but it is certain that the pathogenesis of tumor is very complex and multifactorial. There are physical factors, chemical factors, biological factors, genetic factors, and even spiritual factors. Living habits, such as smoking and drinking, are closely related to the occurrence and development of cancers. The survey found that about a third of the patients who died of cancer were related to smoking, which was the main risk factor of lung cancer. Ingestion of a large amount of liquor can lead to the occurrence of malignant tumors in the mouth, throat, and esophagus. High energy and high fat foods can increase the incidence rate of breast cancer, endometrial cancer, prostate cancer, and colon cancer. Drinking polluted water and eating moldy food can induce liver cancer, esophageal cancer, and gastric cancer.

Recent studies have found that environmental pollution is closely related to the incidence of cancer. For example, the pollution of air, drinking water, and food can cause serious harm to human beings. The carcinogenic substances related to the environment published by the World Health Organization include arsenic, asbestos, benzidine, 4-aminobiphenyl, chromium, diethylstilbestrol, radioactive radon, coal tar, mineral oil, coupled estrogen, and so on. These chemical or physical carcinogens in the environment enter the human body through the body surface, respiratory, and digestive tract to induce cancer. Ultraviolet rays can cause skin cancer, especially in plateau areas. Virus carcinogenesis is a biological factor, including DNA virus and RNA virus. DNA viruses such as Epstein-Barr (EB) virus are associated with nasopharyngeal carcinoma and Burkitt lymphoma, human papillomavirus infection is associated with cervical cancer, and hepatitis B virus is associated with liver cancer. RNA viruses such as T-cell leukemia/lymphoma virus are associated with T-cell leukemia/lymphoma. In addition, bacteria, parasites, and fungi can cause cancer under certain conditions, such as *Helicobacter pylori* infection and gastric cancer. Schistosomiasis is proved to induce bladder cancer. *Aspergillus flavus* and its toxin can cause liver cancer. Studies have found that trauma and local chronic irritation, such as burns, deep scars, and chronic skin ulcers, may also lead to carcinogenesis. Examination instruments in hospitals, such as ionizing radiation, X-ray, radionuclide can cause skin cancer, leukemia, and so on. There are also some drugs that cause cancers, such as cytotoxic drugs, hormones, arsenic agents, immune-suppressants, and so on.

Tumors are related to family and heredity, but they are only a few uncommon tumors with direct heredity, such as familial adenomatous polyps of colon. Most of them have colorectal cancer after the age of 40; mutations in Brca-1 and Brca-2 are associated with breast cancer. Congenital or acquired immunodeficiency is prone to malignant tumors. For example, patients with hepatitis C deficiency are prone to leukemia and lymphohe-matopoietic tumors. The incidence of malignancies in AIDS patients is significantly higher. Abnormal hormone levels in the body are one of the tumor-inducing factors, such as estrogen and prolactin, which are related to breast cancer. Growth hormone can stimulate the development of cancer.

At present, many tumor treatment methods and technologies have been developed, such as surgery, chemotherapy, radiotherapy, biotherapy, hyperthermia, radiofrequency therapy, minimally invasive interventional therapy, and traditional Chinese medicine treatment, but these treatments inflict a lot of pain on the patients. Even so, many tumors are still difficult to treat. The recently developed minimally invasive surgery can greatly reduce the pain of patients and play an important role in the treatment of tumors. In addition, early detection and treatment of cancer are very important, which can greatly reduce the mortality of patients and increase the treatment effect. Chemotherapy is the treatment of cancer with drugs that kill cancer cells. Since the biggest difference between cancer cells and normal cells lies in rapid cell division and growth, the action principle of anticancer drugs is usually to inhibit the growth of cancer cells by interfering with cell division, such as inhibiting DNA replication or preventing chromosome separation. However, most chemotherapy drugs are not specific, so they often damage healthy tissues that need to divide to maintain normal function, such as intestinal mucosal cells. Chemotherapy is often a "combination chemotherapy" that uses two or more drugs at the same time, and the chemotherapy of most patients is carried out in this way.

Therefore, tumor prevention is the most important. The prevention of tumor includes keeping away from various environmental carcinogenic risk factors, preventing infection factors related to tumor incidence, changing bad lifestyle, appropriate exercise, maintaining mental pleasure, and adopting certain medical intervention measures for extremely high-risk groups or precancerous lesions to reduce the risk of tumor incidence. The World Health Organization believes that more than 40% of cancers can be prevented. The occurrence of malignant tumor is the result of long-term interaction between the body and external environmental factors. Therefore, tumor prevention should be incorporated into one's daily life and continue for a long time.

If we can find some natural substances to prevent cancer, it will reduce the incidence rate and mortality of malignant tumors, thereby reducing the harm of malignant tumors to national health and family, as well as the consumption of national medical resources, and reducing the economic burden of family and society caused by malignant tumors. It is found that antioxidants, especially tea polyphenols, may play a role in this regard, which is worthy of in-depth study and discussion.

14.3 Cancer and oxidative stress

Oxidative stress is a physiological state where high levels of reactive oxygen species (ROS) and reactive nitrogen species (RNS) free radicals are generated. Oxidative stress is caused by the imbalance between the generation of ROS, RNS, and the ability of biological systems to respond to these reactive intermediates or the resulting damage. Thus, oxidative stress is accepted as a critical pathophysiological mechanism in different frequent human pathologies, including cancer. In fact, ROS and RNS can cause protein, lipid, and DNA damage, and malignant tumors often show increased levels of DNA base oxidation and mutations. Studies have found that the occurrence, development, harm, and treatment of cancer are related to oxidative stress. Oxidative stress is complex for cancer. Oxidative stress can cause cell mutation and accumulation of DNA damage and lead to cancer. Oxidative stress can also kill cancer cells and play a role in cancer treatment. Many perturbations, such as pathogen infection and inflammation increase oxidative stress and promote the accumulation of DNA mutations, ultimately leading to carcinogenesis. Lipid peroxides, aldehydes, and carboxyl groups can aggravate peroxidation damage. For example, aldehydes formed by microsomal lipid peroxidation can change the function of mitochondria, microtubule, adenylate cyclase activity, and protein and DNA synthesis. Sulfur and amino groups are particularly sensitive to aldehydes. Lipid peroxide malondialdehyde is a mutagen that can crosslink protein and DNA. The influence of cell membrane lipid peroxidation is multifaceted. Cell membrane lipid peroxidation can reduce the coordination binding to receptors and inhibit the activity of ATPase adenylate cyclase. The change of ATP synthesis will affect and limit the total cell metabolism and the repair of DNA breakage. Mitochondrial lipid peroxidation can cause mitochondrial swelling and changes in respiratory function. Lipid peroxidation of nuclear membrane will directly damage DNA, and nuclear cytoplasmic exchange and RNA transport will also be affected. Therefore, lipid peroxidation of all cell membranes will directly and indirectly regulate the carcinogenic process [Marnertt, 1987]. Additionally, several signaling pathways associated with carcinogenesis can control ROS and RNS generation and regulate ROS and RNS downstream mechanisms, which could have potential implications in anticancer research.

14.3.1 *Oxidative stress and various carcinogenic factors*

An association of ROS and RNS generation and human cancer induction has been shown. It appears that oxidative stress may both cause as well as modify the cancer process. Oxidative stress can cause cell and DNA damage and lead to cancer. Many perturbations, such as pathogen infection, inflammation, environmental pollution, intestinal diseases, heavy metals, bacteria, and viruses can all lead to increase oxidative stress and promote the accumulation of DNA mutations, ultimately leading to carcinogenesis. ROS and RNS are induced through a variety of endogenous and exogenous sources. Overwhelming amount of antioxidant and DNA repair mechanisms in the cell by ROS and RNS may result in oxidative stress and oxidative damage to the cell. This resulting oxidative stress can damage critical cellular macromolecules and/or modulate gene expression pathways. Oxidative damage resulting from ROS and RNS generation can participate in all stages of the cancer process. The association between polymorphisms in oxidative DNA repair genes and antioxidant genes (single nucleotide polymorphisms) and human cancer susceptibility has been shown. Transformation of a normal cell to a malignant one requires phenotypic changes often associated with each of the initiation, promotion, and progression phases of the carcinogenic process. Growing evidence supports a role of ROS and RNS-induced generation of oxidative stress in these epigenetic processes. Inadequate or excessive nutrient consumption leads to oxidative stress, which may disrupt oxidative homeostasis, activate a cascade of molecular pathways, and alter the metabolic status of various tissues. Cancer initiation may be modulated by the nutrition-mediated elevation in ROS and RNS levels, which can stimulate cancer initiation by triggering DNA mutations, damage, and pro-oncogenic signaling.

1. Inflammation carcinogenesis and oxidative stress

Extensive research during the past two decades has revealed the mechanism by which continued oxidative stress can lead to chronic inflammation, which in turn could mediate most chronic diseases including cancer etc. Inflammation and oxidative stress are common and co-substantial pathological processes accompanying, promoting, and even initiating numerous cancers. Our study found that only 5 ng/ml inflammatory factor carcinogen phorbol (PMA) caused polymorphonuclear leukocyte (PMN) to produce respiratory burst and release a large number of oxygen free radicals. The superoxide anion free radicals and hydroxyl free radicals produced can be captured by spin trap. The oxygen free radicals produced in this process can also be detected by chemiluminescence. Comparing the analogues of various PMA cancer promoters, it is found that there is a good correlation between their oxygen free radical production and cancer promoting activity. PMA with strong cancer-promoting effect also stimulates PMN to produce more oxygen free radicals, and PMA without cancer-promoting effect cannot stimulate PMN to produce oxygen free radicals. Weak cancer-promoting PMA stimulated PMN to produce less oxygen free radicals. PMA-stimulated PMN for five minutes can break DNA, superoxide dismutase (SOD) can block this break, while dimethyl sulfoxide (DMSO) and mannitol are not very effective, indicating that superoxide anion free radicals are mainly involved in this reaction. The oxygen free radicals produced by PMA-stimulated PMN are also genotoxic to other cells, which can increase the sister chromosome exchange between Chinese hamster ovary (CHO) and Chinese hamster cancer ovary (V79) cells, and SOD can inhibit this mutation [Zhao *et al.*, 1989].

ROS and RNS can also promote the formation of cancer by inducing DNA mutations and pro-oncogenic signaling pathways. Oxidative stress can activate a variety of transcription factors including nuclear factor-kappaB (NF-κB), AP-1, p53, HIF-1α, PPAR-γ, β-catenin/Wnt, and Nrf2. Activation of these transcription factors can lead to the expression of over 500 different genes, including those for growth factors, inflammatory cytokines, chemokines, cell cycle regulatory molecules, and anti-inflammatory molecules. Oxidative stress activates inflammatory pathways, leading to transformation of a normal cell to tumor cell, tumor cell survival, proliferation, chemoresistance, radioresistance, invasion, angiogenesis, and stem cell survival. Many intracellular pathways commonly involved downstream will help maintain and amplify inflammation, oxidative stress, and cancer. Thus, many WNT/β-catenin target genes such as c-Myc, cyclin D1, and HIF-1α are involved in the development of cancers. NF-κB can activate many inflammatory factors such as TNF-α, TGF-β, interleukin-6 (IL-6), IL-8, matrix metalloproteinase (MMP), vascular endothelial growth factor, Cyclooxygenase-2 (COX2), Bcl2, and inducible nitric oxide synthase. These factors are often associated with cancerous processes and may even promote them. ROS, generated by cellular alterations, stimulate the production of inflammatory factors such as NF-κB, signal transducer and activator transcription, activator protein-1, and HIF-α. NF-κB inhibits glycogen synthase kinase-3β (GSK-3β) and therefore activates the canonical extracellular factor WNT pathway. ROS activates the phosphatidylinositol 3 kinase/protein kinase B (PI3K/Akt) signaling in many cancers. PI3K/Akt also inhibits serine/threonine kinase GSK-3β. Many gene mutations of the canonical WNT/β-catenin pathway giving rise to cancers have been reported (CTNNB1, AXIN, APC). Conversely, a significant reduction in the expression of PPARγ has been observed in many cancers. Moreover, PPARγ agonists promote cell cycle arrest, cell differentiation, and apoptosis and reduce inflammation, angiogenesis, oxidative stress, cell proliferation, invasion, and cell migration. All these complex and opposing interactions between the canonical WNT/β-catenin pathway and PPARγ appear to be fairly common in inflammation, oxidative stress, and cancers [Reuter *et al.*, 2010; Vallée & Lecarpentier, 2018]. Inflammation and oxidative stress are carcinogenesis and the key aspects of the carcinogenic process, oxidative stress, and inflammation contribute to this process. The connection between metastasis and oxidative stress are the key players in the tumor microenvironment that leads to inflammation, oxidative stress, and DNA damage. Moreover, oxidative stress and inflammation play important role in cancer and metastasis [Nowsheen *et al.*, 2012].

These suggest that oxidative stress, chronic inflammation, and cancer are closely linked.

2. Environmental carcinogenesis and oxidative stress

Environmental pollutants, such as pesticides, herbicides, additives to food and water, and electromagnetic fields threaten public health by promotion of cancer. ROS and RNS are the more abundant free radicals in nature and have been related with a number of tissue/organ injuries induced by xenobiotics, ischemia, activation of leucocytes, UV exposition, etc. Oxidative stress is caused by an imbalance between ROS and RNS production and a biological system's ability to readily detoxify these reactive intermediates or easily repair the resulting damage. Different lifestyle- and environmental-related factors (including tobacco smoking, diet, alcohol, ionizing radiations, biocides, pesticides, viral infections, etc.) may be pro-carcinogenic. In all these cases, oxidative stress acts as a critical pathophysiological mechanism. Many pollutants cause adverse health outcomes by effects on mitochondrial function to produce oxidative stress through loss of the active site complex for oxidative phosphorylation, thioretinaco ozonide oxygen nicotinamide adenine dinucleotide phosphate, from opening of the mitochondrial permeability transition pore. Glyphosate, fluoride, and electromagnetic fields, are examples of carcinogenic pollutants that promote loss and decomposition of the active site for oxidative phosphorylation, producing mitochondrial dysfunction and oxidative stress. Ionizing radiation has long been known to be carcinogenic, and non-ionizing electromagnetic fields from microwaves, radar, cell phones, and cathode ray screens are carcinogenic. Excessive exposure of skin to UV radiation triggers the generation of oxidative stress, inflammation, immunosuppression, apoptosis, matrix-metalloproteases production, and DNA mutations, leading to the onset of photo ageing and photocarcinogenesis. At the molecular level, these changes occur via activation of several protein kinases as well as transcription pathways, formation of ROS, and release of cytokines, interleukins, and prostaglandins together [Garg *et al.*, 2020].

Benzo(a)pyrene (BP) is a strong polycyclic aromatic hydrocarbon carcinogen. It is a large class of environmental pollutants, mainly from the exhaust gas from industry and automobile and smoking tar. Benzopyrene is converted into bp-7,8-diol-9,10-epoxy by cytochrome P-450 and epoxy hydratase to form a genotoxic carcinogen, but this is not the only way. The one electron oxidation of BP at the C-6 position forms a free radical cation, which is attacked by water electrophiles to obtain a volatile 6-hydroxybp derivative. After molecular rearrangement, it forms a 6-hydroxybp radical, which reacts with oxygen molecules to form BP quinone metabolites, 6-oxygen-bp radical and semi-quinone radical can electrophilically bind to DNA. Under aerobic conditions, superoxide anion radical and hydroxyl radical are also generated to break DNA. SOD can inhibit this effect [Lorentzen, 1979]. Benzene is widely used in rubber, plastics, and petroleum industries. It exists in a wide range of environmental pollutants. A large amount of exposure to benzene can cause aplastic anemia and myeloid leukemia. Benzene can be metabolized into phenol, hydroquinone, catechol, etc., and these substances can easily react with oxygen to generate reactive oxygen free radicals. Microsomes and cytochrome P-450 play an important role in this metabolic process. Most carcinogens are biologically transformed from inert chemicals into highly active substances that can react with cellular components. This activation process often includes free radical intermediates formed by one-electron or two-electron redox. Under aerobic conditions, free radical intermediates, semi-quinone, azo and nitro ions, can transfer an electron to oxygen to form superoxide anion radicals, which further produce hydroxyl radicals and singlet oxygen. These ROS and RNS can damage almost all cell components and form secondary and tertiary free radicals. These primary and secondary free radicals play an important role in carcinogenesis. SOD can block this metabolism, indicating that superoxide anion free radicals are involved in the carcinogenic process of these substances [Snyder *et al.*, 1980]. Peroxides used in some chemical and pharmaceutical industries are also cancer promoters. They can produce oxygen free radicals, such as benzoyl peroxide, lauryl peroxide, sunflower peroxide, isopropyl benzene peroxide and so on. They are all highly active skin cancer promoters; even hydrogen peroxide is an effective cancer promoter in the presence of PMA [Haroz & Thomasson, 1980].

Nitroso compounds widely existing in some pickled foods are another kind of carcinogens. They can cause lung cancer and leukemia. The cytotoxicity, mutagenicity, and carcinogenicity of these compounds are mainly due to the reduction of nitro metabolism to an ionic free radical, and then produce highly reactive hydroxylamine, and can react with oxygen molecules to produce superoxide anion free radical and hydroxyl free radical [Floyd *et al.*, 1978]. Gossypol, the main toxic substance of cottonseed oil, is a cancer promoter of skin cancer and has contraceptive activity. Its mechanism is that SOD and catalase can inhibit its activity by producing oxygen free radicals [Zhao *et al.*, 1989].

Long-term inhalation of asbestos produces pulmonary fibrosis, mesothelial cancer, and bronchial cancer. Asbestos is a carcinogen of mesothelial tissue and fibroblasts, and an auxiliary carcinogen and cancer promoter of bronchial cancer. Asbestos workers and smokers are prone to bronchial cancer, which is 92 times higher in asbestos workers and smokers than in the general population [Mossman *et al.*, 1983]. Treatment of rodent bronchus with asbestos pre-exposed to non-carcinogenic dose of 7,12-methylphenylanthracene has obvious cancer promoting effect, showing typical cancer promoting phenomenon, inducing ornithine decarboxylase, increasing DNA synthesis, proliferation, and tissue deformation of bronchial tissue and epithelial cells. SOD can greatly reduce the cytotoxicity caused by asbestos, while catalase and singlet oxygen scavengers are ineffective, indicating that superoxide anion free radicals play an important role in the cytotoxicity of asbestos. Prolonged exposure to asbestos significantly increases endogenous SOD activity, indicating a stress response. When rat liver or lung microsomes were co-cultured with asbestos, lipid peroxide increased with dose. It is proved that asbestos can catalyze hydrogen peroxide to produce hydroxyl and superoxide anion radicals by spin trapping method. Compared with normal mice, lung lavage fluid macrophages and PMN of asbestos exposed mice were 3–9 times higher. Co-culture of human PMN or guinea pig macrophages with asbestos can produce ROS [Haroz & Thomasson, 1980].

The above research results show that some environmental factors around us may be carcinogenic factors, so we should be particularly careful to avoid their influence, which otherwise will lead to cancer and harm human health.

3. Digestive tract carcinogenesis and oxidative stress

As there are many opportunities for digestive tract to be exposed to various carcinogenic factors, many digestive tract cancers may occur. Oral cancer accounts for 2–3% of all malignancies and is the fifth most common cancer worldwide according to the World Health Organization. A paper studied the oxidative stress based-biomarkers in oral carcinogenesis. Oxidative stress implies a cellular state whereby ROS and RNS production exceeds its metabolism resulting in excessive ROS and RNS accumulation and overwhelmed cellular defenses. Such a state has been shown to be involved in the multistage process of human carcinogenesis (including oral cancer) via many different mechanisms. Amongst them are ROS and RNS-induced oxidative modifications on major cellular macromolecules like DNA, proteins, and lipids with the resulting byproducts being involved in the pathophysiology of human oral malignant and premalignant lesions [Hanafi *et al.*, 2012].

Oxidative stress has been implicated in the pathogenesis of diverse gastrointestinal (GI) diseases including gastroesophageal reflux disease (GERD), gastritis, enteritis, colitis, and associated cancers as well as pancreatitis and liver cirrhosis. Oxidative stress in inflammation-based gastrointestinal tract diseases, include reflux esophagitis, *Helicobacter pylori*-associated gastritis, non-steroidal anti-inflammatory drug-induced enteritis, ulcerative colitis, and associated colorectal cancer. The challenge is that ROS and RNS can contribute to diverse gastrointestinal dysfunction or manifest dual roles in cancer promotion or cancer suppression. ROS and RNS can contribute to diverse gastrointestinal dysfunction, or manifest dual roles in cancer promotion or cancer suppression [Kim *et al.*, 2012]. Short-chain fatty acids in intestinal barrier function, inflammation, oxidative stress, and colonic carcinogenesis play a role. Most *in vitro* and *ex vivo* studies have shown that short-chain fatty acids affect the regulation of inflammation, carcinogenesis, intestinal barrier function, and oxidative stress [Liu *et al.*, 2021]. Alterations to the microbiome caused by infection, diet and/or lifestyle can disturb the gut homeostasis and promote disease, such as inflammatory bowel diseases and cancer. Some bacterial species have been identified and suspected to play a role in colorectal carcinogenesis, such as *Streptococcus bovis, Helicobacter pylori, Bacteroides fragilis, Enterococcus faecalis, Clostridium septicum, Fusobacterium*, and *Escherichia coli*. These factors, such as inflammation, host defenses modulation, and bacterial-derived metabolism, induce oxidative stress and anti-oxidative defenses modulation [Gagnière *et al.*, 2016].

One of the contributory causes of colon cancer is the negative effect of ROS and RNS on DNA repair mechanisms. Several studies have documented the importance of antioxidants in countering oxidative stress and preventing colorectal carcinogenesis. Molecular mechanisms underlying development and progression of gastrointestinal (GI) cancers are mediated by both oxidative stress and microRNAs (miRNAs) involvement. A growing body of evidence has indicated a reciprocal connection between oxidative stress signaling pathways and miRNA regulatory machines in GI cancer development and progression. Since the use of antioxidants is limited owing to the contrasting consequences of oxidative stress signaling in cancer, the discovery of oxidative stress-responsive miRNAs may provide a potential new strategy to overcome oxidative stress-mediated GI carcinogenesis. Given the possible interaction between oxidative stress and miRNAs in GI cancers, the existing evidence on the interaction between oxidative stress and miRNA regulatory machinery and its role in GI carcinogenesis, illustrate the function of miRNAs which target OS systems during homeostasis and tumorigenesis [Akbari *et al.*, 2020].

The above research results show that various carcinogenic factors in the digestive tract, namely the oral cavity and gastrointestinal tract, and the carcinogenic process are closely related to oxidative stress.

4. Roles of oxidative stress in metal carcinogenesis

Occupational and environmental exposures to metals are closely associated with an increased risk of various cancers. Accumulating evidence indicates that ROS and RNS generated by metals play important roles in the etiology of degenerative and chronic diseases. In the mechanism of heavy metal carcinogenesis, metal-generated ROS and RNS are essential. ROS and RNS play two roles in metal carcinogenesis; two stages in the process of metal carcinogenesis differ in the amounts of ROS and RNS activating a dual redox-mediated mechanism. In the early stage of metal carcinogenesis, ROS and RNS act in an oncogenic role. However, in the late stage of metal carcinogenesis, ROS and RNS play an anti-oncogenic role. The dual role of ROS and RNS represents a "double-edged sword" with many possible novel ROS and RNS-mediated strategies in cancer therapy in metal carcinogenesis. KRAS is the most frequently mutated oncogene in tumors. A paper revealed that cobalt-induced oxidative stress contributes to alveolar/bronchiolar carcinogenesis. Mice and rats exposed to cobalt metal dust (CMD) by inhalation developed alveolar/bronchiolar carcinogenesis in a dose-dependent manner. In cobalt metal dust-exposed mice, the incidence of Kras mutations in alveolar/bronchiolar carcinogenesis was 67%, with 80% of those being G to T trans-versions on codon 12, suggesting the role of oxidative stress in the pathogenesis. Oxidative stress increased due to cobalt exposure. In addition, significantly increased 8-oxo-dG adducts were demonstrated in lungs of mice exposed to cobalt metal dust for 90 days. Furthermore, significant alterations were observed in canonical pathways related to MAPK signaling (IL-8, ErbB, Integrin, and PAK pathway) and oxidative stress (PI3K/AKT and Melatonin pathway) in alveolar/bronchiolar carcinogenesis from CMD-exposed mice. Oxidative stress can stimulate PI3K/AKT and MAPK signaling pathways. NADPH oxidoreductase 4 (NOX4) was significantly upregulated only in cobalt metal dust-exposed age-associated B cells (ABCs) and NOX4 activation of PI3K/AKT can lead to increased ROS levels in human cancer cells. Ereg gene encodes a secreted peptide hormone and member of the epidermal growth factor (EGF) family of proteins. The gene encoding Ereg was markedly upregulated in cobalt metal dust-exposed mice. Oncogenic KRAS mutations have been shown to induce EREG overexpression [Ton *et al.*, 2021].

Cadmium (Cd) is a toxic metal, targeting the lung, liver, kidney, and testes following acute intoxication, and causing nephrotoxicity, immunetoxicity, osteotoxicity, and tumors after prolonged exposures. ROS and RNS are often implicated in Cd toxicology. A paper studied the role of oxidative stress in cadmium toxicity and carcinogenesis and found direct evidence for the generation of free radicals in intact animals following acute Cd overload the association of ROS and RNS in chronic Cd toxicity and carcinogenesis. Cd-generated superoxide anion, hydrogen peroxide, and hydroxyl radicals *in vivo* have been detected by the electron spin resonance spectra, which are often accompanied by activation of redox sensitive transcription factors (e.g., NF-kappaB, AP-1, and Nrf2) and alteration of ROS and RNS-related gene expression. It is generally agreed upon that oxidative stress plays important roles in acute Cd poisoning. Acquired apoptotic tolerance renders damaged cells to proliferate with inherent oxidative DNA lesions, potentially leading to tumorigenesis [Liu *et al.*, 2009]. ROS and RNS are generated following Cd overload and play important roles in tissue damage and carcinogenesis.

In addition to cobalt and cadmium, there are also reports that other metals such as excessive iron also play a role in human carcinogenesis. The similarity of genomic alterations between Fe-NTA-induced renal cancer and human cancers suggests that excess iron also plays a role in human carcinogenesis. Furthermore, excess iron is a major pathology in asbestos-induced mesothelioma, including chrysotile [Toyokuni, 2016].

The above research results show that various metal carcinogenic factors are closely related to oxidative stress.

5. Oxidative stress and virus carcinogenesis

Extensive experimental work has conclusively demonstrated experimental data and indirect evidence on promoting the activity of oxidative stress in viral infection and viral. Recent data suggest the involvement of oxidative DNA damage in carcinogenesis and acquired immunodeficiency syndrome (AIDS). Oxidative DNA damage may lead to apoptotic cell death of patients infected with human immunodeficiency virus (HIV) and may influence the progression of AIDS. Oxidants act at several stages in the malignant transformation of cells [Olinski *et al.*, 2002]. A paper has found that oxidative stress is a trigger of hepatitis C and B virus-induced liver carcinogenesis. Virally-induced liver cancer usually evolves over long periods of time in the context of a strongly oxidative microenvironment, characterized by chronic liver inflammation and regeneration processes. They ultimately lead to oncogenic mutations in many cellular signaling cascades that drive cell growth and proliferation. Oxidative stress, induced by hepatitis viruses, therefore is one of the factors that drives the neoplastic transformation process in the liver [Ivanov *et al.*, 2017]. The so-called high-risk human papillomavirus, represent a most powerful human carcinogen. Epithelial tissues, the elective target for human papillomavirus (HPV) infection, are heavily exposed to all named sources of oxidative stress. Two different types of cooperative mechanisms are presumed to occur between oxidative stress and human papillomavirus. The oxidative stress genotoxic activity and the human papillomavirus-induced genomic instability concur independently to the generation of the molecular damage necessary for the emergence of neoplastic clones. This first mode is merely a particular form of co-carcinogenesis; oxidative stress specifically interacts with one or more molecular stages of neoplastic initiation and/or progression induced by the human papillomavirus infection [De Marco, 2013].

The above results indicate that the carcinogenesis of various virus is closely related to oxidative stress.

6. Oxidative stress in cancer radiotherapy

Many facts have proved that radiation can cause cancer. As early as 1902, it was found that radioactive workers easily get skin cancer of the hand and leukemia. The incidence rate of leukemia in Hiroshima atomic bomb in Japan is particularly high within five years. The incidence rate of lung cancer is high in the radioactive elements mineral mining workers due to inhalation of radioactive dust [Wilmer & Schubert, 1981]. The potential cause of radiation-induced biotoxicity is relevant to oxidative stress. Oxidative stress, an accumulation of ROS, disrupts intracellular homeostasis through chemical modification and damages proteins, lipids, and

DNA. Therefore, it results in a series of related pathophysiological changes [Ping *et al.*, 2020].

Fluorescence and ultraviolet light can also cause skin cancer. Skin cancer mainly occurs in exposed parts, such as hands, head, neck, and arms, and skin cancer mainly occurs in outdoor workers. Irradiation of animal skin with ultraviolet light can indeed produce skin cancer. DNA can be damaged by energy during irradiation, but more importantly, irradiation of water produces a series of free radicals such as hydroxyl radicals with high reactivity and hydrated electrons. These free radicals can attack cell components, especially DNA, produce a series of reactants, cause irreparable damage and cause mutations. Oxidative stress in radiotherapy induces toxicity for cancer cells and provides a treatment method for cancer. Oxidative stress, an accumulation of ROS, disrupts intracellular homeostasis through chemical modification and damages proteins, lipids, and DNA. Therefore, it results in a series of related pathophysiological changes [Ivanov *et al.*, 2018].

The above research results show that some carcinogenic factors, such as inflammation, environmental carcinogens, heavy metals, viruses, radiation, and other carcinogenic processes, including oxidative stress, play an important role. Increased ROS and RNS production has been detected in various cancers and has been shown to have several roles, they can activate pro-tumorigenic signaling, enhance cell survival and proliferation, and drive DNA damage and genetic instability.

14.3.2 *Oxidative stress in the treatment of cancer*

Oxidative stress also plays an important role in the treatment of cancer. The regulation of oxidative stress is an important factor in both cancer development and responses to anticancer therapies. Oxidative stress due to imbalance between ROS production and detoxification plays a pivotal role in determining cell fate. In response to excessive ROS and RNS, apoptotic signaling pathway is activated to promote normal cell death. Counterintuitively, ROS and RNS can also promote anti-tumorigenic signaling, initiating oxidative stress-induced cancer cell death. Cancer cells express elevated levels of antioxidant proteins to detoxify elevated ROS and RNS levels, establish a redox balance, while maintaining pro-tumorigenic signaling and resistance to apoptosis. Cancer cells have an altered redox balance to that of their normal counterparts and this identifies ROS and RNS manipulation as a potential target for cancer therapies. The treatment for cancer is generally based on using cytotoxic drugs, radiotherapy, chemotherapy. The modulation of oxidative stress response might represent a potential approach to eradicate cancer in combination with chemotherapies, radiotherapies, as well as immunotherapies. ROS and RNS and oxidative stress play important roles in all these treatments generate. Many signaling pathways that are linked to tumorigenesis can also regulate the metabolism of ROS and RNS through direct or indirect mechanisms. High ROS and RNS levels are generally detrimental to cells, and the redox status of cancer cells usually differs from that of normal cells. Due to metabolic and signaling aberrations, cancer cells exhibit elevated ROS and RNS levels. High ROS and RNS levels may constitute a barrier to tumorigenesis. The discovery of the role of free radicals in cancer has led to a new medical approach. Minimizing oxidative damage may be a significant advancement in the prevention or treatment of these diseases since antioxidants are able to stop the free-radical formation and prevent oxidizing chain reactions. These findings have generated great interest in therapeutic oxidant-based cancer drug development. The design and development of synthetic oxidant compounds, able to increase free radicals, could present a significant therapeutic advance, in particular for treating pathological conditions such as cancer.

1. Oxidative stress in cancer radiotherapy

Radiotherapy is one of the major cancer treatment strategies. Exposure to penetrating radiation causes cellular stress, directly or indirectly, due to the generation of ROS and RNS, DNA damage, and subcellular organelle damage and autophagy. These radiation-induced damage responses cooperatively contribute to cancer cell death, but paradoxically. A paper studied the oxidative stress and radioiodine treatment of differentiated thyroid cancer. It evaluated the oxidative stress level changes using the measurement of malondialdehyde (MDA) concentration in patients with differentiated thyroid cancer undergoing radioactive iodine treatment. Considering the results obtained in the study group, the serum levels of MDA in differentiated thyroid cancer patients were significantly higher compared to the healthy subjects ($p < 0.05$). The MDA concentration was significantly higher on the third day after radioactive iodine ($p < 0.001$) and significantly lower one year after radioactive iodine ($p < 0.05$) in differentiated thyroid cancer patients compared to the baseline concentration. Moreover, the redox stabilization after radioactive iodine treatment in patients with differentiated thyroid cancer during a year-long observation was demonstrated. Accordingly, an increased oxidative stress impact on the related biochemical parameters reflecting the health conditions of the differentiated thyroid cancer patients was determined [Buczyńska *et al.*, 2021].

Dose assessment is an important issue for radiation emergency medicine to determine appropriate clinical treatment. A paper examined the induction of oxidative stress biomarkers in lymphocytes to identify new biomarkers for dosimetry following whole-body irradiation in mice. Radiation-induced oxidative damage is useful not only for radiation dose assessment, but also for evaluation of radiation risks on humans [Shimura *et al.*, 2020]. Tyrosine nitrosylation is a marker of oxidative stress following radiation in malignant tumors in a study. It immune-stained sections from previously radiated vestibular schwannomas with an antibody that recognizes nitrosylated tyrosine residues to assess for ongoing oxidative stress [Robinett *et al.*, 2018]. Another paper revealed that oxidative stress-associated metabolic adaptations regulate radio-resistance in human lung cancer cells. Differential inherent and acquired radio-resistance of human lung cancer cells contribute to poor therapeutic outcome and tumor recurrence after radiotherapy. It exposed human lung cancer cells to a cumulative dose of 40Gy and allowed the radio-resistant (RR) survivors to divide and form macroscopic colonies after each fraction of 5Gy dose. The radio-resistant subline exhibited enrichment of cytosolic cancer cells without specific increase in mitochondrial ROS and RNS levels [Singh *et al.*, 2020].

The above research results show that radiotherapy is one of the main cancer treatment strategies with relative oxidative stress. Since radiation cannot only kill tumor cells, but also cause damage to normal cells, it is necessary to design accurately and aim at the treatment cells to avoid damage to normal tissue cells, which is worthy of in-depth study.

2. Oxidative stress in cancer chemotherapy

Chemotherapy is another one of the main cancer treatment strategies, which can damage cancer cells. However, the peripheral normal cells and tissue lesions also caused by chemotherapy are the side effects of many anticancer chemotherapy. Oxidative stress is the main factor leading to this effect. Bleomycin is mainly used to treat Hodgkin's malignant lymphogranulomatosis and testicular cancer. It can bind to DNA, especially guanine residues, cause single-strand or even double-strand breaks and deoxyribose decomposition to form basic acrolein, and further break to release MDA. The decomposition of DNA by bleomycin requires the existence of transition metal ions and oxygen. It needs to form complexes with transition metal ions to cause DNA decomposition. This complex can be obtained by adding Fe (II) to bleomycin, or by reducing Fe (III)-bleomycin with biological reducing agent ascorbic acid, GSH, or superoxide anion free radical system. In the presence of hydrogen peroxide, this complex can also activate and decompose DNA. Bleomycin and Fe (II) saline solution could produce hydroxyl and superoxide anion radicals. The main side effect of bleomycin is to cause lung injury, increase oxygen concentration, and aggravate lung injury, indicating the involvement of aerobic free radicals [Bickers *et al.*, 1984].

Quinone anticancer drugs may kill malignant cells through redox cycle or semi-quinone reaction with SH group. Quinone redox produces superoxide anion radical and hydrogen peroxide, and then hydroxyl radical can damage DNA. Not only superoxide anion radical, but also semi-quinone can reduce Fe (III). Naphthol can selectively kill tumor cells; it can attack the SH group of cells and generate superoxide anion free radical and hydrogen peroxide through oxidation and reduction, but its mechanism needs to be further studied. Streptomelanin is the first quinone anticancer drug that has been proved to produce oxygen free radicals *in vivo* [Lown *et al.*, 1977]. Anthracycline antibiotics are widely used in the treatment of acute leukemia, thymic cancer, Hodgkin's disease, and sarcoma. The most famous are daunomycin and adriamycin. Like all anticancer drugs, anthracycline antibiotics also have side effects, the most crucial of which is the damage to the heart. After taking it for a few minutes, there are symptoms such as arrhythmia and ECG changes. In severe cases, there can be acute irreversible congestive heart failure, which greatly limits its clinical application. The mechanism of action of anthracycline drugs include a variety of processes. They can insert DNA bases and interfere with DNA replication and RNA transcription. Daunomycin can bind to the cell membrane, change the permeability of the cell membrane to calcium ions and interfere with the electron transfer of myocardial mito-chondria. In the experiments presented in this study we examined the effects of doxorubicin (Adriamycin), daunorubicin, and related quinonoid anticancer agents on superoxide, hydrogen peroxide, and hydroxyl radical production by preparations of beef heart submitochondrial particles. Superoxide anion formation was stimulated from (mean±S.E.) 1.6 ± 0.2 to 69.6 ± 2.7 or 32.1 ± 1.5 nmol × min-1 × mg-1 by the addition of 90 μM doxorubicin or daunorubicin, respectively. Furthermore, H_2O_2 production increased from undetectable control levels to 2.2 ± 0.3 nmol × min-1 × mg-1 after treatment of submitochondrial particles with doxorubicin (200 μM). The hydroxyl radical, or a related chemical oxidant, was also detected after the addition of an anthracycline to this system by both ESR spectroscopy using the spin trap 5,5-dimethylpyrroline-N-oxide and by gas chromatographic quantitation of CH_4 produced from dimethyl sulfox-ide. Hydroxyl radical production, which was iron-dependent in this sys-tem, occurred in a nonlinear fashion with an initial lag phase due to a requirement for H_2O_2 accumulation. These experiments suggest that injury to cardiac mitochondria, which is produced by anthracycline anti-biotics, may result from the generation of the hydroxyl radical during anthracycline metabolism by NADH dehydrogenase. [Doroshow & Davies, 1986].

Actinomycin D is a peptide antibiotic. It is generally believed that its anticancer activity is due to its interaction with DNA bases to prevent RNA synthesis. When co-cultured with microsomes, if NADPH exists, it can be transformed into free radical intermediates, and then O_2 can be reduced to superoxide anion free radicals. Under hypoxic condition, actinomycin D reduced the toxicity of breast cancer in mice, indicating that oxygen free radicals play an important role in the metabolism of actinomycin D and the process of killing tumor cells [Bachur *et al.*, 1978]. Mitomycin C com-bined with other reagents can treat most cancers, and its mechanism is similar to actinomycin D. P-450 system can convert mitomycin C into offensive DNA products, but it is more toxic under hypoxia, indicating that the products reduced by nadph-cytochrome-p450 system are far more toxic to target cells than oxygen free radicals [Sartorilli, 1986].

Adriamycin is another kind of chemotherapy drug. We detected the effects of adriamycin on myocardial homogenate, mitochondria, and sub-mitochondria by ESR technology, and found that semiquinone free radical signal is produced by adriamycin. Promethazine and tanshinone can clear this signal, indicating that the toxicity of adriamycin to myocardium and the protective effect of promethazine and antioxidant Tanshinone on myocardium are related to free radicals [Zhang *et al.*, 1991; Xu *et al.*, 1995].

Chemotherapy will cause oxidative stress and DNA damage of the patients. In this study, stained sections of frontal lobe autopsy tissue were done on cancer patients treated with chemotherapy, cancer patients not treated with chemotherapy, and patients without history of cancer for markers of oxidative stress (nitrotyrosine, 4-hydroxynonenal) and DNA damage. Cancer patients treated with chemotherapy had increased staining for markers of oxidative stress and DNA damage in frontal lobe cortical neurons compared to controls [Torre *et al.*, 2021]. The study highlights the potential relevance of oxidative stress and DNA damage in the pathophysiology and the neurotoxicity of chemotherapy.

Developing resistance to chemotherapeutic drugs are major clinical obstacles in the successful therapeutic strategies to cancer treatment. Acquired cancer chemoresistance is a multifactorial phenomenon involving factors such as tumor type, tumor stage, cellular ROS and RNS level, or ROS and RNS-responsive microRNAs profile. ROS and RNS level could influence the microRNAs expression level, which changes the cellular profile of the content of miRs. Such significant changes in the cellular microRNAs profile generate subsequent biological effects through the regulation of their target genes [Kozak *et al.*, 2020].

The above results indicate that cancer chemotherapy is closely related to oxidative stress. Oxidative stress, directly or indirectly caused by chemotherapeutics as exemplified by doxorubicin, is one of the underlying mechanisms of the toxicity of anticancer drugs in noncancerous tissues, including the heart and brain, but can also cause serious side effects. Injury to non-targeted tissues in chemotherapy often complicates cancer treatment by limiting therapeutic dosages of anticancer drugs and by impairing the quality of life of patients during and after treatment. A comprehensive understanding of the mechanisms of oxidative injury to normal tissue will be essential for the improvement of strategies to prevent or attenuate the toxicity of chemotherapeutic agents without compromising their chemotherapeutic value.

3. Radiation sensitization therapy and oxidative stress

Photodynamic therapy (PDT) is a novel and promising cancer treatment which employs a combination of a photosensitizing chemical and visible light to induce apoptosis in cancer cells. Radiation sensitization therapy is a combination of chemotherapy and radiotherapy. Radiation sensitizer is a kind of chemical reagent, which can enhance the killing effect of radiation on tumor. There are three sensitizers, which act respectively before radiation, during radiation, and after radiation. It is mainly used to synchronize the cell division cycle before radiation. After radiation, it is mainly used to inhibit the process of cell repair. During radiation, it is mainly through the interaction with free radicals produced by radiation. Hematoporphyrin is a photosensitizer for the treatment of tumor, which has attracted extensive attention in recent years. When injected into the body, it can quickly focus on the tumor site and show fluorescence when illuminated with appropriate wavelength, which can be used to diagnose tumors. Irradiation with a certain wavelength can improve the killing effect of irradiation on tumor. We used ESR spin trapping technology to study the free radicals produced by light hematoporphyrin. Figure 14-1 shows the ESR spectrum of free radicals captured by light hematoporphyrin. Through the analysis and calculation of spectral parameters and computer spectral simulation, it is found that there are mainly three spectral components, one of which is hydroxyl radical. The addition of SOD or NaN_3 can reduce this signal, but it cannot disappear completely. It shows

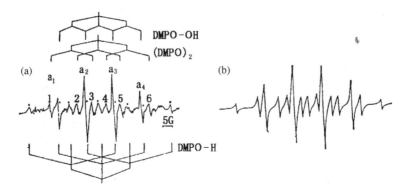

Figure 14-1. (a) ESR spectra and (b) computer simulation spectra of DMPO adducts of free radicals produced by photohemoporphyrin.

that the hydroxyl radical comes partly from the Harper Weiss reaction of superoxide anion radical and partly from the reaction of singlet oxygen and water. The ESR signal of protons is also detected here, which may be produced by the combination of hydrated electrons formed by the interaction between light hematoporphyrin and water molecules and H^+ [Zhao et al., 1985a, 1985b; Zhang et al., 1986a, 1988; Huang et al., 1990].

We also studied the interaction between light hematoporphyrin and liposome membrane. It was found that light hematoporphyrin can significantly increase the permeability of liposome membrane, especially infrared laser and ultraviolet laser. Ultraviolet laser has a greater effect than infrared laser. Illumination of hematoporphyrin can reduce the ESR signal of fatty acid spin marker in liposome membrane, indicating the occurrence of electron transfer. The order of phospholipid molecules in liposome membrane decreased slightly under light. ESR technology is also used to study the singlet oxygen, hydroxyl radical, superoxide anion radical, and hematoporphyrin anion radical produced by Yangzhou hematoporphyrin under light. The study on the photosensitive mechanism of hypocrellin A and hypocrellin B has also obtained very meaningful results [Zhang et al., 1986b, 1991].

Our study found that photodynamic treatment of 2-ba-2-dmhb resulted in a 3–5-fold increase in intracellular ROS in a short time, and we found that most of the ROS came from mitochondria). Sodium azide is a scavenger of singlet oxygen, which can inhibit the increase of mitochondrial ROS, indicating that $^1O_2$ is the main source of mitochondrial reactive oxygen species after photodynamic treatment, which is also confirmed in ESR spin trapping experiment. The increase of mitochondrial calcium concentration or mitochondrial uncoupling agent treatment had little effect on ROS, indicating that the increase of mitochondrial reactive oxygen species induced by photodynamic does not depend on mitochondrial function. Our results clearly show that NO is produced on the mitochondria of MCF-7 cells during 2-butylamine-2-demethoxyhypocrellin (2-BA-2-DMHB) photodynamic treatment (Figs. 14-2e, 14-2g and 14-2h) [Lu et al., 2006, 2009]. ESR method was used to further determine the production of singlet oxygen and the increase of intracellular ROS and NO during photodynamic treatment. ESR method was used to further determine the production of singlet oxygen and the increase of

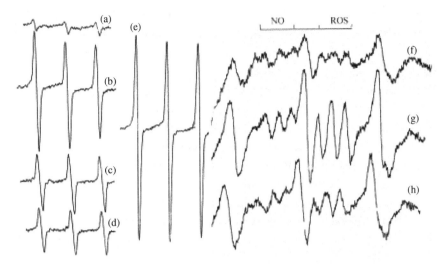

Figure 14-2. ESR spectra of spin trapped singlet oxygen, NO, and ROS. 2-BA-2-DMHB-PDT-induced singlet oxygen production (a) before light, (b) subjected to 20 J cm-2 light, (c) subjected to 40 J cm-2 light, (d) subjected to 40 J cm^{-2} light + 20 mg/ml cell lysis (protein content), and (e) subjected to 40 J cm-2 light + 20 mg/ml permeabilized cell (protein content). PDT-induced intracellular ROS and NO generation for (f) control group, (g) light group, and (h) light + NaN3 pretreatment group.

intracellular reactive oxygen species and NO during photodynamic treatment. The content of NO in cells significantly increased to 160% after 30 minutes of illumination in cells and 2-BA-2-DMHB. Intracellular ROS were measured by fluorescence spectrophotometer. After 2-BA-2-DMHB photodynamic treatment, the fluorescence intensity increased to 350% and the intensity could be basically stable at 500% after 30 minutes [Lu *et al.*, 2006, 2009].

After 12 hours of light treatment, the nuclear morphology showed obvious apoptotic changes, that is, pyknosis or fragmentation of the nucleus. The destruction of tumor tissue caused by photodynamic therapy is mainly due to the oxidative damage caused by singlet oxygen. The role of NO in apoptosis is complex. NO-induced apoptosis is usually associated with the activation of p53 gene, which upregulates cell cycle factors or pro-apoptotic proteins. We summarized the sequence of various apoptosis-related events in the photodynamic treatment mediated by

2-BA-2-DMHB in MCF-7 cells. After receiving red light irradiation, the photosensitizer in the cells produced a large number of $^1O_2$ and other free radicals, increased the level of intracellular reactive oxygen species, led to calcium influx, increased the concentration of intracellular free calcium ions, and activated cellular mitochondrial NOS enzyme. It produced a large amount of NO, promoted the expansion of mitochondria and the release of cytochrome c, and finally led to apoptosis. Pretreatment with sodium azide or ptio can reduce the changes of nuclear morphology, inhibit the up regulation of p53 and p21 genes, and improve cell viability, indicating that reactive oxygen species and NO are very important factors in photodynamic treatment induced apoptosis [Lu *et al.*, 2006, 2009].

The above research results show that radiation sensitizers can enhance the killing effect of radiation n on tumors by producing oxygen and NO free radicals before, during, and after radiation.

4. Oxidative stress in cancer immunotherapies

Immunotherapy (IMT) is now a core component of cancer treatment. It has been well-established that cancer cells are under constant oxidative stress, as reflected by elevated basal level of ROS and RNS, due to increased metabolism driven by aberrant cell growth. Cancer cells can adapt to maintain redox homeostasis through a variety of mechanisms. The prevalent perception about ROS and RNS is that they are one of the key drivers promoting tumor initiation, progression, metastasis, and drug resistance. Based on this notion, numerous antioxidants that aim to mitigate tumor oxidative stress have been tested for cancer prevention or treatment, although the effectiveness of this strategy has yet to be established. In recent years, it has been increasingly appreciated that ROS and RNS have a complex, multifaceted role in the tumor microenvironment, and that tumor redox can be targeted to amplify oxidative stress inside the tumor to cause tumor destruction. Accumulating evidence indicates that cancer immunotherapies can alter tumor redox to intensify tumor oxidative stress, resulting in ROS and RNS-dependent tumor rejection. Immune cells face a harsh metabolic environment that can significantly impair effector function. These tumor-mediated metabolic perturbations include hypoxia, oxidative stress, and metabolites of cellular energetics. Primarily through HIF-1-dependent processes, hypoxia invokes an immunosuppressive phenotype via altered molecular markers, immune cell trafficking, and angiogenesis. Additionally, oxidative stress can promote lipid peroxidation and endoplasmic reticulum stress, all associated with immune dysregulation.

A paper studied genomics to immunotherapy of ovarian clear cell carcinoma. Ovarian clear cell carcinoma (OCCC) is distinctive from other histological types of epithelial ovarian cancer, with genetic/epigenetic alterations, a specific immune-related molecular profile, and epidemiologic associations with ethnicity and endometriosis. These findings allow for the exploration of unique and specific treatments for ovarian clear cell carcinoma. Two major mutated genes in ovarian clear cell carcinoma are phosphatidylino-sitol 3-kinases (PIK3CA) and ARID1A, which frequently coexist with each other. Other genes' alterations also contribute to activation of the PI3K and dysregulation of the chromatin remodeling complex (e.g., ARID1B, and SMARKA4). Although the number of focal copy number variations is small in ovarian clear cell carcinoma, amplification is recurrently detected at chromosome 20q13.2, 8q, and 17q. Both expression and methylation profiling highlight the significance of adjustments to oxidative stress and inflammation. In particular, upregulation of HNF-1β resulting from hypomethylation contributes to the switch from anaerobic to aerobic glucose metabolism. Additionally, upregulation of HNF-1β activates STAT3 and NF-κB signaling and leads to immune suppression via production of IL-6 and IL-8 [Oda *et al.*, 2018]. A paper found that nanoscale coordination polymers induce immunogenic cell death by amplifying radiation therapy-mediated oxidative stress [Huang *et al.*, 2021]. In recent years, it has been increasingly appreciated that ROS and RNS have a complex, multifaceted role in the tumor microenvironment, and that tumor redox can be targeted to amplify oxidative stress inside the tumor to cause tumor destruction. Accumulating evidence indicates that cancer immunotherapies can alter tumor redox to intensify tumor oxidative stress, resulting in ROS and RNS-dependent tumor rejection [Aboelella *et al.*, 2021].

The above results indicate that the treatment of immune tumors is also related to oxidative stress, but the research of mothers is not too clear, and further research is needed. The recent progresses regarding the impact of ROS and RNS on cancer cells and various immune cells in the tumor microenvironment and the emerging ROS- and RNS-modulating strategies can be used in combination with cancer immunotherapies to achieve enhanced antitumor effects.

14.4 Anticancer effects of tea polyphenols

Epidemiological investigation found that drinking tea can reduce the risk of cancer. Scientific research have shown that tea polyphenols have clear anticancer effects in all stages of tumor induction, occurrence, and development, such as in skin cancer, lung cancer, gastrointestinal cancer, and other types of cancer etc. Important progress has been made in the recent years concerning the effects of green and black tea on health. Experimentation with new accurate tools provide useful information about the metabolism of tea components in the body, their mode of action as antioxidants at the cellular level, and their protective role in the development of cancer and other pathologies [Zhao, 2020].

14.4.1 *Epidemiological results of tea drinking to reduce cancer risk*

Epidemiological studies have found that drinking tea can reduce the risk of cancer, including lung cancer, oral cancer, breast cancer, and colon cancer etc. It can delay cancer onset of patients with a history of consuming over 10 cups of green tea per day, and absence of any severe adverse effects among volunteers who took 15 green tea tablets per day (2.25 g green tea extracts, 337.5 mg EGCG, and 135 mg caffeine) for six months. Cancer onset of patients who had consumed over 10 cups of green tea per day was 8.7 years later among females and three years later among males, compared with patients who had consumed under three cups per day [Fujiki *et al.*, 1999]. There are many reports on whether tea can prevent and treat cancer, and there are many detailed clinical experimental results. For example, 26 patients with prostate cancer undergoing radical prostatectomy took 1.3 g of tea polyphenols (containing 800 mg EGCG) every day. The results showed that the value of tumor markers decreased significantly, which provided strong support for the development of EGCG as a drug for the prevention and treatment of prostate cancer [Manjula *et al.*, 2019].

A statistical analysis of green tea and esophageal cancer in 2020, including 16 studies, explored the relationship between green tea intake and the risk of esophageal cancer through meta-analysis. A total of 20 studies were included. It was concluded that green tea can be used as a preventive factor for esophageal cancer [McLarty *et al.*, 2009]. To investigate the association between green tea consumption and esophageal cancer risk through meta-analysis. Another paper searched MEDLINE, Embase, Web of Science and Cochrane Library for studies on the relationship between green tea and esophageal cancer risk. The attributable hazard ratio (ARP) was calculated. A total of 16 studies were included. The highest relative risk of green tea consumption was 0.86. Since consumption of green tea and green tea consumption were negatively correlated with breast cancer risk, green tea consumption could be a protective factor for esophageal cancer [Yu *et al.*, 2020].

The risk of breast cancer is high for women in world. The search was performed using PubMed, Embase, and Web of Science databases. The generalised least square method and constrained cubic spline model were performed to assess the dose-response trends between green tea consumption and BC risk. A total of 16 studies were included and the pooled relative risks was 0.86 (95%CI: 0.75–0.99) for BC risk at the highest versus lowest levels of GT consumption. The results demonstrated that in China, 23.5% of people who do not suffer from breast cancer are associated with the consumption of green tea. Drinking green tea may have a positive effect on reducing breast cancer risk, especially in long-term, high doses [Wang *et al.*, 2020a]. Ovarian cancer accounts for 4% of female malignant tumors in the world. The random effect model was used to measure the comprehensive effect, and the adjusted indirect comparison method was used to compare the effects of green tea and black tea. Linear and nonlinear models were used to explore the dose-response relationship, and a total of 14 studies were included. In the binary meta-analysis stratified by tea type, it was observed that there was a significant negative correlation between ovarian cancer and drinking green tea [Zhang *et al.*, 2018]. A data analysis on the relationship between lung cancer and tea drinking habits of different populations included 42 literatures, 30 case-control studies, 14,578 patients with lung cancer, and 180,574 controls. The 12 cohort studies included 543,825 subjects, of which the 5,085 had lung cancer. The study found that compared with people who do not drink tea, people who drink tea have a lower risk of lung cancer (or = 0.80, 95% confidence interval: 0.73, 0.87). Compared with people who do not drink tea at ordinary times, drinking green tea, black tea, or other tea has protective effects [Guo *et al.*, 2019].

In addition, consumption of tea and non-Hodgkin's lymphoma (RR = 0.61) (95% confidence interval 0.38–0.99) [Mirtavoos-Mahyari *et al.*, 2019] and oral cancer (RR = 0.85) (95% confidence interval 0.75–0.93) [Liu *et al.*, 2016] and colon cancer risk, were reduced by 30–40%, and there was also a negative correlation. Another study shown that daily consumption of a cup of tea was associated with a decrease in risk of bladder cancer. A concentration of 100–400 μM catechins can induce U937 cell apoptosis in a concentration dependent manner, and the ability of different monomers to induce apoptosis depends on the gallic acid group on the B ring and the gallic acid group at the 3'position of the B ring. In other cell lines, EGCG was also found to inhibit the activity of CDK2 and CDK4, induce cell arrest in G1 phase, inhibit DNA synthesis and the amount of cells entering S phase, or induce cell arrest in G2/M phase. These studies show that the regulatory effect of catechins on cell cycle and the ability to induce tumor cell apoptosis are the basis of tea polyphenols' anti-cancer properties [Liu *et al.*, 2016].

Colorectal cancer (CRC) is a multifactorial disease, which is usually induced and developed through complex mechanisms, including the effects of diet and lifestyle, genomic abnormalities, changes in signaling pathways, inflammatory response, oxidative stress, ecological imbalance, and so on. Most animal studies and epidemiological studies have shown that tea polyphenols can prevent and treat colorectal cancer. Tea polyphenols inhibit the growth and metastasis of colorectal cancer by exerting anti-inflammatory, antioxidant or pro-oxidative, and pro-apoptotic effects. These effects are achieved through multiple levels of regulation. Many experiments have shown that TPS can regulate a variety of signal pathways in cancer cells, including mitogen-activated protein kinase pathway, phosphatidylinositol-3 kinase/Akt pathway, and Wnt pathway/β-Catenin pathway and 67 kDa laminin receptor pathway to inhibit proliferation and promote apoptosis. In addition, new research also shows that tea polyphenols can improve the immune system and reduce inflammatory response by regulating the composition of intestinal microbiota, so as to prevent the growth and metastasis of colorectal cancer [Wang *et al.*, 2020a, 2020b]. Black tea polyphenols, mainly consist of high molecular weight species that predominantly persist in the colon. There, they can undergo a wide range of bioconversions by the resident colonic microbiota but can in turn also modulate gut microbial diversity. The impact of back tea polyphenols on colon microbial composition can now be assessed by microbiomics technologies.

However, in a review update, they included in total 142 completed studies (11 experimental and 131 nonexperimental) and two ongoing studies. After stratifying the analysis according to study design, they found strongly conflicting results for some cancer sites: oesophageal, prostate and urinary tract cancer, and leukaemia showed an increased RR in cohort studies and a decreased RR or no difference in case-control studies [Filippini *et al.*, 2020].

The above epidemiological investigation shows that drinking tea is effective in preventing most cancers, but the results of some cancer sites are contradictory, so more-strict epidemiological investigation and clinical verification should be carried out.

14.4.2 *Tea polyphenols prevent various mutations and cancers*

A large number of scientific studies have found that tea polyphenols can prevent various cancer and inhibit the growth of cancer cells and mutations. The administration of green tea, black tea, or EGCG and theaflavin can inhibit the growth of established nonmalignant and malignant tumors in tumor. EGCG and theaflavin demonstrated potential anticancer effects in different preclinical and clinical studies. DNA mutation is a very important step in carcinogenesis and elevated levels of oxidative DNA damage have been monitored in a variety of tumors. Based on recent evidence that tea consumption contributes to a decreased incidence of human carcinomas, a number of investigators have focused on the mechanisms of cancer prevention by tea extracts, especially green tea polyphenols. Green and black tea can inhibit DNA synthesis and enhance apoptosis in both non-malignant and malignant tumors. In the setting of *in vivo* cancer prevention studies, administration of the tea and tea polyphenols at preinitiation stages only showed partial prevention, whereas continuous administration showed potential effect in restriction of carcinogenesis in the body's multiple organs at early premalignant stages throughout the experiment. Similar to different *in vitro* cancer cell models, treatment after initiation stages showed potential therapeutic efficacy *in vivo*. However, the mechanisms of prevention and therapy were found to be similar regardless of tea and its polyphenols. Tea polyphenols mainly serve as antioxidants and induce the detoxification system, thereby inhibiting carcinogen metabolism and cancer initiation. Additionally, they could inhibit self-renewal, proliferation, and survival of the tumor-initiating population in restriction of the carcinogenesis progression from cancer initiation and promotion. This might be a result of the modulation of membrane organization, interaction with DNA/RNA/proteins and epigenetic modifications, as well as regulation of cellular replicative potential by the tea polyphenols.

Most black and green tea extracts strongly inhibited neoplastic transformation in mouse mammary organ cultures, rat tracheal epithelial cells, and human lung tumor epithelial cells. Nearly all tea fractions strongly inhibited benzo[a]pyrene adduct formation with human DNA. A paper studied the inhibition of activator protein 1 activity and cell growth by green tea and black tea polyphenols in H-ras-transformed cells. *Ras* gene mutation, which perpetually turns on the growth signal transduction pathway, occurs frequently in many cancer types. This study investigated the ability of different pure green and black tea polyphenols to inhibit this Ras signaling pathway. The major green tea polyphenols catechins, EGCG, (-)-epigallocatechin (EGC), (-)-epicatechin-3-gallate (ECG), (-)-epicatechin, and their epimers, and black tea polyphenols, theaflavin, theaflavin-3-gallate, theaflavin-3'-gallate, and theaflavin-3,3'-digallate (TFdiG), were compared with respect to their ability to inhibit the growth of 30.7b Ras 12 cells and AP-1 activity. All the tea polyphenols, except (-)-epicatechin, showed strong inhibition of cell growth and AP-1 activity.

EGCG lowered the level of c-jun, whereas TFdiG decreased the level of fra-1. These results suggest that tea polyphenols inhibited AP-1 activity and the mitogen-activated protein kinase pathway, which contributed to the growth inhibition [Chung *et al.*, 1999].

Another paper studied the inhibition of Fms-like tyrosine kinase 3 (FLT3) expression by green tea catechins in FLT3-mutated acute myeloid leukemia cells. Acute myeloid leukemia is a heterogeneous disease characterized by a block in differentiation and uncontrolled proliferation. FLT3 is a commonly mutated gene found in acute myeloid leukemia patients. FLT3-ITD mutation is the most important risk factor for predicting leukemia recurrence. In clinical trials, the presence of a FLT3-ITD mutation significantly correlates with an increased risk of relapse and dismal overall survival. Therefore, activated FLT3 is a promising molecular target for acute myeloid leukemia therapies. This study showed that green tea polyphenols including EGCG, EGC, and ECG suppress the proliferation of acute myeloid leukemia cells. Interestingly, EGCG, EGC, and ECG showed the inhibition of FLT3 expression in cell lines harboring FLT3 mutations. In the THP-1 cells harboring FLT3 wild-type, EGCG showed the suppression of cell proliferation but did not suppress the expression of FLT3 even at the concentration that suppress 100% cell proliferation. Moreover, EGCG-, EGC- and ECG-treated cells showed the suppression of MAPK, AKT, and STAT5 phosphorylation [Ly *et al.*, 2013].

A study tested the anti-mutagenic activities of Chinese green tea water extract, tea polyphenols, and tea catechins (EGCG, ECG, EGC, EC). In the V79 cell forward gene mutation and V79 cell cytokinesis-block micronuclei tests, all samples showed significant inhibitory effects on mitomycin, which indicate their effects on the initiation stage of chemical carcinogenesis. However, none of the catechins and tea polyphenols had effects as strong as the tea water extract on the basis of their relative contents in tea. ECG and EGCG were the most potent in the four catechins. Using the V79 cell metabolic co-operation test as an indicator for the promotion stage, tea water extract and tea polyphenols showed weak inhibitory effects and the individual catechins showed much stronger inhibition [Han, 1997]. Investigation of the mechanisms for inhibiting tumor necrosis factor-induced IL-6 production by which EGCG, ECG, and theaflavin in HGFs found that tumor necrosis factor increased IL-6 production in human gingival fibroblasts in a concentration-dependent manner. EGCG, ECG, and theaflavin prevented tumor necrosis factor-mediated IL-6 production in human gingival fibroblasts. EGCG, ECG, and theaflavin prevented necrosis factor receptor-induced extracellular signal-regulated kinase (ERK), c-Jun N-terminal kinase (JNK), and NF-kB activation in human gingival fibroblasts. Inhibitors of ERK, JNK, and NF-kB decreased tumor necrosis factor-induced IL-6 production. In addition, EGCG, ECG, and theaflavin attenuated tumor necrosis factor receptor expression on human gingival fibroblasts [Hosokawa *et al.*, 2010].

The above research results show that tea polyphenols EGCG, ECG, and theaflavin, mainly as antioxidants, can inhibit the metabolism of carcinogens and the occurrence of cancer, prevent various cancers, and inhibit the mutation of cancer cells.

14.4.3 *Green and black tea and tea polyphenols inhibit tumor growth*

It has been demonstrated in various animal models that the oral administration of green tea extracts in drinking water can inhibit tumor growth. The administration of green tea, black tea, or EGCG inhibited the growth of established nonmalignant and malignant tumors in tumor-bearing mice. Oral administration with black tea inhibited DNA synthesis and enhanced apoptosis in both nonmalignant and malignant tumors in tumor-bearing mice [Conney *et al.*, 1999]. We found green tea polyphenols in combination with copper (II) induced apoptosis in Hela cells [Li *et al.*, 2002].

This study investigated the time-course anticancer effects of EGCG on human ovarian cancer cells to provide insights into the molecular-level understanding of growth suppression mechanism involved in EGCG-mediated apoptosis and cell cycle arrest. EGCG exerts a significant role in suppressing ovarian cancer cell growth. Also, EGCG showed growth inhibitory effects in each cell line in a dose-dependent fashion and induced apoptosis and cell cycle arrest. The cell cycle was arrested at the G(1) phase by EGCG in SKOV-3 and OVCAR-3 cells. In contrast, the cell cycle was arrested in the G(1)/S phase arrest in PA-1 cells. EGCG differentially regulated the expression of genes and proteins (Bax, p21, Retinoblastoma, cyclin D1, CDK4, Bcl-X (L)) more than two-fold, showing a possible gene regulatory role of EGCG. The continual expression in p21WAF1 suggests that EGCG acts in the same way with p53 proteins to facilitate apoptosis after EGCG treatment. Moreover, Bax, PCNA, and Bcl-X are important in EGCG-mediated apoptosis [Huh *et al.*, 2004]. A study investigated the anticancer effects of EGCG in human papillomavirus (HPV)-16 associated cervical cancer cell line, CaSki cells. The growth inhibitory mechanism(s) and regulation of gene expression by EGCG were also evaluated. EGCG showed growth inhibitory effects in CaSki cells in a dose-dependent fashion. Cell cycles at the G1 phase were arrested at 35 μM EGCG, suggesting that cell cycle arrests might precede apoptosis. When CaSki cells were tested for their gene expression using 384 cDNA microarray, an alteration in the gene expression was observed by EGCG treatment. EGCG downregulated the expression of 16 genes over time more than two-fold. In contrast, EGCG upregulated the expression of four genes more than two-fold, suggesting a possible gene regulatory role of EGCG. This data supports that EGCG can inhibit cervical cancer cell growth through induction of apoptosis and cell cycle arrest as well as regulation of gene expression *in vitro* [Ahn *et al.*, 2003].

The above results indicate that tea polyphenols EGCG, and theaflavin, mainly as antioxidants, can induce apoptosis, block cell cycle, regulate gene expression *in vitro*, and inhibit the growth of malignant tumors.

14.4.4 *Tea and tea polyphenol cancer chemopreventive, anti-mutagenic, and anti-proliferative properties*

Tea is one of the most highly consumed beverages worldwide and has been linked to improvements in human health. Tea contains many active components, including tea polyphenols, tea polysaccharides, L-theanine, tea pigments, and caffeine among other common components. Several studies have identified components in tea that can directly or indirectly reduce carcinogenesis with some being used in a clinical setting. Many studies, *in vitro* and *in vivo*, have focused on the mechanisms that functional components of tea utilized to protect against cancer. One particular mechanism is an improvement in antioxidant capacity. Other mechanisms include anti-pathogen, anti-inflammation, and alterations in cell survival pathways. Tumor volume was decreased significantly in mice consuming green tea, and tumor size was significantly correlated with green tea polyphenol content in tumor tissue. There was a significant reduction in hypoxia-inducible factor 1-alpha and vascular endothelial growth factor protein expression. Green tea consumption significantly reduced oxidative DNA and protein damage in tumor tissue as determined by 8-hydroxydeoxyguanosine/deoxyguanosine ratio and protein carbonyl assay, respectively. Methylation is known to inhibit anti-oxidative enzymes such as glutathione S-transferase pi to permit reactive oxygen species promotion of tumor growth. Green tea inhibited tumor 5-cytosine DNA methyltransferase 1 mRNA and protein expression significantly, which may contribute to the inhibition of tumor growth by reactivation of anti-oxidative enzymes.

The chemistry of tea and tea polyphenols, and their antioxidant potential, immune-potentiating properties, and mode of action against various cancer cell lines that showed its potential as a chemopreventive agent against colon, skin, lung, prostate, and breast cancer. In a study, black tea extracts (hot aqueous, polyphenols, and theaflavins) and green tea extracts (hot aqueous, polyphenols, epicatechin, epicatechin gallate, epigallocatechin, and epigallocatechin gallate) were tested in nine standardized cell culture assays for comparative cancer chemopreventive properties. Most black and green tea extracts strongly inhibited neoplastic transformation in mouse mammary organ cultures, rat tracheal epithelial cells, and human lung tumor epithelial cells. Nearly all tea fractions strongly inhibited benzo[a]pyrene adduct formation with human DNA. Induction of phase II enzymes, glutathione-S-transferase, and quinone reductase, were enhanced by nearly all tea fractions, while glutathione was induced by only a few fractions. Ornithine decarboxylase activity was inhibited by nearly all the green tea fractions, but none of the black tea fractions. 12-O-tetradecanoylphorbol-13-acetate-induced free radicals were inhibited by most tea fractions. These results provide strong evidence of both anti-mutagenic, anti-proliferative, and anti-neoplastic activities for both black and green tea extracts [Steele *et al.*, 2000].

Much of the cancer chemo-preventive properties of green tea are mediated by EGCG that induces apoptosis and promotes cell growth arrest, by altering the expression of cell cycle regulatory proteins, activating killer caspases, and suppressing NF-κB activation. Besides, it regulates and promotes IL-23 dependent DNA repair and stimulates cytotoxic T cells activities in a tumor microenvironment. It also blocks carcinogenesis by modulating the signal transduction pathways involved in cell proliferation, transformation, inflammation, and metastasis. Green tea is also supposed to enhance humoral and cell-mediated immunity, decrease the risk of certain cancers, and may have certain advantage in treating inflammatory disorders. Green tea and its components effectively mitigate cellular damage due to oxidative stress [Butt & Sultan, 2009].

Dietary administration of green tea polyphenols inhibited the formation of aberrant crypt foci (ACF) in the colon of azoxymethane (AOM)-treated F344 rats. Herein, we reported cancer-preventive activity of polyphenols using colorectal cancer as an end point. F344 rats were given two weekly injections of AOM, and then maintained on a 20% high-fat diet with or without 0.24% PPE for 34 weeks. In the control group, 83% of rats developed colorectal tumors. Green tea polyphenols treatment significantly increased the plasma and colonic levels of tea polyphenols and decreased tumor multiplicity and tumor size. Histological analysis indicated that polyphenols significantly decreased the incidence of adenocarcinoma and the multiplicity of adenocarcinoma as well as the multiplicity of adenoma. Green tea polyphenols treatment significantly decreased plasma levels of pro-inflammatory eicosanoids, prostaglandin E2, and leukotriene B4. It also decreased β-catenin nuclear expression, induced apoptosis, and increased expression levels of RXRα, β, and γ in adenocarcinomas [Hao *et al.*, 2017].

The above research shows that tea polyphenols and theaflavins significantly inhibits the expression of 5-cytosine DNA methyltransferase 1 mRNA and protein in tumor tissue, which may inhibit tumor growth by reactivating antioxidant enzymes and significantly reduce oxidative DNA and protein damage in tumor tissue. EGCG can change the expression of cell cycle regulating proteins, activate killer caspases, and inhibit NF-κB activation, induce cell apoptosis, and promote cell growth arrest.

14.4.5 *Tea and tea polyphenols inhibit cell hyperproliferation, tumorigenesis, and tumor progression*

Both green and black tea have been shown to inhibit lung tumorigenesis in laboratory animal experiments. Green tea inhibited N-nitrosodiethylamine-induced lung tumor incidence and multiplicity in female A/J mice when tea was given either during the carcinogen treatment period or during the post-carcinogen treatment period. In a separate tumorigenesis model, both decaffeinated black tea and decaffeinated green tea inhibited 4-(methylnitrosamino)-1-(3-pyridyl)-1-butanone (NNK)-induced lung tumor formation. Studies in which tea was administered during different time periods in relation to the NNK suggest that tea can inhibit lung tumorigenesis at both the initiation and promotion stages. The antiproliferative effects of tea may be responsible for these anti-carcinogenic actions. Black tea polyphenol preparations decreased NNK-induced hyperproliferation. Black tea also inhibited the progression of pulmonary adenomas to adenocarcinomas and the formation of spontaneous lung tumors in A/J mice. Growth inhibition by various tea polyphenols has been demonstrated in human lung H661 and H1299 cells [Yang *et al.*, 1998]. Black tea extract enriched in theaflavins inhibited the chymotrypsin-like activity of the proteasome and proliferation of human multiple myeloma cells in a dose-dependent manner. Also, an isolated theaflavin can bind to and inhibit the purified 20S proteasome, accompanied by suppression of tumor cell proliferation, suggesting that the tumor proteasome is an important target whose inhibition is at least partially responsible for the anticancer effects of black tea [Mujtaba & Dou, 2012].

EGCG is widely studied for its anti-cancer properties. EGCG is effective *in vivo* at micromolar concentrations, suggesting that its action is mediated by interaction with specific targets that are involved in the regulation of crucial steps of cell proliferation, survival, and metastatic spread. Recently, several proteins have been identified as EGCG direct interactors. Among them, the trans-membrane receptor 67LR has been identified as a high affinity EGCG receptor. 67LR is a master regulator of many pathways affecting cell proliferation or apoptosis, also regulating cancer stem cells

(CSCs) activity. EGCG was also found to be interacting directly with Pin1, TGFR-II, and metalloproteinases (MMPs) (mainly MMP2 and MMP9), which respectively regulate EGCG-dependent inhibition of NF-κB, epithelial-mesenchimal transaction (EMT), and cellular invasion. EGCG interacts with DNA methyltransferases (DNMTs) and histone deacetylases (HDACs), which modulates epigenetic changes [Negri *et al.*, 2018].

Many studies have shown that tea polyphenols have anticancer activity. In one study, the effects of theaflavin-3-3'- diglyceride (TF), the main theaflavin monomer in black tea, with reducing agent ascorbic acid (AA), and EGCG, the main polyphenol in green tea, and coordinated with reducing agent ascorbic acid, on the viability and cell cycle of human lung adenocarcinoma SPC-A-1 cells, were investigated. The 50% inhibitory concentrations of TF, EGCG, and AA on SPC-A-1 cells were 4.78 μmol/L, 4.90 μmol/L, and 30.62 μmol/L, respectively. When the molar ratio of TF to AA (TF + AA) and EGCG to AA (EGCG + AA) was 1:6, the inhibition rates of SPC-A-1 cells were 54.4% and 45.5%, respectively. Flow cytometry analysis showed that TF + AA and EGCG + AA could significantly increase the number of cells in G(0)/G(1) phase of SPC-A-1 cell cycle, from 53.9% to 62.8% and 60.0%, respectively. At 50% inhibitory concentration, G(0)/G(1) accounted for 65.3%. Therefore, TF + AA and EGCG + AA have a synergistic inhibitory effect on the proliferation of SPC-A-1 cells and can significantly maintain SPC-A-1 cells in G(0)/G(1) phase. The results show that the combination of theaflavin and green tea polyphenols with reducing agent ascorbic acid can improve its anticancer activity [Hudlikar *et al.*, 2019].

This study attempted to identify signaling pathways or target molecules regulated by each of or a mixture of green tea polyphenols, including epicatechin (EC), epicatechin-3-gallate (ECG), epigallocatechin (EGC), and EGCG, in the human lung cancer cell line A549. ECG, EGC, and a catechin mixture, in addition to EGCG, significantly decreased cell viability. In contrast, caspase 3/7 activity, an apoptosis indicator, was specifically induced by EGCG. By conducting a series of luciferase-based reporter assays, we revealed that the catechin mixture only upregulates the p53 reporter. EGCG was a more potent inducer of p53-dependent transcription, and this induction was further supported by the induced level of p53 protein. RNA interference (RNAi)-mediated p53 knockdown completely abolished EGCG-induced apoptosis [Yamauchi *et al.*, 2009].

Prevention of oxidative stress, modulation of carcinogen metabolism, and prevention of DNA damage have been suggested as possible cancer preventive mechanisms for tea and tea polyphenols. We studied DNA oxidative damage mediated by (OP) $2Cu^{2+}$ and the protective effects of four tea polyphenols on DNA oxidative damage by chemiluminescence and ESR methods. At the same time, the different protective mechanisms of tea polyphenols were discussed. All four compounds showed a strong inhibition on signals of chemiluminescence and spin trapping DMSO-OH and excellent protective effects on oxidative damage of DNA. The inhibitory efficiency of the four compound decreases as the following order: ECG>EC>EGCG>EGC. The efficiency of the inhibition could be explained by both the antioxidant capacities and interaction between antioxidants and microenvironment of reaction. Experimental studies have shown that tea polyphenols have clear anticancer effects in all stages of tumor induction, occurrence, and development, as well as skin cancer, lung cancer, gastrointestinal cancer, and other types of cancer. In the induction stage of tumor, tea polyphenols can react with carcinogens activated by phase I enzymes to neutralize them before damaging cells, such as inhibiting the formation of nitrosamines and heterocyclic amines [Nie *et al.*, 2001].

Above experimental studies have shown that tea polyphenols have clear anticancer effects in all stages of tumor induction, occurrence, and development, inhibit cell hyper-proliferation, tumorigenesis, and tumor progression. Tea polyphenols and theaflavins can bind and inhibit the purified 20S proteasome and inhibit the proliferation of tumor cells.

14.4.6 *Tea polyphenols can resist tumor promotion and invasion*

Green tea polyphenols provide protection against tumor promotion and tumor progression in mouse skin. A study evaluated the protective effect of green tea polyphenols against the induction and subsequent progression of papillomas to squamous cell carcinomas (SCCs) in experimental protocols where papillomas were developed with a low or high probability of their malignant conversion. Topical application of green tea polyphenols (6 mg/animal) 30 minutes prior to that of 12-O-tetradecanoylphorbol-13-acetate (TPA) either once a week for five weeks (high risk TPA protocol) or once a week for 20 weeks (low risk TPA protocol) or mezerein (MEZ) twice a week for 20 weeks (high risk MEZ protocol) in 7,12-dimethylbenz[a] anthracene (DMBA)-initiated mouse skin resulted in significant protection against skin tumor promotion in terms of tumor incidence, multiplicity, and tumor volume/mouse at the termination of the experiment at 20 weeks. Green tea polyphenols resulted in significant protection against the malignant conversion of papillomas to SCC in all the protocols employed. At the termination of the experiment at 51 weeks, these protective effects were evident in terms of mice with carcinomas, carcinomas per mouse, and percent malignant conversion of papillomas to carcinomas. The kinetics of malignant conversion suggest that a subset of papillomas formed in the early phase of tumor promotion in all the protocols had a higher probability of malignant conversion into SCCs because all the positive control groups (acetone treated) produced nearly the same number of carcinomas at the end of the progression period. In the GTP-treated group of animals, the number of carcinomas formed was less, which shows the ability of green tea polyphenols to protect against the malignant conversion of papillomas of higher probability of malignant conversion to SCCs [Katiyar *et al.*, 1997].

Prostate cancer is one of the most frequently diagnosed male neoplasia in the Western countries, which is in agreement with this gland being particularly vulnerable to oxidative stress processes, often associated with tumorigenesis. Study showed that polyphenols from green tea could prevent prostate cancer. It showed the beneficial effects of green tea polyphenols in chemoprevention of cancer with particular emphasis on the involvement of Matrix metalloproteinases (MMPs) in prostate cancer. MMP play a crucial role in the development and metastatic spread of cancer. One of the earliest events in the metastatic spread of cancer is the invasion through the basement membrane and proteolytic degradation of the extracellular matrix proteins, such as, collagens, laminin, elastin, and fibronectin, and non-matrix proteins. MMPs are the important regulators of tumor growth, both at the primary site and in distant metastases. Metastatic spread of cancer continues to be the greatest barrier in the prevention or cure of cancer. The recognition that MMPs facilitate tumor cell growth, invasion, and metastasis of cancer has led to the development of MMP inhibitors as cancer therapeutic agents. Understanding the molecular mechanism of metastasis is also crucial for the design and effective use of novel therapeutic strategies to combat metastases [Katiyar, 2006].

A paper studied the inhibitory effect of EGCG on growth and invasion in human biliary tract carcinoma cells. The human biliary tract carcinoma cells (TGBC-2, SK-ChA-1, and NOZC-1) were treated with different doses of EGCG (0 mM, 25 mM, 50 mM, 100 mM, and 200 mM) for 48 hours in cell medium. EGCG and green tea extract showed inhibitory effects on the growth of lung and mammary cancer cell lines. All human biliary tract cancer cells studied showed a significant suppression of cell growth by EGCG treatment in a dose-dependent manner (27.2%, 16.0%, and 10.1%, in TGBC-2, SK-ChA-1, and NOZC-1, respectively, at the dose of 200 mM). EGCE treatment also produced a significant suppression of invasive ability of the carcinoma cells (12.6%, 11.2%, 7.9%, in TGBC-2, SK-ChA-1, and NOZC-1, respectively, at a dose of 100 mM). These data indicated that EGCG might be a potent biological inhibitor of human biliary tract cancers, reducing their proliferative and invasive activities. EGCG was also tested in human pancreatic carcinoma cells. The cells were treated with different doses of EGCG (0 µl/L, 25 µl/L, 50 µl/L, 100 µl/L, and 200 µl/L) for 48 hours in culture medium. The growth of all three pancreatic carcinoma cells was significantly suppressed by EGCG treatment in a dose-dependent manner. EGCG treatment caused significant suppression of the invasive ability of pancreatic carcinoma cells PANC-1, MIA PaCa-2, and BxPC-3 but did not affect the cell cycle protein cyclin D1. Proliferation was inhibited with the addition of EGCG in a dose-dependent manner. EGCG (200 µl/L) produced a profound growth suppression of HLF cells (24.5%). Cell invasion was also inhibited with pre-incubation of 100 µl/L of EGCG (10.2%). In addition to the antitumor effects, neurite-like conformational changes of HLF cells were observed. Addition of EGCG (100 µl/L) showed the expression of RAGE on cell surface in accordance with the morphological changes [Takada *et al.*, 2002a, 2002b, 2002c].

A paper studied the inhibition of cyclin-dependent kinases 2 and 4 activities as well as induction of cyclin-dependent kinase 2 (Cdk) inhibitors p21 and p27 during growth arrest of human breast carcinoma cells by EGCG. DNA flow cytometric analysis indicated that 30 µM of EGCG blocked cell cycle progression at G1 phase in asynchronous MCF-7 cells. In addition, cells exposed to 30 µM of EGCG remained in the G1 phase after release from aphidicolin block. Over a 24-hour exposure to EGCG, the retinoblastoma protein (Rb protein) changed from hyper- to hypo-phosphorylated form and G1 arrest developed. The protein expression of cyclin D1 and cyclin E reduced slightly under the same conditions. Immuno-complex kinase experiments showed that EGCG inhibited the activities of cyclin-dependent kinase 2 (Cdk2) and Cdk4 in a dose-dependent manner in the cell-free system. As the cells were exposed to EGCG (30 µM) over 24 hours, a gradual loss of both Cdk2 and Cdk4 kinase activities occurred. EGCG also induced the expression of the Cdk inhibitor p21 protein and this effect correlated with the increase in p53 levels. The level of p21 mRNA also increased under the same conditions. In addition, EGCG also increased the expression of the Cdk inhibitor p27 protein within six hours after EGCG treatment [Liang *et al.*, 1999].

The above research results show that EGCG and green tea extract can inhibit the growth of lung cancer and breast cancer cell lines, green tea polyphenols can significantly prevent the malignant transformation from papilloma to squamous cell carcinoma, and tea polyphenols can significantly inhibit the invasive ability of cancer cells.

14.5 The mechanism of anticancer effects by tea polyphenols

Tea polyphenols display many anti-carcinogenic properties. Numerous studies have revealed the molecular mechanisms of the anti-cancer effects of tea polyphenols, including their inhibitory effects on cancer cell proliferation, tumor growth, angiogenesis, metastasis, and inflammation as well as inducing apoptosis, signaling, cell cycle, and various malignant behaviors. In addition, they can modulate immune system response and protect normal cells against free radical damage. Their anti-oxidative actions help scavenge ROS and RNS and downregulate NF-κB to produce favorable anti-inflammatory effects. Meanwhile, pro-oxidant actions appear to promote ROS and RNS generation leading to the activation of 5'-AMP-activated protein kinase, which modulates different enzymes and factors with health beneficial roles. Tea polyphenols can inhibit cell-proliferation and cell cycle arrest by suppressing the NF-kB pathway in various cancers. EGCG is effective *in vivo* at micro-molar concentrations, suggesting that its action is mediated by interaction with specific targets that are involved in the regulation of crucial steps of cell proliferation, survival, and metastatic spread. Recently, several proteins have been identified as EGCG direct interactors. Among them, the trans-membrane receptor 67kDa laminin receptor (67LR) has been identified as a high affinity EGCG receptor. 67LR is a master regulator of many pathways affecting cell proliferation or apoptosis and regulating cancer stem cells (CSCs) activity. EGCG was also found to interact directly with Pin1 and metalloproteinases (MMPs), which respectively regulate EGCG-dependent inhibition of NF-kB, epithelial-mesenchimal transaction (EMT), and cellular invasion. EGCG interacts with DNA methyltransferases (DNMTs) and histone deacetylases (HDACs), which modulates epigenetic changes. The bulk of this novel knowledge provides information about the mechanisms of action of EGCG and may explain its onco-suppressive function [Negri *et al.*, 2018]. The most chemical or biochemical reactions of tea polyphenols have concentrated on the B ring of the C6-C3-C6 skeleton. This novel knowledge provides information about the mechanisms of action of tea polyphenols and may explain their onco-suppressive function.

14.5.1 *Inhibitory effect of tea polyphenols on carcinogen-induced cancers*

Some dietary factors, genetic factors, and gene mutations can contribute to the development of cancer. There are also some smoked and barbecued foods that contain some substances which are carcinogens and some chemical radiation or nuclear pollution that are carcinogens. A study shown that green tea, administered as drinking water, inhibits lung tumor development in A/J mice treated with 4-(methylnitrosamino)-1-(3-pyridyl)-1-butanone (NNK), a potent nicotine-derived lung carcinogen found in tobacco. The levels of 8-hydroxydeoxyguanosine (8-OH-dG), a marker of oxidative DNA damage, were significantly suppressed in mice treated with green tea or EGCG. The oxidation products found in black tea, thearubigins and theaflavins, also possess antioxidant activity, suggesting that black tea may also inhibit NNK-induced lung tumorigenesis. Indeed, bioassays in A/J mice have shown that black tea given as drinking water retarded the development of lung cancer caused by NNK [Chung, 1999].

Aflatoxin is one of the pathogenic factors of liver cancer, which is also a carcinogen. The active compounds in tea, particularly tea polyphenols, can directly or indirectly scavenge ROS and RNS to reduce oncogenesis and cancer-ometastasis. Many experiments have demonstrated that tea polyphenols can modulate several signaling pathways in cancer cells, including the mitogen-activated protein kinase pathway, phosphatidylinositol-3 kinase/Akt pathway, Wnt/β-catenin pathway, and 67 kDa laminin receptor pathway, to inhibit proliferation and promote cell apoptosis.

A present study was to compare the antioxidant potential of lipophilic tea polyphenols (LTP) against the one of naturally-occurring water-soluble green tea polyphenols (GTP) in a two-stage model of diethylnitrosamine (DEN)/phenobarbital (PB)-induced hepatocarcinogenesis in Sprague-Dawley rats. Histopathological and electron microscopic examination of liver tissue confirmed the protective effect of LTP on DEN/PB-induced liver damage and pre-carcinogenesis. LTP treatment significantly increased total antioxidant capacity (T-AOC) and glutathione peroxidase (GSH-Px) activity in liver tissues. Immuno-histochemical detection of cellular nuclear factor erythroid-2-related factor-2 (Nrf2) and peroxiredoxin-6 (P6) indicated a downregulation in Nrf2 and upregulation of P6 expression in the liver of LTP-supplemented rats [Zhou *et al.*, 2014].

Research shows that smoking can cause cancer of the mouth, esophagus and lungs, and tea can inhibit these cancers. In this study, when green tea or its main polyphenol EGCG was given, the mice given tobacco nitrosamine 4-(methyl-nitrosamine)-1-(3-pyridyl)-1-butanone (NNK) had significantly less lung tumors than the control group. The active ingredients in tea, especially tea polyphenols, can directly or indirectly ROS and RNS, reduce the occurrence of tumors and oncosuppressis [Yan *et al.*, 2010]. Theaflavins and thearubin are the most abundant polyphenols in black tea, and the understanding of their dose-related effects is limited. In one study, 0.75%, 1.5%, and 3% black tea were used as raw materials to investigate the effects of different doses of black tea extracts on the biochemical parameters and lung carcinogenicity of a/J mice. After pretreatment with tea red polymerized polyphenols, the expression and activity of cytochrome P450 isozymes in liver and lung tissues decreased, and the expression and activity of phase I enzymes in lung tissues induced by carcinogens, and the number and intensity of DNA adducts in liver and lung decreased significantly in a dose-dependent manner. The content of polyphenols in black tea (1.5%, 3%) decreased in dose and relationship with the incidence rate and diversity of lung cancer and was further related to the proliferation and apoptosis of benazepril and nitrosamine models. In conclusion, the dose-dependent chemo-preventive effects of theaflavins- and thearubin-polymerized polyphenols include inhibiting carcinogenesis initiation (inducing the second stage and inhibiting the first stage enzymes induced by carcinogens, resulting in a decrease in DNA adducts) and inhibiting carcinogenesis promotion (reducing cell proliferation and apoptosis, reducing incidence rate, and/or increasing the diversity of lung lesions). Theaflavins- and thearubin-polymerized polyphenols were observed in A/J mice without obvious toxicity [Hudlika *et al.*, 2019].

14.5.2 *Tea polyphenols inhibit angiogenesis, tumor growth*

The development of new blood vessels from a pre-existing vasculature (also known as angiogenesis) is required for many physiological processes including embryogenesis and post-natal growth. However, pathological angiogenesis is also a hallmark of cancer and many ischaemic and inflammatory diseases. Green tea and its polyphenolic substances (like catechins) show chemo-preventive and chemotherapeutic features in various types of cancer and experimental models for human cancers. The tea catechins, including EGCG, have multiple effects on the cellular proteome and signalome. Note that the polyphenolic compounds from green tea can change the miRNA expression profile associated with angiogenesis in various cancer types [Rashidi *et al.*, 2017].

Polyphenols in brewed green tea inhibit prostate tumor xenograft growth by localizing to the tumor and decreasing oxidative stress and angiogenesis. It has been demonstrated in various animal models that the oral administration of green tea (GT) extracts in drinking water can inhibit tumor growth. GT was administered instead of drinking water to male severe combined immunodeficiency (SCID) mice with androgen-dependent human LAPC4 prostate cancer cell subcutaneous xenografts. Tumor volume decreased significantly in mice consuming GT and tumor size significantly correlated with GT polyphenol (GTP) content in tumor tissue. There was a significant reduction in hypoxia-inducible factor 1-alpha and vascular endothelial growth factor protein expression. GT consumption significantly reduced oxidative DNA and protein damage in tumor tissue as determined by 8-hydroxydeoxyguanosine/deoxy-guanosine ratio and protein carbonyl assay, respectively. Methylation is known to inhibit anti-oxidative enzymes such as glutathione S-transferase pi to permit reactive oxygen species promotion of tumor growth. GT inhibited tumor 5-cytosine DNA methyltransferase 1 mRNA and protein expression significantly, which may contribute to the inhibition of tumor growth by the reactivation of anti-oxidative enzymes [Henning *et al.*, 2012]. Tea polyphenols display many anti-carcinogenic properties, including their inhibitory effects on cancer cell proliferation, tumor growth. Tea polyphenols with vitamin C, amino acids, and other micronutrients (EPQ) demonstrated significant suppression of ovarian cancer ES-2 xenograft tumor growth and suppression of ovarian tumor growth from IP injection of ovarian cancer A-2780 cells [Niedzwiecki *et al.*, 2016].

The results suggest that the inhibition of tumor growth is due to GTP-mediated inhibition of oxidative stress and angiogenesis in the tumor. The studies advance our understanding of tumor growth inhibition by brewed GT in an animal model by demonstrating tissue localization of GTPs in correlation with inhibition of tumor growth.

14.5.3 *Tea polyphenols inhibit metastasis*

Metastasis is the deadliest aspect of cancer and results from several interconnected processes including cell proliferation, angiogenesis, cell adhesion, migration, and invasion into the surrounding tissue. The appearance of metastases in organs distant from the primary tumor is the most destructive feature of cancer. Metastasis remains the principal cause of the deaths of cancer patients despite decades of research aimed at restricting tumor growth. Therefore, inhibition of metastasis is one of the most important issues in cancer research. Tea polyphenols, display many anti-carcinogenic properties including their inhibitory effects on angiogenesis and metastasis. Tea polyphenols with vitamin C, amino acids, and other micronutrients (EPQ) demonstrated significant suppression of ovarian cancer ES-2 xenograft tumor and lung metastasis from IP injection of ovarian cancer A-2780 cells [Niedzwiecki *et al.*, 2016]. Several *in vitro*, *in vivo*, and epidemiological studies have reported that the consumption of green tea may decrease cancer risk. (-)-Epigallocatechin-3-gallate, a major component of green tea, has been shown to inhibit tumor invasion and angiogenesis which are essential for tumor growth and metastasis [Khan & Mukhtar, 2010]. Cell experiments revealed that natural nano-vehicles from tea showed strong cyto-toxicities against cancer cells due to the stimulation of ROS and RNS amplification. The increased intracellu-lar ROS and RNS amounts could not only trigger mitochondrial damage, but also arrest cell cycle, resulting in the *in vitro* anti-proliferation, anti-migration, and anti-invasion activities against breast cancer cells. Further mice investigations demonstrated that natural nanovehicles from tea after intravenous (i.v.) injection or oral administration could accumulate in breast tumors and lung metastatic sites, inhibit the growth and metastasis of breast cancer, and modulate gut microbiota [Chen *et al.*, 2022].

The above research results showed that tea polyphenols compounds could inhibit cancer metastasis and the role of main green tea polyphenols on the most common cancer sites of cancer metastasis.

14.5.4 *Tea polyphenols induce apoptosis of cancer cells*

Interestingly, the excessive levels of ROS and RNS induced by consuming tea could induce programmed cell death (PCD) or non-PCD of cancer cells. Apoptosis is programmed cell death and is a kind of cell death with characteristic cell shrinkage, membrane blistering, DNA fragmentation, and the formation of apoptotic bodies. Apoptosis play an important role in homeostasis of living body and is related to various diseases such as can-cer. It was reported that GTP could inhibit grows of cancer by inhibiting urokinate and angiogenesis [Cao & Cao, 1999]. GTP have several hydroxyl groups in their structure. GTP not only act as an antioxidant, but

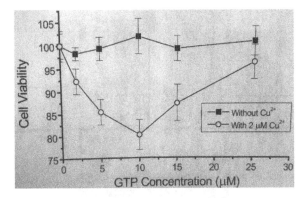

Figure 14-3. The cell death induced by GTP in combination with copper cation in Hela cell. Cells were incubated with GTP or GTP+Cu^{2+} for 14 hours and the viability was assessed by the MTT assay. Data are means±SEM of eight samples.

also can auto-oxidized to form semiquinone free radicals and act as potential pro-oxidant, even generate oxygen free radicals under special conditions, such as in combination with transition metal cation, Cu^{2+} [Li *et al.*, 1999]. Thus, GTP triggers oxidative stress and subsequently, induces cell apoptosis. We studied the apoptosis of Hela cells induced by GTP plus cupper and the mechanisms. Figure 14-3 shows the cell death induced by GTP in combination with copper cation in Hela cell. Cells were incubated with GTP or GTP+Cu^{2+} for one hour and the viability was assessed by the MTT assay. No significant cell death occurred in Hela cells treated by GTP alone, while treatment with 2–15 μM GTP+Cu^{2+} for 14 hours caused significant cell death. Interestingly, GTP at higher concentration (>25 μM) cannot induce cell death in the presence of Cu^{2+}.

During the apoptotic process, the endonuclease was activated, resulting in the internucleosomal fragmentation of DNA and the formation of ~200 bp DNA ladder which can be detected by agarose gel electrophoresis. Figure 14-4 shows that after treatment of Hela cells with 2 μM Cu^{2+} and GTP (7–25 μM) for 14 hours, the extracted DNA shows typical apoptotic features (~200 bp DNA ladder) upon electrophoresis. Higher concentrations of GTP did not lead to DNA fragmentation, nor did Cu^{2+} or GTP alone. This showed that GTP at certain concentrations, with the cooperation with Cu^{2+}, could induce apoptosis of Hela cells (Fig. 14-4).

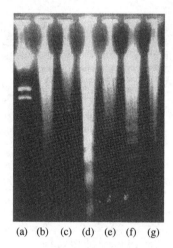

(a) (b)   (c)   (d)   (e)   (f)   (g)

Figure 14-4. Agarose gel electrolysis detection of DNA fragmentation in GTP/Cu^{2+}-teated Hela cells. (a) DNA marker; (b) Lela cells treated with 2 μM Cu^{2+} for 14 hours; (c) Control Hela cells; (d) Hela cells treated with 15 μM Cu^{2+} for 14 hours; (e) Lela cells treated with 3 μM Cu^{2+} for 14 hours; (f) Lela cells treated with 2 μM Cu^{2+} for 14 hours; (g) Lela cells treated with 20 μM Cu^{2+} for 14 hours.

The fragmentation of DNA was also detected by flow cytometry. Flow cytometry DNA analysis was performed in order to evaluate the percentage of apoptotic cells whose DNA content was lower than that of diploid cells [Ni *et al.*, 1996; Xin *et al.*, 2000]. As generally accepted apoptotic cells can be recognized by the decrease of DNA specific fluorescence which is due to DNA degradation and subsequent leakage from the cell. As shown in Figures 14-5(c) and 14-6(b), typical apoptotic peak and electron microscopy photos can be found in the cells treated with 2 μM Cu^{2+} and GTP.

In cells treated with 2 μM Cu^{2+} and low concentration of GTP, morphological alteration characterized apoptosis such as cell shrinkage, chromatin condensation, and formation of apoptotic bodies could be observed by transmission electron microscopy. Figure 14-5(b) shows a typical apoptotic cell. No such morphological alteration occurred in cells treated with Cu^{2+} or GTP alone or with Cu^{2+} plus high concentration of GTP. This strengthens the above conclusion that low concentrations of GTP with the cooperation of Cu^{2+} led to apoptosis in Hela cells.

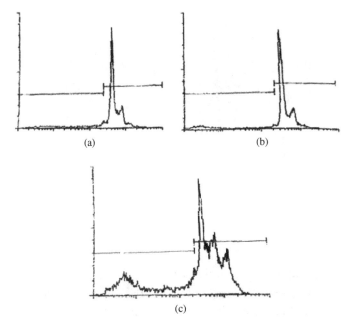

Figure 14-5. Detection of apoptosis by flow cytometry. Hela cells were treated with GTP/Cu^{2+} for 14 hours stained with PI, and the histogram of cellular DNA content was analyzed by flow cytometry. (a) Normal cells; (b) Cells treated with 2 μM Cu^{2+} for 14 hours; (c) Cells treated with 15 μM GTP plus 2 μM Cu^{2+} for 14 hours.

Tea is one of the most popular beverages consumed worldwide. GTP, which constitutes about 30% of dried tea leaves, has been studied extensively. GTP is a potent antioxidant and is expected to be used as a potential drug for diseases associated with the overproduction of reactive oxygen species. Several reports claimed that GTP shows significant preventive and suppressive effect on different tumors of skin, esophagus, liver, and mammary gland. It has been proposed that the antitumor effect of GTP is related with its potent antioxidant capacity. GTP has many hydroxyl groups in its structure, which act as potent hydrogen donors in the elimination of reactive oxygen species. However, the hydroxyl groups make GTP very prone to be auto-oxidized to form various products such as semiquinone free radicals. During the auto-oxidation process, reactive oxygen species might be formed under certain experimental conditions.

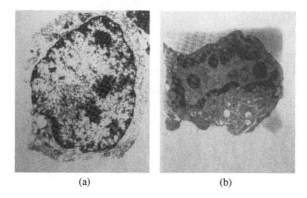

(a)   (b)

Figure 14-6. Ultrastructure of Hela cells. (a) Normal cell; (b) Hela cells treated with 15 μM GTP and Cu^{2+} for 14 hours.

By ESR spin trapping technique, we observed that GTP could generate semiquinone free radicals upon auto-oxidation. In the presence of copper ion, hydroxyl radicals were formed during the auto-oxidation of GTP:

$$GTP + O_2 \rightarrow GTP\text{-}OH + OH$$
$$GTP\text{-}OH + Cu^{2+} \rightarrow GTP\text{-}O\bullet + Cu^+$$
$$GTP\text{-}\bullet O^- + O_2 \rightarrow O_2\cdot + GTP = 0$$
$$2O_2 + 2H^+ \rightarrow H_2O_2 + O_2$$
$$H_2O_2 + Cu^+ \rightarrow \cdot OH + OH^- + Cu^{2+}$$

Hydroxyl radical and hydrogen peroxide formation during copper cation-catalyzed oxidation of phenolic compounds has been observed in both GTP and propyl gallate. Hydroxyl radical and hydrogen peroxide are both potent oxidants and could induce apoptotic cell death in various cell types. This suggested that GTP, as well as other phenolic compounds, may exert pro-oxidant effect under certain condition and may be used as a cytotoxic agent for the induction of cell death. Jacobi and coworkers (1999) have reported that propyl gallate in combination with copper ion induces lipid peroxidation in human fibroblasts and resulted in cell death. In the present investigation, we studied the effect of GTP on Hela cells and found that low concentration of GTP in combination of copper ion could induce apoptotic cell death of Hela cells, while neither GTP nor copper ion alone could induce apoptosis. We also found that the same concentration of other transition metal cations such as Fe^{3+} or Fe^{2+} had no such effect. Interestingly, higher concentration of GTP failed to induce apoptosis in Hela cells in combination with copper ion. This suggests that only under certain experimental conditions do GTP/Cu^{2+} show cytotoxic effect. Under our experimental conditions, the cytotoxic effect of GTP/Cu^{2+} might be due to the formation of reactive oxygen species. When Hela cells were treated with an appropriate concentration of GTP, and copper ion catalyzed the oxidation of GTP and caused the formation of reactive oxygen species, apoptosis in Hela cells was induced. When the concentration of GTP is higher, GTP itself might eliminate the reactive oxygen species formed during the oxidation of GTP, and thus GTP/Cu^{2+} would fail to cause oxidative damage to cells. In this regard, GTP can inhibit cancer by inducing oxidative stress.

In our experiment, we found that incubation with certain concentration of GTP plus copper cation triggered the accumulation of cytosolic free calcium cation. The increase in the cytosolic free calcium cation level is related to the signaling pathways involved in the induction of apoptosis. Under our experimental conditions, neither GTP nor Cu^{2+} alone caused the increase in Ca^{2+} level, suggesting that the accumulation of cytosolic Ca^{2+} is related to the formation of reactive oxygen species formed by GTP plus copper cation. However, it has been reported that oxidative stress may result in the influx of Ca^{2+}, which may play important roles in the activation of the apoptosis process.

The above results indicate that GTP can induce apoptosis of cancer cells, which may open a new way for cancer treatment.

14.6 Conclusion

This chapter discusses that extensive DNA damage caused by ultraviolet radiation, ionizing radiation, environmental factors, and therapeutic carcinogenic factors are all related to oxidative stress. This chapter also discussed the latest knowledge of the anti-cancer effect of tea polyphenols in the prevention and treatment of cancer, especially the molecular mechanism of its effect, such as the prevention and treatment of cancer through the inhibition of angiogenesis, tumor metastasis, inhibition of cancer

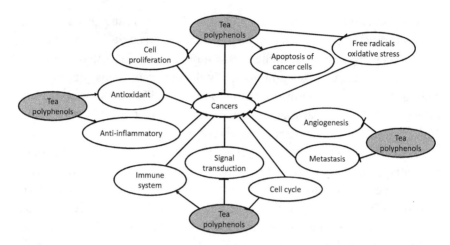

Figure 14-7.   Schematic diagram of mechanism about inhibition effects of tea polyphenols on cancers.

promoting factors, inhibition of cell proliferation, tumor growth, and induction of apoptosis of cancer cells through antioxidant, anti-inflammatory, signal transduction, cell cycle, etc. Tea polyphenols can also change the cell cycle and various malignant behaviors of cancer. In addition, tea polyphenols can regulate the immune system response and protect normal cells from free radical damage. Therefore, tea polyphenols can be the future direction of cancer treatment strategies as shown briefly in Figure 14-7. However, during cancer treatment, since tea polyphenols inhibit the damage of oxidative stress on cancer cells during treatment, which may reduce the therapeutic effect, they should be used to alleviate the side effects caused by the treatment process after cancer treatment. All these need to be verified by strict clinical experiments.

References

Aboelella NS, Brandle C, Kim T, Ding ZC, Zhou G. (2021) Oxidative stress in the tumor microenvironment and its relevance to cancer immunotherapy. *Cancers (Basel)*, **13**(5), 986.

Akbari A, Majd HM, Rahnama R, Heshmati J, Morvaridzadeh M, Agah S, Amini SM, Masoodi M. (2020) Cross-talk between oxidative stress signaling and microRNA regulatory systems in carcinogenesis: focused on gastrointestinal cancers. *Biomed Pharmacother*, **131**, 110729.

Ahn WS, Huh SW, Bae SM, Lee IP, Lee JM, Namkoong SE, Kim CK, Sin JI. (2003) A major constituent of green tea, EGCG, inhibits the growth of a human cervical cancer cell line, CaSki cells, through apoptosis, G(1) arrest, and regulation of gene expression. *DNA Cell Biol*, **22**(3), 217–224.

Bachur NR, Gee MV, Gordon SL. (1978) Enzymatic activation of actinomycin D (ACT-D) to free radical state. *Proc Am Assoc Cancer Res*, **19**, 75–83.

Bickers DR, Dixit R, Mukhtar H. (1984) Enhancement of blemycin-mediated DNA damage by epidermal microsomal systems. *Bichim Biophys Acta*, **781**, 265–273.

Buczyńska A, Sidorkiewicz I, Rogucki M, *et al.* (2021) Oxidative stress and radioiodine treatment of differentiated thyroid cancer. *Sci Rep*, **11**(1), 17126.

Butt MS, Sultan MT. (2009) Green tea: nature's defense against malignancies. *Crit Rev Food Sci Nutr*, **49**(5), 463–473.

Cao Y, Cao R. (1999) Angiogenesis inhibited by drinking tea. *Nature*, **398**, 381.

Chung FL. (1999) The prevention of lung cancer induced by a tobacco-specific carcinogen in rodents by green and black Tea. *Proc Soc Exp Biol Med*, **220**(4), 244–248.

Chung JY, Huang C, Meng X, Dong Z, Yang CS. (1999) Inhibition of activator protein 1 activity and cell growth by purified green tea and black tea polyphenols in H-ras-transformed cells: structure-activity relationship and mechanisms involved. *Cancer Res*, **59**(18), 4610–4617.

Chen Q, Li Q, Liang Y, Zu M, Chen N, Canup BSB, Luo L, Wang C, Zeng L, Xiao B. (2022) Natural exosome-like nanovesicles from edible tea flowers suppress metastatic breast cancer *via* ROS generation and microbiota modulation. *Acta Pharm Sin B*, 907–923.

Conney AH, Lu Y, Lou Y, Xie J, Huang M. (1999) Inhibitory effect of green and black tea on tumor growth. *Proc Soc Exp Biol Med*, **220**(4), 229–233.

De Marco F. (2013) Oxidative stress and HPV carcinogenesis. *Viruses*, **5**(2), 708–731.

Doroshow JH, Davies KJA. (1986) Redox cycling of anthracyclines by cardiac mitochondria. Formation of superoxide anion, hydrogen peroxide and hydroxyl radical. *J Biol Chem*, **261**, 3068–3095.

Fujiki H, *et al.* (1999) Mechanistic findings of green tea as cancer preventive for humans. *Proc Soc Exper Biol Med*, **220**, 225–228.

Filippini T, Malavolti M, Borrelli F, Izzo AA, Fairweather-Tait SJ, Horneber M, Vinceti M. (2020) Green tea (Camellia sinensis) for the prevention of cancer. *Cochrane Database Syst Rev*, **3**(3), CD005004.

Floyd RA, Soong LM, Stuart MA, Reigh DL. (1978) Spin trapping of free radicals produced from nitrosoamine carcinogens. *Photochem photobiol*, **28**, 857–863.

Gagnière J, Raisch J, Veziant J, Barnich N, Bonnet R, Buc E, Bringer MA, Pezet D, Bonnet M. (2016) Gut microbiota imbalance and colorectal cancer. *World J Gastroenterol*, **22**(2), 501–518.

Garg C, Sharma H, Garg M. (2020) Skin photo-protection with phytochemicals against photo-oxidative stress, photo-carcinogenesis, signal transduction pathways and extracellular matrix remodeling—an overview. *Ageing Res Rev*, **62**, 101127.

Guo Z, Jiang M, Luo W, Zheng P, Huang H, Sun B. (2019) Association of lung cancer and tea-drinking habits of different subgroup populations: meta-analysis of case-control studies and cohort studies. *Iran J Public Health*, **48**, 1566–1576.

Han C. (1997) Screening of anticarcinogenic ingredients in tea polyphenols. *Cancer Lett*, **114**(1-2), 153–158.

Hanafi R, Anestopoulos I, Voulgaridou GP, *et al.* (2012) Oxidative stress based-biomarkers in oral carcinogenesis: how far have we gone? *Curr Mol Med*, **12**(6), 698–703.

Hao X, Xiao H, Ju J, Lee MJ, Lambert JD, Yang CS. (2017) Green tea polyphenols inhibit colorectal tumorigenesis in azoxymethane-treated F344 rats. *Nutr Cancer*, **69**(4), 623–631.

Haroz RK, Thomasson J. (1980) Tumor initiating and promoting activity of gossypol. *Toxicol Lett Suppl*, **6**, 72–80.

Henning SM, Piwen Wang, Jonathan Said, *et al.* (2012) Polyphenols in brewed green tea inhibit prostate tumor xenograft growth by localizing to the tumor and decreasing oxidative stress and angiogenesis. *J Nutr Biochem*, **23**(11), 1537–1542.

Hosokawa Y, Hosokawa I, Ozaki K, Nakanishi T, Nakae H, Matsuo T. (2010) Tea polyphenols inhibit IL-6 production in tumor necrosis factor superfamily 14-stimulated human gingival fibroblasts. *Mol Nutr Food Res*, **54**(Suppl 2), S151–S158.

Hudlikar RR, Pai V, Kumar R, Thorat RA, Kannan S, Ingle AD, Maru GB, Mahimkar MB. (2019) Dose-related modulatory effects of polymeric black tea polyphenols (PBPs) on initiation and promotion events in B(a)P and NNK-induced lung carcinogenesis. *Nutr Cancer*, **71**(3), 508–523.

Huh SW, Bae SM, Kim YW, Lee JM, Namkoong SE, Lee IP, Kim SH, Kim CK, Ahn WS. (2004) Anticancer effects of (-)-epigallocatechin-3-gallate on ovarian carcinoma cell lines. *Gynecol Oncol*, **94**(3), 760–768.

Huang N, Chen L, Zhao B-L, Zhang Ji, Xin W. (1990) Photosensitive effect of HPD on artificial membrane. *J Biophys*, **6**, 9–15.

Huang Z, Wang Y, Yao D, Wu J, Hu Y, Yuan A. (2021) Nanoscale coordination polymers induce immunogenic cell death by amplifying radiation therapy mediated oxidative stress. *Nat Commun*, **12**(1), 145.

Khan N, Mukhtar H. (2010) Cancer and metastasis: prevention and treatment by green tea. *Cancer Metastasis Rev*, **29**(3), 435–445.

Ivanov AV, Valuev-Elliston VT, Tyurina DA, Ivanova ON, Kochetkov SN, Bartosch B, Isaguliants MG. (2017) Oxidative stress, a trigger of hepatitis C and B virus-induced liver carcinogenesis. *Oncotarget*, **8**(3), 3895–3932.

Ivanov IV, Mappes T, Schaupp P, Lappe C, Wahl S. (2018) Ultraviolet radiation oxidative stress affects eye health. *J Biophotonics*, **11**(7), e201700377.

Jacobi H, Hinrichsen ML, Wess D, Witte I. (1999) Induction of lipid peroxidation in human fibroblasts by the antioxidant propyl gallate in combination with copper(II). *Toxicol Lett*, **110**(3), 183–190.

Katiyar SK, Mohan RR, Agarwal R, Mukhtar H. (1997) Protection against induction of mouse skin papillomas with low and high risk of conversion to malignancy by green tea polyphenols. *Carcinogenesis*, **18**, 497–502.

Katiyar SK. (2006) Matrix metalloproteinases in cancer metastasis: molecular targets for prostate cancer prevention by green tea polyphenols and grape seed proanthocyanidins. *Endocr Metab Immune Disord Drug Targets*, **6**(1), 17–24.

Kim YJ, Kim EH, Hahm KB. (2012) Oxidative stress in inflammation-based gastrointestinal tract diseases: challenges and opportunities. *J Gastroenterol Hepatol*, **27**(6), 1004–1010.

Kozak J, Jonak K, Maciejewski R. (2020) The function of miR-200 family in oxidative stress response evoked in cancer chemotherapy and radiotherapy. *Biomed Pharmacother*, **125**, 110037.

Li W, Guo Q, Zhao BL, Shen S, Hou J, Xin W. (1999) ESR spectroscopy structural analysis of semiqunne free radicals of green tea polyphenols and their components. *Clin J Magn Reso*, **15**, 507–513.

Li W, Wei T-T, Xin W-J, Zhao B-L. (2002) Green tea polyphenols in combination with copper(II) induce apoptosis in Hela cells. *Res Chem Interm*, **28**, 505–618.

Liang YC, Lin-Shiau SY, Chen CF, Lin JK. (1999) Inhibition of cyclin-dependent kinases 2 and 4 activities as well as induction of Cdk inhibitors p21 and p27 during growth arrest of human breast carcinoma cells by (-)-epigallocatechin-3-gallate. *J Cell Biochem*, **75**(1), 1–12.

Liu J, Qu W, Kadiiska MB. (2009) Role of oxidative stress in cadmium toxicity and carcinogenesis. *Toxicol Appl Pharmacol*, **238**(3), 209–214.

Liu Q, Jiao Y, Zhao Y, Wang YR, Li J, Ma S, *et al.* (2016) Tea consumption reduces the risk of oral cancer: a systematic review and meta-analysis. *Int J Clin Exp Med*, **9**, 2688–2697.

Liu P, Wang Y, Yang G, Zhang Q, Meng L, Xin Y, Jiang X. (2021) The role of short-chain fatty acids in intestinal barrier function, inflammation, oxidative stress, and colonic carcinogenesis. *Pharmacol Res*, **165**, 105420

Lown JW, Sim S, Majumdar KC, Chang RY. (1977) Strand scission of DNA by bound adriamycin and daunorubicin in the presence of reducing agents. *Biochem Biophys Res Commun*, **76**, 705–713.

Lorentzen RJ. (1979) Toxicity of metabolic benzo(a)pyrenediones to cultured cells and the dependence upon molecular oxygen. *Cancer Res*, **39**, 3194–3202.

Lu Z, Nie G, Li Y, Shan S-l, Tao y, Cao Y, Zhang Z, Liu N, Ponka P and Zhao B-L. (2009) Overexpression of mitochondrial ferritin sensitizes cells to oxidative stress via an iron-mediated mechanism. *Antioxid Redox Signal*, **11**, 1791–1803.

Lu Z, Tao Y, Zhou Z, Zhang J, Li C, OU L' Zhao B-L. (2006) Mitochondrial ROS and NO mediated cancer cell apoptosis in 2-BA-2-DMHB photodynamic therapy. *Free Rad Biol Med*, **41**, 1590–1605.

Ly BT, Chi HT, Yamagishi M, Kano Y, Hara Y, Nakano K, Sato Y, Watanabe T. (2013) Inhibition of FLT3 expression by green tea catechins in FLT3 mutated-AML cells. *PLoS One*, **8**(6), e66378.

Manjula Vinayak, Akhilendra K Maurya. (2019) Quercetin loaded nanoparticles in targeting cancer: recent development. *Anticancer Agents Med Chem*, **19**(13), 1560–1576.

Marnertt LJ. (1987) Peroxyl free radicals: potential mediators of tumor initiation and promotion. *Carcinogenesis*, **8**, 1365–3172.

McLarty J, Bigelow RL, Smith M, *et al.* (2009) Tea polyphenols decrease serum levels of prostate-specific antigen, hepatocyte growth factor, and vascular endothelial growth factor in prostate cancer patients and inhibit production of hepatocyte growth factor and vascular endothelial growth factor in vitro. *Cancer Prev Res (Phila Pa)*, **2**(7), 673–682.

Mirtavoos-Mahyari H, Salehipour P, Parohan M, Sadeghi A. (2019) Effects of coffee, black tea and green tea consumption on the risk of non-Hodgkin's lymphoma: a systematic review and dose-response meta-analysis of observational studies. *Nutr Cancer*, **71**, 887–97.

Mossman B, Light W, Chung J. (1983) Asbestos: mechanisms of toxicity and carcinogenicity in the respiratory tract. *Annu Rev Pharmocol Toxicol*, **23**, 595–563.

Mujtaba T, Dou QP. (2012) Black tea polyphenols inhibit tumor proteasome activity. *In Vivo*, **26**(2), 197–202.

Negri A, Naponelli V, Rizzi F, Bettuzzi S. (2018) Molecular targets of epigallo-catechin-gallate (EGCG): a special focus on signal transduction and cancer. *Nutrients*, **10**(12), 1936.

Ni Y-C, Zhao B-L, Hou J-W, Xin W-J. (1996) Protection of cerebellar neuron by Ginkgo-biloba extract against apoptosis induced by hydroxyl radicals. *Neuron Sci Letter*, **214**, 115–118.

Nie G-J, Wei T-T, Zhao B-L. (2001) Polyphenol protection of DNA against damage. *Method Enzym*, **335**, 232–244.

Nowsheen S, Aziz K, Kryston TB, Ferguson NF, Georgakilas A. (2012) The interplay between inflammation and oxidative stress in carcinogenesis. *Curr Mol Med*, **12**(6), 672–80.

Niedzwiecki A, Roomi MW, Kalinovsky T, Rath M. (2016) Anticancer efficacy of polyphenols and their combinations. *Nutrients*, **8**(9), 552.

Oda K, Hamanishi J, Matsuo K, Hasegawa K. (2018) Genomics to immunother-apy of ovarian clear cell carcinoma: Unique opportunities for management. *Gynecol Oncol*, **151**(2), 381–389.

Olinski R, Gackowski D, Foksinski M, Rozalski R, Roszkowski K, Jaruga P. (2002) Oxidative DNA damage: assessment of the role in carcinogenesis, atherosclerosis, and acquired immunodeficiency syndrome. *Free Radic Biol Med*, **33**(2), 192–200.

Ping Z, Peng Y, Lang H, Xinyong C, Zhiyi Z, Xiaocheng W, Hong Z, Liang S. (2020) Oxidative stress in radiation-induced cardiotoxicity. *Oxid Med Cell Longev*, **2020**, 3579143.

Rashidi B, Malekzadeh M, Goodarzi M, Masoudifar A, Mirzaei H. (2017) Green tea and its anti-angiogenesis effects. *Biomed Pharmacother*, **89**, 949–956.

Reuter S, Gupta SC, Chaturvedi MM, Aggarwal BB. (2010) Oxidative stress, inflammation, and cancer: how are they linked? *Free Radic Biol Med*, **49**(11), 1603–1616.

Robinett ZN, Bathla G, Wu A, et al. (2018) Persistent oxidative stress in vestibu-lar schwannomas after stereotactic radiation therapy. *Otol Neurotol*, **39**(9), 1184–1190.

Sartorilli AC. (1986) The role of mitomycin antibiotics in the chemotherapy of solid tumors. *Biochem Pharmcol*, **35**, 67–75.

Shimura T, Nakashiro C, Narao M, Ushiyama A. (2020) Induction of oxidative stress biomarkers following whole-body irradiation in mice. *PLoS One*, **15**(10), e0240108.

Singh B, Patwardhan RS, Jayakumar S, Sharma D, Sandur SK. (2020) Oxidative stress associated metabolic adaptations regulate radioresistance in human lung cancer cells. *J Photochem Photobiol B*, **213**, 112080.

Snyder R, Longacre SL, Whitmer CM, Kocsis JJ, Andrews LS, Lee EW. (1980) Biochemical toxicology of benzene. In *Reviews in Biochemical Toxicology Vol 3*, Hodgson E, Bend JR, Philpot RM (eds.), Elsevier, North Holland New York, 123.

Steele VE, Kelloff GJ, Balentine D, Boone CW, Mehta R, Bagheri D, Sigman CC, Zhu S, Sharma S. (2000) Comparative chemopreventive mechanisms of green tea, black tea and selected polyphenol extracts measured by in vitro bioassays. *Carcinogenesis*, **21**(1), 63–67.

Takada M, Nakamura Y, Koizumi T, Toyama H, Kamigaki T, Suzuki Y, Takeyama Y, Kuroda Y. (2002a) Suppression of human pancreatic carcinoma cell growth and invasion by epigallocatechin-3-gallate. *Pancreas*, **25**(1), 45–48.

Takada M, Nakamura Y, Koizumi T, *et al.* (2002b) Inhibitory effect of epigallo-catechin-3-gallate on growth and invasion in human biliary tract carcinoma cells. *World J Surg*, **26**(6), 683–686.

Takada M, Ku Y, Toyama H, Suzuki Y, Kuroda Y. (2002c) Suppressive effects of tea polyphenol and conformational changes with receptor for advanced gly-cation end products (RAGE) expression in human hepatoma cells. *Hepatogastroenterology*, **49**(46), 928–931.

Ton TT, Kovi RC, Peddada TN, *et al.* (2021) Cobalt-induced oxidative stress contributes to alveolar/bronchiolar carcinogenesis in B6C3F1/N mice. *Arch Toxicol*, **95**(10), 3171–3190.

Torre M, Dey A, Woods JK, Feany MB. (2021) Elevated oxidative stress and DNA damage in cortical neurons of chemotherapy patients. *J Neuropathol Exp Neurol*, **80**(7), 705–712.

Toyokuni S. (2016) The origin and future of oxidative stress pathology: From the recognition of carcinogenesis as an iron addiction with ferroptosis-resistance to non-thermal plasma therapy. *Pathol Int*, **66**(5), 245–259.

Vallée A, Lecarpentier Y. (2018) Crosstalk between peroxisome proliferator-activated receptor gamma and the canonical WNT/beta-catenin pathway in chronic inflammation and oxidative stress during carcinogenesis. *Front Immunol*, **9**, 745.

Wang ST, Cui WQ, Pan D, Jiang M, Chang B, Sang LX. (2020a) Tea polyphenols and their chemopreventive and therapeutic effects on colorectal cancer. *World J Gastroenterol*, **26**(6), 562–597.

Wang Y, Zhao Y, Chong F, Song M, Sun Q, Li T, *et al.* (2020b) A dose-response meta-analysis of green tea consumption and breast cancer risk. *Int J Food Sci Nutr*, 1–12.

Wilmer J, Schubert J. (1981) Mutagenicity of irradited solution of nuclic acid bases and nucleosides in Salmonella typhimurium. *Mutat Res*, **88**, 337–342.

Xin W-J, Wei T-T, Chen C, Ni Y-C, Zhao B-L, Hou J-W. (2000) Mechanisms of apoptosis in rat cerebellar granule cells induced by hydroxyl radicals and effects of Egb761 and its constitutes. *Toxicology*, **148**, 103–110.

Xu T, Jiang W, Li S, Zhao B-L, Xin W, Lin Z. (1995) Protective effect of magnesium ion on adriamycin mitochondrial toxicity. *J Biophys*, **11**, 614–618.

Yan J-Qi, Di X-J, Liu C-Y, *et al.* (2010) Study on the effect of tea filter on eliminating smoking addiction and reducing harm. *Sci China Life Sci*, **53**, 533–541.

Yang CS, Yang GY, Landau JM, Kim S, Liao J. (1998) Tea and tea polyphenols inhibit cell hyperproliferation, lung tumorigenesis, and tumor progression. *Exp Lung Res*, **24**(4), 629–639.

Yamauchi R, Sasaki K, Yoshida K. (2009) Identification of epigallocatechin-3-gallate in green tea polyphenols as a potent inducer of p53-dependent apoptosis in the human lung cancer cell line A549. *Toxicol In Vitro*, **23**(5), 834–839.

Yu Yi, Hailong Liang, Huang Jing, *et al.* (2020) Green tea consumption and esophageal cancer risk: a meta-analysis. *Nutr Cancer*, **72**, 513–521.

Zhang D, Kaushiva A, Xi Y, Wang T, Li N. (2018) Non-herbal tea consumption and ovarian cancer risk: a systematic review and meta-analysis of observational epidemiologic studies with indirect comparison and dose-response analysis. *Carcinogenesis*, **39**, 808–818.

Zhang J, Huang N, Zhao B-L, Chen L, Xin W. (1986a) ESR Study on free radical capture by DMPO in hematoporphyrin biological photosensitive system. *J Chem*, **44**, 627–630.

Zhang J, Huang N, Zhao B-L, Li, Xin W. (1988) Effects of photosensitization of hematoporphyrin derivatives on lipid kinetics and phase diagram of artificial membrane. *Scientific bulletin*, **33**, 1258–1260.

Zhang Y, Zhang Q, Zhao B-L, *et al.* (1991) Scavenging effect of Promethazine on semiquinone free radicals in rat myocardium induced by adriamycin. *Chinese J Pharmacol*, **12**, 20–28.

Zhang Z-Y, Zang L-Y, Sheng P-G and Wang L-H. (1986b) ESR studies of spin trapped radicals in gamma-irradiated polycrystaline pyrimidine nucleosides. *Sci China Chem Ser B*, **24**, 1164–1176.

Zhao B-L. (2020) The pros and cons of drinking tea. *Tradit Med Mod Med*, **3**(3), 1–12.

Zhao B-L, Li X-J, Xin W-J. (1989) ESR study on active oxygen radicals produced in the respiratory burst of human polymophonuclear leukocytes. *Cell Biol Intern Report*, **13**, 529–534.

Zhao B-L, Zhang J-Z. (1984) Free radicals and cancer. *Adv Biochem Biophys*, **11**, 9–14.

Zhao B-L, Huang N, Zhang J, Chen L, Xin W. (1985a) Study on hydroxyl radicals produced by aqueous solution of hematoporphyrin derivatives under illumination by spin trapping technique. Sci Bull, **30**, 1743–1746.

Zhao B-L, Qu B, Zhang C, Zhang J, Xin W. (1985b) Study on the interaction between photoirradiated hematoporphyrin derivatives and liposomes by spin labeled ESR technique. *J Pharm*, **28**, 89–94.

Zhou F, Shen T, Duan T, Xu YY, Khor SC, Li J, Ge J, Zheng YF, Hsu S, DE Stefano J, Yang J, Xu LH, Zhu XQ. (2014) Antioxidant effects of lipophilic tea polyphenols on diethylnitrosamine/phenobarbital-induced hepatocarcinogenesis in rats. *In Vivo*, **28**(4), 495–503.

Chapter 15

Tea Polyphenols and Radiation Damage

Baolu Zhao

Institute of Biophysics, Chinese Academy of Sciences, Beijing, China

15.1 Introduction

Radiation damage is a common problem. The radiation damage caused by the atomic bomb explosion during World War II and the peaceful use of atomic energy in peacetime will also contribute to radiation damage. Radiation-induced skin damage ranges from photoaging and cutaneous carcinogenesis caused by UV exposure, to cutaneous radiation syndrome, a frequently fatal consequence of exposures from nuclear accidents. The major mechanism of skin injury common to these exposures is radiation-induced oxidative stress. We also have relaxation damage caused by using X-ray examination in the hospital, electromagnetic radiation damage from using computers and mobile phones at ordinary times, and ultraviolet (UV) radiation damage due to sunlight. Radiation-induced skin injury remains a serious concerning. As our skin is exposed to the outside, it is often damaged by the sun's UV radiation. Additionally, our vision is obtained by the reflection of light from an object, and therefore, will also be damaged by the sun's UV radiation. Therefore, the focus of this chapter is the protective effect of UV on skin and eye damage and tea polyphenols. There is an urgent need to understand how to protect these radiation injuries. A large number of studies have shown that these radiation injuries are related to oxidative stress injury, and tea and tea polyphenols can reduce these radiation injuries by inhibiting oxidative stress caused by radiation damage. Efforts to prevent or mitigate radiation damage have included the development of antioxidants capable of reducing reactive oxygen species (ROS) and reactive nitrogen species (RNS). These issues will be discussed in this chapter.

15.2 Radiation and health

Radiation is divided into ionizing radiation or non-ionizing radiation. Ionizing radiation has high enough energy to ionize atoms or molecules. The energy of non-ionizing radiation is weaker than that of ionizing radiation. Different non-ionizing radiation can produce different biological effects.

Ionizing radiation (such as X-ray, neutron, proton, α or β particles, γ and X-rays) can damage tissues directly or through secondary reactions. Ionization can cause cancer. The probability of cancer caused by ionizing radiation depends on the radiation dose rate and the sensitivity of organisms receiving radiation. Large doses of radiation can produce visible physical effects in a few days. DNA changes caused by small doses can cause chronic diseases in irradiated people and genetic defects in their offspring. The relationship between the degree of injury and cell healing or death is very complex. Harmful ionizing radiation sources include high-energy X-ray for diagnosis and treatment, radium and other natural radioactive substances (such as radon), sealed cobalt and cesium for cancer treatment, and a large number of artificially generated radioactive substances for medicine and industry. Acute radiation sickness can cause dizziness, fatigue, loss of appetite, digestive tract and brain injury, and serious cases can lead to death. Chronic irradiation can cause chronic radiation-induced diseases, such as chronic skin damage, hematopoietic disorders, cataracts, etc. It can increase the rate of fetal malformation or stillbirth. Chronic radiation can lead to a serious decline in white blood cells, lung cancer, thyroid cancer, breast cancer, bone cancer, and other cancers. Radiotherapy can sometimes cause serious damage to the irradiated organs. It is necessary to pay attention to the impact of radiation on human health and how to protect it.

Electromagnetic waves and particle flows emitted by the sun into space. More than 99% of the solar radiation spectrum is between 0.15–4.0 microns. About 50% of the solar radiation energy is in the visible spectral region (wavelength 0.4–0.76 μm), 7% in the ultraviolet spectral region (wavelength 0.76 μm), and the maximum energy is at the wavelength 0.475 μm. The human body needs sunlight, especially for the synthesis of vitamin D, but the ultraviolet radiation (UVR) of sunlight is the most powerful environmental risk factor for skin cancer. When human skin is exposed to UV radiation, skin tissue will be seriously damaged. The structure of the skin is as follows: proto-protein, elastin, and other fibers form the skeleton that supports the skin. While these elements make the skin look smooth and young, these elements are also vulnerable to UVA and UVB damage.

Ultraviolet rays are also divided into long-wave ultraviolet UVA and short-wave ultraviolet UVB (Fig. 15-1). UVA, a kind of long-wave ultraviolet, can reach the dermis due to its long wavelength, which mainly damages collagen and elastin, makes the skeleton supporting the epidermis disappear, and leads to the uneven collapse of the epidermis, resulting in the formation of skin wrinkles. UVB mainly acts on the epidermis, causing sunburn or melanin deposition, forming spots or skin darkening. UV will produce *in situ* ROS and RNS and matrix metalloproteinases, these factors are the root cause of wrinkles because they destroy the

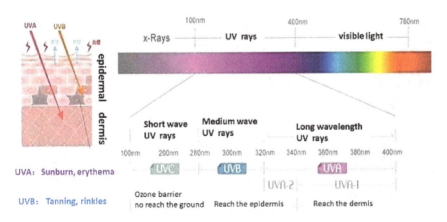

Figure 15-1. Ultraviolet rays are divided into UVA, UVB, and UVC.

collagen matrix of the dermis. Skin is exposed to many environmental factors, including UV, air pollution, and smoking, which can lead to oxidative stress in the skin, leading to premature (external) aging. UVA, UVB, and other factors have the most prominent damage to the skin. Therefore, light protection is the first-line intervention to prevent premature aging and skin cancer. Many studies have shown that tea polyphenols have photo-protective effects, not only through its direct light absorption characteristics, but also through its antioxidant effect (scavenging ROS and RNS) and regulating gene expression, regulation, or cellular inflammatory response induced by ultraviolet light. Intervention studies on human beings with a carotenoid rich diet have shown that it has photo-protective effects on the skin (mainly by reducing the sensitivity to erythema caused by UV radiation) and its beneficial effects in preventing and improving skin aging (improving skin elasticity and hydration, skin texture, wrinkles, and age spots). In addition, tea polyphenols may be helpful in the prevention and treatment of some photosensitive diseases, including erythropoietic proto-porphyrins, delayed cutaneous porphyrins, and polymorphic photo-eruptions. Although UV is considered to be the main pathogenic factor of non-melanoma skin cancer and melanoma, and the photo-protective effect of tea polyphenols is positive, however, it is necessary to study the mechanism of tea polyphenols in skin photo-carcinogenesis. At the molecular level, UV irradiation causes DNA damage such as cyclobutane pyrimidine dimers and photoproducts, which are usually repaired by nucleotide excision repair (NER). Chronic exposure to UV irradiation leads to photoaging, immunosuppression, and ultimately photo-carcinogenesis. Photo-carcinogenesis involves the accumulation of genetic changes, as well as immune system modulation, and ultimately leads to the development of skin cancers. In the clinic, artificial lamps emitting UVB (280–320 nm) and UVA (320–400 nm) radiation in combination with chemical drugs, are used in the therapy of many skin diseases including psoriasis and vitiligo [Matsumura & Ananthaswamy, 2004].

Electromagnetic radiation can be classified according to frequency, from low frequency to high frequency, including radio waves, microwave, infrared, visible light, ultraviolet, (ionizing radiation) X-rays, and gamma rays. People are often exposed to electromagnetic radiation, such as television, mobile phones, and radio transmission systems. The study found that long-term exposure to electromagnetic radiation can also lead to body damage, especially eyes that need attention.

Studies have shown that free radicals are the mediators of radiation damage, and radiation causes excessive production of ROS and RNS in the body. Radiation damage is caused by radiation-induced oxidative stress and inflammation. If some drugs or compounds (such as polyphenols, sulfhydryl compounds) are given before irradiation, the survival rate of animals can be increased. However, there are no preventive drugs available for humans for long time. A large number of studies have found that tea polyphenols have protective effects on radiation damage.

15.3 Radiation and oxidative stress

The detrimental effects of various radiation are mostly mediated via the overproduction of ROS and RNS, especially the hydroxyl radical (•OH). Radiation causes damage to irradiated tissues and tissues that do not receive direct irradiation through a phenomenon called out-of-field effects. This damage through signals such as inflammatory responses can be transmitted to unirradiated cells/tissues and causes many effects such as oxidative damage. The radio-protective and anti-inflammatory effects of antioxidants have been demonstrated in various studies.

Epidemiological, clinical, and biological studies have implicated that solar UV light is a complete carcinogen and repeated exposure can lead to the development of various skin disorders including melanoma and non-melanoma skin cancers. Solar UV radiation-induced immunosuppression is considered to be a risk factor for melanoma and non-melanoma skin cancers. Non-melanoma skin cancer is the most common malignancy in humans and is equivalent to the incidence of malignancies in all other organs combined in the United States. Damaging effects of UV radiation include UV-induced sunburn response, UV-induced immunosuppression, and photoaging of the skin. UVA makes a larger impact on oxidative stress in the skin than UVB by inducing ROS and NOS which damage DNA, protein, and lipids and which also lead to NAD+ depletion, and therefore energy loss from the cell. Lipid peroxidation induces prostaglandin production that in association with UV-induced nitric oxide production causes inflammation. Inflammation drives benign human solar keratosis to undergo malignant conversion into squamous cell carcinoma probably because the inflammatory cells produce ROS and RNS, thus increasing oxidative damage to DNA and the immune system. ROS and RNS appear to cause the increase in mutational burden as solar keratosis progressies into squamous cell carcinoma in humans. UVA is particularly important in causing immunosuppression in both humans and mice, and UV lipid per-oxidation-induced prostaglandin production and UV activation of nitric oxide synthase are important mediators of this event. Other immunosup-pressive events are likely to be initiated by UV oxidative stress. Antioxidants have also been shown to reduce photo-carcinogenesis [Halliday, 2005].

15.3.1 *Ion radiation causes oxidative stress*

On 6 August 1945, the United States dropped an atomic bomb on Hiroshima, Japan. On 15 August, Japanese Emperor Hirohito issued an imperial edict announcing Japan's unconditional surrender and the end of World War II. However, the explosion caused 200,000 deaths, of which light radiation accounted for about 35%, early nuclear radiation accounted for about 5%, and radioactive pollution accounted for about 10%. The intense light wave of the atomic bomb explosion made thousands of people blind. Exposure of astronauts in space to radiation during weightless-ness may contribute to subsequent bone loss. Gamma irradiation of post-pubertal mice rapidly increases the number of bone-resorbing osteo-clasts and causes bone loss in cancellous tissue; similar changes occur in skeletal diseases associated with oxidative stress. Gamma (γ) ray, an elec-tromagnetic radiation, occasionally accompanies the emission of an alpha or beta particle. Irradiation, but not hindlimb unloading, reduced viability and increased apoptosis of marrow cells and caused oxidative damage to lipids within mineralized tissue. Irradiation also stimulated generation of ROS and RNS in marrow cells. Furthermore, injection of alpha-lipoic acid, an antioxidant, mitigated the acute bone loss caused by irradiation. Together, these results showed that disuse and gamma irradiation, alone or in combination, caused a similar degree of acute cancellous bone loss and shared a common cellular mechanism of increased bone resorption. Furthermore, irradiation, but not disuse, may increase the number of osteoclasts and the extent of acute bone loss via increased ROS and RNS production and ensuing oxidative damage, implying different molecular mechanisms [Kondo *et al.*, 2010]. Exposure to such radiation can cause cellular changes such as mutations, chromosome aberration, and cellular damage which depend upon the total amount of energy, duration of exposure, and the dose. Ionizing radiation can impair spermatogenesis and can cause mutations in germ cells. The out-of-field/non-target effect is one of the most important phenomena of ionizing radiation that leads to molecular and cellular damage to distant non-irradiated tissues. The most important concern about this phenomenon is carcinogenesis many years after radiation treatment.

A study evaluated the oxidative damage caused by direct irradiation and out-of-field effects on the lung tissue after pelvic irradiation in rats. The biochemical parameters including malondialdehyde (MDA), glutathione peroxidase (GPx), and superoxide dismutase (SOD) were measured. The results showed that localized irradiation to the lung or pelvis caused an increase in the MDA level. Moreover, pelvis and lung irradiation increased the GPx and SOD activity in the lungs [Ghobadi *et al.*, 2017].

A paper studied oxidative stress and gamma radiation-induced cancellous bone loss with musculoskeletal disuse. It hypothesized that increased oxidative stress mediates radiation-induced bone loss and that musculoskeletal disuse changes the sensitivity of cancellous tissue to radiation exposure. Musculoskeletal disuse by hind-limb unloading (one or two weeks) or total body gamma irradiation (1 or 2 Gy of (137) Cs) of 4-month-old, male C57BL/6 mice each decreased cancellous bone volume fraction in the proximal tibiae and lumbar vertebrae. The extent of radiation-induced acute cancellous bone loss in tibiae and lumbar vertebrae was similar in normally-loaded and hindlimb-unloaded mice. Similarly, osteoclast surface in the tibiae increased 46% as a result of irradiation, 47% as a result of hind-limb unloading, and 64% as a result of irradiation + hind-limb unloading compared with normally-loaded mice. Irradiation, but not hind-limb unloading, reduced viability and increased apoptosis of marrow cells and caused oxidative damage to lipids within mineralized tissue. Irradiation also stimulated generation of reactive oxygen species in marrow cells [Kondo *et al.*, 2010].

In general, type B spermatogonia are sensitive to this type of radiation. A paper studied γ-radiation-induced oxidative stress and apoptosis in rat testis. Rats were divided into groups including C group (control rats) and R (irradiated) group (rats irradiated with γ-radiation). Malondialdehyde

(MDA) is the end product of lipid peroxidation (LPO) and xanthine oxidase (XO). MDA can generate ROS and nitric oxide (NO) in testes homogenate as well as in mitochondrial matrix. The apoptotic markers including DNA-fragmentation (DNAF) in testes homogenate and calcium ions (Ca^{2+}) in mitochondrial matrix were determined. Superoxide dismutase (SOD) and catalase (CAT) activities in testes homogenate, while reduced glutathione "GSH" in nuclear matrix were determined. Also, histopathological examination for testes tissues through electron microscope was studied. Exposure of rats to γ-radiation (R group) increased the levels of MDA, NO, DNAF, Ca^{2+} and XO activity, while it decreased GSH level, SOD, and CAT activities as compared to the C groups; γ-radiation increased oxidative stress (OS), LPO, apoptosis, and induced testes injuries. These results are in agreement with the histopathological examination [Shaban *et al.*, 2017].

A paper studied the radiation-induced oxidative stress at out-of-field lung tissues after pelvis irradiation in rats. In this experiment, SD rats (n = 49) were divided into seven groups (n = 7/each group), including two groups of pelvis-exposed rats (out-of-field groups), two groups of whole body-exposed rats (scatter groups), two groups of lung-exposed rats (direct irradiation groups), and one control sham group. Out-of-field groups were irradiated at a 2 × 2 cm area in the pelvis region with 3 Gy using 1.25 MeV cobalt-60 gamma-ray source, and subsequently, MDA and glutathione (GSH) levels as well as SOD activity in out-of-field lung tissues were measured. Results were compared to direct irradiation, control, and scatter groups at 24 hours and 72 hours after exposure. SOD activity decreased in out-of-field lung tissue 24 hours and 72 hours after irradiation as compared with the controls and scatter groups. GSH level decreased 24 hours after exposure and increased 72 hours after exposure in the out-of-field groups as compared with the scatter groups. MDA level in out-of-field groups only increased 24 hours after irradiation. Pelvis irradiation induced oxidative damage in distant lung tissue that led to a dramatic decrease in SOD activity [Najafi *et al.*, 2016].

Immune system is amongst the most radiosensitive system to radiation-induced cellular and molecular damage. The study found ^{60}Co γ radiation damage in mice can reduce survival time, and body weight, and decrease the number of peripheral blood white blood cells and bone marrow DNA content in irradiated mice. In addition, ^{60}Co γ radiation can significantly damage the antioxidant capacity, active pro-inflammatory cytokines, and destroy the intestinal barrier [Liu *et al.*, 2021].

Radiation-induced skin injury (RSI) refers to a frequently occurring complication of radiation therapy. Skin injury is a major complication during radiation therapy and is associated with oxidative damage to skin cells. Radiation therapy is widely used in the treatment of tumor diseases, but it can also cause serious damage to the body. Exposure to ionizing radiation can cause oxidative damage to human body, leading to various diseases and even death. Radiation-induced cell death is the outcome of oxidative stress caused by free radicals. Ionizing radiation was responsible for augmentation of hepatic oxidative stress in terms of lipid peroxidation and depletion of endogenous antioxidant enzymes. Nearly 90% of patients having received radiation therapy underwent moderate-to-severe skin reactions, severely reducing patients' quality of life and adversely affecting their disease treatment. No gold standard has been formulated for RSIs [Yang *et al.*, 2020].

The above research shows that long-term exposure to ionic radiation produces a large number of ROS and RNS free radical oxidation, which will cause oxidative damage to human body, DNA damage, and even skin cancer, reduce the vitality of bone marrow cells, increase cell apoptosis, and lead to various diseases and even death.

15.3.2 *UV radiation and oxidative stress*

Ultraviolet is divided into long-wave ultraviolet UVA and short-wave ultraviolet UVB. Among them, UVB mainly acts on the epidermis, resulting in sunburn or melanin deposition, forming spots or skin blackening. Ultraviolet UV will produce *in situ* reactive oxygen species and matrix metalloproteinases. These factors are the root cause of wrinkles because they destroy the collagen matrix of the dermis. The human epidermal skin layer is mainly composed of keratinocytes, which is damaged by UV-B radiation-induced intracellular oxidative stress. Epidemiological, clinical, and laboratory studies have shown that solar UVR is associated with a variety of skin diseases, including premature skin aging, melanoma, and non-melanoma skin cancer. UVB radiation-induced inflammatory responses and generation of oxidative stress may have important human health implications. The UVB exposure-induced skin injury and oxidative stress has been associated with a variety of skin disease conditions including photoaging, inflammation, and cancer. Exposure of skin to UV radiation can cause diverse biological effects, including induction of inflammation, alteration in cutaneous immune cells, and impairment of contact hypersensitivity (CHS) responses.

1. UV radiation and oxidative stress cause skin aging and skin diseases.

Skin is the largest human organ, and it provides a first line of defense that includes physical, chemical, and immune mechanisms to combat environmental stress. UV radiation is a prevalent environmental stressor. Radiation-induced skin damage ranges from photoaging and cutaneous carcinogenesis caused by UV exposure to treatment-limiting radiation dermatitis associated with radiotherapy, to cutaneous radiation syndrome, a frequently fatal consequence of exposures from nuclear accidents. The major mechanism of skin injury common to these exposures is radiation-induced oxidative stress. Solar UV radiation-induced ROS and RNS are mainly responsible for the development of photo-aging. ROS and RNS have been shown to play a role in UV-induced skin carcinogenesis. Human skin protects the body from external damage, pathogens, and oxidative stress factors such as UV radiation. Excessive exposure to UV radiation can lead to increased production of free radicals and hence to skin damage such as inflammation, premature skin aging, and skin cancer. Chronic UV radiation exposure-induced skin diseases or skin disorders are caused by the excessive induction of inflammation, oxidative stress, and DNA damage, etc. UVA radiation contributes to skin photoaging. UVA irradiation caused oxidative stress and inflammation in skin. The levels of SOD, MDA, and total anti-oxidative capacity (T-AOC) and RNA (mRNA) and protein levels of the tumor necrosis factor (TNF)-α, interleukin (IL)-1β, and IL-6 were measured. UVA radiation-induced skin injuries and the underlying mechanism involve oxidative stress and inflammatory processes [Li *et al.*, 2016].

Human skin exposed to solar UVR results in a dramatic increase in the production of ROS and RNS. The sudden increase in ROS and RNS shifts the natural balance toward a pro-oxidative state, resulting in oxidative stress. The detrimental effects of oxidative stress occur through multiple mechanisms that involve alterations to proteins and lipids, induction of inflammation, immunosuppression, DNA damage, and activation of signaling pathways that affect gene transcription, cell cycle, proliferation, and apoptosis. Reactive oxygen species produced by UVR, oxidizers, or metabolic processes can damage cells and initiate pro-inflammatory cascades. Immunoblotting and immunofluorescence studies showed that irradiation enhanced the nuclear translocation of nuclear factor kappa B (NF-κB) level, which leads to hepatic inflammation. To investigate further, it found that radiation induced the activation of stress-activated protein kinase/c-Jun NH2-terminal kinase (SAPK/JNK)-mediated apoptotic pathway and deactivation of the NF-E2-related factor 2 (Nrf2)-mediated redox signaling pathway [Khan *et al.*, 2015].

UV and other factors cause the most prominent damage to the skin. UVA, a kind of long-wave ultraviolet, can reach the dermis due to its long wavelength, which mainly damages collagen and elastin, makes the skeleton supporting the epidermis disappear, and leads to the uneven collapse of the epidermis, thus forming skin wrinkles. Due to the short wavelength of UVB, it mainly acts on the epidermis, resulting in skin sunburn or melanin deposition, forming spots or skin blackening. UVC is ultraviolet light with shorter wavelength, which can only reach the skin surface like ozone. Daylight UV induces a characteristic UV-specific mutation, a UV-signature mutation occurring preferentially at methyl-cyclobutane pyrimidine dimers sites, which is also observed frequently after exposure to either UVB or UVA, but not to UVC. UVA has also been suggested to induce oxidative types of mutation, which would be caused by oxidative DNA damage produced through the oxidative stress after the irradiation. Indeed, UVA produces oxidative DNA damage not only in cells but also in skin, which, however, does not seem sufficient to induce mutations in the normal skin genome (Fig. 15-1) [Matsumura & Ananthaswamy, 2004].

Skin damage from exposure to sunlight induces aging-like changes in appearance and is attributed to the UV component of light. Photosensitized production of ROS and RNS by UVA light is widely accepted to contribute to skin damage and carcinogenesis. UVA induced oxidative stress preferentially in mitochondria. Exposing human skin to the blue light contained in sunlight depressed flavin auto-fluorescence, demonstrating that the visible component of sunlight has a physiologically significant effect on human skin. The ROS and RNS produced by blue light is probably superoxide, but not singlet oxygen. These results suggest that blue light contributes to skin aging similar to UVA [Nakashima *et al.*, 2017]. The UVB exposure-induced skin injury and oxidative stress has been associated with a variety of skin disease conditions including photoaging, inflammation, and cancer. It is known that exposure of murine skin to UVB radiation results in cutaneous edema, depletion of the antioxidant-defense system, and induction of ornithine decarboxylase (ODC) and cyclooxygenase activities.

Aging of the skin is also accelerated by UVR exposure, in particular UVA rays that penetrate deep into the epidermis and the dermis where it causes the degradation of collagen and elastin fibers via oxidative stress and activation of matrix metalloproteinases (MMPs). Chronic exposure of solar UV light to human skin results in photoaging. UV-induced oxidative damage and induction of MMP have been implicated in this process. A paper studied UV radiation-induced skin inflammation and oxidative stress in mice. Varied parameters of inflammation and oxidative stress in the skin of mice were evaluated after UV radiation (4.14 J/cm^2). UV radiation induced skin edema, neutrophil recruitment (myeloperoxidase activity and LysM-eGFP+ cells), MMP-9 activity, deposition of collagen fibers, epidermal thickness, sunburn cell counts, and production of pro-inflammatory cytokines (TNF-α, IL-1β, IL-6, and IL-33). Depending on the time point, LXA4 increased the levels of anti-inflammatory cytokines (TGF-β and IL-10). UV radiation-induced oxidative damage returning the oxidative status to baseline levels in parameters such as ferric-reducing ability, scavenging of free radicals, GSH levels, catalase activity, and superoxide anion production. UV radiation-induced gp91phox nicotinamide adenine dinucleotide phosphate (NADPH) oxidase 2 mRNA expression and enhanced nuclear factor erythroid 2-related factor 2 and its downstream target enzyme nicotinamide adenine dinucleotide (phosphate) quinone oxidoreductase mRNA expression [Martinez *et al.*, 2018]. A study assessed the oxidative stress-induced injury caused by UVB irradiation and investigated the molecular mechanisms. *In vitro*, UVB-induced HaCaT cells were collected for the detection of reactive oxygen species, 8-iso-prostaglandin F2α, malondialdehyde. Additionally, the expression level of PI3K, Akt, Nrf2, and heme oxygenase-1 (HO-1) were investigated. All results indicated that the levels of reactive oxygen species, 8-iso-prostaglandin F2α

and malondialdehyde, promoted the UVB exposure-induced expression of PI3K, Akt, Nrf2, and heme oxygenase-1 in HaCaT cells [Zhang *et al.*, 2018]. This *in vivo* pilot study addressed the distribution of radical production in skin types IV and V during irradiation in the UV, visible, and NIR spectral regions, comparing the first results with those of skin type II. In skin types IV–V, most radicals were induced in the visible + NIR region, followed by the NIR and UV regions of the sun spectrum. Significantly ($p \leq 0.05$) more radicals were induced in skin types IV–V than in type II during NIR irradiation, whereas skin types IV–V exhibited significantly less UV-induced radicals ($p \leq 0.01$) than skin type II. All spectral regions (UV, visible, and NIR) caused free radical formation in skin types II and IV–V. After four minutes of solar-simulated exposure (UV-NIR), the radical formation in skin types IV–V was 60% of that in skin type II [Albrecht *et al.*, 2019].

The above research shows that long-term exposure to UV radiation produces a large number of reactive oxygen species free radical oxidation, causing skin photoaging, DNA damage, and even skin cancer.

2. UV radiation oxidative stress causes visual impairment and visual diseases

The aging eye appears to be at considerable risk from oxidative stress. A great deal of research indicates that dysfunctional mitochondria are the primary site of ROS and RNS. More than 95% of O_2 produced during normal metabolism is generated by the electron transport chain in the inner mitochondrial membrane. Mitochondria are also the major target of ROS and RNS. Cataract formation, the opacification of the eye lens, is one of the leading causes of human blindness worldwide, accounting for 47.8% of all causes of blindness. Cataracts result from the deposition of aggregated proteins in the eye lens and lens fiber cell plasma membrane damage, which causes clouding of the lens, light scattering, and obstruction of vision. ROS and RNS-induced damage in the lens cell may consist of oxidation of proteins, DNA damage, and/or lipid peroxidation, all of which have been implicated in cataractogenesis. Excessive exposure to UV radiation may increase oxidative stress of various eye tissues and thus lead to the advancement of serious ocular pathologies. Children are especially vulnerable to UV radiation because of their larger pupils and more transparent ocular media; up to 80% of a person's lifetime exposure to UV radiation is reached before the age of 18 [Ivanov *et al.*, 2018]. UV-induced oxidative stress is related to non-cancer ocular pathologies; various corneal pathologies include cataract, glaucoma, and age-related macular degeneration. As the first blinding disease, cataract causes visual impairment, which has a great impact on the quality of life of cataract patients. Studies have confirmed that UV radiation is closely related to the occurrence of cataract. When the soluble protein in the lens is damaged by ultraviolet light, the activity of insoluble protein in the lens will be reduced, which will also lead to opacification.

Human retina and central nervous system are rich in unsaturated fatty acids. Therefore, free radicals produced by oxidation can easily cause peroxidation damage. Eye diseases associated with age and diabetes cause millions of Americans to suffer from visual impairment and blindness. As the first blinding disease, cataract causes visual impairment, which has a great impact on the quality of life of cataract patients. Studies have confirmed that UV radiation is closely related to the occurrence of cataract. After being damaged by UV rays, a large number of reactive oxygen free radicals are produced to oxidize and damage the lens, resulting in the decrease of soluble protein and the increase of insoluble protein in the lens, and the turbidity of the lens, resulting in visual impairment. At this time, the pupil turns white, what is known as cataract. The retina has the ability to distinguish images. Light is the driving force for the evolution and development of animal vision. The retina is an important nervous system evolved by animals to capture light signals. External light can form objective and real visual images only when it reaches the normal retina [Record & Dreosti, 1998; Katiyar *et al.*, 1999a].

Light is the driving force for the evolution and development of animal vision. High intensity light irradiation will lead to acute retinal damage. Photon excited free radicals will increase the level of retinal lipid peroxidation and lead to retinal degeneration. High intensity light irradiation will lead to acute retinal damage, and photon excited free radicals will increase the level of retinal lipid peroxidation and lead to retinal degeneration [Saric *et al.*, 2016; Sharma *et al.*, 2018]. Besides UV, the visible and near infrared (NIR) regions are also a source of radical production. Half of all free radicals are induced by the visible + NIR region of the solar spectrum in people with skin types I–III.

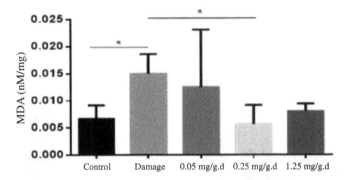

Figure 15-2. Effect of L-arginine antioxidant at different doses on MDA in lens of xenon lamp-irradiated cataract rats. *, $p < 0.05$.

The energy distribution of xenon lamp radiation spectrum is close to that of the sun. We studied the damage effect of oxidative stress induced by xenon lamp radiation on the lens of rats. It can be seen from Figure 15-2 that compared with the control group, the content of MDA in the injury group increased significantly. Low dose of L-arginine antioxidant has a certain inhibitory effect on the increase of MDA caused by UV light injury, while the content of MDA in the medium dose of L-arginine antioxidant formula group decreased significantly compared with the injury group, with significant differences. The results showed that the medium dose of L-arginine antioxidant could significantly reduce the increase of MDA in lens of cataract rats (Fig. 15-2).

Figure 15-3 showed the effect of xenon lamp radiation and different doses of L-arginine antioxidant on retinal cell apoptosis in rats with retinal injury. Compared with the control group, the number of retinal cell apoptosis in the injured group was significantly increased. Compared with the injury group, the number of apoptotic retinal cells of rats in the low-dose and high-dose L-arginine antioxidant formula groups showed a downward trend, but there was no statistical difference, while the number of apoptotic retinal cells of rats in the middle dose L-arginine antioxidant formula group was significantly reduced, almost restored to the level of the control group. It shows that the medium dose of L-arginine antioxidant can significantly alleviate the increase in the number of retinal cell apoptosis caused by xenon lamp radiation strong light injury.

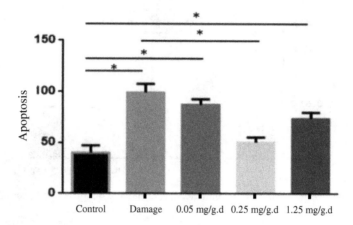

Figure 15-3.   Statistics of the number of apoptosis of retinal cells in rats with xenon lamp radiation-induced retinal damage by different doses of L-arginine antioxidant, for the control group, damage group, low dose 0.05 mg/(g.d) L-arginine antioxidant formula group, medium dose 0.25 mg/(g.d) L-arginine antioxidant group, and high dose 1.25 mg/(g.d) L-arginine antioxidant group. *, $p < 0.05$.

The above research shows that long-term exposure to UV radiation produces a large number of ROS and RNS free radical oxidation, causing retinal damage, damage to the lens, resulting in the reduction of soluble proteins in the lens, the increase of insoluble proteins, lens turbidity, leading to visual impairment.

15.3.3 *Oxidative stress induced by electromagnetic radiation*

Reports on hazardous health effects stemming from exposure to radiofrequency electromagnetic waves (RF-EMW) emitted from cell phones is still controversial. Although the energy of electromagnetic radiation (EMR) is not strong, it can still cause body damage. Several epidemiological studies have shown that exposure to electromagnetic radiation can be harmful to human health. Disturbance in ROS and RNS metabolism caused by RF-EMW and delineate NADH oxidase mediated ROS and RNS formation as playing a central role in oxidative stress (OS) due to cell phone radiation (with a focus on the male reproductive system). Widespread concern continues in the community about the deleterious effects of radiofrequency radiations (with which cell phones operate) on human tissues and the subsequent potential for carcinogenesis. A detailed survey of published studies researching this question was done in preparation of this manuscript. Included in the survey were case reports, *in vitro* studies, population-based retrospective studies and other investigations. A search was conducted on the database of indexed journals for keywords such as "cell phone", "radiation", "cancer", and "radio waves". Guidelines issued by the World Health Organization, federal and technical authorities, Institute of Electrical and Electronic Engineers, and the International Commission for Non-Ionizing Radiation Protection were reviewed. The evaluation of current evidence provided by various studies suggest that the possible carcinogenic potential of radiofrequency radiation is inconclusive. This risk assumes significance in light of the burgeoning number of people who are continually exposed to the high frequency radiation from cell phones and towers that serve as receiving and transmitting stations [Kohli *et al.*, 2009]. The preponderance of published research works over several decades including some with over 10 years of follow up have not demonstrated any significant increase in cancer among mobile phone users. However, the need for caution is emphasized as it may take up to four decades for carcinogenesis to become fully apparent [Abdus-salam *et al.*, 2008]. Latest epidemiological data reveal a significant increase in risk of development of some types of tumors in chronic (over 10 years) users of mobile phone. It was detected a significant increase in incidence of brain tumors (glioma, acoustic neuroma, meningioma), parotid gland tumor, seminoma in long-term users of mobile phone, especially in cases of ipsilateral use (case-control odds ratios from 1.3 up to 6.1). Two epidemiological studies have indicated a significant increase of cancer incidence in people living close to the mobile telephony base station as compared with the population from distant area [Yakymenko & Sidorik *et al.*, 2010].

A study determined the effects of extremely low frequency electromagnetic field (ELF-EMF) on energy metabolism and oxidative stress in *Caenorhabditis elegans* (*C. elegans*). ELF-EMF can enhance worm energy metabolism and elicit oxidative stress, mainly manifesting as ATP and ROS level elevation together with ATP synthase upregulation and ROS-TAC score decrease in young adult *C. elegans* [Sun *et al.*, 2019]. The study found that the number of dead cells in hippocampus of rats induced by electromagnetic radiation increased significantly, and the concentration of malondialdehyde was significantly higher than that of the control group. In addition, the activities of catalase, glutathione peroxidase, and superoxide dismutase in the electromagnetic radiation group were significantly lower than those in the control group [Rasouli Mojez et al., 2021].

Most epidemiological studies have also indicated the overall effect of EMR exposure in premenopausal women, particularly for estrogen receptor positive breast tumors. It has been proposed that chronic exposure to EMR may increase the risk of breast cancer by suppressing the production of melatonin; this suppression may affect the development of breast cancer either by increasing levels of circulation of estrogen or through over production of free oxygen radicals. Melatonin may modulate breast cancer through modulation of enhanced oxidative stress and Ca^{2+} influx in cell lines [Nazıroğlu et al., 2012]. This study examined oxidative parameters and apoptosis induced by EMR in human kidney embryonic cells (HEK293). EMR group had higher malondialdehyde (MDA) level and lower superoxide dismutase (SOD) activity compared with control group. The number of the apoptotic cells and caspase-3 immuno-positive cells at EMR group was increased significantly compared with the control group, whereas Bcl-2 was decreased. EMR caused oxidative stress and apoptotic activation in HEK293 cells [Özsobacı et al., 2020].

During the last couple of decades, there has been a tremendous increase in the use of mobile phones (MP). It has significantly added to the rapidly increasing EMF smog, an unprecedented type of pollution consisting of radiation in the environment, thereby prompting the scientists to study the effects on humans. A study was aimed to examine changes in major parameters (oxidative stress, level of pro-inflammatory cytokines (PICs), hypothalamic-pituitary-adrenal (HPA) axis hormones, and contextual fear conditioning) which are linked to hippocampus directly or indirectly, upon exposure to mobile phone radiofrequency electromagnetic field (MP-RF-EMF) radiation. It was observed that radiation exposure caused significant increase in hippocampal oxidative stress ($p < 0.05$) and elevated level of circulatory PICs viz. IL-1beta ($p < 0.01$), IL-6 ($p < 0.05$), and TNF-alpha ($p < 0.001$) in experimental animals upon exposure to MP-RF-EMF radiation. Adrenal gland weight ($p < 0.001$) and level of stress hormones viz. adrenocorticotropic hormone (ACTH) ($p < 0.01$) and corticosterone (CORT) ($p < 0.05$) were also found to increase significantly in MP-RF-EMF radiation-exposed animals as compared with control [Singh *et al.*, 2020].

The above studies show that exposure to EMF radiation emitted from mobile phones may induce oxidative stress, inflammatory response, and HPA axis deregulation, causing changes in hippocampal functionality.

15.4 Protective effect of tea and tea polyphenols against radiation damage

Several botanical products have been found to exhibit potent antioxidant tea polyphenols capacity and the ability to counteract UV-induced insults to the skin. These natural products exert their beneficial effects through multiple pathways, including some known to be negatively affected by solar UVR. Since natural compounds tea polyphenols are capable of attenuating some of the UV-induced aging effects in the skin, increased attention has been generated in the area of cosmetic sciences. ROS and RNS have been shown to play a role in UV-induced skin carcinogenesis. Tea polyphenols have been reported to possess substantial skin photo-protective effects. Green tea polyphenols reduce experimental skin cancers in mice mainly because of their antioxidant properties [Nichols & Katiyar, 2010]. It has been shown that topical treatment or oral consumption of green tea polyphenols, GTP, inhibit chemical carcinogen- or UVR-induced skin carcinogenesis in different laboratory animal models. Topical treatment of GTP and EGCG or oral consumption of GTP resulted in the prevention of UVB-induced inflammatory responses, immunosuppression, oxidative stress, and several skin disease states.

15.4.1 *Protective effects of tea and tea polyphenols against ultraviolet radiation damage*

Several reports have shown that antioxidants tea polyphenols protect cells against UV irradiation [Stewart *et al.*, 1996; Salucci *et al.*, 2017; Yin *et al.*, 2013]. Efforts to prevent or mitigate radiation damage have included the development of antioxidants capable of reducing ROS and RNS. Green tea is consumed as a popular beverage worldwide particularly in Asian countries like China, Korea, Japan, and India. It contains polyphenolic compounds also known as epicatechins, which are antioxidant in nature. Many laboratories have shown that topical treatment or oral consumption of green tea polyphenols inhibits chemical carcinogen- or UVR-induced skin tumorigenesis in different animal models. Studies have shown that green tea extract also possesses anti-inflammatory activity. These anti-inflammatory and anti-carcinogenic properties of green tea are due to their polyphenolic constituents present therein. Treatment of green tea polyphenols to skin has been shown to modulate the biochemical pathways involved in inflammatory responses, cell proliferation, and responses of chemical tumor promoters as well as UV light-induced inflammatory markers of skin inflammation. Topical treatment with EGCG on mouse skin also results in prevention of UVB-induced immunosuppression and oxidative stress. The protective effects of green tea treatment on human skin are either topically or consumed orally against UV light-induced inflammatory or carcinogenic responses [Zhao, 1999, 2007, 2008, 2016, 2020].

1. Protective effect of tea polyphenols against skin damage caused by ultraviolet radiation

Skin is often exposed to the outside with direct contact with sunlight, therefore it is very easy to cause UVR damage, leading to skin photoaging and skin diseases. Studies have shown that tea polyphenols can be used to treat the damage caused by UVR. The skin disease or skin disease caused by UVR is caused by excessive induced inflammation, oxidative stress, and DNA damage.

(1) Protective effect of tea polyphenols against skin photoaging induced by ultraviolet light

This study investigated the effects of topical application of EGCG to human skin before UV irradiation on UV-induced markers of oxidative stress and antioxidant enzymes. Results showed that the application of EGCG (mg/cm^2 skin) before a single UV exposure of 4× minimal erythema dose markedly decreases UV-induced production of hydrogen peroxide (68–90%, $p < 0.025$–0.005) and nitric oxide (30–100%, $p < 0.025$–0.005) in both epidermis and dermis in a time-dependent manner. EGCG pretreatment also inhibits UV-induced infiltration of inflammatory leukocytes, particularly CD11b (+) cells (a surface marker of monocytes/macrophages and neutrophils), into the skin, which are considered to be the major producers of reactive oxygen species. EGCG treatment was also found to inhibit

UV-induced epidermal lipid peroxidation at each time point studied (41–84%, $p < 0.05$). A single UV exposure of 4× minimal erythema dose to human skin was found to increase catalase activity (109–145%) and decrease glutathione peroxidase (GPx) activity (36–54%) and total glutathione (GSH) level (13–36%) at different time points studied. Pretreatment with EGCG was found to restore the UV-induced decrease in GSH level and afforded protection to the antioxidant enzyme GPx [Katiyar *et al.*, 2001a, 2001b]. This study assessed the protective effect of GTP on these UVB radiation-caused changes in murine skin. Oral feeding of 0.2% green tea polyphenols (wt/vol) as the sole source of drinking water for 30 days to SKH-1 hairless mice followed by irradiation with UVB (900 mJ/cm^2), resulted in significant protection against UVB radiation-caused cutaneous edema ($p < 0.0005$) and depletion of the antioxidant-defense system in epidermis ($p < 0.01$–0.02). The oral feeding of green tea polyphenols also resulted in significant protection against UVB radiation-caused induction of epidermal ornithine decarboxylase ($p < 0.005$–0.01) and cyclooxygenase activities ($p < 0.0001$) in a time-dependent manner [Agarwal *et al.*, 1993].

The photo-protection of tea polyphenols *in vivo* mouse model shown that topical treatment of polyphenols from green tea or EGCG (1 mg/cm^2 skin area) in hydrophilic ointment USP before single (180 mJ/cm^2) or multiple UVB exposures (180 mJ/cm^2, daily for 10 days) resulted in significant prevention of UVB-induced depletion of antioxidant enzymes such as glutathione peroxidase (78–100%, $p < 0.005$–0.001), catalase (51–92%, $p < 0.001$), and glutathione level (87–100%, $p < 0.005$). Treatment of EGCG or GTP also inhibited UVB-induced oxidative stress when measured in terms of lipid peroxidation (76–95%, $p < 0.001$) and protein oxidation (67–75%, $p > 0.001$). Further, treatment of EGCG to mouse skin resulted in marked inhibition of a single UVB irradiation-induced phosphorylation of ERK1/2 (16–95%), JNK (46–100%), and p38 (100%) proteins of MAPK family in a time-dependent manner. Identical photo-protective effects of EGCG or GTP were also observed against multiple UVB irradiation-induced phosphorylation of the proteins of MAPK family *in vivo* mouse skin. Photo-protective efficacy of GTP given in drinking water (0.2%, w/v) was also determined and compared with that of topical treatment of EGCG and GTP. Treatment of GTP in drinking water also significantly prevented single or multiple UVB irradiation-induced depletion of antioxidant enzymes (44–61%, $p < 0.01$–0.001), oxidative stress (33–71%, $p < 0.01$), and phosphorylation of ERK1/2, JNK, and p38 proteins of MAPK family but the photoprotective efficacy was comparatively less than that of topical treatments of EGCG and GTP. Lesser photoprotective efficacy of GTP in drinking water in comparison with topical application may be due to its less bioavailability in skin target cells [Vayalil *et al.*, 2003].

Chronic exposure of solar UV light to human skin results in photoaging. UV-induced oxidative damage and induction of matrix metalloproteinases (MMP) have been implicated in this process. A paper studied the skin delivery of EGCG and hyaluronic acid-loaded nano-transfersomes for antioxidant and anti-aging effects in UV radiation induced skin damage. The work attempts to develop and statistically optimize transfersomes containing EGCG and hyaluronic acid to synergize the UV radiation-protective ability of both compounds, along with imparting antioxidant and anti-aging effects. The optimized transfersomes were found to increase the cell viability and reduce the lipid peroxidation, intracellular ROS, and expression of MMP in HaCaT cells. The optimized transfersomal formulation of EGCG and HA exhibited considerably higher skin permeation and deposition of EGCG than that observed with plain EGCG. The results underline the potential application of the developed transfersomes in sunscreen cream/lotions for improvement of UV radiation-protection along with deriving antioxidant and anti-aging effects [Avadhani *et al.*, 2017]. UV irradiation-induced oxidative damage and induction of MMP might be prevented *in vivo* in mouse skin by oral administration of GTP. GTP was administered in drinking water (0.2%, wt/vol) to SKH-1 hairless mice, which were then exposed to multiple doses of UVB. Treatment of GTP resulted in inhibition of UVB-induced protein oxidation *in vivo* in mouse skin, a hallmark of photoaging, when analyzed biochemically. GTP treatment also inhibited UVB-induced protein oxidation *in vitro* in human skin fibroblast HS68 cells, which supports *in vivo* observations. Moreover, oral administration of GTP also resulted in inhibition of UVB-induced expression of matrix-degrading MMP, such as MMP-2 (67%), MMP-3 (63%), MMP-7 (62%), and MMP-9 (60%) in hairless mouse skin [Vayalil *et al.*, 2004].

These data indicate that the inhibition of UVB radiation-caused changes in these markers of skin aging and tumor promotion in murine skin by GTP may be one of the possible mechanisms of chemo-preventive effects associated with green tea against UVB-induced tumorigenesis and skin ageing.

(2) Protective effect of tea polyphenols against skin inflammation caused by ultraviolet light

Several studies have shown that polyphenolic compounds isolated from green tea provide protection against UVB-induced inflammatory responses and photocarcinogenesis in murine models. Topical application of EGCG (3 mg/mouse), a major polyphenolic component of green tea, before exposing a single low dose UVB (72 mJ/cm^2) to C3H/HeN mice, prevented UVB-induced inhibition of the contact hypersensitivity response and tolerance induction to the contact sensitizer 2,4-dinitrofluorobenzene. Topical application of EGCG before UVB exposure reduced the number of CD11b+ monocytes/macrophages and neutrophils infiltrating into skin inflammatory lesions, which are considered to be responsible for creating the UV-induced immunosuppressive state. In addition, application of EGCG before UVB exposure decreased UVB-induced production of the immunomodulatory cytokine interleukin (IL)-10 in skin as well as in draining lymph nodes (DLN), whereas production of IL-12, which is considered to be a mediator and adjuvant for induction of contact sensitivity, was found to be markedly increased in DLN when compared with UVB alone-exposed mice [Katiyar *et al.*, 1999b].

A study was performed to investigate the possible protective effects of green tea polyphenols UV-C light irradiation-induced cell death in the cultured rat cortical neurons. Protective effects of green tea polyphenols on UV-C light irradiation-induced apoptosis in cortical neurons were demonstrated by testing the content of Bax, which is involved in cell death. The expression of active Bax in cultured rat cortical neurons was inhibited significantly by green tea polyphenols compared to UV irradiation group tested. These results demonstrated that the green tea polyphenols inhibited the active Bax expression, suggesting a neuroprotective effect of green tea polyphenols against the UV-C light irradiation-induced injury on cortical neurons [Liu & Yu, 2008]. It has demonstrated that oral feeding as well as topical application of a polyphenolic fraction isolated from green tea, GTP, affords protection against the carcinogenic effects of UVB (280–320 nm) radiation. GTP could protect against UVB-induced immunosuppression and cutaneous inflammatory responses in C3H mice. Topical application of GTP (1–6 mg/animal), 30 minutes prior to or 30 minutes after exposure to a single dose of UVB (2 kJ/m^2), resulted in significant protection against local (25–90%) and systemic suppression (23–95%) of contact hypersensitivity (CHS) and inflammation in mouse dorsal skin (70–80%). These protective effects were dependent on the dose of GTP employed; increasing the dose (1–6 mg/animal) resulted in an increased protective effect (25–93%). The protective effects were also dependent on the dose of UVB (2–32 kJ/m^2). Among the four major epicatechin derivatives present in GTP, (-)-epigallocatechin-3-gallate, the major constituent in GTP, was found to be the most effective in affording protection against UVB-caused CHS and inflammatory responses [Katiyar *et al.*, 1995].

The above data demonstrate that EGCG protects against UVB-induced immunosuppression and tolerance induction by blocking UVB-induced infiltration of CD11b+ cells into the skin; reducing IL-10 production in skin as well as in DLN while markedly increasing IL-12 production in DLN. Hence, protection against UVB-induced immunosuppression by EGCG may be associated with protection against UVB-induced damages.

2. Protective effect of tea polyphenols against visual impairment caused by ultraviolet radiation

Studies have found that tea polyphenols have protective effects on a variety of visual impairment caused by ultraviolet radiation, such as keratitis, cataract, and macular degeneration.

(1) Protective effect of tea polyphenols against keratitis caused by ultraviolet radiation

UVB radiation from sunlight is a known risk factor for human corneal injury. This study investigated the protective effects of green tea polyphenol EGCG on UVB radiation-induced corneal oxidative damage in male imprinting control region mice. Corneal oxidative damage was induced by exposure to UVB radiation at 560 μW/cm^2. The animals received 0%, 0.1%, and 0.01% EGCG eye drops at a 5 mg/ml dose, twice daily for eight days. UVB radiation caused significant damage to the corneas, including apparent corneal ulceration and severe epithelial exfoliation, leading to a decrease in SOD, catalase, GSH-Px, GSH-Rd, and GSH activity in the cornea. However, the corneal TBARS and protein carbonyls increased compared with the control group. Treatment with EGCG eyedrops significantly ($p < 0.05$) ameliorated corneal damage, increased SOD, catalase, GSH-Px, GSH-Rd, and GSH activity, and decreased the TBARS and protein carbonyls in the corneas compared with the UVB-treated group. EGCG eyedrops exhibit potent protective effects on UVB radiation-induced corneal oxidative damage in mice, due to the increase in antioxidant defense system activity and the inhibition of lipid peroxidation and protein oxidation [Chen *et al.*, 2014]. A study found that ECG dose-dependently attenuated UVB-induced keratinocyte death. Moreover, ECG markedly inhibited UVB-induced cell membrane lipid peroxidation and H_2O_2 generation in keratinocytes, suggesting that ECG can act as a free radical scavenger when keratinocytes were photodamaged. In parallel, H_2O_2-induced activation of ERK1/2, p38, and JNK in keratinocytes could be inhibited by ECG. UVB-induced pre-G1 arrest, leading to apoptotic changes of keratinocytes, were blocked by ECG [Huang *et al.*, 2007].

A paper studied the antioxidant and cyto-protective effects of (-)-epigallocatechin-3-(3′-O-methyl)gallate against UV irradiation. This study examined the antioxidant roles of 3′Me-EGCG in keratinocytes (HaCaT cells). 3′Me-EGCG showed scavenging effects on H_2O_2 in cell and cell-free systems. Under H_2O_2 exposure, 3′Me-EGCG recovered cell viability and increased the expression of heme oxygenase 1 (HO-1). Under UVB and sodium nitroprusside (SNP) exposure, 3′Me-EGCG protected keratinocytes and regulated the survival protein AKT1. By regulating the AKT1/NF-κB pathway, 3′Me-EGCG augmented cell survival and proliferation in HaCaT cells. These results indicate that 3′Me-EGCG exhibits antioxidant properties, resulting in cyto-protection against various external stimuli [Kim *et al.*, 2019].

Above results indicated that UVB radiation caused significant damage to the corneas, including apparent corneal ulceration and severe epithelial exfoliation, EGCG eyedrops exhibit potent protective effects on UVB radiation-induced corneal oxidative damage.

(2) Protective effect of tea polyphenols against cataract caused by ultraviolet radiation

Age related cataract is a major cause of visual loss worldwide, which is the result of opacification of the eye lens proteins. One of the major reasons behind this deterioration is UV-induced oxidative damage. This study reported an investigation of the oxidative stress-induced damage to γB-crystallin under UV exposure. Human γB-crystallin had been expressed and purified from *E. coli*. It was found that ECG has a higher affinity towards the protein compared to EGC. The *in vitro* study of UV irradiation under oxidative damage to the protein in the presence of increasing concentrations of GTPs is indicative of their effective role as potent inhibitors of oxidative damage. Docking analyses show that the GTPs bind to the cleft between the domains of human γB-crystallin that may be associated with the protection of the protein from oxidative damage [Chaudhury *et al.*, 2017].

The transparency of the human eye lens depends on the solubility and stability of the structural proteins of the eye lens, the crystallins. Although the mechanism of cataract formation is still unclear, it is believed to involve protein misfolding and/or aggregation of proteins due to the influence of several external factors such as ultraviolet UV radiation, low pH, temperature, and exposure to chemical agents. This work studied the UV-induced photo-damage (under oxidative stress) of recombinant human γB-crystallin *in vitro* in the presence of the major green tea polyphenol, EGCG. The results showed that EGCG has the ability to protect human γB-crystallin from oxidative stress-induced photo-damage [Chaudhury *et al.*, 2016].

Aggregation of human ocular lens proteins, the crystallins, is believed to be one of the key reasons for age-onset cataract. An article investigated the aggregation propensity of human γB-crystallin in absence and presence of EGCG, *in vitro*, when exposed to stressful conditions. The experimental results have been substantiated by molecular dynamics simulation studies and showed that EGCG possesses inhibitory potency against the aggregation of human γB-crystallin at low pH and elevated temperature [Chaudhury *et al.*, 2018].

Crystallin aggregation in eye lens leads to reduction in lens opacity, which is a condition generally referred to as a cataract. Peptide fragment αA (66–80), derived from αA-crystallin, possesses high aggregation propensity and forms amyloid-like structures. αA (66–80) aggregates are known to interact with soluble crystallins and destabilize native structures that subsequently undergo aggregation. In order to inhibit the aggregation of αA (66–80) peptide, EGCG, a major active constituent of green tea, was employed. The inhibitory potential of EGCG toward αA-crystallin was clearly observed as in the presence of EGCG, the αA (66–80) aggregation was considerably inhibited and the pre-formed fibrillary aggregates of αA (66–80) were found to be disassembled. Therefore, the study suggests that EGCG can be a potential molecule that can prevent the initiation of cataract as well as be helpful in the disease reversal [Kumar *et al.*, 2017].

Oxidative stress damage of eye lens caused by ultraviolet radiation is considered to involve protein misfolding and/or protein aggregation due to the influence of various external factors such as ultraviolet radiation, low pH value, temperature, and exposure to chemical reagents. The level of ROS in eye tissue is regulated by the protection and repair treatment system, therefore, damage of these systems can lead to the formation of senile eye disease and cataract as mitochondrial reactive oxygen species in aging eyes change. Tea polyphenols were found to have antioxidant effect on lens. A study found that tea polyphenols have antioxidant effects on cultured rat lens *in vitro*. Rat lenses were cultured in Dulbecco's Modified Eagle's Medium (DMEM) medium containing H_2O_2 (1 mM) at $37°C$ in CO_2 incubator. Iron and copper salts were added to the medium. At 24 hours or 48 hours, photographs recorded the changes in transparency of rat lens and the lens proteins were analyzed. Tea polyphenols and allitridi were added to the medium with lens in order to test the antioxidant effects. Lens opacities occurred at the equatorial region by 24 hours, progressing to become totally opaque by 48 hours. New bands higher than 43KD were found. In addition, the 30KD band disappeared in the medium containing copper salt. With the addition of tea polyphenols or allitridi to the culture system, all lens remained transparent. Exposure of rat lens to H_2O_2 with iron or copper salt resulted in opacification of lens with crosslinking of crystallins and degradation of lens polypeptides. Addition of tea polyphenols or allitridi prevented the lens from undergoing oxidative stress. After 24 hours, the development of the oxidative modifications of crystallins continued. Comparing catalytic strength, copper ions were stronger than the iron ions. This work reported that the anti-oxidative action of tea polyphenols is strong [Xiang *et al.*, 1997, 2018].

Mitochondrial ROS and RNS are altered in the aging eye along with protective and repair therapeutic systems believed to regulate ROS and RNS levels in ocular tissues. Mitochondria-targeted antioxidants might be used to effectively prevent ROS and RNS-induced oxidation of lipids and proteins in the inner mitochondrial membrane *in vivo*. As a result of the combination of weak metal chelating, •OH and lipid peroxyl radicals scavenging, reducing activities to liberated fatty acid, and phospholipid hydro-peroxides, carnosine and carcinine appear to be physiological anti-oxidants able to efficiently protect the lipid phase of biologic membranes and aqueous environments and act as the anti-apoptotic natural drug compounds. A study showed that EGCG has the ability to protect human γB-crystallin from oxidative stress-induced photo-damage [Chaudhury *et al.*, 2016]. This paper studied the mechanism underlying the UVB irradiation-induced apoptosis of human lens epithelial cells (HLECs), and to investigate the protective effect of EGCG against the UVB-induced apoptosis of human lens epithelial cells. Human lens epithelial cells were exposed to different concentrations of EGCG plus UVB (30 mJ/cm^2). The results revealed that UVB irradiation reduced the mitochondrial membrane potential of HLECs and induced apoptosis. Notably, EGCG significantly attenuated the generation of H_2O_2 and hydroxyl free radicals caused by UVB irradiation in human lens epithelial cells, and significantly increased CAT, SOD, and GSH-Px activities, however, GSH levels were not significantly increased. EGCG also reduced UVB-stimulated Bax, cytochrome c, caspase-9 and caspase-3 expression, and elevated Bcl-2 expression, suggesting that EGCG may possess free radical-scavenging properties, thus increasing cell viability [Wu *et al.*, 2022]. They also investigated the protective effect of EGCG against UVB-induced apoptotic death and the underlying mechanism in human lens epithelial cells. HLECs were exposed to various concentrations of EGCG under UVB (30 mJ/cm^2), and cell viability was monitored. Mitochondrial membrane potential, ROS, and apoptosis were detected. The total antioxigenic capacity was determined and the expression of apoptosis inducing factor and endonuclease G was measured respectively. Moreover, the localisation of AIF and Endo G within cells was further detected by confocal optical microscopy. The results indicated that EGCG could enhance the cell viability and protect against cell apoptosis caused by UVB irradiation in human lens epithelial cells. EGCG could also decrease the UVB-induced generation of ROS and collapse of mitochondrial membrane potential and increase the total antioxidant capacity level. In addition, EGCG could also inhibit the UVB-stimulated increase of apoptosis inducing factor and Endo G expression at mRNA and protein levels and ameliorate the

UVB-induced mitochondria-nuclear translocation of apoptosis inducing factor and endonuclease G [Wu *et al.*, 2021].

This study determined the protective effect of EGCG against H_2O_2-induced apoptotic death and the possible mechanisms involved in human lens epithelial cells. HLEB-3, a human lens epithelial cell line, was exposed to various concentrations of H_2O_2 and EGCG, and subsequently monitored for cell death. The ability of EGCG to block the accumulation of intracellular reactive oxygen species and the loss of mitochondrial membrane potential induced by H_2O_2 was examined. EGCG protected against cell death caused by H_2O_2 in human lens epithelial cells. EGCG reduced the H_2O_2-induced generation of ROS, the loss of mitochondrial membrane potential, and the release of cytochrome c from the mitochondria into the cytosol. EGCG inhibited the H_2O_2-stimulated increase of caspase-9 and caspase-3 expression and the decrease of the Bcl-2/Bax ratio. Moreover, EGCG attenuated the reduced activation and expression of ERK, p38 MAPK, and Akt induced by H_2O_2 [Yao *et al.*, 2008]. Mitochondria are particularly susceptible to oxidative stress, and mitochondrial-dependent apoptosis plays a major role in radiation-induced tissue damage. We reasoned that targeting a redox-cycling nitroxide to mitochondria could prevent reactive oxygen species accumulation, limiting downstream oxidative damage and preserving mitochondrial function. Here, we show that in both mouse and human skin, topical application of a mitochondrially targeted antioxidant prevents and mitigates radiation-induced skin damage characterized by clinical dermatitis, loss of barrier function, inflammation, and fibrosis. Further, damage mitigation is associated with reduced apoptosis, preservation of the skin's antioxidant capacity, and reduction of irreversible DNA and protein oxidation associated with oxidative stress [Brand *et al.*, 2017]. Mitochondria are particularly susceptible to ROS and RNS and subsequent DNA damage as they are a major intracellular source of oxidants.

Therefore, the development of mitochondrially-targeted agents to mitigate mitochondrial oxidative stress and resulting DNA damage is a logical approach to prevent and treat UV-induced lens damage. Mitochondria are particularly susceptible to ROS and subsequent DNA damage as they are a major intracellular source of oxidants. Therefore, the development of mitochondrially targeted agents to mitigate mitochondrial oxidative stress and resulting DNA damage is a logical approach to prevent and treat UV-induced lens damage. Tea polyphenols can pass through the cell membrane and enter mitochondria, which can play an antioxidant role. It is a natural antioxidant worthy of development for prevention of cataract.

(3) Protective effect of tea polyphenols against macular degeneration caused by ultraviolet radiation

Several studies have shown that UVB induces direct DNA damage and oxidative stress in RPE cells by increasing ROS and dysregulating endogenous antioxidants. Activation of different signaling pathways connected to inflammation, cell cycle arrest, and intrinsic apoptosis was reported as well. Besides that, essential functions like phagocytosis, osmoregulation, and water permeability of RPE cells were also affected. EGCG can inhibit UVA-induced RPE cell death. In addition, intracellular H_2O_2 generation in RPE cells irradiated by UVA was inhibited by EGCG in a concentration-dependent manner. EGCG also inhibited UVA-induced extracellular signal-regulated kinase (ERK) and c-jun-NH2 terminal kinase (JNK) activation in RPE cells while a higher concentration of EGCG had an inhibitory effect on UVA-induced p38 activation. Finally, we investigated cyclooxygenase-2 (COX-2) expression in RPE cells exposed to UVA radiation, and EGCG was found to also have inhibited UVA-induced COX-2 expression. Overall, EGCG inhibits UVA-induced H_2O_2 production, mitogen-activating protein kinase activation, and expression of COX-2. Moreover, it enhances RPE cell survival after UVA exposure. This suggests that EGCG is effective in preventing UVA-induced damage in RPE cells and may be suitable for further developments as a chemoprotective factor for the primary prevention of early AMD. TNF-α modulated inflammatory effects in ARPE-19 by induction of ROS and upregulation of ICAM-1 expression [Chan *et al.*, 2008]. Moreover, TNF-α induced phosphor-NFκB nuclear translocation, increased phosphor-NFκB expression and IκB degradation, and increased the degree of monocyte-RPE adhesion. Pretreating the cells with EGCG ameliorated the inflammatory effects of TNF-α.

UVB radiation is part of the spectrum of light produced by the sun. This form of radiation has been implicated as one of the potential etiological factors causing age-related macular degeneration (AMD). Oxidative injury to the retinal pigment epithelium (RPE) has also been thought to play a key role in age-related macular degeneration. This study determined the mechanism by which UVB causes damage to the retinal pigment epithelium cells, whether it occurs through oxidative stress and the mitogen-activated protein kinase (MAPK) pathway and whether the green tea extract, EGCG, has a protective role. Cell viability assays were used to determine the viability of the cells under different conditions. The findings showed that UVB induced apoptosis, which increased intracellular ROS in retinal pigment epithelium cells. Inhibition of c-Jun NH2-terminal kinase (JNK) with a specific inhibitor augmented this apoptosis and anisomycin (an activator of JNK) attenuated this apoptosis. In addition, UVB decreased the phosphorylation of JNK1 and c-Jun. Finally, EGCG reduced the ROS generation and apoptosis, and partially blocked the decreased phosphorylation of JNK1 and c-Jun by UVB irradiation. The findings show that UVB irradiation is able to induce apoptosis in retinal pigment epithelium cells through oxidative stress, while EGCG treatment attenuates this damage [Cao *et al.*, 2012]. GTP effectively suppressed the decrease in viability of the UVB stressed retinal pigment epithelial (RPE) cells and the UVB suppression of surviving gene expression level. GTP alleviated mitochondria dysfunction and DNA fragmentation induced by UVB. GTP effectively suppressed the decrease in viability of the UVB-stressed RPE cells and the UVB suppression of surviving gene expression level [Xu *et al.*, 2010].

Autophagy is an intracellular catabolic process involved in protein and organelle degradation via the lysosomal pathway that has been linked in the pathogenesis of age-related macular degeneration (AMD). UVB irradiation-mediated degeneration of the macular retinal pigment epithelial (RPE) cells is an important hallmark of AMD, along with the change in RPE autophagy. Thus, pharmacological manipulation of RPE autophagy may offer an alternative therapeutic target in AMD. EGCG plays a regulatory role in UVB irradiation-induced autophagy in RPE cells. UVB irradiation results in a marked increase in the amount of LC3-II protein in a dose-dependent manner. EGCG administration leads to a significant reduction in the formation of LC3-II and auto-phagosomes. mTOR signaling activation is required for EGCG-induced LC3-II formation, as evidenced by the fact that EGCG-induced LC3-II formation is significantly impaired by rapamycin administration. Moreover, EGCG significantly alleviates the toxic effects of UVB irradiation on RPE cells in an autophagy-dependent manner [Li *et al.*, 2013].

The above research results show that GTP is a potential candidate for further development of chemical protection factor, which is used to prevent and treat cataract, keratitis, macular degeneration, and other eye injuries.

3. Preventive effect of tea polyphenols against skin cancer caused by ultraviolet radiation

Skin diseases or inflammation, oxidative stress, and DNA damage caused by ultraviolet radiation lead to skin cancer. Studies have found that green tea polyphenols have protective effects on skin inflammation, oxidative stress, and DNA damage caused by ultraviolet light. Laboratory studies on animal models have shown that tea polyphenols can protect the skin from the adverse effects of UV radiation, including the risk of skin cancer. This suggests that tea polyphenols may help supplement the protection of sunscreen and may help treat skin diseases related to inflammation, oxidative stress, and DNA damage caused by solar UVR. *In vitro* and *in vivo* animal and human studies and human clinical trials have shown that tea polyphenols can be used as a pharmacological reagent to prevent melanoma and non-melanoma skin cancer induced by solar UV light. In several mouse skin models, topical application as well as oral consumption of green tea has been shown to afford protection against UVB-induced carcinogenesis and inflammatory responses.

A paper studied the protection by tea against UV-A + B-induced skin cancers in hairless mice. Consumption of tea, especially green tea, has been shown to reduce the incidence of UV-related skin tumors in hairless mice. In separate experiments, there was a significant dose response to black tea as a preventive against UV-related skin lesions, and consumption of black tea was associated with a small but significant reduction in the incidence of papillomas in mice previously exposed to UV radiation [Record & Dreosti, 1998]. This study assessed the effect of oral feeding and topical application of green tea polyphenols on UVB radiation-induced skin carcinogenesis in female SKH-1 hairless mice. Chronic oral feeding of GTP (0.1%, w/v) in drinking water resulted in significantly ($p < 0.01$) lower tumor yield (percent of animals with tumors and number of tumors per mouse) and extended TDT50 ($p < 0.05$), as compared to animals receiving normal drinking water. Topical application of GTP before UVB irradiation also afforded protection against photo-carcinogenesis, however, the protective response was lower than that observed by oral feeding of GTP in drinking water [Wang *et al.*, 1991].

A paper investigated the mechanisms by which GTPs prevent UVB-induced skin cancer in mice. Two groups of 6- to 7-week-old female SKH-1 hairless mice were UVB irradiated (180 mJ/cm^2) three times each week for 24 weeks. One group consumed water and the other, water containing 2 g/L GTPs. The control group drank water and was not exposed to UVB radiation. UVB-induced tumors and skin biopsies from the control group were analyzed. Oral administration of GTPs reduced UVB-induced tumor incidence (35%), tumor multiplicity (63%), and tumor growth (55%). The GTPs+UVB group had reduced expression of the matrix metalloproteinase-2 (MMP-2) and MMP-9, which have crucial roles in tumor growth and metastasis and enhanced expression of tissue inhibitor of MMP in the tumors compared with mice that were treated with UVB alone. The GTPs+UVB group also had reduced expressions of Platelet endothelial cell adhesion molecule-1 (CD31) and vascular endothelial growth factor, which are essential for angiogenesis and inhibited the expression of proliferating cell nuclear antigen in the tumors compared with the UVB group. Additionally, there were more cytotoxic CD8(+) T cells in the tumors of the GTPs+UVB group than in the UVB group and their tumor cells exhibited greater activation of caspase-3, indicating the apoptotic death of the tumor cells [Mantena *et al.*, 2005]. Topical application of EGCG in human skin inhibits UVB-induced infiltration of leukocytes (macrophage/neutrophils), a potential source of generation of ROS and RNS, and generation of prostaglandin (PG) metabolites. In the experiments, it found that topical application of EGCG before UVB exposure decreased UVB-induced erythema. It also found that microsomes from EGCG pretreated human skin and exposed to UVB, compared to UVB exposure alone, significantly reduced PG metabolites, particularly PGE2. The PG metabolites play a critical role in free radical generation and skin tumor promotion in multistage skin carcinogenesis. EGCG pretreated and UVB-exposed human skin contained fewer dead cells in the epidermis with comparison to non-pretreated UVB-exposed skin [Katiyar *et al.*, 1999b].

This study investigated that in human skin, topical application of EGCG, the major polyphenolic constituent in green tea, inhibits UVB-induced infiltration of leukocytes (macrophage/neutrophils), a potential source of generation of ROS, and generation of prostaglandin (PG) metabolites. Human subjects were UVB irradiated on sun-protected skin to four times their minimal erythema dosage and skin biopsies or keratomes were obtained either 24 hours or 48 hours later. This study found that topical application of EGCG (3 mg/2.5 cm^2) before UVB exposure to human skin significantly blocked UVB-induced infiltration of leukocytes and reduced myeloperoxidase activity. These infiltrating leukocytes are considered to be the major source of generation of ROS. In the same set of experiments, we found that topical application of EGCG before UVB exposure decreased UVB-induced erythema [Katiyar *et al.*, 1999a]. UV radiation-induced immunosuppression has been implicated in the development of skin cancers. Green tea polyphenols in drinking water prevented photo-carcinogenesis in the skin of mice. It also studied whether GTPs in drinking water (0.1–0.5%, w/v) prevent UV-induced immunosuppression and potential mechanisms of this effect in mice. GTPs (0.2% and 0.5%, w/v) reduced UV-induced suppression of contact hypersensitivity (CHS) in response to a contact sensitizer in local (58–62% reductions; $p < 0.001$) and systemic (51–55% reductions; $p < 0.005$) models of CHS. Compared with untreated mice, GTP-treated mice (0.2%, w/v) had a reduced number of cyclobutane pyrimidine dimer-positive (CPD(+)) cells (59%; $p < 0.001$) in the skin, showing faster repair of UV-induced DNA damage and had a reduced (two-fold) migration of CPD(+) cells from the skin to draining lymph nodes, which was associated with elevated levels of nucleotide excision repair (NER) genes [Katiyar *et al.*, 2010]. Studies conducted by a group on human skin have demonstrated that GTP prevents UV-B-induced cyclobutane pyrimidine dimers (CPD), which are considered to be mediators of UVB-induced immune suppression and skin cancer induction. GTP-treated human skin prevented penetration of UV radiation, which was demonstrated by the absence of immunostaining for CPD in the reticular dermis. The topical application of GTP or its most potent chemo-preventive constituent EGCG, prior to exposure to UVB, protects against UVB-induced local as well as systemic immune suppression in laboratory animals. Additionally, studies have shown that EGCG treatment of mouse skin inhibits UVB-induced infiltration of CD11b+ cells [Katiyar *et al.*, 2001b].

Topical application of topical application of GTP and EGCG prior to exposure of UVB protects against UVB-induced local as well as systemic immune suppression in laboratory animals, which was associated with the inhibition of UVB-induced infiltration of inflammatory leukocytes. Prevention of UVB-induced suppression of immune responses by EGCG was also associated with the reduction in immunosuppressive cytokine IL-10 production at UV irradiated skin and draining lymph nodes, whereas IL-12 production was significantly enhanced in draining lymph nodes. Antioxidant and anti-inflammatory effects of green tea were also observed in human skin. Treatment of EGCG to human skin resulted in the inhibition of UVB-induced erythema, oxidative stress, and infiltration of inflammatory leukocytes. IL-12 knockout (KO) mice on C3H/HeN background and DNA repair-deficient cells from xeroderma pigmentosum complementation group A (XPA) patients were used in this study. The effect of EGCG was determined on UV-induced suppression of contact hypersensitivity and UV-induced DNA damage in the form of yclobutene pyrimidine dimers (CPD) in mice and XPA-deficient cells using immunohistochemistry and dot-blot analysis. Topical treatment with EGCG prevented UV-induced suppression of the contact hypersensitivity in wild-type (WT) mice but did not prevent it in IL-12 KO mice. Injection of anti-IL-12 monoclonal antibody to WT mice blocked the preventive effect of EGCG on UV-induced immunosuppression. EGCG reduced or repaired UV-induced DNA damage in skin faster in WT mice as shown by reduced number of CPDs(+) cells and reduced the migration of CPD(+) antigen-presenting cells from the skin to draining lymph nodes. In contrast, this effect of EGCG was not seen in IL-12 KO mice [Meeran *et al.*, 2006]. Topical treatment with EGCG prevented UV-induced suppression of the contact hypersensitivity in wild-type mice but did not prevent it in IL-12 KO mice. Injection of anti-IL-12 monoclonal antibody to WT mice blocked the preventive effect of EGCG on UV-induced immunosuppression. EGCG reduced or repaired UV-induced DNA damage in skin faster in wild-type mice as shown by the reduced number of cyclobutane pyrimidine dimers (+) cells and reduced the migration of cyclobutane pyrimidine dimers (+) antigen-presenting cells from the skin to draining lymph nodes. In contrast, this effect of EGCG was not seen in IL-12 KO mice. Further, EGCG was able to repair UV-induced cyclobutane pyrimidine dimers in xeroderma pigmentosum complementation group A proficient cells obtained from healthy person but did not repair in xeroderma pigmentosum A complementation group A-deficient cells, indicating that nucleotide excision repair mechanism is involved in DNA repair [Katiyar *et al.*, 2010].

The above results indicate that oral administration and local application of polyphenol GTP isolated from green tea can prevent the carcinogenic effect of UVB radiation, and its mechanism is mainly to reduce the expression of MMP-2 and MMP-9, inhibit the inflammatory infiltration of ultraviolet induced leukocytes (macrophages/neutrophils), and prevent ultraviolet-induced immunosuppression. The above data also show that EGCG protects against UVB-induced immunosuppression and tolerance induction by blocking UVB-induced CD11b+cells from infiltrating into the skin, reducing the production of IL-10 in the skin and DLN, and significantly increasing the production of IL-12 in DLN. The protection of EGCG against UVB-induced immunosuppression may be related to the protection against UVB induced light damage.

15.4.2 *Protection of tea polyphenols against electromagnetic radiation-induced injury*

With the increasing use of television and mobile phones, human beings are inadvertently exposed to electromagnetic radiation. In the current situation, radiation exposure is inevitable, and mobile phones are inevitable necessities. Now, almost everyone is using mobile phones, but they are still concerned about whether the electromagnetic fields generated by mobile phones will cause damage to human bodies. Human being are inadvertently being exposed to electromagnetic fields (EMF) as its prevalence increases, mainly through mobile phones. Radiation exposure is unavoidable in the current context, with mobile phones being an inevitable necessity. Research show that EMF are really harmful to public health. EMF exposure might affect the apoptotic process *in vitro*, with results depending on the type of modulation, intermittent mode of exposure, and cell model [Manna & Ghosh, 2016]. A paper studied the effect of 4.5 G (LTE Advanced-Pro network) mobile phone radiation on the optic nerve. The results from the experimental group reveled that the axonal diameter and myelin thickness were shown to be lower and the G-ratio was higher than in the sham group. In the experimental group, malondialdehyde level was significantly higher and superoxide dismutase and catalase activities were significantly lower than sham group. There was also a high correlation between VEP wave amplitudes and oxidative stress markers, and a high correlation between visual evoked potential (VEP) wave amplitudes and oxidative stress markers [Özdemir *et al.*, 2021].

Tea polyphenols have neuroprotective effects against mobile phone-induced neuronal damage as well as through the various mechanisms of action that are elicited to invoke the beneficial effects in validated experimental models [Raghu *et al.*, 2022]. A paper studied the potential protection of green tea polyphenols against 1800 MHz electromagnetic radiation-induced injury on rat cortical neurons. The present study was performed to investigate the possible protective effects of green tea polyphenols against electromagnetic radiation-induced injury in the cultured rat cortical neurons. In this study, green tea polyphenols were used in the cultured cortical neurons exposed to 1800 MHz EMFs by the mobile phone. We found that the mobile phone irradiation for 24 hours induced marked neuronal cell death in the MTT (3-(4,5-dimethylthiazole-2-yl)-2,5-diphenyl-tetrazolium bromide) and TUNEL (TdT mediated biotin-dUTP nicked-end labeling) assay, and protective effects of green tea polyphenols on the injured cortical neurons were demonstrated by testing the content of Bcl-2-associated X protein (Bax) in the immunoprecipitation assay and Western blot assay. In our study results, the mobile phone irradiation-induced increase in the content of active Bax were inhibited significantly by green tea polyphenols, while the contents of total Bax had no marked changes after the treatment of green tea polyphenols [Liu *et al.*, 2011].

The above research results show that the electromagnetic radiation used by mobile phones and televisions may have a certain damage effect on the human body, especially the nerves. The tea polyphenols can prevent and protect the body from the damage caused by these electromagnetic radiation.

15.4.3 *Protection of tea polyphenols against ion radiation*

Among the survivors of the Hiroshima atomic bombing, it was found that those who drank tea for a long time had mild radiation sickness and a high survival rate. Tea extract was discovered in the 1950s to eliminate the harm of radioactive strontium to animals. Based on observations of the survival of animals, those in the control group died, while the animals fed with tea survived. Therefore, tea is called the atomic bomb drink. In the 1970s, China also discovered the anti-radiation effect of tea, and some anti-radiation experiments have achieved satisfactory results. Therefore, drinking tea can prevent ion radiation damage.

1. Protective effect of tea polyphenol against γ radiation damage

On one hand, γ-ray has strong penetration, which can be used for flaw detection or automatic control of assembly line in industry. On the other hand, γ-ray is lethal to cells. While it is used to treat tumors in medicine, it can also cause body damage and oxidative stress. Tea polyphenols can protect cells against γ-ray ionizing radiation damage.

A paper studied the efficacy of tea polyphenols against γ-ray radiation-induced hematopoietic and biochemical alterations in beagle dogs. The dogs were exposed to a single acute dose of whole-body γ-radiation (3 Gy) and orally administered tea polyphenols (80 mg kg^{-1} day^{-1} or 240 mg kg^{-1} day^{-1}) for 28 consecutive days. Dogs exposed to γ-radiation alone exhibited typical hematopoietic syndrome. In contrast, irradiated dogs that received tea polyphenols exhibited an improved blood profile with reduced leucopenia, thrombocytopenia (platelet counts), and reticulocyte levels. Tea polyphenols also significantly elevated levels of the endogenous antioxidant enzyme superoxide-dismutase, reduced the increased levels of serum cytokine in response to radiation-induced toxicity, and increased colony-forming units of bone marrow hematopoietic progenitor cells. In addition, tea polyphenols repaired radiation-induced organ damage [Dong *et al.*, 2019]. The submandibular glands showed that the lesions of the tea polyphenols-group were mild; change in apoptosis of the cells was not obvious compared with the 15 Gy gamma rays-group. The cell apotosis was typical after irradiation in the 15 Gy gamma rays-group. Apoptosis index that was detected in the cells of submandibular glands of the tea polyphenols-group decreased compared with the R-group (statistically significant; $p < 0.01$) on the 3rd, 6th, and 30th day after irradiation [Peng *et al.*, 2011]. Tea polyphenols showed the greatest radioprotective effect against radiation-induced changes in hematological parameters (red blood cell count, white blood cell count, and hemoglobin), and maintained unchanged spleen and thymus indices (spleen or thymus weight/body weight × 1000). Tea polyphenols also significantly decreased radiation-induced lipid peroxidation (malondialdehyde levels), elevated endogenous antioxidant enzymes (superoxide dismutase), and reduced the serum cytokines which were elevated in radiation-induced toxicity [Hu *et al.*, 2011].

Under acellular condition of radiation exposure, *Escherichia coli* gene cloning vector pBR322 plasmid DNA was protected by EGCG in a concentration-dependent manner. Treatment of murine splenocytes with EGCG two hours prior to radiation (3 Gy), protected the cellular DNA against radiation-induced strand breaks. EGCG also inhibited γ-radiation-induced cell death in splenocytes. EGCG pretreatment to the cells decreased the radiation-induced lipid peroxidation and membrane damage. The levels of phase II enzymes, glutathione, and lactate dehydrogenase were restored with EGCG treatment prior to radiation. Our results show that pretreatment with EGCG offers protection to pBR322 DNA under acellular condition and normal splenocytes under cellular condition, against γ-radiation-induced damage, and is better radio-protector in comparison to quercetin and vitamin C [Richi *et al.*, 2012]. The radio-protective actions of black tea against the radiation-induced membrane permeability of human erythrocytes were studied. Tea extracts showed potential scavenging of H_2O_2 and NO, appreciable extent of total antioxidant capacity, and effective anti-hemolytic action. Tea extracts (15 μg/mL) significantly ameliorated the radiation-induced increase of the levels of thiobarbituric acid-reactive substances (TBARS, an index of lipid peroxidation) in the RBC membrane ghosts. Stored blood showed higher levels of K^+ ion as compared to the normal blood which was elevated by radiation. Membrane ATPase was inhibited by the exposure to radiation. Treatment of RBCs with the tea extracts (15 μg/ml) prior to the exposure of radiation significantly mitigated these changes in the erythrocyte membranes caused by the lower dose of radiation (4 Gy) as compared to that induced by the higher dose of radiation [Modak *et al.*, 2016]. Black tea extract scavenged free radicals and inhibited Fenton reaction-mediated 2-deoxyribose degradation and lipid peroxidation in a dose-dependent fashion, establishing its antioxidant properties. The radio-protective effects of black tea extract on strand-break induction in pBR322 plasmid DNA were 100 % at 80 μg/ml and higher. In V79 cells, black tea extract was effective in decreasing the frequency of radiation-induced micronucleated cells and the yields of ROS, and in restoring the integrity of cellular mitochondrial membrane potential significantly. Black tea extract exerted maximum protection against radiation-induced damage in V79 at a dose of 5 µg/ml [Pal *et al.*, 2013].

The above research results show that green tea and black tea polyphenols can protect tissues and cells and prevent γ oxidative stress damage caused by radiation.

2. Protective effect of tea polyphenol against X-rays damage

The frequency and energy of X-rays are second only to γ-rays. X-rays have very strong penetrability and are widely used in human fluoroscopy. When exposed to X-rays, the electrons outside the nucleus can be separated from the atomic orbit and ionized. Therefore, it is also an ionic radiation, which will cause body damage and oxidative stress. Tea polyphenols can protect cells from ionizing radiation damage.

This study investigated whether EGCG confers cyto-protection against ionizing radiation. We found that, compared with the control, pretreatment with EGCG significantly enhanced the viability of human skin cells that were irradiated with X-rays, and decreased apoptosis induced by X-ray irradiation. Mito-Tracker assay showed that EGCG suppressed the damage to mitochondria induced by ionizing radiation via upregulation of SOD2. ROS and RNS in immortalized keratinocytes in adult humans HaCaT cells were significantly reduced when pretreated with EGCG before irradiation. Radiation-induced γH2AX foci, which are representative of DNA double-strand breaks, decreased after pretreatment with EGCG. Furthermore, EGCG induced the expression of the cytoprotective molecule heme oxygenase-1 (HO-1) in a dose-dependent manner via transcriptional activation. HO-1 knockdown or treatment with the HO-1 inhibitor tin proto-porphyrin (SnPPIX) reversed the protective role of EGCG, indicating an important role for HO-1 [Zhu *et al.*, 2014]. Tea polyphenols increase X-ray repair cross-complementing protein 1 and apurinic/apyrimidinic endonuclease/redox factor-1 expression in the hippocampus of rats during cerebral ischemia/reperfusion injury. A study employed a rat model of global cerebral ischemia/reperfusion. It demonstrated that intraperitoneal injection of tea polyphenols immediately after reperfusion significantly reduced apoptosis in the hippocampal CA1 region; this effect started six hours following reperfusion. Immunohistochemical staining showed that tea polyphenols could reverse the ischemia/reperfusion-induced reduction in the expression of DNA repair proteins, X-ray repair cross-complementing protein 1, and apurinic/apyrimidinic endonuclease/redox factor-1 starting at two hours. Both effects lasted at least 72 hours. These experimental findings suggest that tea polyphenols promote DNA damage repair and protect against apoptosis in the brain [Wang *et al.*, 2012].

The above research results show that tea polyphenols can prevent ion radiation damage including γ-ray and X-ray, and has obvious protective effect, which can be used for the protection of radiation damage and the treatment of side effects caused by the treatment of γ-ray and X-ray.

15.5 The mechanism of tea polyphenols in preventing radiation damage

These studies show that tea polyphenols can prevent radiation injury especially for skin, eye lens, and retina. Its mechanism is mainly due to the antioxidant properties of tea polyphenols, which can eliminate free radicals produced by radiation damage and inhibit oxidative stress damage to inhibit oxidative stress, reduce the expression of MMP-2 and MMP-9, inhibit radiation-induced leukocyte (macrophage / neutrophil) infiltration, and prevent radiation-induced immunosuppression. In particular, EGCG can significantly reduce the formation of LC3-II and auto-phagosome, and significantly reduce the toxic effect of radiation on cells in an autophagy dependent manner. EGCG can reduce H_2O_2-induced ROS and RNS production, protect mitochondrial intima lipid loss of mitochondrial membrane potential loss of mitochondrial membrane potential, and release of cytochrome c from mitochondria into the cytoplasm. EGCG can protect proteins and inhibit the increase of caspase-9 and caspase-3 expression stimulated by H_2O_2 and the decrease of Bcl-2/bax ratio. In addition, EGCG attenuated the activation and expression of ERK, p38 MAPK, and Akt induced by H_2O_2, and significantly inhibited the increase of active Bax content and apoptosis induced by radiation. EGCG can also inhibit the increase of UVB-stimulated apoptosis inducing factor and endonuclease G expression at the mRNA and protein levels and improve the mitochondrial nuclear translocation of radiation-induced apoptosis inducing factor and endonuclease G. EGCG also reduced the expression of Bax, cytochrome c, caspase-9, and caspase-3 induced by radiation and increased the expression of Bcl-2. EGCG cleared free radicals, prevented radiation-induced apoptosis, and increased cell viability (Fig. 15-4).

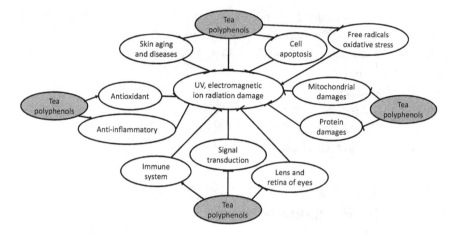

Figure 15-4. The sketch map on the mechanism of preventing effects radiation damage by tea polyphenols.

15.6 Conclusion

The above research results show that γ-ray, X-ray, electromagnetic radiation, and UV radiation, especially the radiation lines with multiple frequencies of sunlight, will produce ROS and RNS in the body, causing oxidative stress and body damage. Tea polyphenols are natural antioxidants, which can remove various kinds of radiation damage and have obvious protective effects. The most important is to prevent skin photo-oxidative damage, skin inflammation, skin aging, and even skin cancer caused by UV light. Tea polyphenols can protect vision damage, including keratitis, cataract, and macular degeneration against UV damage. Tea polyphenols can be used for the prevention and treatment of ionizing radiation damage including γ-rays and X-rays and the treatment of side effects caused by treatment. Of course, it is necessary to strictly determine the dose and duration usage of tea polyphenols to prevent radiation damage in the future to achieve the best effect.

References

Abdus-salam A, Elumelu T, Adenipekun A. (2008) Mobile phone radiation and the risk of cancer; a review. *Afr J Med Med Sci*, **37**(2), 107–118.

Agarwal R, Katiyar SK, Khan SG, Mukhtar H. (1993) Protection against ultraviolet B radiation-induced effects in the skin of SKH-1 hairless mice by a polyphenolic fraction isolated from green tea. *Photochem Photobiol*, **58**(5), 695–700.

Albrecht S, Jung S, Müller R, Lademann J, Zuberbier T, Zastrow L, Reble C, Beckers I, Meinke MC. (2019) Skin type differences in solar-simulated radiation-induced oxidative stress. *Br J Dermatol*, **180**(3), 597–603.

Avadhani KS, Manikkath J, Tiwari M, *et al.* (2017) Skin delivery of epigallocatechin-3-gallate (EGCG) and hyaluronic acid loaded nano-transfersomes for antioxidant and anti-aging effects in UV radiation induced skin damage. *Drug Deliv*, **24**(1), 61–74.

Brand RM, Epperly MW, Stottlemyer JM, Skoda EM, Gao X, Li S, Huq S, Wipf P, Kagan VE, Greenberger JS, Falo LD Jr. (2017) A topical mitochondria-targeted redox-cycling nitroxide mitigates oxidative stress-induced skin damage. *J Invest Dermatol*, **137**(3), 576–586.

Cao G, Chen M, Song Q, Liu Y, Xie L, Han Y, Liu Z, Ji Y, Jiang Q. (2012) EGCG protects against UVB-induced apoptosis via oxidative stress and the JNK1/c-Jun pathway in ARPE19 cells. *Mol Med Rep*, **5**(1), 54–59.

Chan CM, Huang JH, Lin HH, Chiang HS, Chen BH, Hong JY, Hung CF. (2008) Protective effects of (-)-epigallocatechin gallate on UVA-induced damage in ARPE19 cells. *Mol Vis*, **14**, 2528–2534.

Chaudhury S, Bag S, Bose M, Das AK, Ghosh AK, Dasgupta S. (2016) Protection of human γB-crystallin from UV-induced damage by epigallocatechin gallate: spectroscopic and docking studies. *Mol Biosyst*, **12**(9), 2901–2909.

Chaudhury S, Roy P, Dasgupta S. (2017) Green tea flavanols protect human γB-crystallin from oxidative photodamage. *Biochimie*, **137**, 46–55.

Chaudhury S, Dutta A, Bag S, Biswas P, Das AK, Dasgupta S. (2018) Probing the inhibitory potency of epigallocatechin gallate against human γB-crystallin aggregation: spectroscopic, microscopic and simulation studies. *Spectrochim Acta A Mol Biomol Spectrosc*, **192**, 318–327.

Chen MH, Tsai CF, Hsu YW, Lu FJ. (2014) Epigallocatechin gallate eye drops protect against ultraviolet B-induced corneal oxidative damage in mice. *Mol Vis*, **20**, 153–162.

Dong X, Wang D, Han J, Liu P, Hu Y, Luo Q, Guo D, Yan C. (2019) Efficacy of tea polyphenols (TP 50) against radiation-induced hematopoietic and biochemical alterations in beagle dogs. *J Tradit Chin Med*, **39**(3), 324–331.

Ghobadi A, Shirazi A, Najafi M, Kahkesh MH, Rezapoor S. (2017) Melatonin ameliorates radiation-induced oxidative stress at targeted and nontargeted lung tissue. *J Med Phys*, **42**(4), 241–244.

Halliday GM. (2005) Inflammation, gene mutation and photoimmunosuppression in response to UVR-induced oxidative damage contributes to photocarcinogenesis. *Mutat Res*, **571**(1-2), 107–120.

Huang CC, Wu WB, Fang JY, Chiang HS, Chen SK, Chen BH, Chen YT, Hung CF. (2007) (-)-Epicatechin-3-gallate, a green tea polyphenol is a potent agent against UVB-induced damage in HaCaT keratinocytes. *Molecules*, **12**(8), 1845–1858.

Hu Y, Cao JJ, Liu P, Guo DH, Wang YP, Yin J, Zhu Y, Rahman K. (2011) Protective role of tea polyphenols in combination against radiation-induced haematopoietic and biochemical alterations in mice. *Phytother Res*, **25**(12), 1761–1769.

Ivanov IV, Mappes T, Schaupp P, Lappe C, Wahl S. (2018) Ultraviolet radiation oxidative stress affects eye health. *J Biophotonics*, **11**(7), e201700377.

Katiyar SK, Challa A, McCormick TS, Cooper KD, Mukhtar H. (1999a) Prevention of UVB-induced immunosuppression in mice by the green tea polyphenol (-)-epigallocatechin-3-gallate may be associated with alterations in IL-10 and IL-12 production. *Carcinogenesis*, **20**(11), 2117–2124.

Katiyar SK, Matsui MS, Elmers CA, Mukhtar H. (1999b) Polyphenolic antioxidant (-) epigallocatechin-3-gallate from green tea reduce UVB-induced inflammatory responses and infiltration of leukocytes in human skin. *Photochem Photobiol*, **69**, 148–153.

Katiyar SK, Afaq F, Perez A, Mukhtar H.(2001a) Green tea polyphenol (-)-epigallocatechin-3-gallate treatment of human skin inhibits ultraviolet radiation-induced oxidative stress. *Carcinogenesis*, **22**(2), 287–294.

Katiyar SK, Bergamo BM, Vyalil PK, Elmets CA. (2001b) Green tea polyphenols: DNA photodamage and photoimmunology. *J Photochem Photobiol B*, **65**(2–3), 109–114.

Katiyar SK, Vaid M, van Steeg H, Meeran SM. (2010) Green tea polyphenols prevent UV-induced immunosuppression by rapid repair of DNA damage and enhancement of nucleotide excision repair genes. *Cancer Prev Res (Phila)*, **3**(2), 179–189.

Katiyar SK, Elmets CA, Agarwal R, Mukhtar H. (1995) *Photochem Photobiol*, **62**(5), 855–861.

Khan A, Manna K, Das DK, *et al.* (2015) Gossypetin ameliorates ionizing radiation-induced oxidative stress in mice liver--a molecular approach. *Free Radic Res*, **49**(10), 1173–1186.

Kim EM, Hossain MA, Kim JH, Cho JYI. (2019) Antioxidant and cytoprotective effects of (-)-epigallocatechin-3-(3″-*O*-methyl) gallate. *Int J Mol Sci*, **20**(16), 3993.

Kohli DR, Sachdev A, Vats HS. (2009) Cell phones and tumor: still in no man's land. *Indian J Cancer*, **46**(1), 5–12.

Kondo H, Yumoto K, Alwood JS, Mojarrab R, Wang A, Almeida EA, Searby ND, Limoli CL, Globus RK. (2010) Oxidative stress and gamma radiation-induced cancellous bone loss with musculoskeletal disuse. *J Appl Physiol (1985)*, **108**(1), 152–161.

Kumar V, Gour S, Peter OS, Gandhi S, Goyal P, Pandey J, Harsolia RS, Yadav JK. (2017) Effect of green tea polyphenol epigallocatechin-3-gallate on the aggregation of αA (66–80) peptide, a major fragment of αA-crystallin involved in cataract development. *Curr Eye Res*, **42**(10), 1368–1377.

Li CP, Yao J, Tao ZF, Li XM, Jiang Q, Yan B. (2013) Epigallocatechin-gallate (EGCG) regulates autophagy in human retinal pigment epithelial cells: a potential role for reducing UVB light-induced retinal damage. *Biochem Biophys Res Commun*, **438**(4), 739–745.

Li M, Lin XF, Lu J, Zhou BR, Luo D. (2016) Hesperidin ameliorates UV radiation-induced skin damage by abrogation of oxidative stress and inflammatory in HaCaT cells. *J Photochem Photobiol B*, **165**, 240–245.

Liu ML, Wen JQ, Fan YB. (2011) Potential protection of green tea polyphenols against 1800 MHz electromagnetic radiation-induced injury on rat cortical neurons. *Neurotox Res*, **20**(3), 270–276.

Liu ML, Yu LC. (2008) Potential protection of green tea polyphenols against ultraviolet irradiation-induced injury on rat cortical neurons. *Neurosci Lett*, **444**(3), 236–239.

Liu XR, Zhu N, Hao YT, Yu XC, Li Z, Mao RX, Liu R, Kang JW, Hu JN, Li Y. (2021) Radioprotective effect of Whey hydrolysate peptides against γ-radiation-induced oxidative stress in BALB/c mice. *Nutrients*, **13**(3), 816.

Manna D, Ghosh R. (2016) Effect of radiofrequency radiation in cultured mammalian cells: a review. *Electromagn Biol Med*, **35**, 265–301.

Mantena SK, Meeran SM, Elmets CA, Katiyar SK. (2005) Orally administered green tea polyphenols prevent ultraviolet radiation-induced skin cancer in mice through activation of cytotoxic T cells and inhibition of angiogenesis in tumors. *J Nutr*, **135**(12), 2871–2877.

Martinez RM, Fattori V, Saito P, *et al.* (2018) Lipoxin A4 inhibits UV radiation-induced skin inflammation and oxidative stress in mice. *J Dermatol Sci*, **S0923-1811**(18), 30201–30209.

Matsumura Y, Ananthaswamy HN. (2004) Toxic effects of ultraviolet radiation on the skin. *Toxicol Appl Pharmacol*, **195**(3), 298–308.

Meeran SM, Mantena SK, Katiyar SK. (2006) Prevention of ultraviolet radiation-induced immunosuppression by (-)-epigallocatechin-3-gallate in mice is mediated through interleukin 12-dependent DNA repair. *Clin Cancer Res*, **12**(7 Pt 1), 2272–2280.

Modak A, Chakraborty A, Das SK. (2016) Black tea extract protects against -radiation-induced membrane damage of human erythrocytes. *Indian J Exp Biol*, **54**(11), 745–752.

Najafi M, Fardid R, Takhshid MA, Mosleh-Shirazi MA, Rezaeyan AH, Salajegheh A. (2016) Radiation-induced oxidative stress at out-of-field lung tissues after pelvis irradiation in rats. *Cell J*, **18**(3), 340–345.

Nakashima Y, Ohta S, Wolf AM. (2017) Blue light-induced oxidative stress in live skin. *Free Radic Biol Med*, **108**, 300–310.

Naziroğlu M, Tokat S, Demirci S. (2012) Role of melatonin on electromagnetic radiation-induced oxidative stress and Ca2+ signaling molecular pathways in breast cancer. *J Recept Signal Transduct Res*, **32**(6), 290–297.

Nichols JA, Katiyar SK. (2010) Skin photoprotection by natural polyphenols: anti-inflammatory, antioxidant and DNA repair mechanisms. *Arch Dermatol Res*, **302**(2), 71–83.

Özdemir E, Çömelekoğlu Ü, Degirmenci E, Bayrak G, Yildirim M, Ergenoglu T, Coşkun Yılmaz B, Korunur Engiz B, Yalin S, Koyuncu DD, Ozbay E. (2021) The effect of 4.5 G (LTE Advanced-Pro network) mobile phone radiation on the optic nerve. *Cutan Ocul Toxicol*, **40**(3), 198–206.

Özsobacı NP, Ergün DD, Tunçdemir M, Özçelik D. (2020) Protective effects of zinc on 2.45 GHz electromagnetic radiation-induced oxidative stress and apoptosis in HEK293 cells. *Biol Trace Elem Res*, **194**(2), 368–378.

Pal S, Saha C, Dey SK. (2013) Studies on black tea (Camellia sinensis) extract as a potential antioxidant and a probable radioprotector. *Radiat Environ Biophys*, **52**(2), 269–278.

Peng Z, Xu ZW, Wen WS, Wang RS. (2011) Tea polyphenols protect against irradiation-induced injury in submandibular glands' cells: a preliminary study. *Arch Oral Biol*, **56**(8), 738–743.

Raghu SV, Kudva AK, Rajanikant GK, Baliga MS. (2022) Medicinal plants in mitigating electromagnetic radiation-induced neuronal damage: a concise review. *Electromagn Biol Med*, **41**(1), 1–14.

Rasouli Mojez M, Ali Gaeini A, Choobineh S, Sheykhlouvand M. (2021) Hippocampal oxidative stress induced by radiofrequency electromagnetic radiation and the neuroprotective effects of aerobic exercise in rats: a randomized control trial. *J Phys Act Health*. **18**(12), 1532–1538.

Record IR, Dreosti IE. (1998) Protection by tea against UV-A+B-induced skin cancers in hairless mice. *Nutr Cancer*, **32**, 71–75.

Richi B, Kale RK, Tiku AB. (2012) Radio-modulatory effects of green tea catechin EGCG on pBR322 plasmid DNA and murine splenocytes against gamma-radiation induced damage. *Mutat Res*, **747**(1), 62–70.

Salucci S, Burattini S, Buontempo F, Martelli AM, Falcieri E, Battistelli M. (2017) Protective effect of different antioxidant agents in UVB-irradiated keratinocytes. *Eur J Histochem*, **61**, 2784.

Saric S, Sivamani RK. (2016) Polyphenols and sunburn. *Int J Mol Sci*, **17**(9), 1521.

Sharma P, Montes de Oca MK, *et al.* (2018) Tea polyphenols for the prevention of UVB-induced skin cancer. *Photodermatol Photoimmunol Photomed*, **34**(1), 50–59.

Shaban NZ, Ahmed Zahran AM, El-Rashidy FH, Abdo Kodous AS. (2017) Protective role of hesperidin against gamma-radiation-induced oxidative stress and apoptosis in rat testis. *J Biol Res (Thessalon)*, **24**, 5.

Singh KV, Gautam R, Meena R, *et al.* (2020) Effect of mobile phone radiation on oxidative stress, inflammatory response, and contextual fear memory in Wistar rat. *Environ Sci Pollut Res Int*, **27**(16), 19340–19351.

Stewart MS, Cameron GS, Pence BC. (1996) Antioxidant nutrients protect against UVB-induced oxidative damage to DNA of mouse keratinocytes in culture. *J Investig Dermatol*, **106**, 1086–1089.

Sun YY, Wang YH, Li ZH, Shi ZH, Liao YY, Tang C, Cai P. (2019) Extremely low frequency electromagnetic radiation enhanced energy metabolism and induced oxidative stress in Caenorhabditis elegans. *Sheng Li Xue Bao*, **71**(3), 388–394.

Vayalil PK, Elmets CA, Katiyar SK. (2003) Treatment of green tea polyphenols in hydrophilic cream prevents UVB-induced oxidation of lipids and proteins, depletion of antioxidant enzymes and phosphorylation of MAPK proteins in SKH-1 hairless mouse skin. *Carcinogenesis*, **24**(5), 927–936.

Vayalil PK, Mittal A, Hara Y, Elmets CA, Katiyar SK. (2004) Green tea polyphenols prevent ultraviolet light-induced oxidative damage and matrix metalloproteinases expression in mouse skin. *J Invest Dermatol*, **122**(6), 1480–1487.

Wang Z, Xue R, Lei X, Lv J, Wu G, Li W, Xue L, Lei X, Zhao H, Gao H, Wei X. (2012) Tea polyphenols increase X-ray repair cross-complementing protein 1 and apurinic/apyrimidinic endonuclease/redox factor-1 expression in the hippocampus of rats during cerebral ischemia/reperfusion injury. *Neural Regen Res*, **7**(30), 2355–2361.

Wang ZY, Agarwal R, Bickers DR, Mukhtar H. (1991) Protection against ultraviolet B radiation-induced photocarcinogenesis in hairless mice by green tea polyphenols. *Carcinogenesis*, **12**(8), 1527–1530.

Wu Q, Song J, Gao Y, Zou Y, Guo J, Zhang X, Liu D, Guo D, Bi H. (2022) Epigallocatechin gallate enhances human lens epithelial cell survival after UVB irradiation via the mitochondrial signaling pathway. *Mol Med Rep*, **25**(3), 87.

Wu Q, Li Z, Liu D, *et al.* (2021) Epigallocatechin gallate protects the human lens epithelial cell survival against UVB irradiation through AIF/endo G signalling pathways in vitro. *Cutan Ocul Toxicol*, **40**(3), 187–197.

Xiang H, Pan S, Li S. (1997) Study of the antioxidative effect of tea polyphenol and allitridi on cultured rat lens in vitro. *Yan Ke Xue Bao*, **13**(4), 173–176.

Xiang H, Pan S, Li S. (2018) Studies on human fetal lens crystallins under oxidative stress and protective effects of tea polyphenols. *Yan Ke Xue Bao*, **14**(3), 170–175.

Xu JY, Wu LY, Zheng XQ, Lu JL, Wu MY, Liang YR. (2010) Green tea polyphenols attenuating ultraviolet B-induced damage to human retinal pigment epithelial cells in vitro. *Invest Ophthalmol Vis Sci*, **51**(12), 6665–6670.

Yakymenko I, Sidorik E. (2010) Risks of carcinogenesis from electromagnetic radiation of mobile telephony devices. *Exp Oncol*, **32**(2), 54–60.

Yang X, Ren H, Guo X, Hu C, Fu J. (2020) Radiation-induced skin injury: pathogenesis, treatment, and management. *Aging (Albany NY)*, **12**(22), 23379–23393.

Yao K, Ye P, Zhang L, Tan J, Tang X, Zhang Y. (2008) Epigallocatechin gallate protects against oxidative stress-induced mitochondria-dependent apoptosis in human lens epithelial cells. *Mol Vis*, **14**, 217–223.

Yin Y, Li W, Son YO, Sun L, Lu J, Kim D, Wang X, Yao H, Wang L, Pratheeshkumar P. (2013) Quercitrin protects skin from UVB-induced oxidative damage. *Toxicol Appl Pharmacol*, **269**, 89–99.

Zhang B, Zhao Z, Meng X, Chen H, Fu G, Xie K. (2018) Hydrogen ameliorates oxidative stress via PI3K-Akt signaling pathway in UVB-induced HaCaT cells. *Int J Mol Med*, **41**(6), 3653–3661.

Zhao B-L. (1999) Oxygen free radicals and natural antioxidants (First Edition). Science Press, Beijing.

Zhao B-L. (2007) Free radicals and natural antioxidants and health, China Science and Culture Press, Hong Kong.

Zhao B-L. (2008) Nitric oxide free radical, Science Press, Beijing.

Zhao B-L. (2016) Nitric oxide free radical biology and medicine, Science Press, Beijing.

Zhao B-L. (2020) The pros and cons of drinking tea. *Tradit Med Mod Med*, **3**(3), 1–12.

Zhu W, Xu J, Ge Y, Cao H, Ge X, Luo J, Xue J, Yang H, Zhang S, Cao J. (2014) Epigallocatechin-3-gallate (EGCG) protects skin cells from ionizing radiation via heme oxygenase-1 (HO-1) overexpression. *J Radiat Res*, **55**(6), 1056–1065.

https://doi.org/10.1142/9789811274213_0016

Chapter 16

Tea Polyphenols and Osteoporosis and Anemia

Baolu Zhao

Institute of Biophysics, Chinese Academy of Sciences, Beijing, China

16.1 Introduction

Osteoporosis is a systemic bone disease that is prone to fracture due to the decline of bone density and bone quality, the destruction of bone microstructure, and the increase of bone fragility due to various reasons. The most direct cause of osteoporosis is the reduction of calcium absorption and excessive calcium loss. Oxidative stress-induced DNA damage, apoptosis, and cell aging are all the reasons for this tissue dysfunction and bone homeostasis imbalance.

Anemia is the reduction in volume of human peripheral blood erythrocytes, lower than the normal range, resulting in poor blood supply to the heart, such as chest pain, angina pectoris, chest tightness after activity, hypoxia, and severe anemia can also cause anemia heart disease and other symptoms. While anemia is caused by a variety of reasons, the direct cause of anemia is attributed to having too little / due too much loss in iron absorption.

Studies have found that both osteoporosis and anemia are closely related to oxidative stress. Since tea and tea polyphenols can complex metal ions, it is generally believed that drinking tea will lead to the loss of calcium and iron, osteoporosis, and anemia. In fact, drinking tea prevents osteoporosis and anemia [Zhao, 2020]. The inducements of osteoporosis and anemia are very complex, among which oxidative stress injury plays an important role, while tea polyphenols have antioxidant effect and inhibit oxidative stress damage. This chapter will discuss these issues.

16.2 Osteoporosis

Osteoporosis is the second most common disease, only secondary to cardiovascular disease, with the risk of fracture increasing with age. Osteoporosis is caused by an imbalance between osteoblasto-genesis and osteoclast-genesis processes. Osteoclasto-genesis may be enhanced, osteoblasto-genesis may be reduced, or both may be significant decreased. Osteoporosis is a systemic bone disease that is prone to fracture due to the decline of bone mineral density and bone quality, the destruction of bone microstructure, and the increase of bone fragility. Osteoporosis can be divided into primary and secondary. Primary osteoporosis is divided into postmenopausal osteoporosis (type I), senile osteoporosis (type II), and idiopathic osteoporosis (including adolescent type). Postmenopausal osteoporosis generally occurs in women within 5–10 years after menopause. Senile osteoporosis generally refers to the osteoporosis of the elderly after the age of 70. Idiopathic osteoporosis mainly occurs in adolescents.

The etiology of osteoporosis is complex. In addition to primary osteoporosis, which is mainly related to menopause and aging, osteoporosis may also be caused by a variety of diseases. Diabetes (type 1, type 2), hyperparathyroidism, hypogonadism, hyperthyroidism, pituitary prolactinoma, hypopituitarism, systemic lupus erythematosus, rheumatoid arthritis, Sjogren's syndrome, dermatomyositis, chronic kidney disease, leukemia, lymphoma, multiple myeloma, etc. Malabsorption syndrome, after subtotal gastrectomy, chronic pancreatic disease, chronic liver disease, malnutrition, hemiplegia, paraplegia, motor dysfunction, muscular dystrophy, stiff person syndrome and myotonic syndrome, and long-term bed rest, or space travel can lead to osteoporosis. Regular use of glucocorticoids, immune-suppressants, heparin, anticonvulsants, and anticancer drugs can also lead to osteoporosis.

Osteoporosis can cause many symptoms, including pain. Patients can have low back pain or body pain. When the load increases, the pain increases, or the activity is limited. In serious cases, it is difficult to turn over, sit up, and walk. Severe osteoporosis may have shortened height and hunchback and spinal deformation. Vertebral compression fractures can cause thoracic deformities, abdominal compression, and affect cardiopulmonary function. Even fractures without trauma or slight trauma are fragile fractures. The common sites of brittle fractures are chest, lumbar spine, hip, radius, distal ulna, and proximal humerus. Osteoporosis can also lead to skin damage and aging. In osteoporosis, the content of superoxide dismutase (SOD) in skin decreases and the content of MDA increases significantly [Ye *et al.*, 2021]. Osteoporosis can reduce the quality of life of patients because spinal deformation and fracture can cause disability, so that patients have limited activities and are unable to take care of themselves, therefore, increasing the incidence of pulmonary infection and bedsore. It not only decreases the life quality and increases mortality of patients, but also brings a heavy economic burden to individuals, families, and society.

The International Osteoporosis Foundation (IOF) organized the global osteoporosis campaign and launched World Osteoporosis Day (WOD) in 1997. This day is celebrated on 20 October every year to raise people's awareness of the disease prevention. Osteoporosis is the biggest cause of fracture in elderly women. Bones become thin and fragile. Hormone deficiency is the main cause of osteoporosis, so women over 60 are the most vulnerable. To prevent osteoporosis, we need to supply a balanced diet rich in calcium, low salt, and an appropriate amount of protein and vitamin D and pay attention to appropriate outdoor activities. From adolescence, strength exercise, ensuring adequate calcium intake, preventing and actively treating various diseases, especially chronic consumptive diseases, malnutrition, malabsorption, etc., and preventing various gonadal dysfunction diseases and growth and development of diseases are important. It is crucial to avoid long-term use of drugs that affect bone metabolism, so as to obtain the ideal peak bone mass and reduce the risk of osteoporosis in the future. Prevention in adulthood mainly includes two aspects. First, delay the rate and degree of bone loss. For postmenopausal women, the recognized measure is to supply estrogen or estrogen and progesterone mixture as soon as possible. Second, prevent fractures in patients with osteoporosis, by avoiding the risk factors of fractures which can significantly reduce the incidence of fractures.

Studies have found that osteoporosis can lead to oxidative stress, or various factors leading to osteoporosis are related to oxidative stress.

16.3 Osteoporosis and oxidative stress

Oxidative stress has been implicated as a causative factor in many disease states, including diminished bone mineral density in osteoporosis. Osteoporosis is caused by many factors related to oxidative stress, such as osteoporosis caused by estrogen reduction, aging, microgravity, and castration. Inflammation and high reactive oxygen species (ROS) reactive nitrogen species (RNS) enhance osteoclastogenesis while reducing osteoblastogenesis by inducing osteoblast apoptosis and suppressing osteoblastic proliferation and differentiation. Oxidative stress-induced DNA damage, cellular apoptosis, and cellular senescence are all responsible for this tissue dysfunction and the imbalance in the bone homeostasis [Chandra & Rajawat, 2021]. A diet rich in whole plant foods with high antioxidant content along with antioxidant-preserving lifestyle changes may improve bone mineral density and reduce the risk of fragility-related fractures [Kimball *et al.*, 2021].

16.3.1 *Decreased estrogen leads to osteoporosis and oxidative stress*

Estrogen deficiency has been considered the seminal mechanism of osteoporosis in both women and men, but epidemiological evidence in humans and recent mechanistic studies in rodents indicate that aging and the associated increase in ROS and RNS are the proximal culprits. ROS and RNS greatly influence the generation and survival of osteoclasts, osteoblasts, and osteocytes. Estrogen participates in the metabolism of bone calcium. When estrogen level drops rapidly after menopause, calcium metabolism will become abnormal, which will lead to osteoporosis. The estrogen in middle-aged and elderly women, especially before and after menopause, will decline rapidly, leading to the loss of bone in bone parts, which easily leads to fracture. The epidemiological results showed that superoxide dismutase (SOD) in erythrocytes, catalase (CAT), total antioxidant status, hydroperoxides, advanced oxidation protein products (AOPP), MDA, and vitamin B12 (VB12) in plasma/serum were not statistically different between the postmenopausal osteoporosis and control group, whereas significantly increased level of homocysteine and nitric oxide (NO), along with decreased SOD, glutathione peroxidase (GPx), folate, and total anti-oxidant power in plasma/serum were obtained in the postmenopausal osteoporosis group [Zhou *et al.*, 2016]. Another epidemiological result showed that total oxidant status, SOD, hydroperoxides, paraoxonase, NO, and homocysteine were not statistically different between the postmeno-pausal osteoporosis and control groups, whereas significantly increased levels of oxidative stress index, MDA, advanced oxidation protein prod-ucts, and vitamin B12, along with decreased total antioxidant status, total antioxidant power, catalase (CAT), glutathione peroxidase, uric acid, and folate, were detected in the postmenopausal osteoporosis group. Subgroup analysis based on biological samples displayed significantly elevated NO in erythrocyte and hydroperoxides in serum, along with decreased SOD in serum [Zhao *et al.*, 2021]. Loss of estrogens or androgens decreased defense against oxidative stress in bone, and this accounts for increased bone resorption associated with acute loss of these hormones. ROS-activated FoxOs in early mesenchymal progenitors also diverted ss-catenin away from Wnt signaling, leading to decreased osteoblastogenesis. This latter mechanism may be implicated in the pathogenesis of type 1 and 2 diabetes and ROS-mediated adverse effects of diabetes on bone formation. Attenuation of Wnt signaling by the activation of peroxisome proliferator-activated receptor gamma by ligands generated from lipid oxidation, also contributes to the age-dependent decrease in bone forma-tion, suggesting a mechanistic explanation for the link between athero-sclerosis and osteoporosis. Additionally, increased glucocorticoid production and sensitivity with advancing age decreased skeletal hydra-tion and thereby increased skeletal fragility by attenuating the volume of the bone vasculature and interstitial fluid [Manolagas, 2010].

The above research results show that estrogen is involved in bone calcium metabolism. When estrogen level drops rapidly, calcium metabolism will be abnormal, ROS and RNS will increase, and anti-oxidant enzyme SOD will decrease, which will lead to osteoporosis.

16.3.2 *Osteoporosis caused by aging and oxidative stress*

The loss of bone calcium and the decline of bone mineral density, which occur with aging, are natural processes of human bones and cannot be completely reversed. Epidemiological evidence in humans and recent mechanistic studies in rodents indicate that aging and the associated increase in ROS and RNS are the proximal culprits. ROS and RNS greatly influence the generation and survival of osteoclasts, osteoblasts, and osteocytes. Moreover, oxidative defense by the protein expressed by gene (FoxO) transcription factors is indispensable for skeletal homeostasis at any age. Loss of estrogens or androgens decreases defense against oxidative stress in bone, and this accounts for the increased bone resorption associated with the acute loss of these hormones. ROS and RNS-activated FoxOs in early mesenchymal progenitors also divert ss-catenin away from Wnt signaling, leading to decreased osteoblastogenesis [Manolagas, 2010]. Oxidative stress-induced DNA damage, cellular apoptosis, and cellular senescence are all responsible for this tissue dysfunction and the imbalance in the bone homeostasis [Chandra & Rajawat, 2021]. A study indicated that oxidative stress is a major contributor to several alterations observed in age-related conditions (sarcopenia, osteoporosis) and, more significantly, in brain aging, suggesting a pivotal role in the pathogenesis and progression of one of the most dramatic age-related diseases, Alzheimer's disease (AD). 8-hydroxy-2'-deoxyguanosine (OH8dG), a marker of oxidation, increases progressively with aging and more so in AD brain, supporting the hypothesis that oxidative stress, a condition of unbalance between the production of reactive oxygen species and antioxidants, strongly contributes to the high incidence of AD in old age subjects [Mecocci *et al.*, 2018].

The above research results show that the loss of bone calcium, the decrease of bone density with aging, and the increase of ROS and RNS are the main culprits of osteoporosis.

16.3.3 *Osteoporosis caused by microgravity and oxidative stress*

Astronauts lose their bone density in space due to the loss of microgravity. Microgravity-induced bone loss can be modeled by *in vivo* and *in vitro* microgravity and hind-limb suspension (HLS). Hind-limb suspension induced reduction of bone mineral density and preserved bone structure in tibiae and mechanical strength in femurs. Microgravity-induced ROS and RNS formation and enhanced osteoblastic differentiation, downregulated vitamin D receptor (VDR) expression in femurs of rats exposed to HLS and mouse embryonic osteoblast precursor cell line MC3T3-E1 cells exposed to modeled microgravity. HLS induced bone loss in rats possibly via suppressing oxidative stress and upregulating vitamin D receptor expression [Xin *et al.*, 2015]. HLS and rotary wall vessel bioreactor were used to model microgravity *in vivo* and *in vitro*, respectively. Sprague-Dawley rats were exposed to HLS to induce bone loss in this study. HLS induced reduction of bone mineral density, ultimate load, stiffness, and energy in femur and lumbar vertebra, induced augmentation of MDA content and peroxynitrite content, and reduction of total sulfhydryl content in femur and lumbar vertebra. In cultured MC3T3-E1 cells, modeled microgravity-induced ROS formation, reduction of osteoblastic differentiation, increase of ratio of receptor activator of nuclear factor kappa B ligand to osteoprotegerin, inducible nitric oxide synthetase upregulation, and Erk1/2 phosphorylation. In cultured RAW264.7, incubation with HRM aggravated modeled microgravity-induced ROS formation, osteoclastic differentiation, and osteoclastogenesis [Sun *et al.*, 2013]. This study investigated the effects of blockade of IL-6 on bone loss induced by modeled microgravity. Adult male mice were exposed to HLS and treated with IL-6-neutralizing antibody (IL-6 nAb) for four weeks. HLS in mice led to upregulation of IL-6 expression in both sera and femurs. IL-6 nAb treatment in HLS mice significantly alleviated bone loss, evidenced by increased bone mineral density of whole tibia, trabecular thickness and number, bone volume fraction of proximal tibiae, and ultimate load and stiffness of femoral diaphysis. IL-6 nAb treatment in HLS mice significantly enhanced levels of osteocalcin in sera and reduced levels of deoxypyridinoline. In MC3T3-E1 cells exposed to microgravity *in vitro*,

IL-6 nAb treatment increased mRNA expression and activity of alkaline phosphatase, mRNA expression of osteopontin and runt-related transcription factor 2, and protein levels of osteoprotegerin and decreased protein levels of receptor activator of the NF-κB ligand. In RAW254.7 cells exposed to microgravity, IL-6 nAb treatment downregulated mRNA expression of cathepsin K and tartrate-resistant acid phosphatase (TRAP) and reduced numbers of TRAP-positive multinucleated osteoclasts. In conclusion, blockade of IL-6 alleviated the bone loss induced by microgravity [He *et al.*, 2020].

The above results indicate that microgravity induces ROS and RNS formation, enhances osteoblast differentiation, and downregulates the expression of vitamin D receptor (VDR), leading to osteoporosis.

16.3.4 *Orchidectomy causes oxidative stress and increases the incidence of osteoporosis*

Numerous studies support the detrimental effects of oxidative stress and cell senescence on skeletal homeostasis. In a study, the results showed that orchidectomy could induce osteoporosis by generating oxidative stress, DNA damage, osteocyte senescence, and senescence-associated secretory phenotype (SASP), subsequently stimulating osteoblastic bone formation and osteoclastic bone resorption [Chen *et al.*, 2019]. Another research evaluated antioxidant status and reduced osteoporosis in orchidectomized rats. Fifty-six 90-day-old male Sprague-Dawley rats were randomized into two groups: sham-control group (n = 14) and orchidectomized (ORX) group (n = 42). The orchidectomized group was equally divided among the following three treatments: orchidectomy, orchidectomy + 5.0% glycoprotein, and orchidectomy + 10% glycoprotein. At the termination of the study (day 60), all rats were euthanized and the plasma was collected for antioxidant status and indices of bone turnover. Bone quality and mineral contents in the bone, urine, and feces were evaluated. Orchidectomy lowered ($p < 0.05$) antioxidant status, bone quality, and bone mineral contents, and elevated ($p < 0.05$) indices of bone turnover, urinary deoxypyridinoline, and fecal calcium excretion [Deyhim *et al.*, 2008]. Coenzyme Q10 (CoQ10) acts as a scavenger for oxidative stress and protects mitochondrial activity from oxidative damage. However, it is unclear whether CoQ10 has a protective effect on osteoporosis caused by orchiectomy. To investigate suppression effect of antioxidant CoQ10 on osteoporosis in orchiectomized (ORX) mice, ORX mice were supplemented with/without CoQ10, and were compared with each other and with sham-operated mice. Our results showed that CoQ10 could prevent ORX-induced bone loss by inhibiting oxidative stress and cell senescence, subsequently promoting osteoblastic bone formation and inhibiting osteoclastic bone resorption [Wu *et al.*, 2020].

The above research results show that castration can cause oxidative stress and increase the incidence rate of osteoporosis. Antioxidant coenzyme Q10 can promote osteoblast bone formation and inhibit osteoclast bone absorption by inhibiting oxidative stress, thus preventing castration induced bone loss.

Above research results not only reveal the close relationship between oxidative stress and osteoporosis, but also show that antioxidants can inhibit oxidative stress and prevent osteoporosis, which also provides experimental and theoretical basis for clinical application.

16.4 Tea and tea polyphenols prevent osteoporosis

Patients with severe osteoporosis not only need calcium supplements, but also medical treatment, even surgery, according to their condition. It is said that there is a substance called tannic acid in tea, which will affect the absorption of calcium. Another suggests that the caffeine in tea can significantly inhibit the absorption of calcium in the digestive tract and increase the excretion of calcium in urine. Excessive caffeine can cause bone calcium loss. There are many causes of bone loss and osteoporosis, but all are related to oxidative stress. Drinking tea and tea polyphenols can inhibit oxidative stress and prevent bone loss and osteoporosis. A large number of studies have shown that drinking tea and tea polyphenols not only does not cause bone loss and osteoporosis, but also can prevent bone loss and osteoporosis. This includes epidemiological investigation, animal experiment, and its mechanism research.

Osteoporosis is a metabolic disease that affects bone mineral density and thus, compromise the strength of the bones. The most reasonable way to prevent disease through diet. Among the several nutrients investigated, the intake of tea polyphenols seems to influence bone mineral density by acting as free radical scavengers and preventing oxidation-induced damage to bone cells. In addition, the growing understanding of the bone remodeling process supports the theory that inflammation significantly contributes to the etiopathogenesis of osteoporosis. Osteoporosis is a major health problem in the aging population worldwide. Cross-sectional and retrospective evidence indicates that tea consumption may be a promising approach in mitigating bone loss and in reducing risk of osteoporotic fractures among older adults. Tea polyphenols enhance osteoblastogenesis and suppress osteoclastogenesis *in vitro*. Tea drinking has positive effects on bone health and may prevent and treat osteoporosis, especially in older and postmenopausal women. Some studies also suggest that tea intake may protect against bone loss. A diet rich in whole plant foods with high antioxidant content along with antioxidant-preserving lifestyle changes may improve bone mineral density and reduce the risk of fragility-related fractures. While it is not explicitly clear if antioxidant activity is the effector of this change, the current evidence supports this possibility.

16.4.1 *Epidemiological investigation on the prevention of osteoporosis by tea polyphenols*

Several studies have shown beneficial associations between tea consumption and bone mineral density (BMD) and fracture risk. The relationship between tea drinking and bone mineral density in elderly British women was studied. The bone mineral density of lumbar spine, femoral neck, greater trochanter, and ward triangle were measured in 1,256 free-living women aged 65–76 in Cambridge, UK. A self-administered questionnaire was used to evaluate tea drinking, and the women were divided into tea drinkers and non-tea drinkers. There were a total of 1,134 tea drinkers (90.3%) and 122 non-tea drinkers (9.7%). Research results showed that compared with people who do not drink tea, those who drink tea for a long time have higher bone density in lumbar spine, femoral neck, and ward triangle. Those who drank black tea, oolong tea, or green tea for more than 10 years had the highest bone density in the hip. Those who drank tea had significantly higher mean bone mineral density measurements of lumbar spine (0.033 g/cm; $p = 0.03$), greater trochanter (0.028 g/cm; $p = 0.004$), and ward's triangle (0.025 g/cm). Older women who drank tea had higher bone mineral density measurements than did those who did not drink tea. Nutrients found in tea, such as flavonoids, may influence bone mineral density. Tea drinking may protect against osteoporosis in older women [Halder & Bhaduri, 1998; Chisari *et al.*, 2019]. Another random survey was conducted on 623 postmenopausal women of Han nationality who lived in Fuzhou, and 593 cases met the requirements. Grouping: 112 cases in the habitual tea (oolong tea) group and 481 cases in the non-habitual tea group. The bone mineral density of lumbar spine and hip was measured by dual energy X-ray (DXA) bone densitometer. Among 593 postmenopausal women, the tea drinking group accounted for 18.89%, and the non-tea drinking group accounted for 81.11%. Comparing the age, occupation, education level, living habits, fertility, disease factors, and understanding of osteoporosis between the two groups, only the body mass of the two groups was statistically significant ($p < 0.05$). The body mass of the tea drinking group was significantly higher than that of the non-tea drinking group. Using body mass as a covariate, a covariance analysis was conducted to compare the bone mineral density between the two groups. The greater trochanter bone mineral density in the tea drinking group was higher than that in the non-tea drinking group ($p = 0.042$), and the difference was statistically significant. The ward's triangle bone mineral density in the tea drinking group was higher than that in the non-tea drinking group ($p = 0.022$). Comparing the incidence of postmenopausal fractures: tea drinking group and no tea drinking group, the difference was not statistically significant [Chen *et al.*, 2014].

Randomly selected women (n = 1500) aged 70–85 years, participated in a five-year prospective trial to evaluate whether oral calcium supplements prevent osteoporotic fractures in a paper. Impaired hip structure assessed by dual-energy X-ray absorptiometry (DXA) areal bone mineral density is an independent predictor for osteoporotic hip fracture. In the cross-sectional analysis, total hip a bone mineral density was 2.8% greater in tea drinkers than in non-tea drinkers ($p < 0.05$). Tea drinking is associated with preservation of hip structure in elderly women. This finding provides further evidence of the beneficial effects of tea consumption on the skeleton [Devine *et al.*, 2007].

This study analyzed the relationship between oolong tea drinking and bone mineral density in postmenopausal Han Chinese women, while living and diet habits, fertility, disease elements, and other baseline conditions were controlled. One group included 124 cases who routinely drank oolong tea, and the other included 556 who did not drink tea. Data were collected on participant age, lifestyle habits, fertility condition, disease elements, lumbar, and hip bone densities. It was found that the bone densities of the greater trochanteric bone in tea drinkers were higher than that in non-tea drinkers ($p = 0.013$). Similarly, the bone density of Ward's triangular bone in tea drinkers was higher than that in non-tea drinkers ($p = 0.013$). Therefore, the results indicate that oolong tea drinking could help prevent bone loss in postmenopausal Chinese women [Wang *et al.*, 2014, 2018].

Higher bone mineral density is often associated with greater consumption of black tea. A study found that drinking oolong tea is associated with increased calcaneus bone mineral density in postmenopausal women. From an epidemiological survey in Shantou, 476 postmenopausal women aged 40–88 years were enrolled in the study. All women were questioned about their demographic features, lifestyle, health status, types of tea consumed, habit of tea consumption, and habitual dietary intake by use of a structured questionnaire. As compared with non-tea drinkers, oolong tea drinkers had higher calcaneus bone mineral density. In addition, calcaneus bone mineral density was significantly increased for those drinking 1–5 cups/day but not >5 cups/day. No linear increase was observed in calcaneus bone mineral density with increasing years of tea consumption and local polynomial regression fitting showed a parabola-shaped association between years of tea consumption and calcaneus bone mineral density. However, symptoms of osteoporosis did not differ by types of tea consumed. Long-term moderate oolong tea consumption may have beneficial effects on bone health in postmenopausal women in Shantou of southern China [Duan *et al.*, 2020]. Another study examined the associations of tea consumption with hip bone strength in Chinese women. A total of 1,495 Chinese women aged more than 40 years were included. Tea consumption, socio-demographic information, and lifestyle habits were collected by a face-to-face questionnaire. Hip bone mineral density (BMD) and geometric parameters, i.e., cross-sectional area

(CSA), section modulus (Z), and buckling ratio (BR) were recorded. Tea drinkers (n = 732) had approximately 1.9% higher hip bone mineral density ($p < 0.05$) and 3.6% lower buckling ratio ($p < 0.05$) than non-tea drinkers (n = 763). The dose-response relationships of hip bone mineral density, buckling ratio, or cross-sectional area with total tea consumption were identified ($p < 0.05$). Tea drinking was found to be a significant and independent predictor of bone mineral density ($p < 0.05$) or buckling ratio ($p < 0.05$). Tea consumption was associated with increased bone strength in middle-aged and elderly Chinese women [Huang *et al.*, 2018].

This study found that green tea consumption is beneficially associated with the number of remaining teeth. A total of 24,147 people responded and 22,278 valid data were included in the analysis. The average age of participants was 74.2 years, 52.2% had ≥20 teeth and 34.2% drank 2–3 cups of green tea a day. Both higher green tea consumption and a larger social network size were associated with more remaining teeth ($p < 0.001$). The association of green tea was greater among those with smaller social networks ($p < 0.05$) [Manami *et al.*, 2020]. This paper analyzed the cross-sectional data from the Ohsaki Cohort 2006 Study. Usable self-administered questionnaires about green tea consumption and tooth loss were returned from 25,078 persons (12,019 men and 13,059 women) aged 40–64 years in Japan. Consumption of ≥1cups/day of green tea was significantly associated with decreased odds for tooth loss, and the association appeared to fit a threshold model. In men, the multivariate-adjusted ORs for tooth loss with a cut-off point of <20 teeth associated with different frequencies of green tea consumption were 1.00 (reference) for <1 cup/day, 0.82 for 1–2 cups/day, 0.82 for 3–4 cups/day, and 0.77 for ≥5 cups/day. The corresponding data for women and the results for cut-off points of 10 teeth and 25 teeth were essentially the same. The findings indicate an association between green tea consumption and decreased odds for tooth loss [Koyama *et al.*, 2010]

These meta-analysis results provide a potential trend that tea consumption might be beneficial for BMD, especially in the lumbar spine, hip, femoral neck, Ward triangle, and greater trochanter, which might help prevent bone loss. These osteoprotective effects appear to be mediated through antioxidant or anti-inflammatory pathways along with their downstream signaling mechanisms. On 7 April 2000, nature also released news that researchers at Cambridge University announced that drinking tea once a day can strengthen the bones and joints of elderly women. For future studies, preclinical animal studies to optimize the dose of tea polyphenols for maximum osteoprotective efficacy and a follow-up short-term dose-response trial in postmenopausal osteopenic women are necessary to inform the design of randomized controlled studies in at-risk populations. Advanced imaging technology should also contribute to determining the effective dose of tea polyphenols in achieving better bone mass, microarchitecture integrity, and bone strength, which are critical steps for translating the putative benefit of tea consumption in osteoporosis management into clinical practice and dietary guidelines [Shen *et al.*, 2013; Zhang *et al.*, 2017].

The above multiple epidemiological investigations show that drinking tea and tea polyphenols will not only not lead to bone loss, but also can prevent osteoporosis. In addition, the bioavailability, safety, bone turnover markers, muscle strength, and quality of life of tea polyphenols should also be evaluated through clinical trials.

16.4.2 *Cell and animal experiments on the prevention of osteoporosis by tea polyphenols*

Studies of animal and cell also reveal that the intake of tea polyphenols have pronounced positive effects on bone as shown by higher bone mass and trabecular bone volume, number, and thickness and lower trabecular separation via increasing bone formation and inhibition of bone resorption, resulting in greater bone strength. Manny studies have shown that green tea catechins play an important role in maintaining joint and skeletal muscle health.

Alveolar bone resorption is a characteristic feature of periodontal diseases and involves the removal of both the mineral and organic constituents of the bone matrix, which is caused by either multinucleated osteoclast cells or matrix metalloproteinases (MMPs). EGCG has been reported to have inhibitory effects on the activity and expression of MMPs. Treatment with the sonicated *Porphyromonas gingivalis* (*P. gingivalis*) extracts stimulated the expression of MMP-9 mRNA and this effect was significantly reduced by EGCG, whereas the transcription levels of

MMP-2 and MMP-13 were not affected by either the sonicated *P. gingivalis* extracts or EGCG. In addition, EGCG significantly inhibited osteoclast formation in the co-culture system at a concentration of 20 μM. Hence, EGCG may prevent the alveolar bone resorption that occurs in periodontal diseases [Yun *et al.*, 2004].

Recent reports discuss the altered bone homeostasis in cigarette smokers, being a risk factor for osteoporosis and negatively influencing fracture healing. Cigarette smoke is known to induce oxidative stress in the body via an increased production of ROS. These increases in ROS are thought to damage the bone-forming osteoblasts. Naturally occurring polyphenols contained in green tea extract, e.g., catechins, are known to have anti-oxidative properties. Co-, pre-, and post-incubation with green tea extract and catechins significantly reduced ROS formation and thus improved the viability of cigarette smoke medium-treated osteoblasts. Besides GTE's direct radical scavenging properties, pre-incubation with both green tea extract and catechins protected osteoblasts from cigarette smoke medium-induced damage. Inhibition of the anti-oxidative enzyme HO-1 significantly reduced the protective effect of green tea extract and catechins, indicating the key role of this enzyme in the anti-oxidative effect of green tea extract [Holzer *et al.*, 2012].

Recent studies show that green tea polyphenols (GTPs) attenuate bone loss and microstructure deterioration in ovariectomized aged female rats, a model of postmenopausal osteoporosis. This study evaluated the efficacy of GTPs at mitigating bone loss and microstructure deterioration along with related mechanisms in androgen-deficient aged rats, a model of male osteoporosis. Orchidectomy suppressed serum testosterone and tartrate-resistant acid phosphatase concentrations, liver glutathione peroxidase activity, bone mineral density, and bone strength, as well as decreased trabecular bone volume, number, and thickness in the distal femur and proximal tibia and bone-formation rate in trabecular bone of proximal tibia, but increased serum osteocalcin concentrations and bone-formation rates in the endocortical tibial shaft. GTP supplementation resulted in increased serum osteocalcin concentrations, bone mineral density, and trabecular volume, number, and strength of femur, increased trabecular volume and thickness and bone formation in both the proximal tibia and periosteal tibial shaft, decreased eroded surface in the proximal tibia and endocortical tibial shaft, and increased liver glutathione peroxidase activity [Shen *et al.*, 2011]. In addition to tea polyphenols, tea polysaccharide is a major bioactive constituent in tea, despite its profound effects on human health.

A paper also investigated the anti-osteoporotic effects of tea polysaccharide. *In vitro*, tea polysaccharide effects on osteoclastogenesis were examined using osteoclast precursor RAW264.7 cells. For *in vivo* studies, 12-week-old female Wistar rats were divided randomly into a sham-operated group (sham) and four ovariectomized (OVX) subgroups: OVX with vehicle (model) and ovariectomized with low-, medium-, and high-dose tea polysaccharide. Tea polysaccharide inhibited osteoclast differentiation significantly and dose-dependently, and its inhibitory effect was not due to toxicity to RAW264.7 cells. Tea polysaccharide suppressed expression of osteoclastogenesis-related marker genes and proteins significantly. In *in vivo* studies, medium-dose tea polysaccharide treatment ameliorated ovariectomized-induced calcium loss significantly. Low-dose tea polysaccharide treatment decreased the activity of acid phosphatase (ACP) in ovariectomized rats significantly. In addition, tea polysaccharide treatment improved other blood biochemical parameters and femoral biomechanical properties to a certain extent. More importantly, tea polysaccharide treatment ameliorated bone microarchitecture in ovariectomized rats strikingly because of increased cortical bone thickness and trabecular bone area in the femur [Xu *et al.*, 2018].

Bone mineralization refers to the biochemical process in which inorganic minerals are deposited in the organic substance of bone, so that calcium and phosphorus form hydroxyapatite and chelate with organic matter to form bone. This study compared the ability of different tea types to promote mineralization. Saos-2 cells underwent mineralization (five days) in the presence of tea (white tea: WT, green tea: GT, black tea: BT, green tea: GR, or red tea: RR; 1 µg/mL of polyphenols) or control. Total polyphenol content, was highest in GT and BT. The ability of each tea to inhibit DPPH (1,1-Diphenyl-2-picrylhydrazyl radical 2,2-Diphenyl-1-(2,4,6-trinitrophenyl)hydrazyl) also differed (WT, GT>RR) after normalizing for polyphenol quantity. Each tea increased mineralization and differences were observed among types (GT/BT/GR/RR>WT, GT=BT=GR, RR>BT/GT). The mRNA expression of alkaline phosphatase and ectonucleotide pyrophosphatase/phosphodiesterase remained unchanged, whereas osteopontin and sclerostin were reduced in cells treated with tea, regardless of type. At 24-hour and 48-hour postexposure to tea, cell activity was greater in cells receiving any of the teas compared with the vehicle control. Supplementation increased mineralization regardless of tea type with both rooibos teas and black teas, stimulating greater mineralization than WT, whereas green tea is similar to the others. While future study is needed to confirm the *in vivo* effects, the results suggest that consuming any of the teas studied may benefit bone health [McAlpine *et al.*, 2012]. Comparison of black, green and rooibos tea on osteoblast activity found that the lower level of polyphenols resulted in greater mineral content as well as cellular and alkaline phosphatase activity in Saos2 cells. Moreover, this was associated with higher markers of differentiation (osteopontin, sclerostin) and reduced cellular toxicity and pro-inflammatory markers (IL6, TNFα). Green, black, and rooibos teas improved osteoblast activity at the low level and support epidemiological evidence suggesting that tea consumption may benefit bone heath [Nash & Ward, 2016].

This study determined the dose-dependent response of black tea in Saos-2 cells and investigated the changes to several proteins involved in the mineralization process. Mineralization was induced in the presence of BT at concentrations that represent levels likely achieved through daily consumption (0.1, 0.5, 0.75, 1 µg gallic acid equivalents/mL) or through supplementation (2, 5, or 10 µg gallic acid equivalents/mL). Black tea exerted a positive dose-response on bone mineralization, peaking at 1 µg gallic acid equivalents/mL of black tea ($p < 0.05$). Cellular activity was significantly greater than control with exposure to 2–10 µg gallic acid equivalents/mL of black tea (at 24 hours) ($p < 0.05$) and 1–10 µg gallic acid equivalents /mL (at 48 hours) ($p < 0.05$), with a peak at 5 µg gallic acid equivalents/mL at 24 hours and 48 hours ($p < 0.05$). Protein expression of alkaline phosphatase and ectonucleotide pyrophosphatase/ phosphodiesterase-1 were unchanged, whereas a moderate dose of black tea (0.75 µg gallic acid equivalents/mL) resulted in greater expression of osteopontin compared with the highest dose (10 µg gallic acid equivalents/mL) ($p < 0.05$). Doses of BT from 0.5–10 µg gallic acid equivalents/ mL resulted in higher antioxidant capacity compared with control ($p < 0.05$). The results indicate that the higher antioxidant capacity, enhanced cell viability, and upregulated mineralization suggest that consumption of

BT may have a positive effect on BMD at levels obtained through consumption of tea [Cleverdon *et al.*, 2020].

The above research results show that the supplement of tea polyphenols and tea polysaccharide tea polyphenols not only does not lead to bone loss, but also can prevent osteoporosis. Tea polyphenols can reduce bone loss by improving antioxidant capacity, enhancing cell vitality, and increasing bone formation through up regulation of mineralization. At the same time, tea polyphenols can also increase bone absorption through antioxidant capacity.

16.4.3 *The mechanism on the prevention of osteoporosis by tea polyphenols*

From the above introduction, we found that drinking tea not only does not cause osteoporosis, but also can prevent bone cell damage caused by oxidative damage, improve bone mass, and protect bone mineral density, especially for the health of teeth. The incidence rate and etiology of osteoarthritis and muscle reduction are excessive inflammation and oxidative stress, mitochondrial dysfunction, and autophagy. Green tea catechins regulate the expression of miRNAs and reduce chondrocyte death, collagen degradation, and cartilage protection by downregulating inflammatory signal mediators and upregulating the effects of anabolic mediators on joint health By maintaining the dynamic balance of protein synthesis and degradation, green tea catechins promote the synthesis of mitochondrial energy metabolism and have an impact on skeletal muscle health, which is conducive to the stability of muscle internal environment and reduce muscle atrophy. Tea polyphenols in green tea have a certain effect on reducing the incidence rate of osteoporosis. For example, flavonoids can improve bone mass, while polyphenols and tannins in green tea help to improve bone mass. The intake of polyphenols in tea can eliminate inflammation as a free radical scavenger, prevent bone cell damage caused by oxidation, and protect bone mineral density [Chisari *et al.*, 2019].

MMPs play an important role in degeneration of the matrix associated with bone and cartilage. Regulation of osteoclast activity is essential in the treatment of bone disease, including osteoporosis and rheumatoid arthritis. Polyphenols in green tea, particularly EGCG, inhibit MMPs expression and activity. However, the effects of the black tea polyphenol, theaflavin-3,3′-digallate (TFDG), on osteoclast and MMP activity are unknown. TFDG and EGCG affect MMP activity and osteoclast formation and differentiation. TFDG or EGCG (10 μM and 100 μM) was added to cultures of rat osteoclast precursors cells and mature osteoclasts. Numbers of multinucleated osteoclasts and actin rings decreased in polyphenol-treated cultures relative to control cultures. MMP-2 and MMP-9 activities were lower in TFDG- and EGCG-treated rat osteoclast precursor cells than in control cultures. MMP-9 mRNA levels declined significantly in TFDG-treated osteoclasts in comparison to control osteoclasts. TFDG and EGCG inhibited the formation and differentiation of osteoclasts via inhibition of MMPs. TFDG may suppress actin ring formation more effectively than EGCG. Thus, TFDG and EGCG may be suitable agents or lead compounds for the treatment of bone resorption diseases [Oka *et al.*, 2012]. TFDG is extracted from black tea and has strong antioxidant capabilities. TFDG efficiently decreased receptor activator of NF-κB ligand-induced osteoclast formation and ROS generation in a dose-dependent manner, reduced ROS generation by activating nuclear factor erythroid 2-related factor 2 (Nrf2) and its downstream heme oxygenase-1 (HO-1), and also inhibited the mitogen-activated protein kinases (MAPK) pathway. C57BL/6J female mice showed that TF3 markedly attenuated bone loss and osteoclastogenesis in mice. The levels of MDA and superoxide dismutase (SOD) revealed that TFDG increased the expression of Nrf2 and decreased the intracellular ROS level *in vivo*. These findings indicated that TFDG may have the potential to treat osteoporosis and bone diseases related to excessive osteoclastogenesis via inhibiting the intracellular ROS level [Ai *et al.*, 2020].

The *in vitro* effects of tea catechins on osteogenesis have been confirmed in several animal models, as well as in epidemiological observational studies on human subjects. Catechins, the main polyphenols found in green tea with potent anti-oxidant and anti-inflammatory properties, can counteract the deleterious effects of the imbalance of osteoblastogenesis and osteoclastogenesis caused by osteoporosis. Green tea catechins can attenuate osteoclastogenesis by enhancing apoptosis of osteoclasts, hampering osteoclastogenesis, and prohibiting bone resorption *in vitro*. Catechin effects can be directly exerted on pre-osteoclasts/osteoclasts or indirectly exerted via the modulation of mesenchymal stem cells (MSCs)/ stromal cell regulation of pre-osteoclasts through activation of the nuclear factor kB (RANK)/RANK ligand (RANKL)/osteoprotegerin (OPG) system. Catechins also can enhance osteoblastogenesis by enhancing osteogenic differentiation of MSCs and increasing osteoblastic survival, proliferation, differentiation, and mineralization [Huang *et al.*, 2020].

16.5 Anemia

Anemia (anemia) is a common clinical symptom where the volume of human peripheral blood red blood cells decreases below the lower limit of the normal range. Due to the difficulty in determining red blood cell volume, hemoglobin (Hb) concentration is often used in clinical practice. Hematologists in China believe that in the sea level areas of China, adult male Hb<120g/L, adult female (non-pregnant) Hb<110g/L, and pregnant women Hb<100g/L have anemia. There are many types of anemia and reasons for anemia, including: 1) Erythropoietic anemia is an abnormality of hematopoietic cells, bone marrow hematopoietic microenvironment and hematopoietic raw materials affects erythropoiesis and can form erythropoiesis-reducing anemia. 2) Aplastic anemia is a kind of bone marrow hematopoietic failure, which is related to the damage of primary and secondary hematopoietic stem cells. 3) Congenital erythropoietic anemia (CDA) is a refractory anemia caused by benign clonal abnormalities of hereditary erythroid stem progenitor cells. 4) Malignant clonal disease of the hematopoietic system. 5) Anemia is a bone marrow necrosis, bone marrow fibrosis, bone marrow sclerosis, marbling disease, bone marrow metastasis of various extramedullary tumors and various infectious or non-infectious osteomyelitis caused by bone marrow stromal cells and stromal cell damage can affect hematopoiesis due to bone marrow stromal cell damage and abnormal hematopoietic microenvironment. 6) Anemia is caused by abnormal levels of hematopoietic regulatory factors — stem cell factor (SCF), interleukin (IL), granulocyte monocyte colony stimulating factor (GM-CSF), granulocyte colony stimulating factor (G-CSF), erythropoietin (EPO), thrombopoietin (TPO). Platelet growth factor (TGF), tumor necrosis factor (TNF), and interferon (IFN) play a positive and negative role in regulating hematopoiesis. 7) Lymphocyte hyper-function aplastic anemia (AA) is an autoimmune disease, autoimmune hemolytic anemia. 8) Hematopoietic cell apoptosis anemia is caused by myelodysplastic syndrome (MDS). 9) Anemia is caused by folic acid deficiency or utilization disorder, or megaloblastic anemia is caused by absolute or relative deficiency or utilization disorder of folic acid or vitamin B12 caused by various physiological or pathological factors. 10) Iron deficiency and iron utilization disorder anemia are the most common anemia in clinic. 11) Erythrocyte destructive anemia is characterized by abnormalities in erythrocytes, including their membrane, enzymes, globulin, and heme, and the abnormal surrounding environment of red blood cells, immune, vascular, and hemolytic anemia. 12) Hemorrhagic anemia can be divided into acute and chronic according to the blood loss rate.

Although there are many types of anemia and various reasons for anemia, many studies have shown that oxidative stress is closely related to anemia. In addition, drinking tea and tea polyphenols can inhibit oxidative stress and anemia.

16.6 Anemia and oxidative stress

Cell oxidative status, which represents the imbalance between oxidants and antioxidants, is involved in abnormal functions. Under pathological conditions, there is a shift toward the oxidants, leading to oxidative stress, which is cytotoxic, causing oxidation of cellular components that result in cell death and organ damage. Human erythrocytes are organelle-free cells packaged with iron-containing hemoglobin, specializing in the transport of oxygen. With a total number of approximately 25 trillion cells per individual, the erythrocyte is the most abundant cell type not only in blood but in the whole organism. Despite their low complexity and their inability to transcriptionally upregulate antioxidant defense mechanisms, they display a relatively long lifetime of 120 days. The average lifespan of circulating erythrocytes usually exceeds 100 days. Prior to that, however, erythrocytes may be exposed to oxidative stress in the circulation. Recent compelling evidence suggests that oxidative stress is an important perpetrator in accelerating erythrocyte loss in different systemic conditions and an underlying mechanism for anemia associated with these pathological states. The programmed death of erythrocytes relate with Ca^{2+} influx, the generation of ceramide, oxidative stress, kinase activation, and iron metabolism. It is well known that iron metabolism underlies the dynamic interplay between oxidative stress and antioxidants in many pathophysiological processes. Both iron deficiency and iron overload can affect redox state, and these conditions can be restored to physiological conditions using iron supplementation and iron chelation, respectively.

16.6.1 *Thalassemia and oxidative stress*

The thalassemia, formerly called marine anemia, is a group of hereditary hemolytic anemia diseases. This anemia or pathological condition is caused by the absence or insufficient synthesis of one or more globin chains in hemoglobin due to genetic defects. Thalassemia is a wide-spread problem in many parts of the world, and one of the highest incidence areas is in Southeast Asia. It is common in Guangdong, Guangxi, and Sichuan province of China, while sporadic cases are found in provinces and regions to the south of the Yangtze River, and rare cases are found in the north. Oxidative stress and generation of free radicals are fundamental in initiating pathophysiological mechanisms leading to an inflammatory cascade resulting in high rates of morbidity and death from many inherited point mutation-derived hemoglobin opathies. Hemoglobin (Hb) E is the most common point mutation worldwide. The βE-globin gene is found in greatest frequency in Southeast Asia. While HbE by itself presents as a mild anemia and a single gene for β-thalassemia is not serious, it remains unexplained why HbE/β-thalassemia (HbE/β-thal) is a grave disease with high morbidity and mortality. Patients often exhibit defective physical development, severe chronic anemia, and often die of cardiovascular disease and severe infections. Pathophysiological mechanisms are derived from and initiated by the dysfunctional property of HbE as a reduced nitrite reductase concomitant with excess α-chains exacerbating unstable HbE, leading to a combination of nitric oxide imbalance, oxidative stress, and pro-inflammatory events [Hirsch *et al.*, 2017].

Liver is affected by secondary iron overload in transfusions-dependent b-thalassemia patients. The redox iron can generate reactive oxidants that damage biomolecules, leading to liver fibrosis and cirrhosis. Iron chelators are used to treat thalassemia to achieve negative iron balance and relieve oxidant-induced organ dysfunctions. Thalassemia is a hereditary hemolytic anemia caused by mutations in globin genes that cause reduced or complete absence of specific globin chains (commonly, α or β). Although oxidative stress is not the primary etiology of thalassemia, it mediates several of its pathologies. The main causes of oxidative stress in thalassemia are the degradation of the unstable hemoglobin and iron overload, both stimulate the production of excess free radicals. The symptoms aggravated by oxidative stress include increased hemolysis, ineffective erythropoiesis, and functional failure of vital organs such as the heart and liver. The oxidative status of each patient is affected by multiple internal and external factors, including genetic makeup, health conditions, nutrition, physical activity, age, and the environment (e.g., air pollution, radiation). In addition, oxidative stress is influenced by the clinical manifestations of the disease (unpaired globin chains, iron overload, anemia, etc.) [Fibach & Dana, 2019]. Within cells, most of the acquired iron becomes protein-associated, as once released from endocytosed transferrin, it is used within mitochondria for the synthesis of protein prosthetic groups, or it is incorporated into enzyme active centers or alternatively sequestered within ferritin shells. The labile forms of iron infiltrate the mitochondria and damage cells by inducing noxious ROS formation, resulting in heart failure. The oral iron chelator deferiprone, because of its small size and neutral charge, demonstrably enters cells and chelates labile iron, thereby rapidly reducing ROS formation, allowing better mitochondrial activity and improved cardiac function. Deferiprone may also rapidly improve arrhythmias in patients who do not have excessive cardiac iron. It maintains the flux of iron in the direction hemosiderin to ferritin to free iron and it allows clearance of cardiac iron in the presence of other iron chelators or when used alone. To date, the most commonly used chelator combination therapy is deferoxamine plus deferiprone, whereas other combinations are in the process of assessment [Berdoukas *et al.*, 2015].

The prevention and treatment of atrial fibrillation (AF) in β-thalassemia patients make use of iron chelators and/or antioxidants. One of the most important causes of atrial fibrillation is cardiac iron overload and the harmful effects of increased oxidative stress. Iron-induced atrial fibrillation can be reversed by using an intensive iron chelation regimen. The combination of iron chelators with some antioxidants, including N-acetyl-l-cysteine (NAC), vitamin C, and acetaminophen, can lead to improved cardiac protection [Nomani *et al.*, 2019].

The above results show that thalassemia is closely related to oxidative stress injury, and antioxidants can prevent and improve anemia symptoms.

16.6.2 *Chronic kidney disease, anemia, and oxidative stress*

Uremic syndrome of chronic kidney disease (CKD) is a term used to describe clinical, metabolic, and hormonal abnormalities associated with progressive kidney failure. Chronic kidney disease is a debilitating pathology with various causal factors, culminating in end-stage renal disease (ESRD) requiring dialysis or kidney transplantation. Oxidative stress (OS), defined as disturbances in the pro-/antioxidant balance, is harmful to cells due to the excessive generation of highly ROS and RNS. When the balance is not disturbed, oxidative stress has a role in physiological adaptations and signal transduction. However, an excessive amount of ROS and RNS results in the oxidation of biological molecules such as lipids, proteins, and DNA. Oxidative stress has been reported in kidney disease, due to both antioxidant depletions as well as increased ROS and RNS production. The kidney is a highly metabolic organ, rich in oxidation reactions in mitochondria, which makes it vulnerable to damage caused by oxidative stress, and several studies have shown that oxidative stress can accelerate kidney disease progression. In patients at advanced stages of chronic kidney disease (CKD), increased oxidative stress is associated with complications such as hypertension, atherosclerosis, inflammation, and anemia [Daenen *et al.*, 2019]. In chronic kidney disease, anemia and oxidative stress are common features and both are involved in increasing morbidity and mortality. Patients with advanced chronic kidney disease are characterized by an imbalance between pro- and antioxidant factors, and increased oxidative stress has been associated with complications of end-stage renal disease such as atherosclerosis, Beta2-microglobulin amyloidosis, and anemia. Since blood membrane interaction plays a key role in generating oxidative stress, direct free radical scavenging at the membrane site is a logical approach. Antioxidants such as vitamin E work by inhibiting LDL oxidation by oxidants and by limiting cellular response to oxidized LDL, and are potentially useful adjuncts to the usual medical therapy provided to such patients. 8-hydroxy 2'-deoxyguanosine (8-OHdG) of leukocyte DNA has been identified as a surrogate marker of oxidative stress in chronic hemodialysis (HD) patients. In this study, we focused on the determinants of the 8-OHdG level in leukocyte DNA of HD patients and the intracellular production of ROS, H_2O_2, and O_2^-. Treatment with vitamin E dialyzer effectively reduced the 8-OHdG content in leukocyte DNA and suppressed intracellular ROS production of granulocytes [Tarng *et al.*, 2000]. There is data in the literature indicating increased oxidative stress in chronic kidney disease (CKD). Numerous reports have shown the association between uremia and oxidative stress, which increases patients' risk for cumulative injury to multiple organs. Anemia is a common and disabling feature of CKD and seems to be a main cause of oxidative stress; correction of anemia represents an effective approach to reduce oxidative stress and, consequently, cardiovascular risk. The vast majority of patients with CKD seem to be iron-deficient, as evaluated by the usual parameters and by iron staining on bone marrow biopsy, because of multiple forms of interference with all phases of iron metabolism. However, iron deficiency frequently complicates anemia in patients with CKD, and ferrous iron cation is a co-factor that is needed for hydroxyl radical production, which can promote cytotoxicity and tissue injury. This has raised a justifiable concern that prescription of intravenous iron may exacerbate oxidative stress and, hence, endothelial dysfunction, inflammation, and progression of cardiovascular disease, which are widely known consequences of CKD. Correction of anemia represents an effective approach to reduce oxidative stress and, consequently, cardiovascular risk [Garneata, 2008].

CKD affects both brain structure and function. Patients with CKD have a higher risk of both ischemic and hemorrhagic strokes. Nervous system complications occur in every patient with uremic syndrome of CKD. Their pathogenic mechanisms are complex and multiple. They include accumulation of uremic toxins resulting in neurotoxicity, blood-brain barrier injury, neuroinflammation, oxidative stress, apoptosis, brain neurotransmitters imbalance, ischemic/microvascular changes, brain metabolism dysfunction, and erythropoietin and iron deficiency anemia [Hamed, 2019]. Generating ROS and RNS is necessary for both physiology and pathology. An imbalance between endogenous oxidants and antioxidants causes oxidative stress, contributing to vascular dysfunction. The ROS and RNS-induced activation of transcription factors and proinflammatory genes increases inflammation. The effect of ROS and RNS disrupts the excretory function of each section of the nephron. It prevents the maintenance of intra-systemic homeostasis and leads to the accumulation of metabolic products. Renal regulatory mechanisms, such as tubular glomerular feedback, myogenic reflex in the supplying arteriole, and the renin-angiotensin-aldosterone system, are also affected. It makes it impossible for the kidney to compensate for water-electrolyte and acid-base disturbances, which progresses further in the mechanism of positive feedback, leading to a further intensification of oxidative stress [Podkowińska & Formanowicz, 2020]. Renal-specific pathophysiologic derangements, such as oxidative stress, chronic inflammation, endothelial dysfunction, vascular calcification, anemia, gut dysbiosis, and uremic toxins are important mediators. The progression of CKD is closely associated with systemic inflammation and oxidative stress, which are responsible for the manifestation of numerous complications such as malnutrition, atherosclerosis, coronary artery calcification, heart failure, anemia and mineral, and bone disorders, as well as enhanced cardiovascular mortality. Conventional therapy with anti-inflammatory and antioxidative agents has indicated beneficial effects in these disturbances [Rapa *et al.*, 2019].

The above research results show that chronic renal failure anemia leads to inflammation, oxidative stress damage, apoptosis, lipid, protein, and DNA oxidation, and antioxidants can prevent and improve the symptoms of chronic renal anemia.

16.6.3 *Anemia during pregnancy and oxidative stress*

Iron deficiency anemia (IDA) during pregnancy will increase the risk of fetal and neonatal hypoxia, amniotic fluid reduction, stillbirth, premature delivery, neonatal asphyxia, neonatal hypoxic-ischemic encephalopathy, and affect the growth and development of the fetus. Iron deficiency anemia during pregnancy, although associated with disturbances of hematological parameters, is now also considered as a source of oxidative stress. Iron deficiency anemia ranks 9th among 26 diseases with highest burden and Asia bears 71% of this global burden. Iron deficiency anemia can severely impair the outcome of pregnancy. IDA has been shown to cause oxidative stress, which may be exacerbated by oral iron therapy. IDA was shown to increase oxidative stress levels in all the studied organs and in placenta as well as hypoxia and inflammation in placenta. A paper investigated the effect of anemia in pregnancy on oxidative stress and cardiac parameters. A total of 100 patients (pregnant women with anemia n = 34, healthy pregnant women n = 33, non-pregnant control group n = 33) were enrolled. Serum thiol-disulfide and ischemia modified albumin levels, and echocardiographic parameters were compared. A significant positive correlation was determined between thiol levels ($p = 0.041$) in anemia group. Anemia in pregnant women may trigger oxidative stress and increased oxidative stress may be related to changes in cardiac functions [Bozkaya *et al.*, 2021]. IDA in pregnant rats impaired pregnancy outcome, increased the expression of hypoxia and inflammatory markers in the placenta, and increased oxidative stress in dams, fetuses, and placentas [Toblli *et al.*, 2012]. A study aimed to detect any alteration in superoxide dismutase (SOD) and glutathione peroxidase (GSH-Px) enzymes activity in pregnant women with IDA. Levels of GSH-Px and SOD were measured in 156 anemic, pregnant women and compared with similar levels in 20 non-anemic, pregnant women. Activity of SOD was found to be reduced in the anemic group when compared with the control group. It found a non-significant increase in GSH-Px activities in the anemic group. These findings could be explained in terms of oxidative stress under hypoxic condition which preserves the activity of GSH-Px with a decreased activity of SOD. A positive association was seen between IDA during pregnancy and oxidative stress [Khalid *et al.*, 2019]. Iron is important to prevent or treat anemia and is a cofactor of many enzymes in the antioxidant process. Effect of sodium iron ethylenediaminetetraacetate (NaFeEDTA) and ferrous sulfate on iron bioavailability and oxidative stress in anemic pregnant women was evaluated [Han *et al.*, 2011].

The above results show that iron deficiency anemia during pregnancy leads to decreased SOD activity, oxidative stress damage, cell apoptosis, lipid, protein, and DNA oxidation, and antioxidants can prevent and improve the symptoms of iron deficiency anemia during pregnancy.

16.6.4 *Hemolytic anemia and oxidative stress caused by drugs*

Anemia caused by drugs is generally hemolytic anemia, as well as bone marrow hematopoiesis reduction. Hemolytic anemia is an anemia that occurs because the rate of red blood cell destruction increases and exceeds the compensatory capacity of bone marrow hematopoiesis. Drug-induced hemolytic anemia is a very rare but potentially lethal adverse drug reaction, which can take the form of oxidative damage to vulnerable erythrocytes, drug-induced thrombotic microangiopathy, or immune-mediated hemolytic anemia. Drug-induced immune hemolytic anemia occurs in one case per million and can be fatal. Hemolysis presents as acute or chronic anemia, reticulocytosis, or jaundice. The diagnosis is established by reticulocytosis, increased unconjugated bilirubin and lactate dehydrogenase, decreased haptoglobin, and peripheral blood smear findings. Premature destruction of erythrocytes occurs intravascularly or extravascularly. Glucose-6-phosphate dehydrogenase (G6PD) deficiency leads to hemolysis in the presence of oxidative stress [Dhaliwal *et al.*, 2004]. The G6PD enzyme catalyzes the first step in the pentose phosphate pathway, leading to antioxidants that protect cells against oxidative damage. A G6PD-deficient patient, therefore, lacks the ability to protect red blood cells against oxidative stresses from certain drugs, metabolic conditions, infections, and ingestion of fava beans. G6PD deficiency has now migrated to become a worldwide disease. Numerous drugs, infections, and metabolic conditions have been shown to cause acute hemolysis of red blood cells in the G6PD-deficient patient, with the rare need for blood transfusion. Benzodiazepines, codeine/codeine derivatives, propofol, fentanyl, and ketamine were not found to cause hemolytic crises in the G6PD-deficient patient. The most effective management strategy is to prevent hemolysis by avoiding oxidative stressors. Thus, management for pain and anxiety should include medications that are safe and have not been shown to cause hemolytic crises, such as benzodiazepines, codeine/ codeine derviatives, propofol, fentanyl, and ketamine [Elyassi & Rowshan, 2009]. Cases were retrospectively identified using spontaneous notifications collected by the pharmacovigilance center and the results of immuno-hematological investigations were performed by the laboratory of French blood establishment of Lyon between 2000 and 2012. Ten cases were identified and the causal drugs were ambroxol, beta-interferon, cefotetan, ceftriaxone, loratadine, oxacillin, oxaliplatine, piperacillin-tazobactam, pristinamycin, and quinine. The median time to onset of anemia after starting the culprit drug was six days. The median nadir of hemoglobin was 57.9 g/L. The direct antiglobulin test was positive in eight patients. Drug-induced immune haemolytic anemia was considered as definite in five cases with positive drug-induced antibodies [Bollotte *et al.*, 2014]. Drug-induced hemolytic anemia can take the form of oxidative damage to vulnerable erythrocytes (as in G6PD deficiency), drug-induced thrombotic microangiopathy, or immune-mediated hemolytic anemia [Renard & Rosselet, 2017].

The above research results show that drug-induced immune hemolytic anemia can cause oxidative stress injury, leading to glucose-6-phosphate dehydrogenase deficiency symptoms.

16.6.5 *Sideroblastic anemias and oxidative stress*

Sideroblastic anemias appear to result from a disturbance at the interface between mitochondrial function and iron metabolism. A defining feature is excessive iron deposition within mitochondria of developing red cells, the consequences of which are an increase in cellular free radical production, increased damage to proteins, and reduced cell survival. Due to its mitochondrial location, superoxide dismutase (SOD2) is the principal defense against the toxicity of superoxide anions generated by the oxidative phosphorylation. This paper used hematopoietic stem cell transplantation to study blood cells lacking SOD2. The result suggests that oxidative stress and in particular, mitochondrial-derived oxidants, play an important role in the pathogenesis of the human disorder, sideroblastic anemia [Martin *et al.*, 2006]. X-linked sideroblastic anemia with ataxia (XLSA/A) is a rare inherited disorder characterized by mild anemia and ataxia. XLSA/A is caused by mutations in the ABCB7 gene, which encodes a member of the ATP-binding cassette transporter family. Studies in yeast, mammalian cells, and mice have shown that ABCB7 functions in the transport of iron-sulfur (Fe-S) clusters into the cytoplasm. To further investigate the mechanism of this disease, we have identified and characterized the *Caenorhabditis elegans* homologue of the ABCB7 gene, abtm-1. We have studied the function of abtm-1 using mutants and RNAi. abtm-1-depleted animals produce arrested embryos that have morphogenetic defects and unusual premature, putative apoptotic events. abtm-1(RNAi) animals also show accumulation of ferric iron and increased oxidative stress [González-Cabo *et al.*, 2011].

The above research results show that thalassemia, chronic kidney disease, iron-deficiency anemia in pregnancy, drug-induced hemolytic anemia, sideroblastic anemia, and other anemia can indeed produce reactive oxygen species and reactive nitrogen, which is related to oxidative stress injury. Some antioxidants can alleviate and prevent these types of anemia. Oxidative stress is a common denominator in the pathogenesis of many chronic diseases. Oxidative stress activates Ca^{2+}-permeable nonselective cation channels in the cell membrane, thus, stimulating Ca^{2+} entry and subsequent cell membrane scrambling resulting in phosphatidylserine exposure and activation of Ca^{2+}-sensitive K^+ channels leading to K^+ exit, hyperpolarization, Cl-exit, and ultimately cell shrinkage due to loss of KCl and osmotically driven water.

16.7 Tea and tea polyphenols prevent anemia

People also often hear that drinking tea may cause iron loss and anemia. The most checked literatures have reported the opposite results. From the following discussion, we can see that this is not the truth. There are many causes of iron loss and anemia, but they are all related to oxidative stress. Besides not causing iron loss and anemia, drinking tea and tea polyphenols also inhibit oxidative stress and prevent iron loss and anemia. Oxidative stress reduce erythrocyte survival and provide novel insights into the possible use of antioxidants as putative anti-eryptotic and anti-anemic agents in a variety of systemic diseases. Addition of antioxidants to these treatment regimens has been suggested as a viable therapeutic approach for attenuating tissue damage induced by oxidative stress. Notably, many bioactive plant-derived compounds have been shown to regulate both iron metabolism and redox state, possibly through interactive mechanisms. Iron deficiency is a major world health problem, that is, to a great extent, caused by poor iron absorption from the diet. Several dietary factors can influence this absorption. Absorption enhancing factors are ascorbic acid and meat, fish, and poultry; inhibiting factors are plant components in vegetables, tea, and coffee (e.g., polyphenols, phytates), and calcium. A large number of studies have shown that drinking tea and tea polyphenols can prevent anemia.

16.7.1   *Epidemiological studies about the effects of tea polyphenols on anemia*

Epidemiological and clinical studies have shown that drinking tea and tea polyphenols only do not inhibit the absorption of iron and lead to anemia, but also can prevent anemia symptoms caused by various reasons.

The average erythrocyte hemoglobin concentration is suitable for the diagnosis and treatment of various anemia diseases. The decrease of mean erythrocyte hemoglobin concentration is mainly seen in small cell hypochromic anemia. To determine the prevalence of anemia and the relationship between tea-drinking and anemia in reproductive married women in rural China, a cross-sectional study was carried out in four rural communities in Deqing county, Zhejiang Province, China. A total of 1,425 reproductive married women at the ages of 20–49 years participated in this study and had a satisfactory measure of hemoglobin. Among the 1,425 subjects, the average concentration of hemoglobin were 114.7 ± 17.0 g/L, the prevalence of anemia were 63.3%, and most were mild to moderate anemia (Such prevalence were 63.5%, 63.2%, and 63.4% respectively at the ages of 20–30 years, 31–40 years, and 41–49 years.). Subjects that drink tea were higher in average concentration of hemoglobin than those who do not drink tea ($p = 0.001$). There were significant associations between tea-drinking and average concentration of hemoglobin anemia [Fu *et al.*, 2009]. Therefore, suggesting that tea-drinking can prevent anemia.

A paper investigated a sample of African adults participating in the cross-sectional Transition and Health during Urbanization of South Africans (THUSA) study in the North West Province, South Africa. Data were analyzed from 1,605 healthy adults aged 15–65 years by demographic and Food Frequency Questionnaire (FFQ), anthropometric measurements, and biochemical analyses. The main outcome measures were hemoglobin (Hb) and serum ferritin concentrations. No associations were seen between black tea consumption and concentrations of serum ferritin (men $p = 0.059$; women $p = 0.49$) or Hb (men $p = 0.33$; women $p = 0.49$) [Hogenkamp *et al.*, 2008]. These results demonstrate that iron deficiency and iron-deficiency anemia is not significantly explained by black tea consumption in an adult population in South Africa. The randomized clinical study showed that green tea improved the effect of deferoxamine on β efficacy of iron overload in patients with thalassemia; green tea has strong antioxidant and metal chelating effects. The purpose of this study was to investigate the effect of green tea on deferoxamine chelation therapy β effect of iron overload in patients with thalassemia. Blood iron, ferritin, total iron binding capacity, lipid peroxidation, and hematological indexes were measured. Thirty days after treatment with green tea, the levels of iron ($p < 0.001$), ferritin ($p < 0.01$), lipid peroxidation ($p < 0.001$), and leukocyte ($p < 0.05$) in the blood of patients with thalassemia decreased significantly, and the level of iron binding ability increased significantly ($p < 0.001$), while other hematological indexes did not change significantly [Badiee *et al.*, 2015]. It suggested that green tea can be used as a supplement to deferoxamine treatment in patients with thalassemia.

The majority of Indian women have poor dietary folate and vitamin B12 intake resulting in their chronically low vitamin status, which contributes to anemia and the high incidence of folate-responsive neural-tube defects (NTDs) in India. Since tea, the second most common beverage worldwide (after water), is consumed by most Indians every day and appeared an ideal vehicle for fortification with folate and vitamin B12, a paper determined if daily consumption of vitamin-fortified tea for two months could benefit young women of childbearing-age in Sangli, India. Most women had baseline anemia with low-normal serum folate and below-normal serum vitamin B12 levels. After a cup of tea every day for two months, women exhibited significant increases in mean differences in pre-intervention versus post-intervention serum folate levels of 8.37 ng/mL ($p < 0.05$) and 6.69 ng/mL ($p < 0.05$), respectively. Tea is an outstanding scalable vehicle for fortification with folate and vitamin B12 in India, and has the potential to help eliminate hematological and neurological complications arising from inadequate dietary consumption or absorption of folate and vitamin B12 [Vora *et al.*, 2021].

Sickle cell disease (SCD) is an inherited disorder caused by mutations in the structure of the β-globin sequence (HBB), which encodes hemoglobin subunit β. The incidence is estimated to be between 300,000–400,000 neonates globally each year, the majority in sub-Saharan Africa. A variable fraction of dense, dehydrated erythrocytes with high Hb concentration is seen in the blood of patients with sickle cell disease; these dense cells play an important role in the pathophysiology of the vaso-occlusive events of sickle cell disease, due to their higher tendency to polymerize and sickle. Sickle cell dehydration is due to loss of K^+, Cl^-, and water; the two major determinant pathways of dehydration of sickle erythrocytes are the Ca^{2+}-activated K^+ channel (IK1 or Gardos channel) and the K-Cl cotransport (KCC). Hemoglobin molecules that include mutant sickle β-globin subunits can polymerize; erythrocytes that contain mostly hemoglobin polymers assume a sickled form and are prone to hemolysis and anemia. With storage of sickle cells at 4°C for six days, the cells started to undergo spontaneous dehydration when incubated at 37°C. This paper studied the effects of green tea extract on anion transport and sickle cell dehydration *in vitro*. Green tea extract or tea polyphenols) effectively inhibited *in vitro* dehydration of sickle red blood cells induced by K-Cl cotransport or red cell storage. For K-Cl cotransport induced by 500 mM urea, 0.3 mg/ml EGCG almost completely inhibited dehydration, and 6 mg/ml AGE inhibited dehydration to 30% of the control level [Ohnishi *et al.*, 2001].

Fanconi anemia (FA) is a genetic disorder featuring chromosomal instability, developmental defects, progressive bone marrow failure, and predisposition to cancer. The incidence rate in Asian population is 1/160000, and the ratio of male to female is about 1.2:1. In addition to typical aplastic anemia such as bleeding tendency, these patients are also accompanied by multiple developmental abnormalities (brown pigmentation of skin, bone deformity, sexual hypoplasia, etc.). Besides the predominant role in DNA damage response and/or repair, many studies have linked FA proteins to oxidative stress. Oxidative stress, defined as imbalance in pro-oxidant and antioxidant homeostasis, has been considered to contribute to disease development, including Fanconi anemia [Li & Pang, 2014]. A paper investigated the effect of a nutrient mixture containing ascorbic acid, lysine, proline, and green tea extract on Fanconi anemia human fibroblast cell lines FA-A:PD20 and FA-A:PD220 on matrix met-alloproteinase expression, invasion, cell proliferation, morphology, and apoptosis. The data demonstrated that the nutrient mixture inhibited matrix metalloproteinase expression, invasion, and induced apoptosis [Roomi *et al.*, 2013].

Available data suggest that polyphenols from tea can inhibit iron absorption from ferric sodium EDTA (NaFeEDTA) [Lazrak *et al.*, 2021]. Specific attention was paid to the effects of tea on iron absorption. Tea has been shown to be a potent inhibitor of non-heme iron absorption. This study investigated the effect of a 1-hour time interval of tea consumption on non-heme iron absorption in an iron-containing meal in a cohort of iron-replete, non-anemic female subjects with the use of a stable isotope (^{57}Fe). This study showed that tea consumed simultaneously with an iron-containing porridge meal led to decreased non-heme iron absorption and that a 1-hour time interval between a meal and tea consumption attenuates the inhibitory effect, resulting in increased non-heme iron absorption. Mean fractional iron absorption was found to be significantly higher (2.2%) when tea was administered 1 hour post-meal (TM-3) than when tea was administered simultaneously with the meal (TM-2) ($p = 0.046$) [Ahmad *et al.*, 2017]. Recommendations with respect to tea consumption (when in a critical group) include consume tea between meals instead of during the meal and to simultaneously consume ascorbic acid and/or meat, fish, and poultry.

To evaluate the effect of tea drinking on the occurrence of microcytic anemia in infants, we studied 122 healthy infants who underwent routine blood counts at the age of 6–12 months. An overall high frequency of anemia (Hb less than 11 gm/dl; 48.4%), microcytosis (MCV less than 70 Mm^3; 21.3%), and microcytic anemia (19%) was found in the whole group. The percentage of tea drinking infants with microcytic anemia (32.6%) was significantly higher than that of the non-tea drinkers (3.5%). Based on our findings, we do not recommend giving tea to infants whose main source of iron is from milk, grains, vegetables, or medicinal sources [Merhav *et al.*, 1985].

Above epidemiological and randomized clinical studies in different countries show that tea and tea polyphenols can inhibit the absorption of iron and cause anemia and prevent anemia caused by various reasons, such as sickle cell disease and Van Kony's anemia, and improve the symptoms of anemia.

16.7.2 *Animal experiments about the effects of tea polyphenols on anemia*

Iron deficiency leading to anemia is a major world health problem, that is, to a great extent, caused by poor iron absorption from the diet. A paper studied the effects of green tea on iron accumulation and oxidative stress in livers of iron-challenged thalassemic mice. Green tea catechins exhibit anti-oxidation, the inhibition of carcinogenesis, the detoxification of CYP2E1-catalyzed HepG2 cells and iron chelation, and effectiveness of GT in iron-challenged thalassemic mice. Heterozygous BKO type-thalassemia (BKO) mice (C57BL/6) experienced induced iron overload by being fed a ferrocene-supplemented diet (Fe diet) for eight weeks, and by orally being given green tea extract (300 mg/kg) and deferiprone (DFP) (50 mg/kg) for a further eight weeks. Green tea inhibits or delays the deposition of hepatic iron in regularly iron-loaded thalassemic mice effectively. Green tea catechins prevented the iron-induced generation of free radicals via Haber-Weiss and Fenton reactions, and consequently liver damage and fibrosis [Saewong *et al.*, 2010]. Beta-thalassemia patients suffer from secondary iron overload caused by increased iron absorption and multiple blood transfusions. Excessive iron catalyzes free-radical formation, causing oxidative tissue damage. Non-transferrin bound iron (NTBI) detected in thalassemic plasma is highly toxic and chelatable. EGCG and ECG bound chemical Fe^{3+} and chelated the NTBI in a time- and dose-dependent manner. They also decreased oxidative stress in iron-treated erythrocytes. EGCG and ECG could be natural iron chelators that efficiently decrease the levels of NTBI and free radicals in iron overload [Thephinlap *et al.*, 2007]. Combined chelation with green tea would be investigated in beta-thalassemia patients with iron overload in the future.

This study monitored the growth, trace element status, including hematological parameters of weanling rats given either water, 1% black tea, 1% green tea, or 0.2% crude green tea extract as their sole drinking fluid while consuming diets containing either adequate or low amounts of iron. With the exception of manganese, none of the trace elements studied (iron, copper, zinc, and manganese) or the hematological indices measured were affected by the type of beverage supplied, even though the polyphenol extract was shown to chelate metals *in vitro* and all the animals fed the low iron diet were shown to be anemic. There appeared to be an effect of black and green teas on manganese balance in both the first and last weeks of the study. A lower level of brain manganese was associated with green tea consumption, and a higher level of this element in the kidneys of animals fed black tea [Record *et al.*, 1996]. Rats fed diets containing 2.31% green or black tea or given fluid tea had elevated hematocrits but experienced minimal changes in tissue iron levels or in iron absorption [Greger & Lyle, 1988]. The results demonstrate that both black and green teas and a green tea polyphenol extract do not represent a risk to animals consuming the beverages as their sole fluid intake with respect to iron availability.

A paper studied the effects of dietary tea polyphenols on heme iron absorption in a dose-dependent manner in human intestinal Caco-2 cells and found that EGCG markedly inhibited intestinal heme iron absorption by reducing the basolateral iron export in Caco-2 cells. It found that polyphenolic compounds significantly inhibited heme-[55]Fe absorption in a dose-dependent manner. The addition of ascorbic acid did not modulate the inhibitory effect of dietary polyphenols on heme iron absorption when the cells were treated with polyphenols at a concentration of 46 mg/L. A [55]Fe absorption study was conducted by adding various concentrations of EGCG and green tea using Caco-2 intestinal cells. Polyphenols were found to inhibit the transepithelial [55]Fe transport in a dose-dependent manner. The addition of ascorbic acid offset the inhibitory effects of polyphenols on iron transport. Ascorbic acid modulated the transepithelial iron transport without changing the apical iron uptake and the expression of ferroportin-1 protein in the presence of EGCG. The polyphenol-mediated apical iron uptake was inhibited by membrane impermeable Fe^{2+} chelators ($p < 0.001$), but at a low temperature (4°C), the apical iron uptake was still higher than the control values at 37°C ($p < 0.001$) [Ma *et al.*, 2011; Kim *et al.*, 2011].

These animal experiment results suggest that green tea and black tea, and tea polyphenols enhance the apical iron uptake partially by reducing the conversion of ferric to ferrous ions and possibly by increasing the uptake of polyphenol-iron complexes via the energy-independent pathway. The present results indicate that the inhibitory effects of dietary polyphenols on iron absorption can be offset by ascorbic acid.

16.7.3 *Mechanism for the effects of tea polyphenols on anemia*

Many studies explored the scavenging property of black tea and catechins, the major flavonols of tea leaves, against damage by oxidative stress. Studies have shown that oxidative stress is closely related to various anemia, and the prevention of anemia caused by various causes by tea and tea polyphenols is also related to the inhibition of oxidative stress. Oxidant and free radical-generating system were used to promote oxidative damage in erythrocytes. Investigations revealed that using phenyl-hydrazine as well as other oxygen-generating systems (hydrogen peroxide, iron with hydrogen peroxide) increased lipid peroxidation of erythrocyte. It has further been observed that not only lipid peroxidation, but also phenyl-hydrazine, causes significant elevation in methemoglobin formation, catalase activity, and turbidity, in the above system, which are the typical characteristics of hemolytic anemia. However, exogenous administration of green tea leaf extract and ascorbic acid as natural antioxidants and free radical scavengers were shown to separately prevent increased lipid peroxidation caused by phenyl-hydrazine, though the degree of protection is more in case of green tea leaf extract than ascorbic acid. Oxidative damage *in vivo* due to hemolytic disease may be checked to some extent by using natural antioxidants. A study found that three months of green tea extract supplementation resulted in decreases in body mass index, waist circumference, and levels of total cholesterol, low-density cholesterol, and triglyceride. Increases in total antioxidant level and in zinc concentration in serum were also observed. Glucose and iron levels were lower in the green tea extract group than in the control, although HDL-cholesterol and magnesium were higher in the green tea extract group than in the placebo group. At baseline, a positive correlation was found between calcium and body mass index, as was a negative correlation between copper and triglycerides. After three months, a positive correlation between iron and body mass index and between magnesium and HDL-cholesterol, as well as a negative correlation between magnesium and glucose, were observed [Biswas *et al.*, 2005; Suliburska *et al.*, 2012]. Human red blood cell (rbc) was taken as the model and the oxidative damage was induced by a variety of inducers, e.g., phenyl-hydrazine (PHX), Cu^{2+}-ascorbic acid, and xanthine/xanthine oxidase systems. Lipid peroxidation of pure erythrocyte membrane and of whole red blood cell could be completely prevented by black tea extract. Similarly, black tea provided total protection against degradation of membrane proteins. The fluidity studies showed considerable disorganization of its architecture that could be restored back to normal on addition of black tea or free catechins. Black tea extract in comparison to free catechins seemed to be a better protecting agent against various types of oxidative stress [Halder and Bhaduri, 1998]. Microangiopathic hemolytic anemia occurs when the red cell membrane is damaged in circulation, leading to intravascular hemolysis and the appearance of schistocytes. It is greatly influenced by the shape and deformability of RBCs, which can be affected by oxidative stress induced by different drugs and diseases leading to anemia. Many studies found the effect of aqueous tea extracts on lipid peroxidation and alpha and gamma tocopherols concentration in the oxidative damage of human rbc. All tea extracts at level of 4 g/150 mL of water significantly decreased concentration of MDA. The extract of green tea in comparison to black and white tea extracts at the same levels seems to be a better protective agent against oxidative stress. The antioxidant synergism between components extracted from leaves of green tea and endogenous alpha tocopherol in the oxidative damage of red blood cells was observed. All tea extracts did not protect against decrease of gamma tocopherol in human erythrocytes treated with cumene hydroperoxide [Gawlik *et al.*, 2007].

Therefore, the prevention and improvement of anemia caused by various reasons by tea and tea polyphenols are related to the inhibition of oxidative stress injury.

16.8 Conclusion

Through the above discussion, there are many reasons for osteoporosis and anemia, but most of these effects are related to oxidative stress. At the same time, we can see that drinking green tea or black tea and tea polyphenols not only has no effect on osteoporosis and anemia but can prevent osteoporosis and anemia. Although in some specific conditions, drinking tea and supplementing tea polyphenols cannot achieve ideal results, and even have negative effects. For example, the concentration of methemoglobin was increased significantly when challenged with tea extracts and EGC. Plasma hemoglobin levels were higher in G6PD-deficient samples after exposure to tea extracts, EGCG, EGC, and gallic acid, compared with those in normal blood. Tea extracts and polyphenols significantly altered the oxidative status of G6PD-deficient erythrocytes *in vitro* as demonstrated by the decrease of GSH and increase of GSSG, methemoglobin, and plasma hemoglobin. Our data cautions against the excessive consumption of concentrated tea polyphenolic products by G6PD-deficient subjects [Ko *et al.*, 2006].

Although the mechanisms of tea and tea polyphenols to prevent and improve osteoporosis and anemia caused by various reasons are different, there is a common mechanism, which is related to the inhibition of oxidative stress damage through relevant signals (Fig. 16-1).

Therefore, we do not need to worry about the effects of drinking tea on osteoporosis and anemia. Of course, when we drink tea or increase the antioxidant capacity by eating tea polyphenols, we need to pay attention

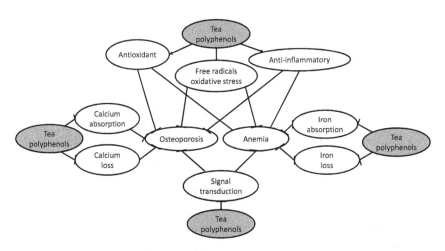

Figure 16-1. Schematic diagram of the mechanism of tea polyphenols inhibiting osteoporosis and anemia.

to the ways, methods, and doses before we can get good results but regardless, will not lead to the opposite results.

References

Ahmad Fuzi SF, Koller D, Bruggraber S, Pereira DI, Dainty JR, Mushtaq S. (2017) A 1-h time interval between a meal containing iron and consumption of tea attenuates the inhibitory effects on iron absorption: a controlled trial in a cohort of healthy UK women using a stable iron isotope. *Am J Clin Nutr*, **106**(6), 1413–1421.

Ai Z, Wu Y, Yu M, Li J, Li S. (2020) Theaflavin-3,3′-digallate suppresses RANKL-induced osteoclastogenesis and attenuates ovariectomy-induced bone loss in mice. *Front Pharmacol*, 11, 803.

Badiee MS, Nili-Ahmadabadi H, Zeinvand-Lorestani H, Nili-Ahmadabadi A. (2015) Green tea consumption improves the therapeutic efficacy of deferoxamine on iron overload in patients with β-thalassemia major: a randomized clinical study. *Biol Forum*, **7**(2), 383–387.

Berdoukas V, Coates TD, Cabantchik ZI. (2015) Iron and oxidative stress in cardiomyopathy in thalassemia. *Free Radic Biol Med*, **88**(Pt A), 3–9.

Biswas S, Bhattacharyya J, Dutta AG. (2005) Oxidant induced injury of erythrocyte-role of green tea leaf and ascorbic acid. *Mol Cell Biochem*, **276**(1–2), 205–210.

Bollotte A, Vial T, Bricca P, Bernard C, Broussolle C, Sève P. (2014) Drug-induced immune hemolytic anemia: a retrospective study of 10 cases. *Rev Med Interne*, **35**(12), 779–789.

Bozkaya VÖ, Oskovi-Kaplan ZA, Erel O, Keskin LH. (2021) Anemia in pregnancy: it's effect on oxidative stress and cardiac parameters. *J Matern Fetal Neonatal Med*, **34**(1), 105–111.

Chandra A, Rajawat J. (2021) Skeletal aging and osteoporosis: mechanisms and therapeutics. *Int J Mol Sci*, **22**(7), 3553.

Chen Ke, Xie Lihua, Lai Yulian, *et al.*, (2014) Study of the relationship between tea drinking and bone mineral density in postmenopausal Han women. *Cell Biochem Biophys.* **70**(2), 1289–1293.

Chisari E, Shivappa N, Vyas S. (2019) Polyphenol-rich foods and osteoporosis. *Curr Pharm Des*, **25**(22), 2459–2466.

Chen L, Wang G, Wang Q, Liu Q, Sun Q, Chen L. (2019) N-acetylcysteine prevents orchiectomy-induced osteoporosis by inhibiting oxidative stress and osteocyte senescence. *Am J Transl Res*, **11**(7), 4337–4347.

Cleverdon RE, McAlpine MD, Ward WE. (2020) Black tea exhibits a dose-dependent response in saos-2 cell mineralization. *J Med Food*, **23**(9), 1014–1018.

Daenen K, Andries A, Mekahli D, Van Schepdael A, Jouret F, Bammens B. (2019) Oxidative stress in chronic kidney disease. *Pediatr Nephrol*, **34**(6), 975–991.

Deyhim F, Mandadi K, Patil BS, Faraji B. (2008) Grapefruit pulp increases anti-oxidant status and improves bone quality in orchidectomized rats. *Nutrition*, **24**(10), 1039–1044.

Dhaliwal G, Cornett PA, Tierney LM Jr. (2004) Hemolytic anemia. *Am Fam Physician*, **69**(11), 2599–2606.

Duan P, Zhang J, Chen J, Liu Z, Guo P, Li X, Li L, Zhang Q. (2020) Oolong tea drinking boosts calcaneus bone mineral density in postmenopausal women: a population-based study in southern China. *Arch Osteoporos*, **15**(1), 49.

Devine A, Hodgson JM, Dick IM, Prince RL. (2007) Tea drinking is associated with benefits on bone density in older women. *Am J Clin Nutr*, **86**(4), 1243–1247.

Elyassi AR, Rowshan HH. (2009) Perioperative management of the glucose-6-phosphate dehydrogenase deficient patient: a review of literature. *Anesth Prog*, **56**(3), 86–91.

Fibach E, Dana M. (2019) Oxidative Stress in beta-Thalassemia. Mol Diagn Ther. **23**(2):245–261.

Fu C, Wei G, Wang F, Wang X, Zhu J, Song J, Chen Y, Jiang Q. (2009) Relationship between tea-drinking and anaemia of rural reproductive married women in a community-based cross-sectional study in China. *Wei Sheng Yan Jiu*, **38**(6), 709–711.

Garneata L. (2008) Intravenous iron, inflammation, and oxidative stress: is iron a friend or an enemy of uremic patients? *J Ren Nutr*,**18**(1), 40–45.

Gawlik M, Czajka A. (2007) The effect of green, black and white tea on the level of alpha and gamma tocopherols in free radical-induced oxidative damage of human red blood cells. *Acta Pol Pharm*, **64**(2), 159–164.

González-Cabo P, Bolinches-Amorós A, Cabello J, Ros S, Moreno S, Baylis HA, Palau F, Vázquez-Manrique RP. (2011) Disruption of the ATP-binding cassette B7 (ABTM-1/ABCB7) induces oxidative stress and premature cell death in Caenorhabditis elegans. *J Biol Chem*, **286**(24), 21304–21314.

Greger JL, Lyle BJ. (1988) Iron, copper and zinc metabolism of rats fed various levels and types of tea. *J Nutr*, **118**(1), 52–60.

Halder J, Bhaduri AN. (1998) Protective role of black tea against oxidative damage of human red blood cells. *Biochem Biophys Res Commun*, **244**(3), 903–907.

Hamed SA. (2019) Neurologic conditions and disorders of uremic syndrome of chronic kidney disease: presentations, causes, and treatment strategies. *Expert Rev Clin Pharmacol*, **12**(1), 61–90.

Han XX, Sun YY, Ma AG, Yang F, Zhang FZ, Jiang DC, Li Y. (2011) Moderate NaFeEDTA and ferrous sulfate supplementation can improve both hematologic status and oxidative stress in anemic pregnant women. *Asia Pac J Clin Nutr*, **20**(4), 514–520.

He B, Yin X, Hao D, Zhang X, Zhang Z, Zhang K, Yang X. (2020) Blockade of IL-6 alleviates bone loss induced by modeled microgravity in mice. *Can J Physiol Pharmacol*, **98**(10), 678–683.

Hirsch RE, Sibmooh N, Fucharoen S, Friedman JM. (2017) HbE/β-thalassemia and oxidative stress: the key to pathophysiological mechanisms and novel therapeutics. *Antioxid Redox Signal*, **26**(14), 794–813.

Hogenkamp PS, Jerling JC, Hoekstra T, Melse-Boonstra A, MacIntyre UE. (2008) Association between consumption of black tea and iron status in adult Africans in the North West Province: the THUSA study. *Br J Nutr*, **100**(2), 430–437.

Holzer N, Braun KF, Ehnert S, *et al.* (2012) Green tea protects human osteoblasts from cigarette smoke-induced injury: possible clinical implication. *Langenbecks Arch Surg*, **397**(3), 467–474.

Huang H, Han GY, Jing LP, Chen ZY, Chen YM, Xiao SM. (2018) Tea consumption is associated with increased bone strength in middle-aged and elderly Chinese women. *J Nutr Health Aging*, **22**(2), 216–221.

Huang HT, Cheng TL, Lin SY, Ho CJ, Chyu JY, Yang RS, Chen CH, Shen CL. (2020) Osteoprotective roles of green tea catechins. *Antioxidants (Basel)*, **9**(11), 1136.

Khalid S, Shaikh F, Imran-Ul-Haq HS. (2019) Oxidative stress associated with altered activity of glutathione peroxidase and superoxide dismutase enzymes with IDA during pregnancy. *Pak J Pharm Sci*, **32**(1), 75–79.

Kim EY, Ham SK, Bradke D, Ma Q, Han O. (2011) Ascorbic acid offsets the inhibitory effect of bioactive dietary polyphenolic compounds on transepithelial iron transport in Caco-2 intestinal cells. *J Nutr*, **141**(5), 828–834.

Kimball JS, Johnson JP, Carlson DA. (2021) Oxidative stress and osteoporosis. *J Bone Joint Surg Am.* **103**(15), 1451–1461.

Ko CH, Li K, Ng PC, Fung KP, Li CL, Wong RP, Chui KM, Gu GJ, Yung E, Wang CC, Fok TF. (2006) Pro-oxidative effects of tea and polyphenols, epigallocatechin-3-gallate and epigallocatechin, on G6PD-deficient erythrocytes in vitro. *Int J Mol Med*, **18**(5), 987–994.

Koyama Y, Kuriyama S, Aida J, Sone T, Nakaya N, Ohmori-Matsuda K, Hozawa A, Tsuji I. (2010) Association between green tea consumption and tooth loss: cross-sectional results from the Ohsaki Cohort 2006 Study. *Prev Med*, **50**(4), 173–179.

Lazrak M, El Kari K, Stoffel NU, Elammari L, Al-Jawaldeh A, Loechl CU, Yahyane A, Barkat A, Zimmermann MB, Aguenaou H. (2021) Tea consumption reduces iron bioavailability from NaFeEDTA in nonanemic women and women with iron deficiency anemia: stable iron isotope studies in Morocco. *J Nutr*, **151**(9), 2714–2720.

Li J, Pang Q. (2014) Oxidative stress-associated protein tyrosine kinases and phosphatases in Fanconi anemia. *Antioxid Redox Signal*, **20**(14), 2290–2301.

Ma Q , Kim E-Y, Lindsay EA, Han O. (2011) Bioactive dietary polyphenols inhibit heme iron absorption in a dose-dependent manner in human intestinal Caco-2 cells. *J Food Sci*, **76**(5), H143–H150.

Manami Hoshi , Jun Aida , Taro Kusama, *et al.* (2020) Is the association between green tea consumption and the number of remaining teeth affected by social networks? A cross-sectional study from the Japan gerontological evaluation study project. *Int J Environ Res Public Health*, **17**(6), 2052.

Manolagas SC. (2010) From estrogen-centric to aging and oxidative stress: a revised perspective of the pathogenesis of osteoporosis. *Endocr Rev*, **31**(3), 266–300.

Martin FM, Bydlon G, Friedman JS. (2006) SOD2-deficiency sideroblastic anemia and red blood cell oxidative stress. *Antioxid Redox Signal*, **8**(7–8), 1217–1225.

McAlpine MD, Gittings W, MacNeil AJ, Ward WE. (2012) Black and green tea as well as specialty teas increase osteoblast mineralization with varying effectiveness. *J Med Food*, **24**(8), 866–872.

Mecocci P, Boccardi V, Cecchetti R, Bastiani P, Scamosci M, Ruggiero C, Baroni M. (2018) A long journey into aging, brain aging, and Alzheimer's Disease following the oxidative stress tracks. *J Alzheimers Dis*, **62**(3), 1319–1335.

Merhav H, Amitai Y, Palti H, Godfrey S. (1985) Tea drinking and microcytic anemia in infants. *Am J Clin Nutr*, **41**(6), 1210–1213.

Nash LA, Ward WE. (2016) Comparison of black, green and rooibos tea on osteoblast activity. *Food Funct*, **7**(2), 1166–1175.

Nomani H, Bayat G, Sahebkar A, Fazelifar AF, Vakilian F, Jomezade V, Johnston TP, Mohammadpour AH. (2019) Atrial fibrillation in β-thalassemia patients with a focus on the role of iron-overload and oxidative stress: a review. *J Cell Physiol*, **234**(8), 12249–12266.

Ohnishi S T, Ohnishi T, Ogunmola G B. (2001) Green tea extract and aged garlic extract inhibit anion transport and sickle cell dehydration in vitro. *Blood Cells Mol Dis*, **27**(1), 148–157.

Oka Y, Iwai S, Amano H, Irie Y, Yatomi K, Ryu K, Yamada S, Inagaki K, Oguchi K. (2012) Tea polyphenols inhibit rat osteoclast formation and differentiation. *J Pharmacol Sci*, **118**(1), 55–64.

Podkowińska A, Formanowicz D. (2020) Chronic kidney disease as oxidative stress- and inflammatory-mediated cardiovascular disease. *Antioxidants (Base l)*, **9**(8), 752.

Rapa SF, Di Iorio BR, Campiglia P, Heidland A, Marzocco S. (2019) Inflammation and oxidative stress in chronic kidney disease-potential therapeutic role of minerals, vitamins and plant-derived metabolites. *Int J Mol Sci*, **21**(1), 263.

Record IR, McInerney JK, Dreosti IE. (1996) Black tea, green tea, and tea polyphenols. Effects on trace element status in weanling rats. *Biol Trace Elem Res*, **53**(1–3), 27–43.

Renard D, Rosselet A. (2017) Drug-induced hemolytic anemia: pharmacological aspects. *Transfus Clin Biol*, **24**(3), 110–114.

Roomi MW, Roomi NW, Bhanap B, Niedzwiecki A, Rath M. (2013) Repression of matrix metalloproteinases and inhibition of cell invasion by a nutrient mixture, containing ascorbic acid, lysine, proline, and green tea extract on human Fanconi anemia fibroblast cell lines. *Exp Oncol*, **35**(1), 20–24.

Saewong T, Ounjaijean S, Mundee Y, Pattanapanyasat K, Fucharoen S, Porter JB, Srichairatanakool S. (2010) Effects of green tea on iron accumulation and oxidative stress in livers of iron-challenged thalassemic mice. *Med Chem*, **6**(2), 57–64.

Shen CL, Cao JJ, Dagda RY, Tenner TE Jr, Chyu MC, Yeh JK. (2011) Supplementation with green tea polyphenols improves bone microstructure and quality in aged, orchidectomized rats. *Calcif Tissue Int*, **88**(6), 455–463.

Shen CL, Chyu MC, Wang JS. (2013) Tea and bone health: steps forward in translational nutrition. *Am J Clin Nutr*, **98**(6 Suppl), 1694S–1699S.

Suliburska J, Bogdanski P, Szulinska M, Stepien M, Pupek-Musialik D, Jablecka A. (2012) Effects of green tea supplementation on elements, total antioxidants, lipids, and glucose values in the serum of obese patients. *Biol Trace Elem Res*, **149**(3), 315–322.

Sun Y, Shuang F, Chen DM, Zhou RB. (2013) Treatment of hydrogen molecule abates oxidative stress and alleviates bone loss induced by modeled microgravity in rats. *Osteoporos Int*, **24**(3), 969–978.

Tarng DC, Huang TP, Liu TY, Chen HW, Sung YJ, Wei YH. (2000) Effect of vitamin E-bonded membrane on the 8-hydroxy 2′-deoxyguanosine level in leukocyte DNA of hemodialysis patients. *Kidney Int*, **58**(2), 790–799.

Thephinlap C, Ounjaijean S, Khansuwan U, Fucharoen S, Porter JB, Srichairatanakool S. (2007) Epigallocatechin-3-gallate and epicatechin-3-gallate from green tea decrease plasma non-transferrin bound iron and erythrocyte oxidative stress. *Med Chem*, **3**(3), 289–296.

Toblli JE, Cao G, Oliveri L, Angerosa M. (2012) Effects of iron deficiency anemia and its treatment with iron polymaltose complex in pregnant rats, their fetuses and placentas: oxidative stress markers and pregnancy outcome. *Placenta*, **33**(2), 81–87.

Vora RM, Alappattu MJ, Zarkar AD, Soni MS, Karmarkar SJ, Antony AC. (2021) Potential for elimination of folate and vitamin B(12) deficiency in India using vitamin-fortified tea: a preliminary study. *BMJ Nutr Prev Health*, **4**(1), 293–306.

Wang G, Liu G, Zhao H, Zhang F, Li S, Chen Y, Zhang Z. (2014) Oolong tea drinking could help prevent bone loss in postmenopausal Han Chinese women. *Cell Biochem Biophys*, **70**(2), 1289–1293.

Wang G, Liu LH, Zhang Z, Zhang F, Li S, Chen Y, Zhao H. (2018) Retraction note to: oolong tea drinking could help prevent bone loss in postmenopausal Han Chinese women. *Cell Biochem Biophys*, **76**(1–2), 325.

Wu X, Liang S, Zhu X, Wu X, Dong Z. (2020) CoQ10 suppression of oxidative stress and cell senescence increases bone mass in orchiectomized mice. *Am J Transl Res*, **12**(8), 4314–4325.

Xin M, Yang Y, Zhang D, Wang J, Chen S, Zhou D. (2015) Attenuation of hindlimb suspension-induced bone loss by curcumin is associated with reduced oxidative stress and increased vitamin D receptor expression. *Osteoporos Int*, **26**(11), 2665–2676.

Xu H, Yin D, Liu T, Chen F, Chen Y, Wang X, Sheng J. (2018) Tea polysaccharide inhibits RANKL-induced osteoclastogenesis in RAW264.7 cells and ameliorates ovariectomy-induced osteoporosis in rats. *Biomed Pharmacother*, **102**, 539–548.

Ye R, Wang HL, Zeng DW, Chen T, Sun JJ, Xi QY, Zhang YL. (2021) GHRH expression plasmid improves osteoporosis and skin damage in aged mice. *Growth Horm IGF Res*, 60-61, 101429.

Yun JH, Pang EK, Kim CS, Yoo YJ, Cho KS, Chai JK, Kim CK, Choi SH. (2004) Inhibitory effects of green tea polyphenol (-)-epigallocatechin gallate on the expression of matrix metalloproteinase-9 and on the formation of osteoclasts. *J Periodontal Res*, **39**(5), 300–307.

Zhang ZF, Yang JL, Jiang HC, Lai Z, Wu F, Liu ZX. (2017) Updated association of tea consumption and bone mineral density: a meta-analysis. *Medicine (Baltimore)*, **96**(12), e6437.

Zhao B-L. (2020) The pros and cons of drinking tea. *Tradit Med Mod Med*, **3**(3), 1–12.

Zhao F, Guo L, Wang X, Zhang Y. (2021) Correlation of oxidative stress-related biomarkers with postmenopausal osteoporosis: a systematic review and meta-analysis. *Arch Osteoporos*, **16**(1), 4.

Zhou Q, Zhu L, Zhang D, Li N, Li Q, Dai P, Mao Y, Li X, Ma J, Huang S. (2016) Oxidative stress-related biomarkers in postmenopausal osteoporosis: a systematic review and meta-analyses. *Dis Markers*. **2016**, 7067984.

https://doi.org/10.1142/9789811274213_0017

Chapter 17

Antisepsis and Preservative Functions of Tea Polyphenols

Baolu Zhao

Institute of Biophysics, Chinese Academy of Sciences, Beijing, China

17.1 Introduction

Many foods, such as salad oil, lard, and other edible oils, have a high fat content. There are also many high-fat foods in the animal body tissue, such as savings meat and its products. These high-fat products are easily oxidized and deteriorated in light, humidity, temperature, and environment. The application of synthetic antioxidants in food has greatly reduced the oxidative corruption rate of these foods, but it also brings a sensitive problem, that is, its safety. Another listed food antioxidant is vitamin E, which is also in insufficient amounts. The main reason for this is because it is not soluble in water and therefore, it does not enter foods easily, so its healthcare effect is also limited.

The oxidation mechanism of oil has been clearly understood. It is considered that oxygen molecule captures unsaturated fatty acids and hydrogen at the olefin position, which initiates a chain reaction to automatically oxidize fats into oxides, and the further pyrolysis products have peculiar smell aldehydes and acids. Tea polyphenols (TP), the most active constituents of tea, are considered natural food additives. Tea polyphenols can provide hydrogen, and its redox potential is low. It is also a high-quality antioxidant material, with broad-spectrum bactericidal effect and free radical combination to produce all kinds of catechin-free radicals with weak activity, which can be oxidized and polymerized into theaflavin polymers to weaken the chain reaction of free radicals and terminate the chain reaction to inhibit the automatic oxidation of fatty acids. Tea polyphenols are among the most abundant functional compounds in tea. They exhibit strong antioxidant, anti-inflammatory, and anti-cancer effects. As functional food components, tea polyphenols have strong antioxidant and physiological activities for human body. A main active ingredient of green tea, epigallocatechin-3-gallate (EGCG), is a polyphenol that possesses antioxidant, antimicrobial, anti-proliferative, and free radical scavenging effects.

A large number of literature reports that oxidative stress is closely related to food corruption. Tea and tea polyphenols can play a great role in anti-corrosion and preservation. These will be discussed in this chapter.

17.2 Food corruption and preservation

Oils, meat, fruits, vegetables, and aquatic products are easy to deteriorate at room temperature. Especially under the conditions of light, humidity, and high temperature, it is easier to oxidize and deteriorate. The proteins in food is decomposed by the protein decomposition enzyme produced by spoilage bacteria, then degraded to low molecular compounds, in turn generating various toxic substances and unpleasant odors. The acid value, carbonyl value, and thiobarbituric acid (TBA) value of edible oil are very high due to long-term exposure to oxygen, sunlight, microorganisms, and enzymes in the air. The acids, aldehydes, ketones, and various oxides produced by oil rancidity not only change the sensory properties of oil, but also have adverse effects on the body.

Low temperature preservation and high temperature sterilization, dehydration and drying preservation, food pickling (including salt pickling and sugar pickling) and smoking preservation, high-pressure processing, ionizing radiation, pulsed electric field and ultraviolet radiation, food radiation preservation, and the use of vacuum technology to preserve food are effective methods to prevent food corruption. Due to the different physiological characteristics of oils, meat, fruits, vegetables, and aquatic products, scientific preservation methods need to be adopted. Although these methods can achieve certain results, there are also many problems. Since microbial food spoilage is a complex phenomenon, which is related to the succession of specific spoilage organisms over a period of time, it needs in-depth research to find a more appropriate method.

The proteins, fats, and carbohydrates of body tissues deteriorate due to bacterial action. Massive necrosis of local tissues of the body is caused by putrefactive bacterial infection. All kinds of bacteria existing in the mouth, respiratory tract, and digestive tract can invade blood vessels and lymphatic vessels and grow and reproduce in large numbers in the body. Bacteria *in vitro* can also invade the human body and reproduce, leading to the corruption of body tissues. We also need to study and find more appropriate methods to deal with it.

Natural food antibacterial agents are bioactive compounds that can inhibit the growth of microorganisms related to food spoilage or food-induced diseases. Food preservative additive is a natural or synthetic substance, which can delay food degradation caused by microbial growth, enzyme activity, and oxidation. Until recently, the use of synthetic additives in food has become more and more common. However, in recent years, synthetic additives have not been widely accepted by consumers because synthetic additives have adverse effects on their health. Therefore, consumers are increasingly inclined to natural additives. Studies have found that tea and tea polyphenols are very good natural antibacterial agents, which can be used for the preservation of oils, meat, fruits and vegetables, aquatic products, and human tissues. Tea polyphenols have excellent antioxidant properties, broad-spectrum bactericidal effect, and high safety. In 1992, relevant departments of Chinese government approved tea polyphenols to be used in the statutory scope of food additives and formulated industrial quality standards. At present, tea polyphenols are widely used in many foods as preservative additive.

17.3 Spoilage and oxidative stress

A few years after it was found that a layer of walnut oil on the surface of water consumed a lot of air, became sticky, and emitted a strange smell, a famous chemist explained this phenomenon and proposed that the absorption of oxygen by walnut oil revealed an important chemical reaction known today as lipid peroxidation, which is the oxidative deterioration of unsaturated fatty acids. Lipid peroxidation exists in the storage of fats, oils, and creams, and in the manufacture and use of paints, plastics, and rubbers. Later, people studied the basic reaction of lipid peroxidation. Oxidative stress caused by reactive oxygen (ROS) and reactive nitrogen (RNS) play an important role in food spoilage [Nieto *et al.*, 2017]. Metabolic processes in muscle tissue *in vivo* result in the production of ROS an RNS and oxidative compounds including superoxide anions and nitric oxide (NO). High oxygen modified atmosphere packaging is a commonly applied method to prolong the minimum shelf life of fresh (red) meats. Upon spoilage, changes of the initial oxygen concentration and microbiome composition can be observed. All bacteria were able to consume dissolved oxygen and all strains showed significant growth enhancement in the presence of heme, indicating respiratory activity. Coupling of respiration and fermentation via regeneration of NADH can be a competitive advantage for meat-spoiling bacteria, resulting in a higher cell count and possibly accelerated spoilage [Zhao, 1999, 2007].

17.3.1 *Oil corruption and oxidative stress*

The process of oil peroxidation is a chain reaction that produces and participates in ROS and RNS free radicals, which can be divided into chain starting, chain expansion, and chain termination. In each step, ROS and RNS free radical intermediates are produced and participated. Lipid peroxy radicals can pump hydrogen from another adjacent lipid molecule to form new lipid radicals. This forms a cycle, which is the extended stage of lipid peroxidation chain reaction. The formed lipid radical reacts with oxygen to form another lipid peroxy radical, so lipid peroxidation forms a chain reaction. Aliphatic peroxy radicals can also form epoxy compounds and inner epoxy compound radicals, which can finally break into various aldehydes and short chain hydrocarbons. Unsaturated lipid peroxidation, which usually occurs simultaneously with oxidative stress, promotes the progress of redox imbalance through chain reaction [Zhang *et al.*, 1992a, 1992b]. Docosahexaenoic acid (DHA) is extremely sensitive to oxidative damage due to the existence of six double bonds between carbon atoms in its polyene chain (c=c). Malondialdehyde (MDA), 4-hydroxy-nonenal (HNE), and the F2-isoprostane 15(S)-8-iso-prostaglandin F2α (15(S)-8-iso-PGF2α) are the best investigated products of lipid peroxidation. MDA, HNE, and 15(S)-8-iso-PGF2α are produced from polyunsaturated fatty acids (PUFAs) by both chemical reactions and reactions catalyzed by enzymes. 15(S)-8-iso-PGF2α and other F2-isoprostanes are derived exclusively from arachidonic acid (AA). The number of PUFAs that may contribute to MDA and HNE is much higher. [Tsikas, 2017].

Affected by oxygen, water, light, heat, and microorganisms, the oil is gradually hydrolyzed or oxidized to deteriorate and rancidity, which decomposes the neutral oil into glycerol and fatty acid or breaks the unsaturated chain in the fatty acid to form peroxide, and then successively decomposes into low-grade fatty acids, aldehydes, ketones, and other substances, resulting in odor and peculiar smell. Some rancidity products also have carcinogenic effect. The rancidity of oil destroys the vitamins contained in it and damages the enzyme system of the body, such as succinate oxidase, cytochrome oxidase, and so on. The free fatty acids generated after oil hydrolysis, especially the double chain position of unsaturated free fatty acids, can be easily oxidized to form peroxides. Among these peroxides, a small amount of odor oxides with ring structure combined with ozone have very unstable properties and are easy to be decomposed into aldehydes, ketones, and small molecular fatty acids. A large number of hydro-peroxides, due to their unstable nature and susceptibility to decomposition, can also polymerize and cause oil rancidity. Moreover, the rancidity will also quickly turn other free fatty acid molecules into hydro-peroxides by chain reaction due to the formation of hydro-peroxides. The result is the accumulation of small molecular substances such as aldehydes, ketones, and acids in oil, showing strong bad flavor and certain physiological toxicity, further worsening the sensory quality of food and increasing the burden of detoxification function of human liver. This oxidative rancidity can occur in oils and fats in most foods. A paper studied the effect of oxygen concentration in modified atmosphere packaging on color and texture of beef patties cooked to different temperatures. Patties were made from raw minced beef after storage for six days in modified atmosphere to study the combined effect of oxygen concentration and cooking temperature on hardness and color. Increased oxygen concentrations generally led to larger ($p < 0.01$) thiobarbituric acid-reactive substances (TBARS) values, greater ($p < 0.01$) loss of free thiols, and more formation of cross-linked myosin heavy chain. Hardness of cooked patties was generally lower ($p < 0.01$) without oxygen. Premature browning of cooked patties was observed already at a relative low oxygen concentration of 20% [Bao *et al.*, 2016].

Oxidized oil can destroy the fluidity and integrity of cell membrane and affect the enzyme activity, resulting in the imbalance of normal physiological and biochemical functions of cells, the destruction of normal structures of tissues and organs, and a variety of pathological symptoms. When the oil is rancid, the composition of oil acid changes, and the saturated fatty acids such as linoleic acid and linolenic acid, which are essential oil acids in the diet, are destroyed, in particular, the relative proportion of oil acid in fish oil decreases significantly. At the same time, the digestibility of oxidized oil decreased, because oxidized oil inhibited the activity of pancreatic lipase. The formation of highly active free radicals in the process of oil oxidation can destroy vitamins, especially fat-soluble vitamins such as vitamin A, vitamin E, and vitamin D, resulting in vitamin deficiency. This paper studied the different oxidative status and expression of calcium channel components in stress-induced dysfunctional chicken muscle. The results indicated that meat production of ROS and RNS increased significantly after transportation ($p < 0.05$) and thiobarbituric acid reactive substance values and carbonyl contents increased significantly in the test group ($p < 0.05$) [Xing *et al.*, 2017].

The above research shows that a large amount of ROS and RNS free radicals will be produced in the process of oil preservation and corruption, which will lead to the decomposition of oil to produce some toxic substances.

17.3.2 *Oxidative stress caused by reactive oxygen ROS in the meat food spoilage*

The onset of oxidation in meat postmortem is well known to produce off-odors, discoloration, and unacceptable flavors associated with rancidity. Oxidation during the immediate postmortem period appears to inhibit tenderization during aging, probably through an inhibitory effect of oxidation on the calpain enzyme. Oxidation of meat occurs under postmortem conditions and is inevitable. The oxidation includes the biochemical changes in meat, leading to changes in color pigments and lipids. As a consequence, color deteriorates and undesirable flavors and rancidity develop in meat, thereby impacting on consumer appeal and satisfaction. Deoxynivalenol produced by various *Fusarium* mold species can induce cytotoxicity and oxidative stress in the gastrointestinal tracts of humans and farm animals. It was found that deoxynivalenol significantly decreased transepithelial electrical resistance and triggered oxidative stress [Pomothy *et al.*, 2021].

Lipofuscin is one of the indicators of oxidative stress. To elucidate the role of oxidative stress in the development of wooden breast, this study investigates lipofuscin accumulation in various parts of wooden breast muscles. In muscles affected by wooden breast, vacuolated muscle fibers were observed, and connective tissues appearing like perimysium were expanded with fibrosis. TBARS value and accumulation of lipofuscin were significantly higher in the wooden breast than in the normal breasts. A lot of lipofuscin granules were localized in the cytoplasm of collapsed muscle fibers of the wooden breast. The cranial portion of the wooden breast showed the highest shear force. The cranial position had a large amount of connective tissue and lipofuscin granules. The results of the present study strongly suggest that high oxidative stress, especially with a significant accumulation of lipofuscin, is associated with the development of wooden breasts [Hasegawa *et al.*, 2021]. A paper studied the relationship between oxygen concentration, shear force, and protein oxidation in modified atmosphere packaged pork. Pork loins were stored at 5°C for 14 days to investigate the effect of oxygen concentration in the modified atmosphere containing 0–80% O_2, 20% CO_2, and balanced with N_2. The results showed that shear force and thiobarbituric acid-reactive substances (TBARS) values increased with increasing oxygen concentration. Protein oxidation, when measured as loss of free thiol groups, was greater in meat packaged under oxygen (20–80%). Myosin heavy chain (MHC) cross-linking, another marker of protein oxidation, was greater in MAP with 80% oxygen than 0% and 20% oxygen [Bao & Ertbjerg, 2015]. This paper studied the effects of high oxygen and vacuum retail ready-packaging formats on lamb loin and topside eating quality. Lamb steaks from semi-membranosus and longissimus thoracis et lumborum (LTL) muscles were allocated to three different packaging treatments — Darfresh® vacuum skin packaging (VSP), Darfresh® Bloom packaging (80% O_2:20% CO_2; Hi-Ox-DB), or high oxygen modified atmosphere packaging (80% O_2:20% CO_2; high-oxygen modified atmosphere packaging (Hi-Ox-MAP), and stored in simulated retail display for five or 10 days. Hi-Ox-MAP and Hi-Ox-DB samples had lower tenderness, flavor, juicinesst, and overall liking scores and higher TBARS values, compared to VSP. Hi-Ox-MAP samples deteriorated in juiciness and flavor between five and 10 days. Hi-Ox-MAP LTL samples had a lower myofibrillar fragmentation index. The LTL exhibited greater desmin degradation and reduced desmin crosslinking relative to the semimembranosus [Frank *et al.*, 2017].

This work investigated the role of ROS-generating systems on the softening of the pale, soft, and exudative-like (PSE-like) rabbit meat during aging. The meat samples had higher values in peroxides value, thiobarbituric acid-reactive substances, metmyoglobin percentage, ferrylmyoglobin content, non-heme iron content, hydroxyl radical content, and ROS concentration compared with the normal ones, suggesting that PSE-like incubation could activate lipid-oxidizing system, myoglobin-mediated oxidation system, together with metal-catalyzed oxidation system. Higher protein carbonyl content was observed in PSE-like meat, along with a significant loss in sulfhydryl group. The results suggested that more serious protein degradation occurred in PSE-like meat [Wang *et al.*, 2022]. This paper studied the oxidative processes in muscle systems and fresh meat. The oxidative process of muscle tissues will vary according to an animal's immunity status, temperament, and ability to cope with stress, with all these affected by nutrition, genetics, management practices, and environmental conditions (hot and cold seasons). The isoprostanes, in the context of complex biochemical reactions, relate to oxidative processes that take place in the biological systems of live animals (*in vivo*) and subsequently in meat (*in vitro*) [Bekhit *et al.*, 2013]. This study tested whether this reduction releases reactive oxygen species that may lead to lipid oxidation in minced meat under two different storage conditions. Krebs cycle substrates (KCS) combinations of succinate and glutamate increased peroxide-forming potential (1.18–1.32 mmol peroxides/kg mince) and

TBARS (0.30–0.38 mg malondialdehyde (MDA) equivalents/kg mince) under low oxygen storage conditions. Both succinate and glutamate were metabolized. Moreover, under high oxygen (75%) storage conditions, KCS combinations of glutamate, citrate and malate increased peroxide-forming potential (from 1.22–1.29 mmol peroxides/kg) and TBARS (from 0.37–0.40 mg MDA equivalents/kg mince [Yi *et al.*, 2015].

This study was conducted to determine whether circulating concentrations of blood isoprostanes can be used as an effective biomarker in lambs to predict degradation of color and/or lipid stability in meat. This included lipid oxidation levels in muscle assessed by thiobarbituric acid reactive substances and meat redness. Lambs that consumed the commercial feedlot pellet had lower muscle vitamin E level ($p < 0.01$) and higher level of polyunsaturated fatty acids (PUFA) ($p < 0.001$) compared with lambs finished on annual ryegrass or lucerne. Lipid oxidation levels were greatest for lambs finished on the feedlot ration, lowest in lambs finished on the ryegrass diet, and intermediate for lambs finished on lucerne and ryegrass-feedlot combination ($p < 0.01$). After eight weeks of feeding, the blood isoprostane concentration was positively correlated with the lipid oxidation of meat displayed for 72 hours in simulated retail conditions ($p < 0.01$) [Ponnampalam *et al.*, 2017].

A paper studied revealed that the high-oxygen modified atmosphere packaging system induces lipid and myoglobin oxidation and protein polymerization. Beef steaks from longissimus lumborum, semimembranosus, and adductor muscles were cut at 24 hours postmortem and randomly assigned to either high-oxygen modified atmosphere packaging (HiOx-MAP; 80% O_2, 20% CO_2) or vacuum (VAC), and displayed for nine days at 1°C. HiOx-MAP packaged beef steaks had a rapid increase in lipid oxidation and a decrease in color stability during display. The steaks in HiOx-MAP had significantly lower tenderness and juiciness scores and higher off-flavor scores compared to steaks in VAC. HiOx-MAP condition did not affect the postmortem degradation of troponin-T or desmin. Furthermore, autolysis of micro-calpain was not influenced by packaging. These results suggest that the HiOx-MAP system may negatively affect meat quality characteristics by inducing lipid and myoglobin oxidation and crosslinking/aggregation of myosin by protein oxidation [Kim *et al.*, 2010]. This paper studied the influence of dietary antioxidants and quality of oil on the oxidative and physicochemical properties of chicken broiler breast and thigh meat stored in either an oxygen-enriched (HiOx: 80% O_2/20% CO_2) or an air-permeable polyvinylchloride (PVC) packaging system during retail display at 2–4°C for up to 14 days and 7 days, respectively. In both packaging systems, lipid oxidation TBARS was inhibited by up to 65% and 57% in chicken breast and thigh, respectively, with an antioxidant-supplemented diet compared to those without. In both breast and thigh samples, protein sulfhydryls and water-holding capacity (purge loss) were better protected by the antioxidant dietary treatment, regardless of oil quality. Thigh muscles had up to seven-fold greater TBARS formation and more myosin heavy chain losses compared to breast samples. Antioxidant supplementation was more protective against lipid oxidation and water-holding capacity in the group fed on high-oxidized oil compared to those fed on low-oxidized oil. The results suggest that dietary antioxidants can minimize the negative impact of oxidized oil on broiler meat quality, and this protection was more pronounced for thigh than breast muscle, indicating inherent variations between muscle fiber types [Delles *et al.*, 2016].

Ferrous had strong pro-oxidant effects. This paper showed that the pro-oxidant effects of ferrous iron, hemoglobin, and ferritin in oil emulsion and cooked-meat homogenates are different from those in raw-meat homogenates. Flax oil was blended with maleate buffer to prepare an oil emulsion and determine the mechanisms of various iron forms on the catalysis of lipid peroxidation. Results showed that ferrous iron and hemoglobin had strong pro-oxidant effects, but ferritin became pro-oxidant only when ascorbate was present. Hemoglobin and ferritin had no pro-oxidant effect in raw-meat homogenates. The status of heme iron and the released iron from hemoglobin had little effect on the pro-oxidant effect of hemoglobin in oil emulsion and cooked meat homogenate systems. The pro-oxidant effect of ferrous iron in oil emulsion and cooked-meat homogenates disappeared in the presence of superoxide ($\bullet O_2^-$), H_2O_2, or xanthine oxidase systems. In raw-meat homogenates, however, ferrous had strong pro-oxidant effects even in the presence of $\bullet O_2^-$, or H_2O_2 [Ahn & Kim, 1998].

The above research shows that the process of meat preservation and corruption will produce a large number of active oxygen free radicals, which will lead to the decomposition of meat, not only causing meat deterioration and producing some toxic substances.

17.3.3 *Oxidative stress caused by reactive nitrogen in the food spoilage*

Preservation of meat with nitrite or nitrate has become important to mankind in controlling meat spoilage and in producing safe and palatable meat products with good keeping properties even at ambient temperature. Nitric oxide (NO) is a free radical that is constantly produced or released throughout the body by diverse tissues and is known to influence proteolytic activity in human and rodent skeletal muscle as well as being involved in regulation of calcium homeostasis in the muscle cell. The influence of nitric oxide on the development of meat tenderness has been studied through postmortem manipulation and also through *in vivo* studies. The effect of NO on meat tenderness is postulated to be via its regulatory effects on the proteins calpain, cathepsins, ryanodine receptor channel in the sarcoplasmic reticulum (SR), and the sarcoplasmic-endoplasmic release calcium ATPase in the SR. NO is an oxidant. Although the effects of NO on effector proteins can be distinguished from a direct oxidation reaction, it is proposed that postmortem meat tenderization is influenced by skeletal muscle's release of NO pre-slaughter and the oxidation of proteases postmortem. Progress has been made, including studies by manipulating the NO levels in muscle cells, suggesting possible effects in the pre-slaughter and post-slaughter environment. NO has potential effects on the meat quality of beef, lamb, chicken, and pork muscles. However, it has been difficult to determine the exact mechanism(s) of NO action as it has variable effects on meat quality including tenderness, water holding capacity, and color. It is speculated that NO and protein S-nitrosylation may be involved in muscle to meat conversion through the regulation of postmortem biochemical pathways including glycolysis, Ca^{2+} release, proteolysis, and apoptosis. NO is an antioxidant in processed meat and may be a possible modulator of transmetalation reactions in meat. NO and NO donor NOR-3 have dual effect on μ-calpain activity, autolysis, and proteolytic ability [Warner *et al.*, 2005].

A work investigated the effects of NO and its induced protein nitros-ylation on calpain-1 activation and protein proteolysis in beef during postmortem aging. Five semimembranosus muscles were removed from a cattle carcass. Beef samples were incubated with one of following treat-ments for 24 hours at 4°C: control (normal saline), NO donor, or nitric oxide synthase (NOS) inhibitor. Results showed that GSNO decreased and L-NAME increased the extent of calpain-1 autolysis at day 1. Degradation of desmin and troponin-T was increased by L-NAME while decreased by GSNO. These results suggest that NO could regulate cal-pain-1 autolysis and its proteolysis activity during postmortem aging in beef SM muscle [Zhang *et al.*, 2018]. This study investigated the bio-chemical changes of nitric oxide synthase (NOS) in pork skeletal muscles during postmortem storage. Results showed that all three muscles exhib-ited NOS activity until day 1, while SM muscle retained NOS activity after three days of storage. The content of nNOS in SM muscle was stable across three days of storage while decreased intensity of nNOS was detected at day 1 and day 3 of aging in PM and LT muscles due to the degradation of calpain [Liu *et al.*, 2015]. A research study examined the influence of NO on calpain activation, protein proteolysis, and oxidation in post-mortem pork. Higher levels of protein oxidation were observed after samples were incubated with NO donor compared to treatment with NOS inhibitor ($p < 0.05$) [Li *et al.*, 2014].

As NO is postulated to be a mediator of the effects of pre-slaughter stress on meat quality, the aims of this experiment were to investigate the effects of modulating NO pharmacologically on meat quality of sedentary lambs. The principal outcomes of the experiment were that L-NAME inhibited proteolysis and reduced tenderness in the semimembranosus. These data indicate that events pre-slaughter that affect NO synthesis can influence meat tenderness, potentially via altered muscle metabolism or modulation of proteolytic enzymes [Cottrell *et al.*, 2015]. Five bovine semimembranosus muscles were incubated with three treatments, includ-ing S-nitrosoglutathione (GSNO, nitric oxide donor), normal saline, and Nω-nitro-L-arginine methyl ester hydrochloride (L-NAME, nitric oxide synthase inhibitor). The results showed that the level of protein S-nitrosylation was improved by GSNO treatment and reduced by

L-NAME treatment ($p < 0.05$). Compared to the control, GSNO treatment had higher shear force while L-NAME treatment presented lower shear force at day 7 postmortem ($p < 0.05$). In addition, μ-calpain autolysis, myofibrillar protein, and desmin degradation were reduced by GSNO treatment and accelerated by L-NAME treatment ($p < 0.05$). Therefore, it can be speculated that protein S-nitrosylation could affect beef tenderization by regulating the autolysis of μ-calpain and the degradation of myofibrillar proteins [Hou *et al.*, 2020].

The above research shows that a large number of active nitrogen free radicals will be produced in the process of food and meat preservation and corruption. NO is a free radical, which is constantly produced or released by various tissues in the body, which can lead to the decomposition of food oils to produce some toxic substances.

17.3.4 *Blood preservation and oxidative stress*

Our studies have found that during blood preservation, living cells will be damaged by oxidation; at 37°C for 24 hours, lipid peroxidation increased by about 25 times and protein oxidation increased by about six times. The lipid peroxidation of cell membrane is the most obvious, and the result is to affect the fluidity of cell membrane. We used fatty acid spin labeling technique to study the changes of erythrocyte membrane fluidity during storage. The results showed that during 35 days of blood preservation, the superficial fluidity of erythrocyte membrane lipids decreased and the deep fluidity increased. The superficial phase transition temperature of membrane lipids decreased significantly, and there were two phase transition points (T1 and T2) in the deep layer. T1 and T2 gradually approached and finally fused into a phase transition point during blood preservation. Another target site of oxidative damage is membrane protein. We studied the conformational changes of erythrocyte membrane protein during blood preservation with maleimide spin markers. The results showed that with the extension of blood preservation, the ESR spectrum of the sulfhydryl binding site of the labeled cell membrane protein had a rapid reduction and then gradually increased process. This reflects the biphasic change of the microviscosity of the environment around the spin marker, which first decreases and then increases, indicating that during blood preservation, the conformation of erythrocyte membrane protein becomes loose or the protein is depolymerized, which is also closely related to the change of cell membrane lipid fluidity [Shi *et al.*, 1995a, 1995b].

This study was to define the associations between blood donor body mass index (BMI) and red blood cell (RBC) measurements of metabolic stress and hemolysis. Evaluations in 18 donors revealed that BMI was significantly ($p < 0.05$) and positively associated with storage and osmotic hemolysis. A BMI of 30 kg/m^2 or greater was also associated with lower post-transfusion recovery in mice 10 minutes after transfusion ($p = 0.026$). It revealed that BMI was a significant modifier for all hemolysis measurements, explaining 4.5%, 4.2%, and 0.2% of the variance in osmotic, oxidative, and storage hemolysis, respectively. In this cohort, obesity was positively associated ($p < 0.001$) with plasma ferritin (inflammation marker). Metabolomic analyses on RBCs from obese donors (44.1 ± 5.1 kg/m^2) had altered membrane lipid composition, dysregulation of antioxidant pathways (e.g., increased oxidized lipids, methionine sulfoxide, and xanthine), and dysregulation of nitric oxide metabolism, as compared to RBCs from non-obese (20.5 ± 1.0 kg/m^2) donors [Hazegh *et al.*, 2021].

From the above discussion, it can be seen that oxidative stress caused by ROS and RNS and plays a very important role in the process of food preservation and corruption. In order to preserve food and prevent corruption, we need to pay attention to and solve the problems of oxidative stress caused by ROS and RNS in the process of food preservation and corruption.

17.4 Inhibitory effect of tea polyphenols on the spoilage of foods and tissues

Although there are many factors causing food corruption, many studies have shown that tea and tea polyphenols can play a role in preventing and reducing spoilage and keeping fresh of food and tissues. A large number of literature reports that tea can scavenge ROS and RNS and inhibit oxidative stress and tea polyphenols play a key role in anti-corrosions and preservations.

17.4.1 *Inhibitory effect of tea polyphenols on food spoilage and growth of microorganisms in food*

Microorganisms are widely distributed in nature. Food will inevitably be polluted by different types and quantities of microorganisms. When the environmental conditions are appropriate, they will grow and reproduce rapidly, resulting in food corruption and deterioration, which not only reduces the nutritional and hygienic quality of food, but also may endanger human health. Therefore, controlling food corruption and food preservation is of great practical significance both in theory and in practice. Tea is one of the most widely consumed beverages in the world and known for its antimicrobial activity against many microorganisms. Preliminary studies have shown that tea polyphenols can inhibit the growth of a wide range of Gram-positive bacteria. Meat contains 50–75% moisture, 14–21% protein, and pH 5.9~6.5, which is an excellent food with rich chemical composition, especially conducive to the development of spoilage bacteria. Under the influence of various microorganisms, meat spoils very rapidly. The deterioration of meat mostly begins on the surface and then gradually goes deep. The aerobic and anaerobic spoilage of meat are different. The pathogens of aerobic corruption are various cocci and bacilli, among which the most active are ordinary bacilli, fluorescent bacilli, and fecal bacilli, while *Escherichia coli*, *Bacillus subtilis*, and potato bacilli cause aerobic corruption. Determination of the effects of tea polyphenol on physicochemical parameters, microbiological counts, and biogenic amines in dry-cured bacons at the end of ripening, showed that plant polyphenols and α-tocopherol significantly decreased pH, thiobarbituric acid reactive substances content, and total volatile basic nitrogen (TVBN) compared with the control ($p < 0.05$). Microbial counts and biogenic amine contents in dry-cured bacons were affected by tea polyphenol being the most effective ($p < 0.05$) in reducing aerobic plate counts, *Enterobacteriaceae*, *Micrococcaceae*, yeast, and molds, as well as in inhibiting the formation of putrescine, cadaverine, tyramine, and spermine [Wang *et al.*, 2015].

1. Inhibitory effect of tea polyphenols on the growth of microorganisms
Spore-forming bacteria are an aggravating problem for the food industry due to spore formation and their subsequent returning to vegetative state

during food storage, thus posing spoilage and food safety challenges. This paper analyzed the effect of tea compounds: gallic acid, EGCG, Teavigo (>90% epigallocatechin gallate), and theaflavin 3,3′-digallate on spore germination and outgrowth and subsequent growth of vegetative cells of *Bacillus subtilis*. Gallic acid most strongly reduced the ability to grow out. All compounds, in particular theaflavin 3,3′-digallate (TFDG), clearly affected the growth of emerging vegetative cells. Gallic acid most strongly reduced the ability to grow out. Additionally, all compounds, especially TFDG, clearly affected the growth of emerging vegetative cells [Pandey *et al.*, 2015]. A work studied the mechanism of the antibacterial action of tea polyphenols such as catechins and theaflavins against *Bacillus coagulans* (*B. coagulans*), and the interaction of EGCG or TFDG with the surface of *B. coagulans* cells was investigated. The antibacterial activities of EGCG and TFDG against *B. coagulans* cells were measured by counting the number viable cells after the mixing with each polyphenol. Bactericidal effect of TFDG was shown at the concentration of greater than or equal to 62.5 mg/l. However, at the same concentration, EGCG did not have this effect. The activity of the glucose transporters of the cells decreased 40% following the treatment with TFDG of 62.5 mg/l. However, this decrease was only slight in case of EGCG. The results suggest that the direct interaction between membrane proteins and TFDG is an important factor in the antibacterial activity of polymerized catechins, affecting their functions and leading to cell death [Sato *et al.*, 2020]. The studies of the effects of EGCG on typical spoilage bacterium, *Pseudomonas fluorescens* (*P. fluorescens*), have found that most of the differentially expressed genes (DEGs) involved in iron uptake, anti-oxidation, DNA repair, efflux system, cell envelope, and cell-surface component synthesis were significantly upregulated by EGCG treatment, while most genes associated with energy production were downregulated. These transcriptomic changes are likely to be adaptive responses of *P. fluorescens* to iron limitation and oxidative stress, as well as DNA and envelope damage caused by EGCG. These results may ultimately contribute to the optimal application of green tea polyphenols in food preservation [Liu *et al.*, 2017].

The above results indicate that tea polyphenols and theaflavins can prevent oxidative stress and food corruption in food preservation by inhibiting the growth of microorganisms.

2. Inhibitory effect of tea polyphenols on oil corruption

During the storage period, the oil will deteriorate, become viscous, and produce strange smell, especially under the conditions of high temperature and high humidity. Peroxide is the initial product of oil oxidative rancidity. Edible fats and fat-containing products undergo oxidation, both during production and storage, causing a sequence of unfavorable changes. Therefore, the production of peroxide in oil is often used as the beginning of oil oxidative rancidity, which is usually expressed by peroxide value (POV). The POV value of fresh vegetable oil is usually less than 10 mg equivalent per kilogram. When the POV value reaches 30–40 mg equivalent per kilogram, it has obvious putrid taste. When POV is lower than 20, the oil can be safely used in the feed, and the peroxide value is below 50, which will not affect the feed intake of animals, but it may affect the feed conversion efficiency and cause toxicity to animals. When POV is higher than 200, the toxicity of oil will increase sharply (significantly changed into hepatomegaly of experimental animals), which will significantly reduce the availability of oil. Enrichment of lipids with plant polyphenols can profitably influence their oxidative stability and additional introduction to human body can also decrease the degenerative diseases morbidity [Jiang *et al.*, 2021]. It is found that tea polyphenols can effectively inhibit the oxidative deterioration of oil.

It was found that pure winter butter was more stable than pure butter from summer season in Rancimat test conditions ($p < 0.05$). Summer season butter oxidative stability was highest in sample with addition of green tea extract: 71.22 hours for Rancimat and 81.23 hours for Oxidograph test. Best anti-oxidative activity in winter butter showed green tea extract, where induction period was 66.5 hours for Rancimat and 64.0 hours for Oxidograph test. Furthermore, rosemary extract and tocopherol showed strong anti-oxidative activity, weaker however than green tea extract. Butylated hydroxytoluene (BHT), a strong synthetic antioxidant, showed much lower activity [Gramza-Michalowska *et al.*, 2007].

It was found that the oxidative deterioration of lard could be significantly inhibited by treating lard with different concentrations of tea polyphenols. When peanut oil, palm oil, oil extraction, and soybean oil were treated with 0.02% tea polyphenols at 63°C for four days, the POV of peanut oil decreased from 48.8 to 11.0 and the inhibition rate was 77.4, the POV of palm oil decreased from 6.4 to 3.3, the inhibition rate was 60%, the POV of rapeseed oil decreased from 44.4 to 17.8, the inhibition rate was 41.6%, and the POV of soybean oil decreased from 48.3 to 28.2 and the inhibition rate was 41.6%. Lard was treated with 0.02%, 0.04%, and 0.08% tea polyphenols, and the POV changes of the treated and control samples were measured. The results showed that the POV of lard stored at room temperature for 60 days exceeded the national edible oil standard (< 20), and the POV of tea polyphenols treated for 120 days and 150 days was still lower than the national edible oil standard (< 10) [Tu & Wang, 2003].

The above research shows that tea polyphenols have strong antioxidant activity in the oil system, and tea polyphenols can effectively inhibit the oxidative and deterioration of oil.

3. Inhibitory effect of tea polyphenols on meat spoilage

Meat is rich in protein and oil, which is the best medium for microorganisms and the target of attack, so it is most likely to decay during storage. Tea polyphenols have good antiseptic effect and antioxidant effect, so it is a very effective choice for meat preservation.

This work studied the effects of four natural extracts from tea in pork patties. During 20 days of storage in modified atmosphere packs at 2°C, pH, color, lipid oxidation, and microbial spoilage parameters of raw minced porcine patties were examined and compared with a synthetic antioxidant (BHT) and control batch. Due to their higher polyphenol content, tea extracts were the most effective antioxidants against lipid oxidation and limiting color deterioration. In addition, natural extracts led to a decrease of total viable counts (TVC), lactic acid bacteria (LAB), *Pseudomonas*, and psychotropic aerobic bacteria compared to the control [Lorenzo et al., 2014].

This study examined the preservative properties of tea polyphenols for *Collichthys* fish ball in well storage. Vacuum-packed *Collichthys* fish balls were treated with tea polyphenols and stored at 0°C for 17 days. Results confirmed that the dominant bacteria in *Collichthys* fish balls are the genera *Serratia* and *Pseudomonas*. Total viable counts dropped two orders of magnitude in *Collichthys* fish balls with 0.25 g kg^{-1} TP compared with the control. The advantages of total volatile basic nitrogen value, 2-thiobarbituric acid value, and texture value were clearly observed, whereas pH and whiteness value exhibited no significant decrease for the group treated with 0.25 g kg^{-1} TP. More than 0.25 g kg^{-1} TP added could retain excellent fish ball characteristics in terms of sensory assessment after 17 days. The shelf life of *Collichthys* fish balls supplemented with tea polyphenols can be prolonged for an additional six days in good condition at 0°C storage. [Yi *et al.*, 2011].

Another paper investigated the quorum sensing (QS) system of *Shewanella baltica* (*S. baltica*) and the anti-QS related activities of green tea polyphenols (GTP) against spoilage bacteria in refrigerated large yellow croaker. GTP at sub-inhibitory concentrations interfered with AI-2 and DKPs activities of *S. baltica* without inhibiting cell growth and promoted degradation of AI-2. The GTP treatment inhibited biofilm development, exopolysaccharide production, and swimming motility of *S. baltica* in a concentration-dependent manner. In addition, GTP decreased extracellular protease activities and trimethylamine production in *S. baltica*. GTP repressed the luxS and torA genes in *S. baltica*, which agreed with the observed reductions in quorum sensing (QS) activity and the spoilage phenotype. EGCG significantly inhibited AI-2 activity of S. *baltica* [Zhou *et al.*, 2015]. These findings strongly suggest that green tea extract could be developed as a new QS inhibitor for seafood preservation to enhance shelf life. Lamb leg chops were sprayed with 0.005%, 0.05%, 0.5%, 5% (p/v) green tea extracts and displayed under retail conditions for 13 days. The extracts showed a concentration-dependent action; the minimum concentration of tea polyphenols which significantly reduced lipid oxidation was 2.08 mg GAE/100cm^2 of meat. Both 0.5% green tea extracts limited color deterioration and reduced metmyoglobin formation. The green tea extracts showed no antimicrobial effect, exceeding microbial counts of 7logCFU/cm^2 at 13 days of display. Sensory analyses determined that none of the extracts added herb odors or flavors to lamb [Bellés *et al.*, 2017]. Evaluation of the antimicrobial effects of tea polyphenols on changes in microbiota composition and quality attributes in silver carp fillets stored at 4°C found that tea polyphenols treatment was found to be effective in enhancing sensory quality, inhibiting microbial growth, and attenuating chemical quality deterioration. Meanwhile, the composition of microbiota of silver carp fillets was investigated using culture-dependent and culture-independent methods. Initially, compared to the control, tea polyphenols obviously decreased the relative abundance of *Aeromonas*, which allowed *Acinetobacter* and *Methylobacterium* to become the dominant microbiota in TP treated fillets on day 0. Tea polyphenols improved the quality of fillets during chilled storage, which was mainly due to their modulating effects on microbiota that resulted in the change in pattern and process of spoilage in fillets [Jia *et al.*, 2018].

A study investigated the melanosis, quality attributes, and bacterial growth of freeze-chilled Pacific white shrimp (*Litopenaeus vannamei*) during six days of chilled storage, as well as the preservative effects of tea polyphenol on shrimp. The results showed that freeze-chilled storage retarded the growth of bacteria and the accumulation of putrescine in shrimp. The growth of spoilage bacteria *Photobacterium* and *Shewanella* were inhibited. However, freeze-chilled storage aggravated melanosis and lipid oxidation. The total volatile basic nitrogen (TVB-N) slightly accumulated in the thawed shrimp. The incorporation of tea polyphenol preserved freeze-chilled shrimp. Melanosis and lipid oxidation of shrimp were alleviated and the accumulation of biogenic amines, TVB-N, hypoxanthine riboside, and hypoxanthine were retarded. Meanwhile, the growth of spoilage bacteria *Pseudoalteromonas*, *Photobacterium*, *Psychrobacter*, and *Carnobacterium* were inhibited [Li *et al.*, 2022].

A research study revealed the quality changes of soy protein-based meat analogues at 4°C, and to investigate the efficacy of antimicrobial packaging on maintaining the qualities of meat analogues during 10 days of storage. Cinnamaldehyde (CI) or tea polyphenol were incorporated in polylactic acid (PLA), polybutylene adipate (PBAT), and starch blends by extrusion technique to prepare antimicrobial packaging. Results suggest that PLA/PBAT-CI film successfully prevents moisture from evaporation, maintains the texture properties, and ensures the quality and safety of meat analogues [Wang *et al.*, 2022]. Slightly acidic electrolyzed water (SAEW) is often used as a disinfectant in beef preservation to ensure microbiological safety. However, it ineffectively inhibits lipid oxidation. Therefore, the combination of SAEW and tea polyphenol was tested to inhibit lipid oxidation and microbial growth in beef preservation. SAEW and tea polyphenol were selected as the optimum sanitizer and antioxidant, respectively. Results revealed that the required quality standard of beef treated with SAEW-TP was prolonged by approximately nine days at 4°C, and this treatment had greater antimicrobial and antioxidant effects than did the single treatment [Bing *et al.*, 2022]. This study evaluated the antimicrobial and preservative effects of the combinations of nisin (NS) tea polyphenols on pasteurized chicken sausage. TP had the largest inhibitory effect on *P. aeruginosa* with a clear zone diameter of 18.2 mm. These results indicate that the combination of NS tea polyphenol could be used as a natural preservative to efficiently inhibit the growth of microorganisms in pasteurized chicken sausage and improve its safety and shelf life [Sun *et al.*, 2021].

A paper evaluated the antimicrobial activities of tea polyphenols and chitosan, and their combinations against both Gram-positive bacteria (GPB) and Gram-negative bacteria (GNB). Results showed that the MIC of nisin was 2.44–1250 mg/L for GPB and reached 5000 mg/L for GNB. The minimum inhibitory concentration (MICs) of tea polyphenols were 313–625 mg/L for GNB, and 156–5000 mg/L for GPB, respectively. These results indicated that tea polyphenols exhibited inhibitory effects against both GPB and GNB. By using the optimum treatment, the shelf life of chilled mutton was extended from 6–18 days at 4°C in the preservative film packages. These results indicate that the TP could be used as preservatives to efficiently inhibit the growth of spoilage microorganisms and pathogens in meat, thus improving the safety and shelf life of chilled mutton [He *et al.*, 2016]. GRA and GTE through vacuum impregnation on the quality retention and microflora of refrigerated grass carp fillets were studied. Generally, the quality degradation of carp fillets was remarkably alleviated using coatings when compared to the control. As suggested by microbial enumeration and high-throughput sequencing, protective coatings were conductive to inhibit bacteria growth, especially spoilage bacteria of *Pseudomonas*. As a result, the indicator related to bacteria such as total volatile basic nitrogen (TVB-N) and K value had lower levels in coating groups than that in control [Zhao *et al.*, 2021].

This paper studied the effects of TP on the post-mortem integrity of large yellow croaker (*Pseudosciaena crocea*) fillet proteins. TP are known to be important for the postmortem deterioration of fish muscle and can enhance food quality. To know the influence of TP on the status of large yellow croaker muscle proteins, control and treated fillets (0.1% TP, 0.2% TP and 0.3% TP, w/v) were analyzed periodically for myofibrillar protein functional properties (Ca^{2+}-ATPase activity, surface hydrophobicity, total sulfhydryl content, emulsion stability index, and rheological behavior). Degradation of the myofibrillar protein myosin could be clearly observed; several proteins were also observed to vary in abundance following post-mortem storage of 25 days. The study offers evidence that TP have an effective impact on muscle protein integrity post-mortem [Zhao *et al.*, 2013].

This study evaluated the antioxidant properties of a complexes between TP and hydroxypropyl-β-cyclodextrin (HP-β-CD) and the effects on myofibrillar protein (MP) from lamb tripe under oxidative conditions. Results showed that the inclusion complex could effectively inhibit protein oxidation, which can provide a reference for the application of polyphenols in meat products and the improvement of protein properties. The addition of an appropriate concentration (5–105 µmol/g) of TP/HP-β-CD inclusion complex decreased the carbonyl content, hydrophobicity, and protein aggregation of MP from lamb tripe, whereas it increased the sulfhydryl content. This improved antioxidant activity and bioavailability of the inclusion complexes will be beneficial for its potential applications in food [Zhang *et al.*, 2018].

The above research shows that tea polyphenols have strong antioxidant activity in meat preservation, and tea polyphenols can effectively inhibit the oxidative deterioration of meat.

17.4.2 *Inhibitory effect of tea polyphenol packaging films on food spoilage*

The study found that not only tea polyphenols, but also edible packaging film of tea polyphenols, can directly inhibit the spoilage of bacteria on preserved food. As a bioactive extract from tea leaves, tea polyphenols are safe and natural. Its excellent antioxidant and antibacterial properties are increasingly regarded as a good additive for improving degradable food packaging film properties. Addition of tea polyphenols could impart antioxidant and antibacterial properties to active packaging films and act as a crosslinking agent to improve other physical and chemical properties of the film, such as mechanical and barrier properties.

A paper studied the preservation effect of meat product by natural antioxidant tea polyphenols. 1% of tea polyphenol, chitosan solution, and potassium sorbate were used as film-forming materials to coat-chilled mutton. Total coliforms, total volatile basic nitrogen (TVB-N) value, and pH value were determined and used as the mutton fresh-keeping indexes. The results showed that after day 12 at the end of the storage, mutton coated with tea polyphenol had the best effects comparing chitosan solution and potassium sorbate. The pH value of mutton coated by tea polyphenol was 6.0 and total volatile basic nitrogen and the total coliforms were both significantly lower than the meat coated by chitosan solution and potassium sorbate. Additionally, mutton coated by tea polyphenol accorded with the requirements of national standards about fresh meat quality. The results indicated that the tea polyphenol film was the most suitable film on chilled mutton coating preservation among the three chemicals used in this research [Wang *et al.*, 2016]. Another study was to investigate the effect of chitosan (CH) film incorporated with tea polyphenol on quality and shelf life of pork meat patties stored at $4 \pm 1°C$ for 12 days. The microbiological, physicochemical (pH, thiobarbituric acid-reactive substances (TBARS) values, and metmyoglobin (MetMb)), and sensory qualities were measured. A microbiological shelf-life extension of six days was achieved for chitosan and chitosan-tea polyphenol treatment groups when compared to the control group. Wrapping with chitosan-tea polyphenol composite film tended to retard the increases in TBARS values and metmyoglobin content. The results indicated that chitosan-tea polyphenol composite film could be a promising material as a packaging film for extending the shelf life of pork meat patties. The addition of tea polyphenols improved the efficiency of the active packaging film in food preservation applications, which accelerates the process of replacing the traditional plastic-based food packaging with active packaging film [Qin *et al.*, 2013]. A study evaluated the effect of a sodium alginate coating infused with tea polyphenols on the quality of fresh Japanese sea bass fillets. Sodium alginate and tea polyphenols are natural preservatives commonly used in the food industry, including the production of fish products. The effect of sodium alginate coating infused with tea polyphenols on the quality of fresh Japanese sea bass fillets was evaluated over a 20-day period at 4°C. Sodium alginate (1.5%, w/v) or tea polyphenols (0.5%, w/v) treatment alone and the sodium alginate coating infused with tea polyphenols all reduced microbial counts, with the sodium alginate-tea polyphenols providing the greatest effect. Fish fillet samples treated with sodium alginate-tea polyphenols had significantly lower levels of total volatile basic nitrogen, lipid oxidation, and protein decomposition during the storage period, relative to the remaining treatments. The samples treated with sodium alginate-tea polyphenols had the highest sensory quality rating as well [Nie *et al.*, 2018].

These results show that, compared with the pure poly-lactic acid (PLA) membrane, adding tea polyphenols and chitosan (CS) significantly increases the heat-sealing strength, water vapor permeability, and solubility of the composite membrane. When the composite membrane was used for the preservation of cherries, it was found that the composite membrane with the mass ratio of TP to chitosan of 3:7 can decrease the rotting rate and mass loss rate significantly, postpone the consumption of soluble solids and vitamin C, maintain the quality of the cherries, and extend the shelf life, thus proving its potential for application in food packaging [Ye *et al.*, 2018]. Edible packaging films using polymer for food preservation have been developed for a long time. A paper studied the fabrication and characterization of tea polyphenol-loaded pullulan-carboxymethyl cellulose sodium (CMC) electro-spun nanofiber for fruit preservation. In this study, the effects of different concentrations (0.5%, 1%, 1.5%, w/v) of tea polyphenols incorporated into pullulan-carboxymethylcellulose sodium (Pul-CMC) solutions on electro-spun nanofiber films were evaluated. Fruit packaging potential was evaluated using strawberry. The pullulan-CMC-TP nanofibers significantly decreased weight loss and maintained the firmness of the strawberries and improved the quality of the fruit during storage [Shao *et al.*, 2018].

The above research shows that using tea polyphenols as additive for packaging film performance can more conveniently and effectively play an antioxidant and antibacterial roles in meat preservation and can effectively inhibit food oxidative deterioration.

17.4.3 *Effects of green tea polyphenol on tissue preservation*

Ideally, tissues should be transplanted immediately from the donor to the recipient. However, this is not always possible, and the problem of tissue preservation needs to be addressed to ensure successful transplantation. Therefore, it is essential to develop storage solutions that can maintain the viability and function of the tissues or organs for longer periods. Generally, mammalian cells and living tissues can be cryopreserved in a frozen state at very low temperatures over a long storage term. The survival rate of cell suspensions is often acceptable, however, living tissues suffer a variety of injuries. The number of tissue or organ transplants has increased substantially in recent years with the advances in surgical methods and the development of immunosuppressive agents. Many studies have shown that tea and tea polyphenols not only play an important role in preventing and reducing food corruption and preservation, but also play an important role in preventing and reducing human tissue necrosis and preservation such as blood, blood vessels, sperm, nerves, skin, teeth, and other tissues. Green tea polyphenols have been recently reported to promote the preservation of tissues, such as blood vessels, corneas, nerves, islet cells, articular cartilage, and myocardium, at room temperature. These findings indicate the possibility of a new method of tissue banking without freezing. It demonstrated that the addition of polyphenols extracted from green tea to conventional cell culture medium and tissue compatible liquid, can control cell proliferation and preserve tissues for several months at ordinary room temperature, including such tissues as blood vessels, cartilage, islet cells, and corneas.

1. Green tea polyphenol in preserving the human vein

The potential role of green tea polyphenol (GTP) in preserving the human saphenous vein was investigated under physiological conditions. The vein segments were incubated for 1 day, 3 days, 5 days, 7 days, and 14 days, either after four hours of treatment with 1.0 mg/ml GTP or in the presence of GTP at the same concentration. When the veins were not treated with GTP, the viability of the endothelial cells was significantly reduced with respect to the progress in the culture time, and none of the cells expressed eNOS after five days. Furthermore, severe histological changes and structural damage were observed in the non-treated veins. In contrast, incubating the veins after four hours of GTP treatment significantly prevented these phenomena. The cellular viability of the GTP-treated vein was approximately 64% after seven days and eNOS expression was maintained up to 40%, compared to that of the fresh vein. The histological observations showed that the vasculature was quite similar to that of the fresh vein. When incubated with GTP, the vein could also be preserved for one week under physiological conditions retaining both its cellular viability (61%) and eNOS expression level (45%) and maintaining its venous structure without any morphological changes. These results demonstrate that GTP treatment may be a useful method for preserving the human saphenous vein [Han *et al.*, 2004]. The injurious effects of reactive oxygen species on venous tissues and the potential protective role played by GTP on human saphenous veins were investigated. Oxidative stress was induced exogenously in the vein segments, either by adding 0.8 M or 1.6 M of H_2O_2, or by using 80 U/L or 160 U/L of xanthine oxidase in the presence of xanthine (0.5 mM). The H_2O_2-induced alterations were prevented by pre-incubating the veins with either 0.5 mg/ml or 1.0 mg/ml of GTP for one hour. When the oxidative stress was induced by xanthine oxidase, cellular viability and venous structure were preserved at the same polyphenol concentrations [Han *et al.*, 2004]. Rat aortas was preserved in a medium containing various amino acids and glucose (DMEM) solution containing polyphenols extracted from green tea leaves. The preserved aortas retained original structures and mechanical strength and were devoid of any undesirable cell secretions for over a month under physiological conditions. In addition, aortas from Lewis rats preserved for a month and transplanted to allogenic ACI rats completely avoided rejection by the host, suggesting that the polyphenols have immunosuppressive actions on the aortic tissues [Hyon *et al.*, 2006].

These results demonstrate that GTP can act as a biological antioxidant and protect veins from oxidative stress-induced toxicity. Polyphenol treatment of aortic tissue transplant can maintain its viability for extended periods of time either before or after transplantation, and the method can be applicable to other transplantation situations.

2. Green tea polyphenol in preserving the human nerves

In this study, allogeneic-transplanted peripheral nerve segments preserved for one month in a green tea polyphenol solution at 4°C could regenerate nerves in rodents, demonstrating the same extent of nerve regeneration as isogeneic fresh nerve grafts. The present study investigated whether the same results could be obtained in a canine model.

Successful nerve regeneration was observed in the green tea polyphenol-treated nerve allografts when transplanted in association with a therapeutic dose of the immunosuppressant FK506. The data indicated that green tea polyphenols can protect nerve tissue from ischemic damage for one month. However, the effects of immune suppression seem insufficient to permit allogeneic transplantation of peripheral nerves in a canine model [Nakayama *et al.*, 2010]. Sciatic nerve segments, 20 mm long, were harvested from Lewis rats and treated in three different ways before transplanting to recipient Lewis rats to bridge sciatic nerve gaps created by removal of 15-mm-long nerve segments. Peripheral nerve segments can be successfully preserved for one month using green tea polyphenol [Ikeguchi *et al.*, 2003]. A study produced a 1.0 cm sciatic nerve defect in rats and divided the rats into four treatment groups: autograft, fresh nerve allograft, green tea polyphenol-pretreated (1 mg/mL, 4°C) nerve allograft, and irradiation-pretreated nerve allograft (26.39 Gy/min for 12 hours; total 19 kGy). The circumference and structure of the transplanted nerve in rats that received autografts or green tea polyphenol-pretreated nerve allografts were similar to those of the host sciatic nerve. Compared with the groups that received fresh or irradiation-pretreated nerve allografts, motor nerve conduction velocity in the autograft and fresh nerve allograft groups was greater, more neurites grew into the allografts, Schwann cell proliferation was evident, and many new blood vessels were observed. In addition, massive myelinated nerve fibers formed, and abundant microfilaments and microtubules were present in the axoplasm. Our findings indicate that nerve allografts pretreated by green tea polyphenols are equivalent to transplanting autologous nerves in the repair of sciatic nerve defects and promoting nerve regeneration. Pretreatment using green tea polyphenols is better than pretreatment with irradiation [Zhou *et al.*, 2015]. In a study, the permeability of nerve tissue to polyphenol solution was investigated using canine sciatic nerve segments stored in 1 mg/ml polyphenol solution for one week and in DMEM for the subsequent three weeks. Electron microscopy revealed that the Schwann cell structure within 500–700 μM of the perineurium was preserved, but cells deeper than 500–700 μM were badly damaged or had disappeared. The infiltration limit for polyphenol solution into neural tissue is inferred to be 500–700 μM [Matsumoto *et al.*, 2005].

The above studies confirm the nerve cell viability of the nerve segment preserved by tea polyphenols. Morphologically, nerve regeneration was similar to that of fresh isografts and superior to that of grafts stored with solution alone. Moreover, the electrophysiological results were equal to those of fresh isografts. Tea polyphenol has the potential to be used for peripheral nerve storage and could be useful for routine peripheral nerve banking.

3. Green tea polyphenol in preserving semen

Once semen leaves the human body's natural environment for half an hour, it will lose the ability to fertilize, which is very unfavorable during IVF. This study evaluated the quality of canine semen after preservation with diluents containing vitamin C and polyphenol at 5°C for four weeks. It investigated the effects of vitamin C combined with polyphenol supplementation on chilled semen quality. The addition of vitamin C (0.5 mM or 1 mM) with 0.75 mg/mL polyphenol to semen extender provided significantly higher percentages of sperm motility and viability during cold storage compared to unsupplemented semen. Supplementation of 0.5 mM vitamin C plus polyphenol yielded the highest percentages of sperm motility and viability. However, there was no beneficial effect on the plasma membrane and acrosomal integrity of the spermatozoa [Wittayarat *et al.*, 2012]. Preservation of rat aortic tissue transplant with green tea polyphenols sperm cell membrane protein (IZUMO1) was recently found to be a crucial mediator in the interaction and fusion with eggs, indicating an important role in assuring the favorable outcome from long-term preservation of chilled semen. A paper studied the effects of green tea polyphenols and α-tocopherol on the quality of chilled cat spermatozoa and IZUMO1 expression during long-term preservation. It investigated whether supplementation of chilled semen extender with green tea polyphenols together with α-tocopherol would provide synergistic effects to prolong sperm survival and maintain IZUMO1 protein stability in cat spermatozoa. Sperm samples were collected from the cat epididymis before being diluted with semen extender containing various concentrations of α-tocopherol (0 µg/ml, 2.5 µg/ml, 5 µg/ml, and 7.5 µg/ml) and 0.75 mg/ml green tea polyphenols and cooled to 4°C. One sample without antioxidants served as a control. Sperm characteristics and IZUMO1 protein expression were investigated before and after chilling at 3 days, 6 days, 9 days, 12 days, and 15 days. Sperm IZUMO1 protein was markedly conserved by supplementation of 5 µg/ml α-tocopherol together with 0.75 mg/ml green tea polyphenols up to 12 days in cold storage [Wittayarat *et al.*, 2022].

The above results indicate that green tea polyphenols can protect sperm characteristics and protein integrity during long-term freezing, and can significantly improve sperm vitality and survival rate during cold storage.

4. Green tea polyphenol in preserving skins

Ultraviolet (UV) radiation causes skin injury and inflammation resulting in impaired immune response and increased risk of skin cancer. It has been shown that GTP enhanced intracellular antioxidant defense and promoted the downregulation of proapoptotic genes, and they could be used to protect against the damage induced by UV irradiation. After eight hours of incubation at 50°C with carboxymethyl cellulose sodium (CMC-Na), 93% of GTPs was preserved, while in the absence of CMC-Na, only 61% of GTPs was preserved. Topical treatment of emulsified GTP effectively inhibited acute UVB-induced infiltration of inflammatory cells, increase of skin thickness, oxidative stress such as depletion of antioxidant enzymes and lipid oxidation, and induced nuclear accumulation of Nrf2 in the mice skin. We also discovered the ability of GTP to simultaneously trigger an accumulation of nuclear Nrf2 and export nuclear Bach1 [Li *et al.*, 2016]. This study examined the effects of EGCG regarding skin preservation. Skin sample biopsy specimens measuring 1 × 1 cm from GFP rats were held in sterile containers with 50 ml preserving solution at 4°C and 37°C for up to about eight weeks. Periodically, some of the preserved skin specimens were directly examined histologically and others were transplanted into nude mice. Histological examinations of skin preserved at 4°C revealed a degeneration of the epidermal and dermal layers from five weeks in all groups. In the groups preserved at 37°C, degeneration and flakiness of the epidermal layer were demonstrated starting at two weeks preservation regardless of addition of EGCG. After 2–7 weeks of preservation, the rat skin grafted to nude mice in the EGCG groups stored at 4°C showed successful engraftment. However, grafts preserved at 4°C without EGCG and at 37°C did not demonstrate GFP-positive keratinocyte or fibroblasts [Kawazoe *et al.*, 2008]. Skin grafts can be preserved by cryopreservation and refrigerated storage at 4°C. EGCG enhances the viability of stored skin grafts and also extends the storage time up to seven weeks at 4°C. EGCG is the main polyphenol component in green tea, which has strong antioxidant, antibacterial, anti-proliferation, and free radical scavenging effects. In the EGCG group, the graft showed higher integrity in the epidermis and dermal matrix. In the EGCG groups, the grafts showed higher integrity in the epidermal layer and dermal matrix [Kim *et al.*, 2009].

The present findings suggest the future clinical usefulness of EGCG for skin preservation without freezing, however, the mechanism by which EGCG promotes skin is through its antioxidant, antibacterial, anti-proliferation, and free radical scavenging effects.

5. Green tea polyphenol in preserving teeth

Many dental diseases are related to bacterial inflammation, leading to tooth damage. The current research confirms that polyphenols can reduce the growth of cariogenic bacteria. Furthermore, they can decrease the adherence of bacteria to the tooth surface and improve the erosion-protective properties of the acquired enamel pellicle. Tea polyphenols, especially, have the potential to contribute to an oral health-related diet [Flemming *et al.*, 2021]. A paper studied the effect EGCG on maintaining the periodontal ligament cell viability of avulsed teeth. Avulsed tooth can be completely recovered if the periodontal ligament (PDL) of tooth is maintained. A study evaluated the effect of EGCG on avulsed-teeth preservation of Beagle dogs for a period of time. The atraumatically extracted teeth of Beagle dogs were washed and preserved with 0/10/100 μM of EGCG at the time of immediate, period 1 (four days in EGCG-contained media and additional one day in EGCG-free media), period 2 (eight days in EGCG-contained media and additional two days in EGCG-free media) and period 3 (12 days in EGCG-contained media and additional two days in EGCG-free media). From the results, the immediately analyzed group presented the highest cell viability and the rate of living cells on teeth surface decreased dependent on the preservation period [Jung *et al.*, 2011]. Thirty freshly extracted single-rooted human teeth with closed apices were randomly assigned to three experimental groups with 10 samples per group and immersed in one of the storage media: EGCG, Hank's balanced salt solution (HBSS), or milk for two hours. The PDL cells were dissociated by an enzyme treatment with collagenase and trypsin. The cells were then labeled with 0.4% Trypan blue for the determination of viability. The result showed that EGCG group had the highest percentage of cell viability, followed by HBSS, and milk group, in descending order [Chen & Huang, 2012].

It was found that salivary pellicle modification with polyphenol-rich teas and natural extracts to improve protection against dental erosion. It performed two experiments: one with teas (green tea, black tea, peppermint tea, rosehip tea, and negative control) and other with natural extracts (grape seed, grapefruit seed, cranberry, propolis, negative control), where the negative control was deionized water. A total of 150 enamel specimens were used (n = 15/group). Results suggested that green tea, black tea, grape seed extract, and grapefruit seed extract were able to modify the salivary pellicle and improve its protective effect against enamel erosion, but rosehip tea and cranberry extract caused erosion [Niemeyer *et al.*, 2021]. These results suggest that EGCG could maintain PDL cell viability of extracted tooth, therefore, could be a perfect additive for tooth preservation and be able to postpone the period of tooth storage.

The above results indicate that TP can reduce the growth of cariogenic bacteria. TP can reduce the adhesion of bacteria to the tooth surface, improve the erosion protection of the acquired enamel membrane, prolong the retention period of teeth, and may promote healthy to oral diet. *In situ* and clinical studies need to be expanded and supplemented in order to make a significant contribution to additional preventive measures in caries prevention.

6. Green tea polyphenol in preserving plant samples

Tea polyphenols can not only protect animal tissues and samples, but also be used to preserve plant samples. In this work, tea polyphenols preserved different types of plant tissue for ultrastuctural and immunocytochemical studies. It also explored oolong tea extract as an alternative for uranyl acetate for the staining of plant samples. It obtained excellent preservation of cell ultrastructure when samples were embedded in epoxy resin, and of cell antigenicity, when embedded in low viscosity hydrophilic acrylic white resin. Furthermore, oolong tea extract successfully replaced uranyl acetate as a counterstain on ultrathin sections and for in-block staining. These novel protocols reduce the time spent at the bench, and improve safety conditions for the investigator. The preservation of the cell components when following these approaches is of high quality [Carpentier *et al.*, 2012]. Oolong tea demonstrated good potentials as a counterstain on ultrathin sections. In addition, sample preparation time was significantly shortened and simplified using LR-White resin. This novel protocol reduced the time for preparing plant samples, and hazardous reagents in traditional method (acetone) were also replaced by less toxic ones (ethanol and oolong tea extract) [He *et al.*, 2018].

The above results show that tea polyphenols, especially oolong tea polyphenols, can not only be used for food preservation to prevent food decay, but also for plant samples.

17.5 Deodorization effect of tea polyphenols

The odor of environment and animals is mainly formed by three types of substances: the first type is nitrogen-containing substances such as methylamine, ammonia, histamine, nicotine; the second category is sulfur compounds, such as ammonia sulfide, sulfur alcohol and sulfur ether; the third type is the odor produced by acids, aldehydes, ketones and esters, and sometimes high concentration alcohol esters and other odor substances. The commonly used deodorization methods at world are the use of surfactants, activated carbon adsorption, and chemical solvent dissolution. These methods can really remove the odor substances in the environment, but they cannot change the intestinal function of animals, and even some methods can lead to secondary pollution of the environment. As a food antioxidant extracted from tea, tea polyphenols have obvious effects on improving animal intestines and refreshing oral cavity. In addition, tea polyphenols can effectively remove a variety of odor substances. Catechin, the main component of tea polyphenols, is an effective component with odor elimination effect, and has been used to eliminate bad breath, deodorize food, and deodorize air freshener.

17.5.1 *Deodorization effect of tea polyphenol on gastrointestinal odor*

The proportion of beneficial bacteria and harmful bacteria in the digestive tract determines intestinal health. Harmful bacteria can cause intestinal diseases and produce ammonia, sulfur dioxide, amine, indole, acid and other odorous substances through the decomposition of intestinal nutrients. The test proves that tea polyphenols can not only kill harmful bacteria, but also activate beneficial bacteria in the intestine, such as *Lactococcus*, *Bifidobacterium*, and *Lactobacillus*, thereby improving the condition of intestinal microflora. Tea polyphenols are used to treat constipation, control intestinal microbiota, and improve intestinal environment. Fifteen elderly patients aged 51–93 years (with the same food intake) who took a total dose of 300 mg tea polyphenols, measured that the number of lactic acid bacteria in the intestinal tract increased significantly, while the number of *Enterobacteriaceae* bacteria decreased significantly. In the course of administration, the number of *Clostridium* showed a downward trend, and the total number of bacteria, *Enterobacteriaceae* bacteria and fungi decreased significantly. The concentration of ammonia in feces decreased significantly and the total amount of organic acids increased significantly. The concentration of hydrogen sulfide first increased and then decreased. Cresol, ethanol, indole, and methyl indole compounds decreased significantly; the value also shows a downward trend. Other experiment indicated that the deodorization effect was significantly increased with the increase of pH value, in particular, the deodorization effect of green tea was greater than that of black tea. After feeding animals with tea polyphenols, the total number of bifidobacteria in the intestinal tract was determined. It was measured that the total number of bifidobacteria in the intestinal tract was still 16% higher than the total number before drug administration two weeks after drug withdrawal, indicating that tea polyphenols were beneficial to the growth of bifidobacteria in the intestinal tract of animals [Yao *et al.*, 1995].

The pig small intestine will produce an unpleasant smell after being stored at 20°C for 0.5–1 day, mainly due to the presence of volatile components such as methyl mercaptan and ethanol. Green tea and brown sugar can reduce this smell. After the hens were fed with 0.1–1.0% green tea powder in the feed for 20–30 days, the eggs produced had no unpleasant fishy smell, the quality of eggs was improved significantly, and they were more resistant to storage than ordinary eggs. When chickens were fed with 3% tea powder, the freshness and tenderness of the chicken eggs were not only better, but also the content of vitamin C and vitamin E was higher than that of the control. When the tea residue containing 5% tea polyphenols was added to the chicken feed, the odor of the chicken house was significantly reduced, indicating that tea polyphenols played a regulatory role in the intestinal metabolism of chickens. In Japan, adding tea polyphenols to pet feed can reduce the odor of pet feces. It is reported that some functional foods containing tea polyphenols can also reduce "fart odor".

17.5.2 *Deodorization effect of tea polyphenol on halitosis*

The so-called halitosis refers to the unpleasant breath. It can seriously affect people's social interaction and mental health. The WHO has classified bad breath as a disease. A survey shows that the prevalence of halitosis in China is 27.5% and 50% in Western countries. There are several reasons for halitosis. The first is due to poor gastrointestinal function. For example, stomach disease, gastric acid reflux, and food indigestion. Moreover, many digestive diseases can cause halitosis, including gastrointestinal motility disorder, indigestion, and malabsorption, which can lead to severe halitosis. If *Helicobacter pylori* infection is combined, this kind of halitosis will be further aggravated and will produce halitosis. Some systemic diseases can also cause halitosis. Dental diseases include dental caries and periodontal infection, which are closely related to bacterial infection. Bacterial infection can decompose food residues in the mouth and produce odor. As natural polyphenols have been known to have the deodorizing activity, the deodorizing properties and mechanisms of action of polyphenols, the main constituents of green tea extract (GTE), black tea extract (BTE), and grape seed extract (GSE), against volatile sulfur compounds (VSCs) in kimchi were investigated. Since tea polyphenols have been shown to have antimicrobial and deodorant effects, we investigated whether green tea powder reduces VSCs in mouth air and compared its effectiveness with that of other foods which are claimed to control halitosis. Immediately after administering the products, green tea showed the largest reduction in concentration of both H_2S and CH_3SH gases, especially CH_3SH which also demonstrated better correlation with odor strength than H_2S, however, no reduction was observed at one hour, two hours, and three hours after administration. Chewing gum, mints, and parsley-seed oil product did not reduce the concentration of VSCs in mouth air at any time. Toothpaste, mints, and green tea strongly inhibited VSCs production in a saliva-putrefaction system, but chewing gum and parsley-seed oil product could not inhibit saliva putrefaction. Toothpaste and green tea also demonstrated strong deodorant activities *in vitro*, but no significant deodorant activity of mints, chewing gum, or parsley-seed oil product were observed. The results suggested that green tea was very effective in reducing oral malodor temporarily because of its disinfectant and deodorant activities, whereas other foods were not effective [Lodhia *et al.*, 2008].

Methylmercaptan, the main component of halitosis, is caused by the metabolism of oral microorganisms. Many studies have proved that tea extracts can effectively inhibit the generation of this unpleasant smell. Catechins were added to chewing gum, saliva was collected after chewing, and the content of methyl mercaptan was determined *in vitro*. The results showed that catechins had a greater inhibitory effect on halitosis than sodium copper chlorophyllin. During the process of chewing gum containing catechins, the content of methyl mercaptan in the mouth decreased significantly. The deodorizing effect of 10 mg catechin was 62%. If the EGCG with the strongest deodorization ability is used, the effect of 6 mg sodium copper chlorophyllin can be achieved with only 3 mg. When 0–1.0 mg tea polyphenols were added to saliva containing 1 ml egg group acid in advance for culture, methyl mercaptan was inhibited. The addition of 0.5% catechin to gum sugar can significantly inhibited methyl mercaptan. This led to the development of a deodorizing gum containing 0.1% tea extract in Japan. After eating 10–20 capsules, more than 85% of the bad breath will be eliminated within 90 minutes. The addition of 0.1% green tea extract to liquor and beverage has deodorization function. With 5 mg/ml tea polyphenol, the growth of *Porphyromonas gingivalis* can be significantly reduced, while the minimum inhibitory concentration against

Porphyromonas, Actinomyces mucilaginosus, Actinomyces neisseria, Fibrophilia, Fusobacterium nucleatum, Pseudomonas fluorescens, Actinomyces concomitant, and other bacteria are only 10 mg/L. Once the reproduction speed of these micro-organisms decreases, the amount of product secreted will naturally decrease, and then the oral odor will also decrease. Another test also further confirmed this mechanism. They rinsed their mouth with 100 ml (5.0 mg/ml) of green tea extract for 30 minutes and 60 minutes, both of which could inhibit caries bacteria, plaque formation, and oral cavity cleaning. After eating garlic, Chinese people often chew dry tea to eliminate its odor. At the concentration of 2 mg/ml, there is only light and refreshing tea taste, and at the concentration of 1 mg/ml, there is basically no astringent taste, which is easy to be accepted by consumers and promoted. Therefore, it is recommended to prepare a dosage form with a concentration of 1–2 mg/ml that can remain in the mouth for 2–5 minutes.

17.5.3 *Deodorization effect of tea polyphenol on environmental odor*

Tea extracts have different effects on eliminating different odor substances. The effect of green tea extract on trimethylamine, ammonia, methyl mercaptan, n-butyric acid and acetaldehyde is as follows: trimethylamine and ammonia are the best, while p-methylmercaptan, n-butyric acid, and acetaldehyde have the same effect, but also have good effect. The scavenging effects of 0.2%, 1.0% tea extract and 0.2% sodium copper chlorophyllin on trimethylamine, ammonia, methyl mercaptan, and tobacco gas were determined. It was found that the effect of tea extract on these four substances was better than that of sodium copper chlorophyllin. Comparing the deodorization ability of tea deodorant with that of air deodorization products on the market, it was found that tea extract had the best deodorization efficiency for sulfur dioxide and ammonia. The decaffeinated wastewater from steaming green tea is used for concentration and the aqueous solution of 0.5% catechin has a certain effect on reducing the odor in pet food processing plants. When 5–20% polyphenol solution is used as indoor deodorant, it can effectively remove phenylmercaptan, amine, sulfur dioxide, and other sulfides in the air. It was found that polyphenols, the main components of green tea extract (GTE) and black tea extract (BTE), could deodorize volatile sulfur compounds (VSC) in pickles. The deodorizing activity toward VSCs was found to be in the following order: GTE (37.6–73.8%) > BTE (28.4–60.3%). This was attributed to the high phenolic. Particularly, the hydroxyl groups in the polyphenols showed deodorizing activity against VSCs via a sulfur-capture reaction [Jeong *et al.*, 2021].

17.5.4 *Scavenging effect of tea polyphenols on formaldehyde*

In recent years, the problem of formaldehyde (FA) has attracted more and more attention around us. FA has brought a lot of trouble to our lives. With the societal development and decorating of new homes, its use is inevitable. The toxicity of FA has always been a concern. The harm of formaldehyde can be divided into several kinds. A small amount of formaldehyde can be eliminated by our body through autoimmunity. While on one hand, a large amount of formaldehyde is also used for food preservation, on the other hand, formaldehyde residues can cause a lot of harm to the human body. If the formaldehyde content exceeds the standard, it will cause a series of adverse reactions, such as dizziness, cough, and red rash on the body.

A study discovered that the the exogenous and endogenous formaldehyde in squid causes serious health threats. Results revealed that 3:2 (v/v) tea polyphenols and chitosan can achieve formaldehyde removal rate of >85%. Apart from the capability of removing formaldehyde, the scavenger derived from natural food exhibited a significant preservation effect (extension of preservation time up to 40%) on squid during chilled storage, and was safe and environmentally friendly [Tang *et al.*, 2022].

EGCG, an active substance in tea polyphenols, has been shown to demonstrate physiological protective functions by in both epidemiological and zoological studies, particularly in the nervous system. The study described here, aims to explore whether EGCG can alleviate the neurotoxic effects induced by formaldehyde. After 14 days of exposure to 3 mg/m^3 formaldehyde, mice exhibited significant cognitive impairment. In the FA group, a significant increase in iNOS level compared with the control group was observed. The reduced GSH level was significantly decreased.

The levels of IL-1β, TNF-α, and Caspase-3 were raised, indicating significant neuronal damage. After administering EGCG as a protective agent, all the above observed changes were reversed, and the protective effect of EGCG became gradually evident in the 20–500 mg/kg range. EGCG could activate the Nrf2 signaling pathway, thus alleviating the oxidative damage caused by formaldehyde [Huang *et al.*, 2019].

17.6 The mechanism of preventing corruption by the deodorization of tea polyphenols

The anti-corruption and deodorization effects of tea polyphenols is due to their unique structure. The inhibitory effect of tea polyphenols on microbial corruption and growth can inhibit the ROS and RNS free radicals and oxidative stress produced during the corruption process of food, oil, meat, and biological samples caused by corruption. Tea polyphenols can be combined with odor substances to directly eliminate odor. Odor substances are generally divided into three categories: sulfur compounds, nitrogen compounds, and acids and alcohols. Methylmercaptan and nitrogen-containing malodorous substance trimethylamine are the main components of halitosis, and catechins have deodorizing effect on these two substances. The mechanism of catechin-eliminating trimethylamine, hydrogen sulfide, and indole odor was studied by gas chromatography. The results showed that the deodorization effect EGCG > EGC > ECG > EC and the deodorization rate gradually increased with time. After anti-corruptive and deodorization, the light absorption value of catechin decreased and new spots appeared on paper chromatography. It was believed that catechin reacted with odor components to generate new substances. Phenyl of flavonoid substances can combine with NH of odor substances, then phenolic groups can react with SH and NH of process odor substances resulting from the corruption in condensation, overlap, and recombination, and catechins can react with odor substances in esterification, transesterification, and other reactions. In addition, the elimination of odor substances by tea polyphenols is also related to electric charge. The nitrogen atom on ammonia, methylamine, and trimethylamine has a pair of lone electrons with a certain negative charge, while the B-ring hydroxyl of tea polyphenols can provide hydrogen protons with a positive charge, so these smelly substances can pair with tea polyphenols to eliminate odor.

Tea polyphenols can inhibit the generation of odor substances resulting from the corruption by inhibiting the activity of enzymes and antibacterial. The inhibition of tea polyphenols on methane production may be realized through the inhibition of microorganisms and their enzymes. Eight 30-day old piglets were fed with basic diet and 0.2% tea polyphenol enriched diet respectively for two weeks. The results showed that the number of lactic acid rod curd significantly increased (measured on the 7th and 14th day of feeding respectively) compared with that before adding tea polyphenol, while the number of bacteria and mycelia decreased. During the feeding period of tea polyphenols, the detection rate of lecithinase of *Clostridium* decreased. After treatment, the ammonia concentration in the feces decreased significantly, and the phenol, p-cresol, and fecal phenol in the feces decreased significantly, while the short chain fatty acid, acetic acid, and milk brain increased significantly. At the same time, the pH decreased, and the fecal odor decreased significantly. It can be seen that tea polyphenols play the anti-corruptive and deodorizing role by regulating intestinal flora and inhibiting odorizing bacteria. The ultimate root cause of halitosis and sweat odor is also the result of microorganisms and their enzymes decomposing organic substances to produce volatile odor substances. Tea polyphenols have broad-spectrum inhibition on microorganisms, which may inhibit or eliminate sweat odor. The anti-corruptive and deodorization effect of tea polyphenols can be formulated into a deodorant such as foot-odor inhibitor. When tea polyphenols and their oxidation products are used as dyes to dye bandages, cloth, and other fabrics, they can be used to produce deodorant and bacteriostatic products.

The order of the anti-corruptive and deodorant activity of the four catechins is EGCG > ECG > EGC > EC, indicating that the deodorant activity is related to the number of hydroxyl groups in catechins. The greater the number of hydroxyl groups, the stronger the anti-corruptive and deodorant ability. The degree of polymerization of tea polyphenols in tea samples is closely related to deodorization. To study the anti-corruptive and deodorization effect of various teas, black tea, green tea, oolong tea, and Baozhong tea were tested for their anti-corruptive and deodorization effect. Among them, black tea > oolong tea > Baozhong tea > green tea has the same change trend as the fermentation degree of the four teas and the polymerization degree of polyphenols. Although the content of catechins and tea polyphenols in green tea extracted with methanol is the highest, its anti-corruptive and deodorization effect is the worst. The anti-corruptive and deodorization ability is closely related to the degree of polymerization of tea polyphenols. Reducing the content of polyphenols in tea extracts with polyvinylpyrrolidone will reduce its anti-corruptive and deodorization effect. With the increase of polyvinylpyrrolidine, the total amount of phenolic compounds in green tea and black tea extracts decreased significantly, and the anti-corruptive and deodorization effect of green tea was lower than that of black tea, indicating that black tea and green tea had different anti-corruptive and deodorization modes. With the increase of pH, the anti-corruptive and deodorization effect is significantly enhanced, especially for green tea. The ability of tea to remove protein can be used as an indicator of anti-corruptive and deodorization ability. All four kinds of tea have the ability to precipitate protein. Black tea is the strongest, while green tea is the weakest. The coagulability of protein in tea extract also plays a role in anti-corruptive and deodorization. The coagulation force of protein in tea extract also plays a role in deodorization.

The anti-corruptive and deodorization mechanism of green tea extract can be summarized as follows:

1. The phenolic hydroxyl of tea polyphenols can be condensed with sulfhydryl, amino, ammonia, and other groups generated from the corruption process to neutralize and eliminate the odor free in the environment.
2. Tea polyphenols can improve the microbial environment of gastrointestinal tract, activate the growth of bifidobacteria, inhibit the growth of harmful bacteria, and play a role in regulating metabolites.
3. Tea polyphenols can inhibit the activity of some aminotransferases and reduce the production of ammonia compounds resulting from the corruption process.

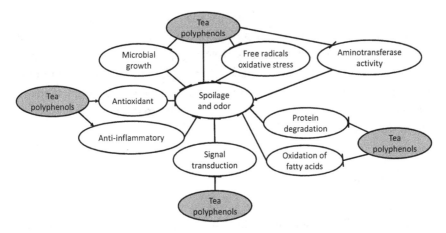

Figure 17-1. Schematic diagram about the mechanism of antisepsis and deodorization of tea polyphenols.

4. The excellent broad-spectrum antibacterial effect of tea polyphenols inhibits the microbial reproduction on the food surface, thus controlling the odor products of protein and other macromolecular substances degradation caused by microbial spoilage products.
5. The active electrophilic effect of tea polyphenols stops the chain oxidation reaction of fatty acids and inhibits the fat degradation caused by their putrefaction process.

The mechanism of antisepsis and deodorization of tea polyphenols can be shown in Figure 17-1.

17.7 Conclusion

From above discussion, we can know that spoilage and odor is closely related to oxidative stress and tea polyphenols play an important role in preservation, anti-corrosion, and deodorization due to their antioxidant properties. It does not only inhibit the growth of some bacteria that produce spoilage, eliminate ROS and RNS free radicals and oxidative stress generated in the process of spoilage, but also have remarkable antiseptic effect on the spoilage of various substances, such as oils, animal, and plant foods, as well as preserve the body tissues and remove the peculiar smell of the body and environment. Since tea polyphenols are natural antioxidants, they have no side effects. Research on its preservation mechanism will play an important role in preservation, anti-spoilage, and deodorization in the future.

References

Ahn DU, Kim SM. (1998) Prooxidant effects of ferrous iron, hemoglobin, and ferritin in oil emulsion and cooked-meat homogenates are different from those in raw-meat homogenates. *Poult Sci*, **77**(2), 348–355.

Bao Y, Ertbjerg P. (2015) Relationship between oxygen concentration, shear force and protein oxidation in modified atmosphere packaged pork. *Meat Sci*, **110**, 174–179.

Bao Y, Puolanne E, Ertbjerg P. (2016) Effect of oxygen concentration in modified atmosphere packaging on color and texture of beef patties cooked to different temperatures. *Meat Sci*, **121**, 189–195.

Bekhit AEA, Hopkins DL, Fahri FT, Ponnampalam EN. (2013) Oxidation processes in muscle systems and fresh meat: sources, makers, and remedies. *Compr Rev Food Sci Food Saf*, **12**(5), 565–597.

Bellés M, Alonso V, Roncalés P, Beltrán JA. (2017) Effect of borage and green tea aqueous extracts on the quality of lamb leg chops displayed under retail conditions. *Meat Sci*, **129**, 153–160.

Bing S, Zang Y, Li Y, Zhang B, Mo Q, Zhao X, Yang C. (2022) A combined approach using slightly acidic electrolyzed water and tea polyphenols to inhibit lipid oxidation and ensure microbiological safety during beef preservation. *Meat Sci*, **183**, 108643.

Carpentier A, Abreu S, Trichet M, Satiat-Jeunemaitre B. (2012) Microwaves and tea: new tools to process plant tissue for transmission electron microscopy. *J Microsc*, **247**(1), 94–105.

Chen H, Huang B. (2012) (–)-Epigallocatechin-3-gallate: a novel storage medium for avulsed teeth. *Dent Traumatol*, **28**(2), 158–160.

Cottrell JJ, Ponnampalam EN, Dunshea FR, Warner RD. (2015) Effects of infusing nitric oxide donors and inhibitors on plasma metabolites, muscle lactate production and meat quality in lambs fed a high quality roughage-based diet. *Meat Sci*, **105**, 8–15.

Delles RM, True AD, Ao T, Dawson KA, Xiong YL. (2016) Fibre type-dependent response of broiler muscles to dietary antioxidant supplementation for oxidative stability enhancement. *Br Poult Sci.*, **57**(6), 751–762.

Flemming J, Meyer-Probst CT, Speer K, Kölling-Speer I, Hannig C, Hannig M. (2021) Preventive applications of polyphenols in dentistry — a review. *Int J Mol Sci*, **22**(9), 4892.

Frank DC, Geesink G, Alvarenga TIRC, Polkinghorne R, Stark J, Lee M, Warner R. (2017) Impact of high oxygen and vacuum retail ready packaging formats lamb loin and topside eating quality. *Meat Sci*, **123**, 126–133.

Gramza-Michalowska A, Korczak J, Regula J. (2007) Use of plant extracts in summer and winter season butter oxidative stability improvement. *Asia Pac J Clin Nutr*, **16**(Suppl 1), 85–88.

Han DW, Park YH, Kim JK, Lee KY, Hyon SH, Suh H, Park JC. (2004) Effects of green tea polyphenol on preservation of human saphenous vein. *J Biotechnol*, **110**(2), 109–117.

Hasegawa Y, Kawasaki T, Maeda N, Yamada M, Takahashi N, Watanabe T, Iwasaki T. (2021) Accumulation of lipofuscin in broiler chicken with wooden breast. *Anim Sci J*, **92**(1), e13517.

Hazegh K, Fang F, Bravo MD, *et al.* (2021) Blood donor obesity is associated with changes in red blood cell metabolism and susceptibility to hemolysis in cold storage and in response to osmotic and oxidative stress. *Transfusion*, **61**(2), 435–448.

He L, Zou L, Yang Q, Xia J, Zhou K, Zhu Y, Han X, Pu B, Hu B, Deng W, Liu S. (2016) Antimicrobial activities of nisin, tea polyphenols, and chitosan and their combinations in chilled mutton. *J Food Sci*, **81**(6), M1466–M1471.

He X, Guo F, Liu B. (2018) Oolong tea and LR-White resin: a new method of plant sample preparation for transmission electron microscopy. *J Microsc*, **270**(2), 244–251.

Hou Q, Zhang CY, Zhang WG, Liu R, Tang H, Zhou GH. (2020) Role of protein S-nitrosylation in regulating beef tenderness. *Food Chem*, **306**, 125616.

Huang J, Lu Y, Zhang B, Yang S, Zhang Q, Cui H, Lu X, Zhao Y, Yang X, Li R. (2019) Antagonistic effect of epigallocatechin-3-gallate on neurotoxicity induced by formaldehyde. *Toxicology*, **412**, 29–36.

Hyon SH, Kim DH, Cui W, Matsumura K, Kim JY, Tsutsumi S. (2006) Preservation of rat aortic tissue transplant with green tea polyphenols. *Cell Transplant*, **15**(10), 881–883.

Ikeguchi R, Kakinoki R, Okamoto T, Matsumoto T, Hyon SH, Nakamura T. (2003) Green tea polyphenol: an experimental study in rats. *Exp Neurol*, **184**(2), 688–696.

Jiang Y, Su M, Yu T, Du S, Liao L, Wang H, Wu Y, Liu H. (2021) Quantitative determination of peroxide value of edible oil by algorithm-assisted liquid interfacial surface enhanced Raman spectroscopy. *Food Chem*, **344**, 128709.

Jia S, Huang Z, Lei Y, Zhang L, Li Y, Luo Y. (2018) Application of Illumina-MiSeq high throughput sequencing and culture-dependent techniques for the identification of microbiota of silver carp (Hypophthalmichthys molitrix) treated by tea polyphenols. *Food Microbiol*, **76**, 52–61.

Jeong S, Lee HG, Cho CH, Yoo S. (2021) Deodorization films based on polyphenol compound-rich natural deodorants and polycaprolactone for removing volatile sulfur compounds from kimchi. *J Food Sci*, **86**(3), 1004–1013.

Jung IH, Yun JH, Cho AR, Kim CS, Chung WG, Choi SH. (2011) Effect of (–)-epigallocatechin-3-gallate on maintaining the periodontal ligament cell viability of avulsed teeth: a preliminary study. *J Periodontal Implant Sci*, **41**(1), 10–16.

Kawazoe T, Kim H, Tsuji Y, Morimoto N, Hyon SH, Suzuki S. (2008) Green tea polyphenols affect skin preservation in rats and improve the rate of skin grafts. *Cell Transplant*, **17**(1-2), 203–209.

Kim H, Kawazoe T, Matsumura K, Suzuki S, Hyon SH. (2009) Long-term preservation of rat skin tissue by epigallocatechin-3-o-gallate. *Cell Transplant*, **18**(5), 513–519.

Kim YH, Huff-Lonergan E, Sebranek JG, Lonergan SM. (2010) Effects of lactate/phosphate injection enhancement on oxidation stability and protein degradation in early postmortem beef cuts packaged in high oxygen modified atmosphere. *Meat Sci*, **86**(3), 852–858.

Li H, Jiang N, Liu Q, Gao A, Zhou X, Liang B, Li R, Li Z, Zhu H. (2016) Topical treatment of green tea polyphenols emulsified in carboxymethyl cellulose protects against acute ultraviolet light B-induced photodamage in hairless mice. *Photochem Photobiol Sci*, **15**(10), 1264–1271.

Li YP, Liu R, Zhang WG, Fu QQ, Liu N, Zhou GH. (2014) Effect of nitric oxide on mu-calpain activation, protein proteolysis, and protein oxidation of pork during post-mortem aging. *J Agric Food Chem*, **62**(25), 5972–5977.

Li Y, Lei Y, Tan Y, Zhang J, Hong H, Luo Y. (2022) Efficacy of freeze-chilled storage combined with tea polyphenol for controlling melanosis, quality deterioration, and spoilage bacterial growth of Pacific white shrimp (Litopenaeus vannamei). *Food Chem*, **370**, 130924.

Liu X, Shen B, Du P, Wang N, Wang J, Li J, Sun A. (2017) Transcriptomic analysis of the response of Pseudomonas fluorescens to epigallocatechin gallate by RNA-seq. *PLoS One*, **12**(5), e0177938.

Liu R, Li YP, Zhang WG, Fu QQ, Liu N, Zhou GH. (2015) Activity and expression of nitric oxide synthase in pork skeletal muscles. *Meat Sci*, **99**, 25–31.

Lodhia P, Yaegaki K, Khakbaznejad A, Imai T, Sato T, Tanaka T, Murata T, Kamoda T. (2008) Effect of green tea on volatile sulfur compounds in mouth air. *J Nutr Sci Vitaminol (Tokyo)*, **54**(1), 89–94.

Lorenzo JM, Sineiro J, Amado IR, Franco D. (2014) Influence of natural extracts on the shelf life of modified atmosphere-packaged pork patties. *Meat Sci*, **96**(1), 526–534.

Matsumoto T, Kakinoki R, Ikeguchi R, Hyon SH, Nakamura T. (2005) Optimal conditions for peripheral nerve storage in green tea polyphenol: an experimental study in animals. *J Neurosci Methods*, **145**(1-2), 255–266.

Nakayama K, Kakinoki R, Ikeguchi R, Yamakawa T, Ohta S, Fujita S, Noguchi T, Duncan SF, Hyon SH, Nakamura T. (2010) Storage and allogeneic transplantation of peripheral nerve using a green tea polyphenol solution in a canine model. *J Brachial Plex Peripher Nerve Inj*, **5**, 17.

Nieto G, Martínez L, Castillo J, Ros G. (2017) Hydroxytyrosol extracts, olive oil and walnuts as functional components in chicken sausages. *J Sci Food Agric*, **97**(11), 3761–3771.

Nie X, Wang L, Wang Q, Lei J, Hong W, Huang B, Zhang C. (2018) Effect of a sodium alginate coating infused with tea polyphenols on the quality of fresh Japanese sea bass (Lateolabrax japonicas) fillets. *J Food Sci*, **83**(6), 1695–1700.

Niemeyer SH, Baumann T, Lussi A, Meyer-Lueckel H, Scaramucci T, Carvalho TS. (2021) Salivary pellicle modification with polyphenol-rich teas and natural extracts to improve protection against dental erosion. *J Dent*, **105**, 103567.

Pandey R, Ter Beek A, Vischer NO, Smelt JP, Kemperman R, Manders EM, Brul S. (2015) Quantitative analysis of the effect of specific tea compounds on germination and outgrowth of Bacillus subtilis spores at single cell resolution. *Food Microbiol*, **45**(Pt A), 63–70.

Pomothy JM, Gatt K, Jerzsele Á, Gere EP. (2021) The impact of quercetin on a porcine intestinal epithelial cell line exposed to deoxynivalenol. *Acta Vet Hung*, **68**(4), 380–386.

Ponnampalam EN, Hopkins DL, Giri K, Jacobs JL, Plozza T, Lewandowski P, Bekhit A. (2017) The use of oxidative stress biomarkers in live animals (in vivo) to predict meat quality deterioration postmortem (in vitro) caused by changes in muscle biochemical components. *J Anim Sci*, **95**(7), 3012–3024.

Qin YY, Yang JY, Lu HB, Wang SS, Yang J, Yang XC, Chai M, Li L, Cao JX. (2013) Effect of chitosan film incorporated with tea polyphenol on quality and shelf life of pork meat patties. *Int J Biol Macromol*, **61**, 312–316.

Sato J, Nakayama M, Tomita A, Sonoda T, Miyamoto T. (2020) Difference in the antibacterial action of epigallocatechin gallate and theaflavin 3,3′-di-o-gallate on Bacillus coagulans. *J Appl Microbiol*, **29**(3), 601–611.

Shao P, Niu B, Chen H, Sun P. (2018) Fabrication and characterization of tea polyphenols loaded pullulan-CMC electrospun nanofiber for fruit preservation. *Int J Biol Macromol*, **107**(Pt B), 1908–1914.

Shi H, Wang J, Zhao B-L, Xin W. (1995a) ESR study of erythrocyte membrane lipid membrane protein interaction in patients with renal anemia. *J Biochem Biophys*, **27**, 323–327.

Shi H-L, Zhao B-L, Xin W-J. (1995b) Scavenging effects of Baicalin on free radicals and its protection on erythrocyte membrane from free radical injury. *Biochem Molec Biol Intern*, **1**, 981–994.

Sun K, Wang S, Ge Q, Zhou XI, Zhu J, Xiong G. (2021) Antimicrobial and preservative effects of the combinations of nisin, tea polyphenols, rosemary extract, and chitosan on pasteurized chicken sausage. *J Food Prot*, **84**(2), 233–239.

Tang Y, Chen X, Yao H, Xie J, Shi W, Lu Y, Deng S, Tao N, Xu C. (2022) Development of a bifunctional edible coating for formaldehyde scavenging and preservation of aquatic products. *J Sci Food Agric*, **102**(5), 1958–1967.

Tsikas D. (2017) Assessment of lipid peroxidation by measuring malondialdehyde (MDA) and relatives in biological samples: analytical and biological challenges. *Anal Biochem*, **524**, 13–30.

Tu Y, Wang Y. (2003) The fresh-keeping effect of tea polyphenols. In *Tea Polyphenol Chemistry*, Yang X, Wang Y, Chen L (eds.), Shanghai Science and Technology Press.

Wang L, Xu J, Zhang M, Zheng H, Li L. (2022) Preservation of soy protein-based meat analogues by using PLA/PBAT antimicrobial packaging film. *Food Chem*, **380**, 132022.

Wang WD, Sun YE. (2016) Preservation effect of meat product by natural antioxidant tea polyphenol. *Cell Mol Biol (Noisy-le-grand)*, **62**(13), 44–48.

Wang Y, Li F, Zhuang H, Li L, Chen X, Zhang J. (2015) Effects of plant polyphenols and alpha-tocopherol on lipid oxidation, microbiological characteristics, and biogenic amines formation in dry-cured bacons. *J Food Sci*, **80**(3), C547–C555.

Warner RD, Dunshea FR, Ponnampalam EN, Cottrell JJ. (2005) Effects of nitric oxide and oxidation in vivo and postmortem on meat tenderness. *Meat Sci*, **71**(1), 205–217.

Wittayarat M, Kimura T, Kodama R, Namula Z, Chatdarong K, Techakumphu M, Sato Y, Taniguchi M, Otoi T. (2012) Long-term preservation of chilled canine semen using vitamin C in combination with green tea polyphenol. *Cryo Letters*, **33**(4), 318–326.

Wittayarat M, Panyaboriban S, Kupthammasan N, Otoi T, Chatdarong K. (2022) Effects of green tea polyphenols and alpha-tocopherol on the quality of chilled cat spermatozoa and sperm IZUMO1 protein expression during long-term preservation. *Anim Reprod Sci*, **237**, 106926.

Xing T, Zhao X, Wang P, Chen H, Xu X, Zhou G. (2017) Different oxidative status and expression of calcium channel components in stress-induced dysfunctional chicken muscle. *J Anim Sci*, **95**(4), 1565–1573.

Yao SG, *et al.* (1995) J of Korea society of food and nutrtion. **24**, 293.

Yi G, Grabež V, Bjelanovic M, *et al.* (2015) Lipid oxidation in minced beef meat with added Krebs cycle substrates to stabilise colour. *Food Chem*, **187**, 563–571.

Ye J, Wang S, Lan W, Qin W, Liu Y. (2018) Preparation and properties of polylactic acid-tea polyphenol-chitosan composite membranes. *Int J Biol Macromol*, **117**, 632–639.

Yi S, Li J, Zhu J, Lin Y, Fu L, Chen W, Li X. (2011) Effect of tea polyphenols on microbiological and biochemical quality of Collichthys fish ball. *J Sci Food Agric*, **91**(9), 1591–1597.

Wang Z, Zhou H, Zhou K, Tu J, Xu B. (2022) An underlying softening mechanism in pale, soft and exudative — like rabbit meat: the role of reactive oxygen species — generating systems. *Food Res Int*, **151**, 110853.

Zhang C, Liu R, Wang A, Kang D, Zhou G, Zhang W. (2018) Regulation of calpain-1 activity and protein proteolysis by protein nitrosylation in postmortem beef. *Meat Sci*, **141**, 44–49.

Zhang J, Zhao B-L, Xin W. (1992a) Using ESR spin trapping technique to study the free radicals produced by fe2+ induced peroxylinoleic acid. *J Biochem Biophys*, **24**, 247–250.

Zhang J, Zhao B-L, Guo Y, Xin W. (1992b) ESR Study on the starting mechanism of linoleic acid peroxidation by cu/vitamin C. *J Biophys*, **8**, 492–500.

Zhao B-L. (1999) Oxygen free radicals and natural antioxidants (First Edition). Science Press, Beijing.

Zhao B-L. (2007) Free radicals and natural antioxidants and health, China Science and Culture Press, Hong Kong.

Zhao J, Lv W, Wang J, Li J, Liu X, Zhu J. (2013) Effects of tea polyphenols on the post-mortem integrity of large yellow croaker (Pseudosciaena crocea) fillet proteins. *Food Chem*, **141**(3), 2666–2674.

Zhao W, Yu D, Xia W. (2021) Vacuum impregnation of chitosan coating combined with water-soluble polyphenol extracts on sensory, physical state, microbiota composition and quality of refrigerated grass carp slices. *Int J Biol Macromol*, **193**(Pt A), 847–855.

Zhou SH, Zhen P, Li SS, Liang XY, Gao MX, Tian Q, Li XS. (2015) Allograft pretreatment for the repair of sciatic nerve defects: green tea polyphenols versus radiation. *Neural Regen Res*, **10**(1), 136–140.

Chapter 18

Cessation and Detoxification Effect of Tea Polyphenols on Tobacco Smoke

Baolu Zhao

Institute of Biophysics, Chinese Academy of Sciences, Beijing, China

18.1 Introduction

As a special consumer's goods, tobacco industry has become an important part of the national economy of all countries in the world and one of the important sources of finance of all countries in the world. However, cigarette smoking can cause a series of serious diseases. The harm of cigarette smoking to health has become a major public health problem faced by mankind. It is universally acknowledged that cigarette smoking is harmful to health, and quitting cigarette smoking is well recommended. However, due to the physiological and psychological enjoyment of cigarette smoking, cigarette smoking is addictive, which also makes it difficult to quit cigarette smoking. Epidemiological research show that cigarette smoking not only causes bronchial and lung injury, but also lead to the most terrible human disease — cancer and cardiovascular disease, with the highest mortality. Now, all over the world, there is a cigarette smoking cessation campaign and the signs of no cigarette smoking in public places can be seen everywhere. Nitrosamine and benzopyrene are the two most

927

carcinogenic substances in cigarette smoking tar. Most people who die from lung cancer are caused by nitrosamine and benzopyrene generated form cigarette smoking.

In the past, people have always believed that the toxicity of cigarette smoking comes from nicotine, but this is not-entirely correct. Cigarette smoking is a very complex combustion process. There are a large number of free radicals in the gas phase and tar of cigarette smoking. These harmful substances cause oxidative stress and attack cell components directly and indirectly, which may be the key cause of various diseases. Free radicals, tar, and carbon monoxide are the most toxic in the large number of harmful substances produced in the process of tobacco ignition. How to reduce the harm of cigarette smoking toxic substances to smokers has become an urgent task for tobacco scientists, medical workers, and technicians.

Many studies have shown that tea and its effective components, tea polyphenols and theanine, can eliminate harmful free radicals produced by tobacco combustion and oxidative stress and various diseases are caused by tobacco combustion. This chapter discusses the free radicals in cigarette smoking, oxidative stress, and their pathogenic mechanism, and discusses the methods to eliminate harmful free radicals in cigarette smoking, prevent damage of harmful free radicals to human health by tea polyphenols, and quit smoking.

18.2 The harm of cigarette smoking

The warning "tobacco seriously damages your health" from the US Department of health is written on Marlboro cigarette boxes in the United States, a clear message of the harm that cigarette smoking has to one's health. The warning "cigarette smoking causes lung cancer, heart disease, emphysema and may complex pregnancy" written on the cigarette box in Canada indicates the health complications that can arise from smoking. In addition, western countries have recently set off a huge campaign to encourage cessation of cigarette smoking. For example, cigarette smoking is prohibited in all public places and offices, and violators will be fined a heavy sum of money. Additionally, some companies even set heavy punishment of dismissal to employees who smoke in the office. The promulgation and implementation of the tobacco convention adopted by the World Health Organization of the United Nations further shows that cigarette smoking is harmful to health. Cigarette smoking has been linked to many life-threatening diseases including heart disease, cancer, and chronic obstructive pulmonary disease [Chung *et al.*, 1999; Zhao, 1988, 1989; Gu *et al.*, 2009]. There are about 1.25 billion smokers in the world and 5 million die every year because of smoking-related diseases.

Through epidemiological investigation and scientific experiments, it is indeed found that cigarette smoking can not only cause bronchitis and lung injury, but also lead to the most terrible human disease, cancer and heart disease, with the highest mortality. What substance in cigarette smoking causes these terrible diseases? Recent studies have shown that cigarette smoking contains a large number of free radicals, which can directly or indirectly attack and damage cells, leading to many disease [Guo *et al.*, 2007].

18.2.1 *Cigarette smoking, free radicals, and carcinogenesis*

Among the multiple components of tobacco smoke, 20 carcinogens caused lung tumors in laboratory animals or humans and are, therefore, likely to be involved in lung cancer induction. Of these, polycyclic aromatic hydrocarbons and the tobacco-specific nitrosamine 4-(methylnitrosamino)-1-(3-pyridyl)-1-butanone are likely to play major roles. Epidemiological investigation shows that about 30% of all cancer deaths are caused by cigarette smoking, and smokers die of lung cancer more than 10 times that of non-smokers. If they do not smoke, it is estimated that 80% of lung cancer deaths can be avoided. An estimated 50–70% of oral cancer deaths are related to cigarette smoking, the risk of laryngeal cancer in heavy smokers is 20–30 times than that of non-smokers while the incidence rate of esophageal cancer is 4–10 times higher than that of non-smokers. The reason is probably free radicals produced by cigarette smoking. The toxicity of free radicals produced by cigarette smoking can be divided into direct and indirect toxicity to cause cancer. The so-called direct toxicity is that some free radicals directly produced in cigarette smoking can damage cell components, causing mutation and cancer formation, such as polycyclic aromatic hydrocarbons (PAHs) free radicals in tar phase, which are a kind of direct carcinogen. Quinone/semiquinone radical (Q•/QH•) free radicals in tar phase are easy to self-oxidize to form reactive oxygen species (ROS) free radicals. Many studies have shown that reactive oxygen free radicals can cause cell damage and participate in the formation and development of cancer. It can also directly bind to cell DNA, leading to cell transformation. Alkyl and alkoxy radicals in the gas phase of cigarette smoking are highly reactive, which can cause lipid peroxidation of cell membrane, damage cell membrane, and lead to a variety of diseases and cancers. The so-called indirect toxicity refers to substances in the tar phase and gas phase of cigarette smoking that can cause polymorphonuclear leukocytes to aggregate and activate in the lung and trachea, produce respiratory burst, and release a large number of ROS and reactive nitrogen species (RNS) free radicals. These ROS and RNS free radicals and the free radicals directly produced in cigarette smoking can damage cells together, leading to the occurrence and development of cancer. Tobacco smoke carcinogens interact with DNA and cause genetic changes-mechanisms that are reasonably well understood and the less well-defined relationship between exposure to specific tobacco smoke carcinogens and mutations in oncogenes and tumor suppressor genes. Tobacco smoke generates free radicals and oxidative damage associate with carcinogenesis [Zhao, 1988, 1994; Zhang *et al.*, 2019].

18.2.2 *Cigarette smoking, free radicals, and myocardial-vascular disease*

Studies have shown that cigarette smoking is a major factor in myocardial infarction. It can act alone or in synergy with other factors such as hypertension and hypercholesterolemia. The risk of local myocardial ischemic heart disease in smokers has been found all over the world, which is directly proportional to the number of smokers. The mortality of local myocardial ischemic heart disease in smokers who quit cigarette smoking is lower than that of continuous smokers. It is generally believed that CO plays is a major role in the pathogenesis of myocardial ischemic heart disease. CO is not only an important component of cigarette smoking gas, but also an index to test cigarettes. Animal experiments show that the cholesterol content in aorta increases after exposure to low concentration CO, which may lead to atherosclerosis. Cigarette smoking can significantly reduce the rate of myocardial mitochondrial respiration and oxidative phosphorylation, mitochondrial edema, abnormal size, outer cell membrane damage, and increase lipid content and autophagy lysosome activity. More and more experiments show that oxygen free radicals play a very important role in the occurrence and development of heart disease. It is likely that CO causes myocardial ischemic heart disease through the mechanism of oxygen free radical. Inhalation of CO mainly causes myocardial hypoxia. Under hypoxia, ATP is transformed into xanthine and hypoxanthine, and xanthine dehydrogenase is transformed into xanthine oxidase. When cigarette smoking is stopped, the supply of oxygen is restored, and xanthine oxidase immediately catalyzes xanthine to produce uric acid. At the same time, a large number of superoxide anion free radicals are produced. In the presence of metal ions, superoxide anion disproportionation generates hydrogen peroxide and •OH radicals which is generated by Fenton reaction, these oxygen radicals further react with unsaturated fatty acids to form lipid radicals. The primary and secondary free radicals produced above are very active and can directly or indirectly damage cell components, leading to the occurrence and development of heart disease. In the gas phase of cigarette smoking, there are alkoxy and alkperoxy groups. They can react with trachea or lung cells or pass through lung cell membrane to cause damage to blood components, so they are associated with atherosclerosis. Q•/QH•, the oxygen free radical produced by self-oxidation and the reactive oxygen free radicals produced by the aggregation and activation of PMN cells in the lung caused by cigarette smoking also play key roles in the occurrence and development of heart disease. In the ischemic myocardium, the concentration of glutathione and the activities of SOD and glutathione peroxidase decrease, which reduces the ability of cardiomyocytes to resist oxygen free radical injury, coupled with the increased production of free radicals caused by cigarette smoking, these factors contribute to the increase risk of heart damage [Zhao 1989a, 1989b, 1996; Zhao *et al.*, 1989a, 1989b, 1996].

18.2.3 *Cigarette smoking, free radicals, and lung diseases*

Oxygen free radicals and their metabolites, collectively described as ROS and RNS, have been implicated in the pathogenesis of many diseases. The pulmonary system is particularly vulnerable to ROS and RNS-induced injury because of its continuous exposure to toxic pollutants from a wide variety of sources in the ambient air. ROS and RNS-induced lung injury at different target levels may contribute to similar patterns of cell injury and alterations at the molecular level by initiation, propagation, and auto-catalytic chain reactions. Intracellular signals, activation and inactivation of enzymes, stimulation, secretion, and release of pro-inflammatory cytokines, chemokines, and nuclear factor activation and alterations are also common events. Understanding the interactions of these intricate mechanistic events is important in the prevention and amelioration of lung injury that results from acute and chronic exposures to toxins in ambient air, especially cigarette smoking. Cigarette smoke contains a large variety of compounds, including many oxidants and free radicals that are capable of initiating or promoting oxidative damage. Also, oxidative damage may result from ROS and RNS generated by the increased and activated phagocytes following cigarette smoking. *In vitro* studies are generally supportive of the hypothesis that cigarette smoke can initiate or promote oxidative damage. However, information obtained from *in vivo* studies is inconclusive. Contrary to expectations, the levels of lipid peroxidation products were found to be decreased or unchanged in the lungs of chroni-cally smoked rats. Metabolic adaptation, such as accumulation of vitamin E in the lung, and increased activities of superoxide dismutase in alveolar macrophages and pulmonary tissues of chronically smoked animals may enable smoked subjects to counteract oxidative stress and to resist further damage to smoke exposure [Chow, 1993].

α-1-antiproteinase is produced from the liver and released into extra-cellular fluid to inhibit several proteolytic enzymes. If their activity is not inhibited, it will hydrolyze elastin in the lung. Elastin is the main compo-nent in lung elastic fibers, which is very important for lung respiration. Congenital people with α-1-antiproteinase deficiency often suffer from emphysema and lose vital capacity. Epidemiological statistics show that even if α-1-antiproteinase functions normally, cigarette smoking can also easily infect a person and cause emphysema. The reason may be that ciga-rette smoking causes pulmonary edema due to inactivation of α-1-antiprotease, which prevents it from inhibiting the elastase released by neutrophils. Sometimes the antiprotease activity in the lung lavage fluid of smokers is lower than normal, therefore, cigarette smoking makes neutrophils and megaphils aggregate and activate in the lung and release elastase. At the same time, cigarette smoking causes α inactivation of 1-antiproteinase to become unable to inhibit these elastases, which may be an important cause of lung injury caused by cigarette smoking. Various oxygen free radicals produced by smoking can inactivate α-1-antiprotein, and the reaction of NO- and NO_2-generated nitrite can also cause the denaturation of α-1-antiprotease.

The above results suggest a mechanism of lung injury and lung disease caused by smoking gas-phase free radicals. After cigarette smoking gas-phase substances enter the lungs, they first cause lipid peroxidation of lung cells, at the same time, make macrophages gather in the lungs. In this way, the oxidized cell membrane phospholipids can stimulate the respiratory burst of macrophages gathered in the lungs to produce a large number of ROS and RNS free radicals, further causing lung cell peroxidation damage, producing a vicious circle, and leading to the occurrence and development of the disease.

18.3 Cigarette smoking and oxidative stress

Many studies have shown that cigarette smoke can cause oxidative stress. Oxidative stress further damages cell membrane, leading to a series of cell damage and the occurrence of various serious diseases.

18.3.1 *Cigarette smoking generates harmful free radicals*

Cigarette smoking is a very complex physical and chemical process. A burning cigarette is a micro chemical plant that can produce thousands of substances, including a large number of free radicals. These free radicals are distributed in the tar and gas phase of cigarette smoking and play different roles in the pathogenesis. The process of cigarette smoking is a process of generating a large number of new compounds from the combustion of the raw materials that make up cigarettes. The formation of these compounds changes with the combustion in different parts of cigarettes and different temperatures. A burning cigarette can reach a maximum temperature of about 900°C, while the smoke from the back of the cigarette is only 50–80°C (Fig. 18-1). There is a temperature gradient from the

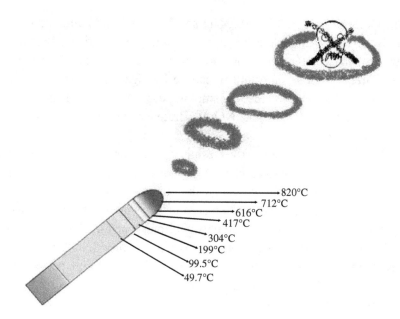

Figure 18-1.   Combustion process and temperature distribution of cigarette smoking.

burning cone to the incombustible cut tobacco. The main components of cigarettes are carbohydrates (48.3%), non-aliphatic organic acids (11.7%), oxygenates (11.1%), total resin (9.8%), and ash (13.2%). If it burns completely above 600°C, it will mainly produce H_2O, CO_2, and NO_2. However, since the oxygen has been basically exhausted on the burning cone surface, the smoking process is not a complete combustion process. When hot air generated by combustion passes through the cut tobacco, distillation and a series of redox processes occurs. When this process is at 450°C, the tobacco is cooked and at 300°C, various vapors from the tobacco condense to form flue gas. Some tobacco components are particularly easy to generate polycyclic aromatic hydrocarbons and olefins at high temperature, while some components are transferred into flue gas intact. Usually, substances with high volatility can be directly distilled out and transferred into the flue gas, while substances with low volatility can enter the flue gas after chemical changes occur through thermal cracking. Some substances are oxidized at the beginning of combustion, but are reduced after entering the flue gas flow, such as CO_2 into CO. In this complex process, thousands of substances are produced. With the hot flue gas flow moving along the smoke column, the temperature rapidly decreases from 800°C to the temperature of the surrounding environment in a few hundredths of a second, forming aerosols and producing a large number of nuclear particles. These include incomplete combustion of organic matter, small fragments, carbon, ash, ionized molecules, or nonvolatile matter splashed from the combustion zone. The diameter of these particles is 0.1–1 mm and they account for 8% of the flue gas. In this complex redox process, electrons are easily converted between these substances to form a variety of stable and unstable products, free radicals, which are distributed in the gas phase of cigarette smoking, and tar [Pryor 1983; Zhao *et al.*, 1990].

We used ESR spin capture technique to study the free radicals produced during tobacco combustion. The distance from the burning point of cigarette to the spin trapping solution is about 300 mm, which is equivalent to the distance from the burning point of cigarette to human lung cells. Alkoxy radical is a short-lived radical and its lifespan is less than 1 second. Free radicals with such a short lifespan cannot decay over such a long distance. Therefore, it is conceivable that alkoxy free radicals are formed instantaneously before entering the trap (or lung cells). Pryor *et al.* [1984] put forward a steady-state hypothesis of the formation of cigarette smoking gas-phase free radicals and confirmed it using some model compounds. They believe that the free radicals in the gas phase of cigarette smoking are constantly forming and decaying. The initial combustion of cigarettes produces nitric oxide, which is not highly reactive with most organic substances, but reacts easily with oxygen in the flue gas stream to produce nitrogen dioxide, which reacts easily with olefins in the flue gas to produce alkyl free radicals. Alkyl radicals are very easily oxidized to form alkoxy radicals. They used nitric oxide/isopropyl/air mixture to pass through the PBN solution of benzene to obtain ESR spectra similar to cigarette smoking gas-phase radicals (Fig. 18-2) [Zhao *et al.*, 1990; Church & Pryor 1985; Halpern & Knieper, 1985].

$$NR + O_2 \text{ --------> } NO$$
$$2NO + O_2 \text{ ---------->} 2NO_2$$
$$NO_2 + R \text{ ------------> } RO_2 + R\bullet$$
$$R\bullet + O_2 \text{ ------------> } ROO\bullet$$
$$RO_2\bullet + NO \text{ --------> } RO\bullet + NO_2$$

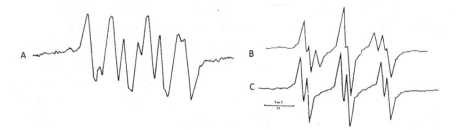

Figure 18-2. Free radicals directly produced by smoking and released by pulmonary inflammatory macrophages due to smoking. (A) ESR spectum of free radicals trapped by DMPO; (B) PBN in one minute; (C) PBN in five minutes.

Since the particles in cigarette smoking smoke are greater than 0.1 μM, 99% tar in cigarette smoking can be collected with a standard filter paper (Cambridge), including several particularly stable free radicals, which can be directly observed by electron spin resonance spectroscopy (ESR). Through ESR spectrum analysis, it was found that the stable free radicals mainly come from quinone/semiquinone radical (Q•/QH•). The concentration of polycyclic aromatic hydrocarbons, carbon, and phosphorus radicals is relatively low in tar, accounting for only 15%, especially the latter two types of radicals. The concentration of free radicals in each cigarette tar is about 6×10^{14} free radicals. Polycyclic aromatic hydrocarbon free radical is a kind of direct carcinogen. Although its ESR signal should not be saturated, it should not be observed at room temperature. The main free radical component in tar, Q•/QH•, is about 85% and its ESR signal is easy to be saturated and easy to observe at low temperature. It is a very important free radical, which can easily produce oxygen free radical by self-oxidation, resulting in a series of toxicological reactions [Zhao, 1988, 1994].

Cigarette smoking generates many free radicals in the gas phase. Most of the free radicals in the gas phase of cigarette smoking are transient unstable free radicals, which cannot be observed directly by ESR spectrometer. It is necessary to capture and convert the unstable free radicals into a spin adduct that can be detected by ESR spectrometer. We used spin trapping agents PBN and DMPO to capture free radicals in the gas phase of cigarette smoking, mainly alkoxy (RO•) and alkane (R•) radicals (Fig. 18-2). These free radicals are constantly formed in the popular process of the airflow from cigarette smoking combustion. Most of the free radicals in the cigarette smoking gas phase are transient unstable free radicals, mainly nitric oxide, nitrogen dioxide, and RO• and R• radicals. Firstly, nitrogen-containing substances oxidize and generate a large amount of NO during cigarette smoking combustion and generate more reactive NO_2 free radicals when encountering oxygen. It can react with olefins generated by cigarette smoking combustion to produce alkane free radicals R•, which can react with O_2 to form alkane peroxy free radicals ROO• and react with NO to form alkoxy radical. These free radicals will react with cellular components when they encounter cells. They not only react with lipid peroxidation of cell membrane, but also oxidize proteins and nucleic acids. It is precisely because of the high reactivity of free radicals and the damage to cells that allows them to cause various diseases [Pryor, 1983, 1984; Zhao, 1988, 1994].

18.3.2 *Cigarette smoking causes lipid peroxidation*

Why does cigarette smoking cause lung injury, cancer, and heart disease? This is a very complex problem. In addition to the reasons analyzed above, there are two points that need to be explained in depth. Cigarette smoking destroys the balance of free radicals in the body. The regulation and modification of low-density lipoprotein (LDL) by free radicals is an important factor in the formation of atherosclerosis. An experiment found that eating too much fat and high cholesterol substances can increase LDL in the body, while cigarette smoking can promote the peroxidation of LDL and the phagocytosis of LDL by macrophages. In an experiment, LDL from a person who smoked 4–7 cigarettes continuously within 90 minutes was separated by blood sampling, the test found that the lipid peroxide of LDL increased by 2–4 times and the phagocytosis of macrophages on LDL increased by 1.5–2 times. If antioxidants were taken before cigarette smoking, LDL peroxidation and phagocytosis of LDL by phagocytes decreased by 50% and 70%, respectively. On the contrary, if you continuously eat fish oil with high fat and cholesterol substances before cigarette smoking and take 10 mg every day, the lipid peroxide in the body will increase by 34–41% and the phagocytosis of macrophages on LDL will increase by 65–73%. If 400 mg vitamin E is also taken while eating fish oil per day, the increase of lipid peroxide and phagocyte phago-cytosis of LDL *in vivo* will be significantly reduced [Zhao, 1988, 1989; Zhao *et al.*, 1995; Zhao & Zhang, 1994].

The cigarette smoking was passed through liposomes and the lipid hydroperoxide TBA reactants and conjugated dienes were measured. With the extension of cigarette smoking time, TBA reactants and conjugated dienes in cell membrane increased, which showed that cigarette smoking gas-phase free radicals could cause lipid peroxidation of cell membrane. The free radical oxidation produced by cigarette smoking consumes a lot of antioxidants in the body. Scientific research shows that the levels of antioxidants such as vitamin C and vitamin E in smokers' plasma and lung lavage fluid are significantly lower than those in non-smokers because the free radicals produced by cigarette smoking oxidized antioxidants in the body. We know that the gas phase of cigarette smoking contains a lot of nitric oxide, nitrogen dioxide, and alkoxy free radicals, which are all highly oxidizing. When they enter body fluid and blood, they will oxidize cell components and antioxidants in body fluid. In addition, cigarette smoking can stimulate the respiratory burst of macrophages to produce more oxygen free radicals. These free radicals are highly-oxidizing oxidants which can react with antioxidants in the body to oxidize and con-sume them. Therefore, the levels of antioxidants in body fluid and plasma of smokers will be reduced, and the ability of the body to produce oxygen free radicals will be strengthened. The oxidation and reduction system in the body should be balanced in order to maintain a normal state of health. Once the antioxidant system in the body is damaged, the antioxidants are reduced, and the system in the body is activated to produce too many oxygen free radicals, which further loses the balance between the produc-tion and elimination of oxygen free radicals in the body, resulting in the damage of oxygen free radicals to cellular components and the occurrence of diseases [Zhao, 1988, 1989; Zhao *et al.*, 1995].

We studied the lipid peroxidation of cell membrane by using ESR spin trapping technology and it was found that with the extension of cigarette smoking time, the captured lipid free radicals increased at the same time [Yan *et al.*, 1991a, 1991b, 1991c]. The gas phase of cigarette smoking contains a large number of oxidizing substances, including alkoxy, alkper-oxy, NO_2, and other free radicals. We directly captured the lipid free radi-cals produced by lipid peroxidation of cell membrane caused by cigarette smoking gaseous substances and detected the final product of lipid peroxidation malondialdehyde (MDA). It was found that with the extension of cigarette smoking time, the levels of lipid free radicals and lipid peroxidation also increased, which itself would cause the damage of cell membrane. More seriously, lipid peroxidation membrane can make macrophages gather and activate in the lung, produce respiratory burst, and release a large number of ROS and RNS free radicals. We used spin trapping technology to capture oxygen free radicals produced by lipid peroxidation film-stimulated PMN respiratory burst caused by cigarette smoking gaseous substances (Fig. 18-2). These oxygen free radicals can cause damage to cell components alone, or they can also produce toxic effects on cells together with the free radicals produced by cigarette smoking. The study of lung cell membrane shows that free radicals in cigarette smoking not only change the fluidity of cell membrane, but also change the conformation of membrane protein.

18.3.3 *Cigarette smoking causes oxidative stress by respiratory burst of macrophages*

A major toxic target of cigarette smoking is lung injury, resulting in bronchitis and lung cancer. Since gaseous substances of cigarette smoking has direct contact with lung cells, cigarette smoking can cause macrophages to gather and activate in the lung, produce respiratory burst, release a large number of ROS and RNS free radicals, and cause more serious damage to lung cells. We studied the effect of cigarette smoking gaseous substances on macrophage respiratory burst and found that cigarette smoking gaseous substances can directly affect the process of macrophage respiratory burst, and can also induce macrophage respiratory burst by oxidizing cell membrane. Smoking stimulates macrophages to release ROS and RNS free radicals and leads to oxidative stress in two ways, one is direct and the other is indirect.

1. Oxidative stress on the respiratory burst of macrophages induced by cigarette smoking gaseous substances

Cigarette smoking gas-phase free radicals indirectly stimulate macrophages to produce free radicals. After smoking, macrophages cannot detect any ESR signal from macrophages without PMA stimulation, which shows cigarette smoking gaseous substances cannot directly stimulate macrophages to produce oxygen free radicals. When polymorphonuclear leukocytes were stimulated with PMA after cigarette smoking, aerobic free radicals were produced, but the signal intensity decreased significantly. Moreover, with the extension of cigarette smoking time, oxygen free radicals gradually decreased. If we follow the kinetics of oxygen free radicals produced by macrophages, we can find that the kinetics of oxygen free radicals produced by macrophages after cigarette smoking is significantly different from the macrophages without cigarette smoking. The oxygen free radicals produced by the former increased gradually with the extension of cigarette smoking time and then decreased gradually. Later, it reached the maximum at the beginning and then gradually decreased, indicating that cigarette smoking gaseous substances changed the respiratory burst process of macrophages [Yan *et al.*, 1992].

2. Lipid peroxidation of polymorphonuclear leukocytes induced by cigarette smoking vapor free radicals

After cigarette smoking, the level of lipid peroxidation of human polymorphonuclear leukocytes was measured. The results showed that compared with the control, the lipid peroxidation reactants increased significantly, and gradually increased with the extension of cigarette smoking time and there was a good correlation, indicating that the polymorphonuclear leukocytes did have lipid peroxidation after cigarette smoking. Smoking gas phase contains a lot of nitrogen dioxide and oxygen center free radicals, which should be highly oxidizing and would lead to lipid peroxidation, while cigarette smoking solid phase mainly contains quinone free radicals, which should be highly reducing. However, some experiments have found that both cigarette smoking solid-phase and gas-phase substances show oxidation [Yang *et al.*, 1992a, 1992b, 1992c, 1993a, 1993b, 1993c].

Our results show that cigarette smoking gaseous substances are oxidizing to polymorphonuclear leukocytes. Combined with the results obtained from liposomes in the previous section, it can be confirmed that cigarette smoking gaseous substances are highly oxidizing. This may explain why cigarette smoking alone cannot stimulate polymorphonuclear leukocytes to produce respiratory burst, but affect the dynamic process of respiratory burst. As MDA produced by lipid peroxidation can crosslink membrane proteins and change their conformation, this results in the change or even loss of function. NADPH oxidase, which is responsible for respiratory burst, is located on the cell membrane and may be the first target of lipid peroxidation.

Since cigarette smoking gaseous substances cannot directly stimulate the respiratory burst of polymorphonuclear leukocytes to produce oxygen free radicals, it will stimulate polymorphonuclear leukocytes to produce respiratory burst through other ways. The above results show that cigarette smoking gaseous substances can cause liposome lipid peroxidation. Lipid peroxidation cannot stimulate polymorphonuclear leukocytes. Therefore, we studied the stimulating effect of lipid peroxidated liposomes on rat granulocytes. Liposomes with concentration of 1 mg/ml were incubated at 37°C after cigarette smoking for 20 seconds. With the extension of incubation time, the concentration of lipid peroxidation increased. If the liposome is incubated with rat granulocytes at 37°C and then stimulated by PMA, as demonstrated by lucigenin-amplified chemiluminescence and electron spin resonance (ESR) spin-trapping with 5,5-dimethyl-1-pyrroline N-oxide (DMPO), the oxygen free radicals produced by DMPO capture are significantly higher than those produced by granulocytes in the control group. With the extension of incubation time, the effect first strengthens, reaches the maximum value at 15 minutes, and then begins to weaken. When the incubation time is greater than 25 minutes, the production of oxygen free radicals by granulocytes began to decrease [Yang *et al.*, 1993b]. The content of lipid peroxide increased with the increase of concentration. The liposomes treated with gas phase smoke were incubated with rat granulocytes at 37°C for 15 minutes and then stimulated with PMA to test the oxygen free radicals produced by respiratory burst. It was found that the oxygen free radicals produced increased significantly and increased with the increase of lipid peroxide concentration compared with the control group. When the lipid concentration was 1 mg/ml, the oxygen free radical reached the maximum, and then began to decrease with the increase of concentration. When the lipid concentration was greater than 15 mg/ml, this peroxidated liposome showed inhibitory effect on the production of oxygen free radicals by rat granulocytes.

ROS and RNS free radicals, contained in cigarette smoke and compromised phagocytic antimicrobial activities including those of polymorphonuclear leukocytes (PMNs), have been implicated in the pathogenesis of severe cigarette smoke-related pulmonary disorders. In cigarette smoke-exposed buffer solutions, $\bullet O_2^-$ was the predominant generated reactive oxygen species. When PMNs were incubated in this buffer, phorbol 12-myristate 13-acetate (PMA)-stimulated active oxygen production and coupled O_2 consumption were strongly impaired without appreciably affecting PMN viability (1-minute exposure inhibited active oxygen production by 75%). Superoxide dismutase (SOD) totally protected and an iron chelator, diethylenetriaminepentaacetic acid (DETAPAC), also protected the cigarette smoke-exposed PMNs, suggesting that generated $\bullet O_2^-$ was an initiating factor in the impairment and $\bullet OH$ generation was a subsequent injurious factor. Pretreatment of PMNs with antioxidants such as alpha-tocopherol and dihydrolipoic acid (DHLA) was partially protective. The results suggest that 1) $\bullet O_2^-$ is probably generated in the upper and lower respiratory tract lining fluid when they come in contact with cigarette smoke, 2) such generated $\bullet O_2^-$ can primarily impair PMN capabilities to generate ROS and RNS, and 3) these effects may contribute to the pathogenesis of cigarette smoke-related lung diseases [Tsuchiya *et al.*, 1992; Tsuchimy *et al.*, 1993].

There may be many reasons for the inhibitory effect of peroxidized liposomes on oxygen free radicals produced by granulocyte respiratory burst. One may be due to the scavenging effect of unsaturated lipids on oxygen free radicals produced by granulocytes and another reason may be that too many lipids are inserted into the granulocyte membrane, resulting in the change of membrane structure and affecting the function of granulocytes. With the extension of incubation time between liposomes and granulocytes treated with cigarette smoke, the oxidized lipid enters the granulocyte membrane and causes lipid peroxidation of granulocyte membrane, resulting in a large number of intermediates, damage of proteins and enzymes in the cell membrane, and destruction of membrane structure. The destruction of NADPH oxidizing enzyme will directly affect the respiratory burst and produce ROS and RNS free radicals.

18.3.4 *Cigarette smoking causes oxidative stress damage on cell membrane*

In order to study the damage and pathogenesis of cigarette smoking gas-phase free radicals to human health, we studied the damage of cigarette smoking gas-phase free radicals to cell membrane. A series of studies from liposome artificial membrane to lung cell membrane were carried out by ESR spin labeling technology. It was found that cigarette smoking gas-phase free radicals can indeed damage cell membrane phospholipids and proteins.

Using ESR spin labeling technique, we studied the effect of cigarette smoking gas-phase free radical on cell membrane fluidity. Two spin markers were used to label the polar end and hydrophobic end of cell membrane respectively. After cigarette smoking, their ESR spectra were measured, and the order parameters and rotation correlation time were calculated from the spectra. The results showed that the sequence parameters and rotation correlation time decreased with the extension of cigarette smoking time, that is, the fluidity of cell were changed [Yang *et al.*, 1992b].

In order to determine the effects of cigarette smoke exposure on the physical properties of cells, nuclear magnetic resonance (NMR) water-proton relaxation time (which measures the intracellular water organization) and ESR spin labeling (which measures membrane order) measurements were performed on cultured Jurkat T cells exposed to cigarette smoke. NMR spin-lattice relaxation time (T1) decreased with cigarette smoke exposure in a dose-dependent fashion. A significantly depressed T1 value was obtained even when cigarette smoke was delivered through a filter. Cell viability was not affected in this condition. Superoxide dismutase (SOD) prevented the depression of T1 value. These results suggest that superoxide radicals or subsequently generated species contained in the gas phase of cigarette smoke increase the intracellular water organization in viable cells. Cigarette smoke exposure also increased the ESR membrane order parameter of nitroxide spin label. These physical characteristic changes may be important in cigarette smoke-induced cell responses and cytopathology [Tsuchiya *et al.*, 1992; Tsuchimy *et al.*, 1993].

The sulfhydryl group of rat lung cell membrane protein was labeled with maleimide spin labe. It was found that with the extension of cigarette smoking time, the ESR spectrum intensity and immobilization ratio gradually decreased, indicating that cigarette smoking gas-phase free radical changed the conformation of the binding position of lung cell membrane protein and made the sulfhydryl binding position of membrane protein more exposed [Yan *et al.*, 1991c, 1992].

It can be seen from the above results that cigarette smoking gas-phase free radicals caused the oxidative fracture of unsaturated phospholipids in cell membrane, the changes in composition of membrane lipids, which then led to the increased fluidity of membrane. This may be because liposomes are used in the experiment. Therefore, the generated malondialdehyde does not have the problem of coupling with proteins. Lipid peroxidation degrades membrane lipids, resulting in increased membrane fluidity and changes in the conformation of the protein.

18.4 Tea and tea polyphenols reduce the harm of tobacco

The beverage tea, from the top leaves of the plant *Camellia sinensis*, is one of the most widely used beverages in the world, second only to water. Black and green tea have mostly similar effects. The active components are polyphenols, mainly epigallocatechin gallate in green tea, and the tea leaf polyphenol oxidase-mediated oxidation to oolong and black tea, yielding other polyphenols, theaflavin, and thearubigins. Many studies have shown that tea and tea polyphenols can reduce the harm of tobacco and prevent diseases caused by cigarette smoking. Drinking tea is a traditional habit in Southeast Asia, which has a history of thousands of years. It is said that "a cup of tea and a cigarette are better than living immortals". According to the survey, drinking tea can prevent cancer and many other diseases. Tea contains a lot of tea polyphenols, vitamin C, trace elements and other substances beneficial to human health. Tea and tea polyphenols are scavengers of ROS and RNS free radicals and blockers of lipid peroxidation. They can prevent the oxidative toxicity of cigarette smoking free radicals on cell components, protect cell membrane phospholipids and proteins from cigarette smoking damage, and supplement the loss of antioxidants caused by cigarette smoking, Tea and tea polyphenols can remove excess free radicals, restore the balance of redox destroyed by cigarette smoking, and prevent the occurrence of heart disease, cancer, and other diseases [Diana & Pryor, 1993; Zhao and Zhang, 1995; Yang *et al.*, 1992b, 1992c].

18.4.1 *Protective effect of tea polyphenols on cigarette smoking-induced injury*

Tea polyphenols have scavenging effect of on ROS and RNS free radicals and the prevention and benefits to human health and may also have therapeutic effects on diseases. We discuss the protective effect of tea polyphenols on cell membrane lipid peroxidation caused by free radicals in cigarette smoking. Many cancers are caused by lifestyle elements. Tea can inhibit cancer in the oral cavity, esophagus, and lung caused by cigarette and tobacco use. The developmental aspects and growth of cancers through promotion are decreased by tea. The regular use of a widely available, tasty, inexpensive beverage, tea, has displayed valuable preventive properties in chronic human diseases.

1. Protective effect of tea polyphenols on cell membrane injury induced by cigarette smoking

The measurement by spin capture technology and ESR spectroscopy shows that tea polyphenols and tea polyphenol filter can selectively reduce the harmful gas-phase free radicals produced by cigarette smoking, protect cells and animals, and reduce the toxicity of cigarette smoking. We studied the protective effect of tea polyphenols on lung cell membrane injury induced by cigarette smoking gaseous substances in rats. Tea polyphenols were added into rat lung cells in advance. Under the same cigarette smoking time, the degree of lipid peroxidation of rat lung cells decreased with the increase of tea polyphenol concentration, indicating that tea polyphenols have a significant inhibitory effect on lipid peroxidation of rat lung cells caused by cigarette smoking gaseous substances. With the increase in the concentration of tea polyphenols, the order parameters at the polar end of cell membrane gradually rise. When the concentration of tea polyphenols is 0.1 mg/ml, the order parameters almost return to the value with no-cigarette smoking time, indicating that tea polyphenols can protect rat lung cell membrane against damage by cigarette smoke [Yang *et al.*, 1992; Zhang *et al.*, 1996].

Chinese hamster lung V79 cells were cultured artificially. The cigarette smoke filtered by two layers of Cambridge filter paper was

Table 18-1.   Effect of cigarette smoke on biophysical characteristics of V79 cell membrane.

Smoking time (seconds)	Order parameter (S)	Rotation correlation time (τ_C)	S/W
0	0.657 ± 0.001	8.326 ± 0.366	0.370 ± 0.006
10	0.651 ± 0.005	9.001 ± 0.864	0.383 ± 0.009
20	0.639 ± 0.005	8.373 ± 0.688	0.392 ± 0.012
30	0.634 ± 0.006	8.344 ± 0.630	0.406 ± 0.019

Note: compared with the control, the $p < 0.05$.

introduced into the cells and the control sample was introduced with the same flow of air. After cigarette smoking, the degree of lipid peroxidation of V79 cell membrane was measured by TBA method. The fluidity of cell membrane was tested by spin labeling and ESR of fatty acids. The sequence parameter S and rotation correlation time τ_C were calculated. At the same time, the sulfhydryl group of cell membrane protein was labeled with maleimide spin labels and the strong and weak immobilization ratio was measured and calculated [Zhang *et al.*, 1996]. With the extension of cigarette smoking time, the number of V79 living cells decreased and the content of TBA reactant increased, indicating that the degree of lipid peroxidation of V79 cell membrane increased. The sequence parameter S (polar end) and rotation correlation time τ_C (hydrophobic end) calculated by the atlas marked by polar end and hydrophobic end are shown in Table 18-1. It can be seen that the sequence parameter S of polar end decreases with the extension of cigarette smoking time, while the rotation correlation time τ_C of hydrophobic end does not change significantly, indicating that cigarette smoking smoke increases the fluidity of shallow layer of cell membrane and has little effect on deep layer. The strong immobilized and weak immobilized components ratio (S/W) calculated from the ESR spectrum of maleimide labeled cells is shown in Table 18-2. It can be seen that S/W increases with the increase of cigarette smoking time, indicating that the structure at the sulfhydryl binding site on the membrane protein becomes compact, which shows that cigarette smoking smoke changes the conformation of the membrane protein [Zhao *et al.*, 1990].

The above research results have shown that tea polyphenols can protect the cell membrane damage caused by smoking.

Table 18-2. Protective effect of tea polyphenols on V79 cell membrane.

	TBA	Survival rate (%)	Order parameter	Rotation correlation time	S/W
Control smoke	2.75 ± 0.35	60 ± 8	0.650 ± 0.002	7.715 ± 0.148	0.409 ± 0.005
Tea polyphenols	1.52 ± 0.25	80 ± 9	0.651 ± 0.002	6.226 ± 0.054	0.394 ± 0.004

Note: compared with the control, the $p < 0.05$.

2. Protective effect of tea polyphenols on animals of toxicity induced by cigarette smoking

We studied the effect of tea polyphenol filter on acute and chronic toxicity of cigarette smoking. The results showed that tea polyphenols had a good protective effect against the two kinds of toxicity of cigarette smoking [Zhao *et al.*, 1990].

(1) Protective effect of tea polyphenols on acute toxicity induced by smoking in animals

Ten mice were placed in a $35 \times 35 \times 20$ cm poison box. A small hole was opened on both sides of the box for ventilation and the flue cigarette smoking gas was introduced into the box. Each cigarette lasted for 2–3 minutes and operated continuously until the mice were poisoned to death. The exposure pressure of CO_2 and O_2 is within the normal range. The experimental results showed that the tea polyphenol filter could significantly prolong the life of mice in the box of cigarette smoking gas (Table 18-3).

Histopathological examination also shows protective effect of tea polyphenols against the cigarette smoking. In the cigarette smoking control group, 80% of the mice had obvious congestion in lung tissue, pulmonary hemorrhage, dilation and congestion of renal interstitial small vessels, mild dilation and bleeding of hepatic lobular central vein or interlobular vein, and no abnormality in heart and spleen. About 40% of the mice in the tea polyphenol filter cigarette group also had interstitial small blood vessels in the lung tissue, renal interstitial small blood vessels dilated and congested, liver lobular central vein or interlobular vein slightly dilated and bled, and there was no abnormality in the heart and spleen.

Table 18-3. Effects of cigarette smoke on survival time of mice.

Groups	Number	Body weight	Survival time	Toxicity reduction rate
Control smoke	10	28.5 ± 1.9	60.0 ± 18.4	
Experimental group	10	27.6 ± 1.5	96.0 ± 9.7	69.5%
Control smoke	10	19.0 ± 1.9	52.0 ± 9.7	
Experimental group	10	19.5 ± 1.5	101.0 ± 1.3	93.3%

Table 18-4. Effect of tea polyphenol filter cigarette smoke on micronucleus rate of mouse bone marrow cells.

Group	Number	Micronucleus rate	Reduce
Control group	20	1.85 ± 1.42	
Cigarette smoke	20	14.60 ± 7.06	
Tea polyphenol	20	5.50 ± 2.10	60%

The above research institutions show that tea polyphenols have a protective effect on acute toxicity caused by smoking.

(2) Protective effect of tea polyphenols on chronic toxicity induced by smoking in animals

The changes of micronucleus rate of bone marrow cells in animals reflect mutation and carcinogenesis. We observed the effect of tea polyphenol filter on the micronucleus rate of bone marrow cells in cigarette smoking animals. There were 30 male mice in each group. They were normal group, control cigarette group, and tea polyphenol filter cigarette group. All groups adopted passive cigarette smoking, one by one, smoking for 20 minutes a day for 75 consecutive days. All animals were killed 24 hours after the last smoking, and the bone marrow was taken to count the polychromatic erythrocytes and cells with micronucleus. Each animal counted 1000 and the micronucleus percentage was calculated ($< 5\%$). At the same time, the ratio of polychromatic erythrocytes to normal erythrocytes was calculated. The experimental results show that tea polyphenol filter cigarette has a significant protective effect on the micronucleus rate of mouse bone marrow cells against cigarette smoking (Table 18-4).

From the above results it can be seen that tea polyphenols can remove the harmful free radicals produced by cigarette smoking, inhibit the lipid peroxidation of cell membrane caused by cigarette smoking gaseous substances, and protect cells and animals from acute and chronic toxicity caused by free radicals produced by smoking.

3. The detoxification effect of the tea on lung damage caused by cigarette smoking

Tea, the most widely consumed beverage, is a source of compounds with anti-oxidative, antimicrobial, anti-mutagenic, and anti-carcinogenic properties. Lung cancer is the most common cause of cancer deaths in both men and women worldwide. Over 1 million people around the world are likely to be killed by lung cancer due to increased tobacco smoking. Therefore chemo-preventive intervention using black tea and its active components may be a viable means to reduce lung cancer death. In the present investigation, it used benzo αpyrene (BP) to induce lung carcinogenesis in mice for the assessment of potential apoptosis-inducing and proliferation-suppressing effects of theaflavins and epigallocatechin gallate, active components of black tea. Hyperplasia, dysplasia, and carcinoma *in situ* evident in the carcinogen control group on the 8th, 17th, and 26th weeks respectively, were effectively reduced after treatment with theaflavins and epigallocatechin gallate. Significant reduction in number of proliferating cells and increased number of apoptotic cells was also found on the 8th, 17th, and 26th week of treatment with theaflavins and epigallocatechin gallate in BP-exposed mice.

Smoking increases DNA methylation and DNA damage, and DNA damage acts as a vital cause of tumor development. The DNA methyltransferase 3B (DNMT3B) enhances promoter activity and methylation of tumor suppressor genes. However, tea polyphenols may inhibit DNMT activity. A study designed a case-control study to evaluate the joint effects of smoking and green tea consumption, with miR-29b and DNMT3B mRNA expression in lung cancer development. A total of 132 lung cancer patients and 132 healthy controls were recruited to measure miR-29b and DNMT3B mRNA expression in whole blood. Results revealed that lung cancer patients had lower miR-29b expression (57.2 versus 81.6; $p = 0.02$)

and higher DNMT3B mRNA expression (37.2 versus 25.8; $p < 0.001$) than healthy controls. Compared to non-smokers with both higher miR-29b and lower DNMT3B mRNA expression, smokers with both low miR-29b and higher DNMT3B mRNA expression had an elevated risk of lung cancer development (OR 5.12, 95% CI 2.64–9.91). Interactions of smoking with miR-29b or DNMT3B mRNA expression in lung cancer were significant. Interaction of green tea consumption with miR-29b expression and DNMT3B mRNA expression in lung cancer was also significant. The study suggests that smokers and green tea nondrinkers with lower miR-29b expression and higher DNMT3B mRNA expression are more susceptible to lung cancer development [Huang *et al.*, 2022].

Animals were randomly divided into the following groups (n = 10) for treatment with different chemicals — control group: mice were injected with 0.9% physiological saline, s.c.; nicotine group: mice were injected with nicotine (0.5 mg/kg.d,); nicotine + GTP group: mice were lavaged with green tea polyphenols (250 mg/kg.d). Specimens of lungs from each animal of all groups were examined under the optical microscope for pathological changes. A variety of pathological alterations were discovered in the lungs of the nicotine group. Thickness of bronchi capillary walls with signs of neutrophil and lymphocyte infiltration was observed. There was an increase of inflammatory exudate and lympho-proliferation in the bronchial lumen, an increased broken alveoli. Tea filters significantly reduced these pathological changes. Abscess and low fiber proliferation were found in the lung tissue in four rats. Only one rat showed the pathological alterations and one rat showed minor pathological changes in the lungs (moderate neutrophil and lymphocyte infiltration of alveoli), and in three animals, almost normal lung tissue was observed from the tea filter group. One rat in normal filtration group showed slight pathological changes [Yan *et al.*, 2010]. Many studies have shown that tea polyphenols have anticancer activity. In one study, the effects of theaflavin-3-3'-diglyceride (TF), the main theaflavin monomer in black tea, with reducing agent ascorbic acid (AA) and epigallocatechin-3-gallate (EGCG), the main polyphenol in green tea, and coordinated with reducing agent ascorbic acid, on the viability and cell cycle of human lung adenocarcinoma SPC-A-1 cells were investigated. The 50% inhibitory concentrations of

TF, EGCG, and AA on SPC-A-1 cells were 4.78 μmol/L, 4.90 μmol/L, and 30.62 μmol/L, respectively. When the molar ratio of TF to AA (TF + AA) and EGCG to AA (EGCG + AA) was 1:6, the inhibition rates of SPC-A-1 cells were 54.4% and 45.5%, respectively. Flow cytometry analysis showed that TF + AA and EGCG + AA could significantly increase the number of cells in G(0)/G(1) phase of SPC-A-1 cell cycle, from 53.9% to 62.8% and 60.0% respectively. At 50% inhibitory concentration, G(0)/G(1) accounted for 65.3%. Therefore, it can synergistically inhibit the proliferation of SPC-A-1 cells. The results show that the combination of L-theaflavin and green tea polyphenols with reducing agent ascorbic acid can improve its anticancer activity [Li *et al.*, 2010].

Above study results suggest detoxification effect of the tea on lung damage caused by cigarette smoking.

4. Tea polyphenols prevent chronic disease caused by tobacco products

Studies show that cigarette and tobacco use leads to cancer in the oral cavity, esophagus, and lung, but this can be inhibited by tea. In this study, mice administered a tobacco nitrosamine, 4-(methylnitrosamino)-1-(3-pyridyl)-1-butanone (NNK), developed significantly fewer lung tumors than controls when given green tea or its major polyphenol, EGCG. Tea suppressed the formation of 8-hydroxydeoxyguanosine (8-OHdG), a marker of oxidative DNA damage, in the lung DNA of mice given NNK. Gastric cancer, caused by a combination of *Helicobacter pylori* and salted foods, is lower in tea drinkers. Western nutritionally-linked cancers of the breast, colon, prostate, and pancreas can be inhibited by tea. The formation of genotoxic carcinogens for these target organs during the cooking of meats, heterocyclic amines, and their effects were decreased by tea. Tea inhibited the formation of reactive oxygen species and radicals and induced cytochromes P450 1A1, 1A2, and 2B1, and glucuronosyl transferase. The higher formation of glucuronides represents an important mechanism in detoxification [Weisburger & Chung, 2002].

The active components of tea are polyphenols, mainly EGCG in green tea, and the tea leaf polyphenol oxidase-mediated oxidation to oolong and black tea, yielding other polyphenols, theaflavin, and thearubigins. The preventive effect of tea on chronic diseases caused by tobacco products depends on 1) its action as an antioxidant, 2) the specific induction of detoxifying enzymes, 3) its molecular regulatory functions on cellular growth, development, and apoptosis, and 4) a selective improvement in the function of the intestinal bacterial flora. The oxidation of LDL cholesterol, associated with a risk for atherosclerosis and heart disease, is inhibited by tea.

5. Effect of tea polyphenols on oxidative damage and apoptosis in bronchial epithelial and fibroblasts cells induced by cigarette smoke

A study determined the effect of tea polyphenols (TP) on oxidative damage and apoptosis in human bronchial epithelial cells induced by low-dose cigarette smoke condensate (CSC). Concentration of intracellular ROS in the CSC group and CSC + TP group was significantly higher than that in the control group ($p < 0.01$), while concentration of intracellular ROS in the CSC + TP group was significantly lower than that in the CSC group ($p < 0.01$). Apparent DNA breakage of the tail belt appeared in the CSC group, while only a small amount of DNA breakage of the tail belt appeared in the CSC + TP group. Compared with the control group, Bcl-2 mRNA expression was reduced and Bax mRNA expression was increased in the CSC group (all $p < 0.01$). Compared with the CSC group, Bcl-2 mRNA expression was increased and Bax mRNA expression was reduced in the CSC + TP group (all $p < 0.01$). Ratio of Bcl-2 mRNA/Bax mRNA in the CSC group and CSC + TP group was significantly lower than that in the control group (all $p < 0.01$). These results shown that TP can antagonize CSC-induced airway epithelial cell apoptosis through the effective removal of ROS, promoting Bcl-2 mRNA expression and inhibiting the expression of Bax mRNA [Li *et al.*, 2010].

Cell culture and animal model studies indicate that (-)-epigallocatechin-3-gallate (EGCG), the major polyphenol present in green tea, possesses potent anti-inflammatory and antiproliferative activity capable of selectively inhibiting cell growth and inducing apoptosis in cancer cells without adversely affecting normal cells. A study demonstrate that EGCG pretreatment (20–80 μM) of normal human bronchial epithelial cells (NHBE) resulted in significant inhibition of cigarette smoke condensate (CSC)-induced cell proliferation. Nuclear factor-kappaB (NF-κB) controls the transcription of genes involved in immune and inflammatory responses. In most cells, NF-κB prevents apoptosis by mediating cell survival signals. Pretreatment of NHBE cells with EGCG suppressed CSC-induced phosphorylation of IkappaBalpha, and activation and nuclear translocation of NF-κB/p65. NHBE cells transfected with a luciferase reporter plasmid containing an NF-κB-inducible promoter sequence showed an increased reporter activity after CSC exposure that was specifically inhibited by EGCG pretreatment. It also showed that pretreatment of NHBE cells with EGCG resulted in a significant downregulation of NF-κB-regulated proteins cyclin D1, MMP-9, IL-8, and iNOS. EGCG pretreatment further inhibited CSC-induced phosphorylation of ERK1/2, JNK, and p38 MAPKs and resulted in a decreased expression of PI3K, AKT, and mTOR signaling molecules [Syed *et al.*, 2007].

Tobacco use has been identified as the most important environmental risk factor for periodontitis. This study investigated the effect of green tea EGCG on the nicotine-induced toxic and inflammatory responses in oral epithelial cells and gingival fibroblasts. Results indicated that nicotine caused a dose-dependent loss of viability in both epithelial cells and fibroblasts. A mixture of nicotine and actinomycetem lipopolysaccharide demonstrated additive instead of synergistic effects on loss of cell viability. Pretreatment of cells with EGCG efficiently neutralized the nicotine-induced toxic effects in epithelial cells and fibroblasts. This study suggests that EGCG, the major polyphenol in green tea, may represent a novel preventive/therapeutic agent for smoking-related periodontitis.

The oxidative effects of cigarette smoke on the human skin were investigated. A remarkable increase in the conversion ratio of squalene (SQ) to squalene monohydroperoxide (SQHPO) due to exposure to cigarette smoke was observed. The results showed that cigarette smoke caused lipid peroxidation. It also found that the addition of chain-breaking-type antioxidants, such as oolong tea extract, inhibited the peroxidation. When cultured human skin fibroblasts were exposed to cigarette smoke, this increased the intensity, suggesting that cigarette smoke caused oxidation in cultured human skin fibroblasts. When the cultured human skin fibroblasts were treated with antioxidants such as glutathione, thiotaurine, hypotaurine, and ascorbic acid there was little increase, meaning that oxidation had been prevented in the human skin fibroblasts. Cigarette smoke had an oxidative effect on SQ, cultured human skin fibroblasts, and the surface of the human skin. The application of antioxidants prevented the cigarette smoke-induced oxidation [Egawa *et al.*, 1999].

The above results indicate that tea polyphenols can protect fibroblasts and epithelial cells against oxidative damage and apoptosis induced by cigarette smoke.

6. Inhibitory effect of green tea on airway neutrophil elastase and matrix metalloproteinase-12 in the upregulation of antioxidant activity of cigarette smoke

Chinese green tea may have potential effect on lung oxidative stress and proteases/anti-proteases in smoking. Chinese green tea may have potential effects on lung oxidative stress and protease/anti-protease in cigarette smoking rat model. Rats were exposed to air or 4% cigarette smoke plus 2% lung green tea or oral gavage. Serine proteases, matrix metalloproteinases (MMPs) and their endogenous inhibitors in bronchoalveolar lavage fluid and lung tissue were determined by gelatin/casein Zymogram and biochemical analysis. The consumption of green tea significantly reduced the increase of lipid peroxidation marker malondialdehyde (MDA) in bronchoalveolar lavage fluid and lung induced by cigarette smoke, and the concentration and activity of neutrophil elastase induced by cigarette smoke were significantly higher than those in the control group ($p < 0.05$). At the same time, the activity of matrix metalloproteinase in bronchoalveolar lavage fluid and lung tissue in cigarette smoke exposure group increased significantly. After green tea treatment, the activity of MMP returned to the air exposure group, but the activity of tissue metalloproteinase inhibitor (TIMP)-1 induced by cigarette smoke decreased, and the consumption of green tea did not reverse. Taken together, these data support the existence of local oxidative stress and protease/anti-protein imbalance in the respiratory tract after cigarette smoke exposure. Green tea can reduce oxidative stress and protease/anti-protein imbalance through its biological antioxidant activity [Chan *et al.*, 2012]. Green tea consumption significantly decreased cigarette smoke-induced elevation of lung lipid peroxidation marker, MDA, and cigarette smoke-induced upregulation of neutrophil elastase (NE) concentration and activity along with that of α(1)-antitrypsin (α(1)-AT) and secretory leukoproteinase inhibitor in bronchoalveolar lavage and lung. In parallel, significant elevation of MMP-12 activity was found in bronchoalveolar lavage and lung of the cigarette smoke-exposed group, which returned to the levels of sham air-exposed group after green tea consumption but not cigarette smoke-induced reduction of tissue inhibitor of metalloproteinase (TIMP)-1 activity, which was not reversed by green tea consumption [Chan *et al.*, 2012].

The above results indicate that tea polyphenols can protect neutrophil elastase and matrix metalloproteinase-12 against oxidative damage and apoptosis induced by cigarette smoke.

7. Inhibitory effect of tea polyphenols on benzopyrazine and nitrosamine in lung cancer

A growing body of evidence from studies in laboratory animals indicates that green tea protects against cancer development at various organ sites. We have previously shown that green tea, administered as drinking water, inhibits lung tumor development in A/J mice treated with 4-(methylnitrosamino)-1-(3-pyridyl)-1-butanone (NNK), a potent nicotine-derived lung carcinogen found in tobacco. The inhibitory effect of green tea has been attributed to its major polyphenolic compound, EGCG. The studies underscore the importance of the antioxidant activity of green tea and EGCG for their inhibitory activity against lung tumorigenesis. Unlike green tea, the effect of black tea on carcinogenesis has been scarcely studied, even though the worldwide production and consumption of black tea far exceeds that of green tea. The oxidation products found in black tea, thearubigins, and theaflavins, also possess antioxidant activity, suggesting that black tea may also inhibit NNK-induced lung tumorigenesis. There are two kinds of strong carcinogens, benzopyrene and nitrosamines, in tobacco barbecue. There is limited understanding of the dose-related effects of thearubin-polymerized polyphenols, the most abundant polyphenol in black tea. One study used 0.75%, 1.5%, and 3% black tea as raw materials to explore the effects of different doses of black tea extract on biochemical parameters and lung carcinogenicity in a/J mice. The expression and activity of cytochrome P450 isozymes in liver and lung tissues decreased after pretreatment with thearubin-polymerized polyphenols, while the expression and activity of phase I enzymes induced by carcinogens in lung tissues, the number and intensity of DNA adducts in liver and lung decreased significantly, which was dose-dependent. Black tea (1.5%, 3%)-derived PBPs showed dose-mediated decrease in lung tumor incidence and multiplicity, which was further correlated with different molecular markers such as cell proliferation and apoptosis in B(a)P and NNK model. In conclusion, the dose-dependent chemo-preventive effects of thearubin-polymerized polyphenols include inhibiting the initiation of carcinogenesis (inducing phase II and inhibiting carcinogen induced phase I enzymes, resulting in the reduction of DNA adducts) and inhibiting the promotion of carcinogenesis (reducing cell proliferation and apoptosis, reducing the incidence and/or increasing the diversity of lung lesions), and thearubin-polymerized polyphenols were observed in a/J mice, There is no obvious toxicity [Hudlika *et al.*, 2019]. These studies also showed that green tea polyphenol EGCG inhibited the tobacco-specific nitrosamine NNK-induced lung tumorigenesis, probably due to its antioxidant property. These studies provide for the first time evidence for the involvement of free radicals in nitrosamine tumorigenesis. The reduced levels of oxidative lesions in lung as a result of EGCG treatment may be related to its ability to reduce reactive oxygen species and/or to chelate iron ion, resulting in a decreased production of hydroxyl radicals [Chung *et al.*, 1993; Chung 1999].

18.5   Detoxification effect of tea filter on toxicity caused by cigarette smoking

Previous studies have shown that tea components, green tea polyphenols can remove tar, free radicals, nitrosamines, α-pyrene, benzo-α-anthracene, chrysene, and total polycyclic aromatic hydrocarbons generated during cigarette smoking [Yang *et al.*, 1993b, 1993c; Gao *et al.*, 2001]. We further studied the effect of tea filter on the acute toxicity of cigarette smoke and found that the compound (semi tea filter/ semi cellulose acetate) and the whole tea filter increased the survival time of animals respectively, and the acute toxicity of cigarette smoke was significantly reduced. In the control group (cigarettes with normal filter tips), most of the mice observed obvious congestion and hemorrhage in the lung tissue. The tea filter reduced the number of mice with these pathological significantly decrease. We studied the effect of tea filters on the incidence of micronucleus in polychromatic erythrocytes (PCE) as a measure of the mutagenicity rate of rats exposed to cigarette smoke for 75 days and as a toxicity indicator. When rats were exposed to cigarette smoke for a long time, the incidence of PCE micronucleus significantly increased, according to the mutagenic activity of cigarette smoke. When using cigarettes with tea filter, the incidence of micronucleus in PCE was inhibited compared with rats who inhaled cigarette smoke with normal filter. Therefore, tea filters significantly reduce the mutagenicity of cigarette smoke.

18.5.1 *Detoxification effect of tea filter on acute toxicity by cigarette smoking*

We studied the efficacy of tea filters on the acute toxicity of cigarette smoke and found that the survival time of animals respectively increased by 32.2% and 60% by complex (half tea filter/half cellules acetate) and full tea filters and the acute toxicity of cigarette smoke was significantly reduced. In the control group (cigarette with normal filters), marked congestion and hemorrhage in lung tissue was observed in 80% of mice. The tea filters reduced the number of mice with these pathological changes to 40% (Table 18-5).

Table 18-5. Effect of tea filter on survival time of mice induced by cigarette smoke.

Group	Number of animals	Weight (g)	Survival time (min)	Increased (%)
Control	8	28.5 ± 1.9	11.5 ± 1.8	
Compound tea filter cigarette	8	27.6 ± 1.5	14.9 ± 2.2	32.2%
Full tea filter cigarette	8	29.5 ± 1.5	18.0 ± 1.9	60.0%

18.5.2 *Detoxification effect of tea filter on mutagenicity caused by cigarette smoking*

We investigated the effect of a tea filter on the incidence of micronuclei in polychromatic erythrocytes (PCE) as a measure of the mutagenicity ratio and as an indicator of toxicity in rats exposed to cigarette smoke for 75 days. The incidence of micronuclei in PCE significantly increased when rats were chronically exposed to cigarette smoke, in accordance with a mutagenic activity of cigarette smoke. When a cigarette with a tea filter was used, the incidence of micronuclei in PCE was inhibited by about 46% as compared to the rats which inhaled smoke from cigarettes with normal filters. Hence there was a significant reduction in the mutagenicity of cigarette smoke by tea filters (Table 18-6).

18.5.3 *Effects of tea filter on carboxyhaemoglobin (COHb) in mouse blood*

Carboxyhaemoglobin (COHb) is a stable complex of carbon monoxide and hemoglobin that forms in red blood cells when carbon monoxide is inhaled. Tobacco smoking, through carbon monoxide inhalation, raises the blood levels of COHb [Cardoso *et al.*, 1991; Cendon *et al.*, 1997], causing cardiovascular and cerebrovascular damage and diseases, e.g., neurasthenia, myocardium damage, and atherosclerosis. In order to study the effect of the tea filter on the toxicity of CO generated from cigarette smoking, we measured the COHb levels in the blood. It was found that the

Table 18-6. Effect of tea filter tip on micronucleus rate and COHb change of erythroblast in rat bone marrow induced by cigarette smoke.

Group	Number of animals	Micronucleus rate	Effect (%)	COHb (mg/mL)	Effect (%)
Control	5	1.85 ± 1.42		1.27 ± 0.66	
Cigarette	5	$14.55 \pm 7.06^{\#}$	+687	$8.40 \pm 0.42^{\#}$	+561
Tea filter cigarette	5	$5.68 \pm 2.10^{*}$	−60	$3.98 \pm 0.99^{*}$	−53

#, $p < 0.05$, compared with non-smoking control group; *, $p < 0.05$, compared with smoking control smoke.

COHb levels in the blood of the mice exposed to smoke from cigarettes with a normal filter were increased about 561% compared with the mice without such exposure, while the COHb levels in the blood of the mice exposed to smoke from cigarettes with a tea filter were decreased about 53% compared with the mice exposed to smoke from cigarettes with normal filters. These results suggest that the tea filter inhibited the COHb levels generated in cigarette smoking and prevented cardiovascular and cerebrovascular diseases caused by cigarette smoking (Table 18-6).

18.5.4 *Inhibitory effect of tea filter on cigarette smoking-induced lung injury in animals*

All experimental animals were examined by pathological sections and light microscopy of lung tissue. The results showed that there were many pathological changes in the lung tissue of rats in the control smoking group, with obvious alveolar rupture and bronchial wall thickening, accompanied by neutrophil lymphocyte infiltration. Abscess and low-grade fibrosis and proliferation were found in the lung tissue of four rats in the control smoking group. However, in the tea filter group, only one showed these pathological changes, one showed minor pathological changes (moderate neutrophil lymphocyte infiltration), and the other three showed almost no difference from normal lung tissue.

18.5.5 *The mechanism of the inhibiting effects of tea polyphenols on tobacco addiction*

From the above discussion it can be concluded that tea polyphenols in tea can inhibit the ROS and RNS free radicals, oxidative stress injury caused by smoking, especially the acute and chronic toxicity, and lung injury caused by smoking. Tea polyphenols have an inhibitory effect on the inflammation induced by cigarette smoke, which leads to medium particle aggregation and the oxidative damage of elastase and matrix metallopro-teinase-12. Tea polyphenols can prevent various cancers caused by cell mutation from smoking and effectively eliminate the harm of smoking. Tea itself can also filter out carcinogens produced by smoking, such as

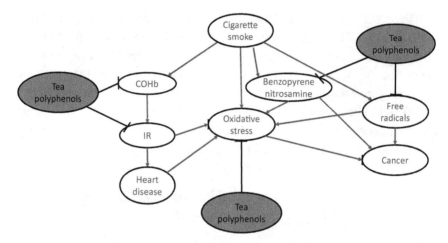

Figure 18-3.   The mechanism schematic diagram about the reduced effect of tea and tea polyphenols on the harm of cigarette smoke.

nitrosamines α-Pyrene, benzo α-Anthracene, chrysene, and total polycyclic aromatic hydrocarbons produced during smoking. Tea can inhibit the level of COHb produced by smoking and prevent cardiovascular and cerebrovascular diseases caused by smoking to protect human health (Fig. 18-3).

18.6 Cessation effect of tea and L-theanine on cigarette smoke

Although the campaign of quitting cigarette smoking has been rising wave by wave in western countries and the number of smokers has decreased sharply, the number of smokers has not decreased and is still increasing among young people. Cigarette smoking can cause a series of serious diseases. The harm of cigarette smoking to health has become a major public health problem faced by mankind. Though scientific research and medical workers have studied various methods, the effect is still not ideal, and tobacco control has become a worldwide problem. Moreover, due to the addictive nature of nicotine, quitting smoking remains extremely difficult. Despite all efforts, currently available smoking cessation methods produce only modestly successful rates with frequent relapse. In addition, they are often perceived as being inconvenient and lead to a wide variety of side effects [Balbani *et al.*, 2005; Ray *et al.*, 2009]. Therefore, the need to develop smoking cessation strategies with improved efficacy and fewer side effects remains a high public health priority. In order to reduce the toxicity of cigarette smoking to human beings, it is necessary to look into low free radical and low toxicity cigarettes. Epidemiological and experimental evidence suggests that drinking tea is adversely associated with lowering the risk of cancer [Jankan *et al.*, 1997], hyperglycemia [Mackenzie *et al.*, 2007], and other diseases [Guo *et al.*, 2007]. In addition, our work has shown that tea components protect cells from cigarette smoke-induced toxicity [Gao *et al.*, 2001; Zhao *et al.*, 1989a, 1989b].

In order to find a new solution to this problem, we developed a tea filter, which can significantly eliminate addiction to smoking dependence, and preliminarily explored its mechanism. Human tests showed that smoking using the tea filters significantly decreased the cigarette consumption of volunteer smokers. The first batch of clinical trials found that when cigarette smoking volunteers used tea filters for two months, the amount of cigarette smoking decreased by about 52%, of which 31% decreased to 0%. Another batch of clinical trials found that after using tea filter for three months, the smoking number of cigarette smoking volunteers significantly decreased respectively in the first, second, and third months. In the last month, the daily smoking number decreased to very seldom. Further study showed that L-theanine, exerted an effect similar to the nicotine acetylcholine receptor inhibitor. Animal experiments showed that L-theanine in tea filter could significantly inhibit the conditioned place preference caused by nicotine in mice, which was similar to the effect of nicotine acetylcholine receptor (nAChRs) inhibitors. It was found that L-theanine-treated animals could inhibit the upregulation of the expression of three nicotine acetylcholine receptor subunits caused by nicotine, and the increase of dopamine release was significantly inhibited. Tea filter can also significantly reduce the harmful substances produced by cigarette smoking and reduce the acute toxicity and chronic carcinogenicity of animals caused by smoking. This work found new substances to inhibit tobacco and nicotine addiction — tea filter and L-theanine, which provided a new strategy to overcome the harm of cigarette smoking [Yan *et al.*, 2010; Di *et al.*, 2012].

18.6.1 *Human test*

A human test for the cessation effect of a tea filter on smoking was performed. In one trial, 100 healthy male cigarette smokers were screened using the standard exclusion/inclusion criteria. Thirty of the volunteers were excluded and 70 of them were double blinded, placebo-controlled and randomized into two groups (smoking with tea filters or regular filters). In another trial, 70 healthy male cigarette smokers were screened and 59 volunteers with a longer smoking history and a stronger desire to quit smoking were tested using the tea filter for three months. Smoking history, including assessment of nicotine dependence, was evaluated at the screening visit.

It was found that, healthy male cigarette smokers who consumed approximately 14 cigarettes per day on average were recruited and randomly divided into two groups (double-blinded, placebo-controlled): smoking with tea filters or with regular filters. After using the tea filter for one month, the daily average cigarette consumption decreased by about 43% in the tea filter group. By contrast, no change in average daily cigarette consumption was detected in the control group using regular filters. Due to the reduction of cigarette consumption, the levels of exhaled carbon monoxide and urine cotinine content in the volunteers who smoked with tea filters were significantly decreased by about 52.6% and 26.3%, respectively. The test for the tea filter group was followed for an additional month. After using the tea filter for two months, the daily average cigarette consumption was decreased by about 56.5%. In another trial, we tested the effect of the tea filter on heavier smokers who had a stronger desire to quit smoking. A total of 59 healthy male cigarette smokers were recruited and tested to smoke with tea filters for three months. The result showed that their daily average cigarette consumption respectively decreased by about 48%, 83%, and 91% after using the tea filter for one month, two months, and three months, suggesting that the tea filter is effective for smoking cessation (Figs. 18-4, 18-5). The efficacy of the tea

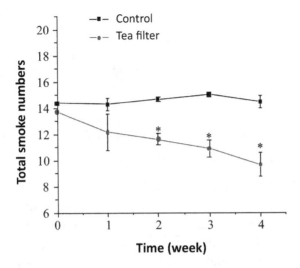

Figure 18-4. Tea filter significantly reduced the smoking number of volunteers smoking. Average daily smoking numbers of each volunteer in weeks 0–4. *, $p < 0.05$, compared with week 0.

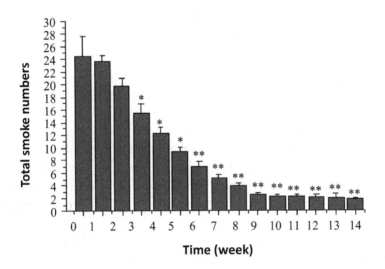

Figure 18-5. Tea filter significantly reduced the smoking number of volunteers. The tea filter was used for three months, and the average daily smoking numbers of each volunteer in 0–14 weeks; *, $p < 0.05$, compared with 0 week; **, $p < 0.01$, compared with 0 week.

filter on smoking cessation is better than many other methods reported. In addition, most subjects described that sputum and their smoking-related symptoms were reduced compared to the control group. Physical examinations of the subjects did not reveal any apparent side-effects [Yan *et al.*, 2010].

18.6.2 *Effect of tea filter on the rewarding effect for animals induced by nicotine*

In order to determine which component in the tea filter has the abstinence effect on cigarette smoking, we conducted animal experiments. The abstinence effects of the main components of tea, tea polyphenols, caffeine, and L-theanine on nicotine-induced animal addiction were tested. The conditioned place preference (CPP) paradigm is a measure of nicotine reinforcement. Mice were randomly divided into the following groups (n = 10) for treatment with different chemicals — control: mice were abdominal injected with 0.9% physiological saline; nicotine: mice were abdominal injected with nicotine (0.5 mg/kg.d); nicotine + GTP: administration of green tea polyphenols to mice by gavage (250 mg/kg.d) and injected with nicotine; nicotine + CF: mice were abdominal injected with caffeine (2 mg/kg.d) and nicotine; nicotine + TH1: mice were injected with L-theanine-low (250 mg/kg.d) and nicotine; nicotine + TH2: mice were abdominal injected with L-theanine-high (500 mg/kg.d) and nicotine; nicotine + DHβE: mice were abdominal injected with dihydro-β-erythroidine hydrobromide (DHβE) (2.0 mg/kg.d.) and nicotine [Di *et al.*, 2012]. Mice were daily abdominal injected with nicotine (0.5 mg/kg.d) or physiological saline for one week. Different compounds were administered 15 minutes before nicotine injection. The result showed that nicotine addiction was induced by daily injection of nicotine (0.5 mg/kg) to the mice for seven days, while different compounds isolated from the tea filter were administrated 15 minutes before each nicotine injection. The results revealed that L-theanine (500 mg/kg), an amino acid derivative component of tea, had a similar effect in mice to DHβE, an inhibitor of nAChR, but green tea polyphenols (250 mg/kg) and caffeine (2 mg/kg) had no effect on nicotine induced reinforcement

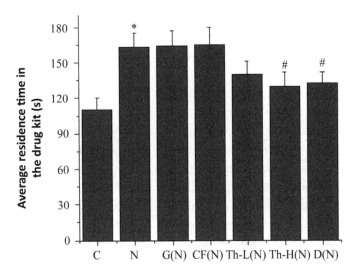

Figure 18-6. Effect of different components in tea on nicotine-induced conditioned place preference. C: control group; N: nicotine group (0.5 mg/kg.d); G(N): tea polyphenols (250 mg/kg) pretreatment group; CF (N): caffeine (2 mg/kg) pretreatment group; Th-L(N): pretreatment group with low dose of L-theanine (250 mg/kg); Th-H(N): pretreatment group with high dose of L-theanine (500 mg/kg); D(N): nicotine acetylcholine receptor inhibitor (2 mg/kg) pretreatment group. The results are expressed by the average residence time in the drug kit, n = 10. Multivariate ANOVA analysis, *, $p < 0.05$, compared with the control group; #, $p < 0.05$, compared with nicotine group.

in the animals. The inhibition effects of L-theanine appeared to be time- and dose-dependent. While administration of L-theanine (250 mg/kg and 500 mg/kg) for seven days respectively inhibited nicotine addiction about 25% and 50% respectively, the inhibition of nicotine addiction was respectively about 90% and 95% after two weeks of L-theanine treatment for both doses (Fig. 18-6).

18.6.3 *Effects of L-theanine on the expression of the nicotine receptor (nAChR) in mouse brains*

Subtypes of neuronal nicotinic acetylcholine receptors (nAChRs) are differentially sensitive to upregulation by chronic nicotine exposure *in vitro*. α3β4-like binding represented >10% of total in 15 out of the 33 regions surveyed [Nguyen *et al.*, 2003]. It has been shown that nicotine treatment increases the expression of nAChR while inhibition of nAChR and its related processes causes smoking cessation [Xiu *et al.*, 2009; Lape *et al.*, 2008]. To study the cessation mechanisms of L-theanine on nicotine dependence, we next investigated if L-theanine caused nicotine cessation by affecting nicotine-induced expression of nAChR using a mouse model. After daily injection with nicotine for two weeks, the protein levels of nAChR subunits α4, α7, and β2 were increased in mouse brains. L-theanine pretreatment significantly inhibited the induction of nAChR subunits in the brain. Therefore, the cessation effect of L-theanine on nicotine dependence may be attributed to its inhibition on nicotine-induced expression of nAChR subunits (Fig. 18-7).

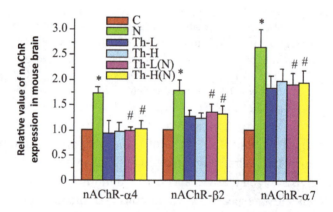

Figure 18-7. Effect of L-theanine and nicotine on the expression of three subunits of nicotinic acetylcholine receptor (nAChRs) in the ventral tegmental area of mice. Western blot analysis of statistical results, C: control group; N: nicotine group (0.5 mg/kg.d); Th-L: L-theanine low dose (250 mg/kg) treatment group; Th-H: L-theanine high dose (500 mg/kg) treatment group; Th-L(N): pretreatment group with low dose of L-theanine (250 mg/kg); Th-H(N): pretreatment group with high dose of L-theanine (500 mg/kg), n = 4. Multivariate ANOVA, *, $p < 0.05$, compared with the control group; #, $p < 0.05$, compared with nicotine group.

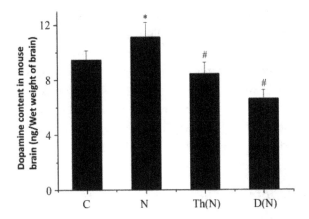

Figure 18-8. Effect of L-theanine on nicotine-induced brain dopamine content in mice. The results were expressed in the amount of dopamine per mg of mouse brain wet weight. C: control group; N: nicotine group (0.5 mg/kg.d); Th(N): L-theanine (250 mg/kg) pretreatment group; D(N): pretreatment group with nicotine acetylcholine receptor inhibitor (2 mg/kg), n = 3. Multi-factor analysis of data. Difference analysis, *, $p < 0.05$, compared with the control group; #, $p < 0.05$, compared with nicotine group.

18.6.4 *Effects of L-theanine on dopamine release in mouse brains*

Dopamine (DA) is the most abundant catecholamine neurotransmitter in the brain. As a neurotransmitter, DA regulates various physiological functions of the central nervous system. The increase of DA release is a significant reward process caused by nicotine [Benwell *et al.*, 1992]. We examined if L-theanine had any effect on the DA release induced by nicotine in mice. After nicotine injection, the levels of DA were significantly increased in the striatum of the mouse brains. Pre-treating animals with L-theanine before nicotine injection significantly reduced the elevation of DA induced by nicotine (Fig. 18-8).

18.7 The mechanism of cessation effect of L-theanine on cigarette smoke

Based on these basic research on nicotine addiction mechanism and our previous research results, we further studied the mechanism of smoking cessation effects by L-theanine in cell and animal systems. L-theanine and nicotinic acetylcholine receptor inhibitor can inhibit the enhancement effect and cell excitation caused by nicotine. We used mice and SH-SY5Y cells *in vitro* as models to study the protein expression level of three nicotinic acetylcholine receptor subunits and tyrosine hydroxylase expression level and the changes of dopamine release level and c-FOS protein expression level were used to study the mechanism of L-theanine on nicotine enhancement effect and cell excitability [Di *et al.*, 2012].

18.7.1 *L-theanine reduces the rewarding effects of nicotine*

The effect of L-theanine on nicotine-induced rewarding was studied using the CPP test in a mouse model. Mice were treated with nicotine (0.5 mg/kg.d) for 14 days to induce the rewarding effects. As shown in Figure 18-9, the duration that the mice spent in the drug-paired compartment increased

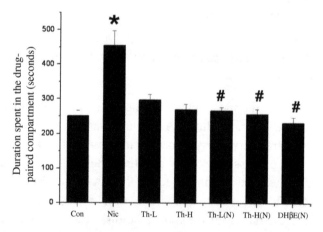

Figure 18-9. Effects of L-theanine pretreatment on reinforcing effect induced by nicotine. The mice were injected with nicotine or different concentrations of L-theanine in physiological saline every day for 14 days. L-theanine were administered 15 minutes before nicotine injection. The nicotine reinforcing effects on the mice were examined on the 17th day by conditioned place preference (CPP) test. Con: mice were treated with physiological saline; Nic: mice were treated with nicotine; TH-L(N): mice were treated with nicotine and L-theanine (250 mg/kg/day); TH-H(N): mice were treated with nicotine and L-theanine (500 mg/kg/day); DHβE(N): mice were treated with nicotine and DHβE. The results are presented as mean ± SEM, n = 8. *, $p < 0.05$ compared with control group; #, $p < 0.05$ compared with nicotine group.

significantly after the treatment when compared with the control group. The preference for the drug-paired compartment induced by nicotine treatment was significantly inhibited when mice were pretreated with L-theanine (250 mg/kg.d or 500 mg/kg.d) or the inhibitor of nicotine receptor (DHβE) before nicotine treatment, the results suggested that both dosages of L-theanine and DHβE pretreatment inhibited the CPP formation induced by nicotine. To further test whether L-theanine treatment reduces the CPP that has been already formed, mice were first injected with nicotine (0.5 mg/kg.d) for 14 days to induce the CPP preference, then treated with L-theanine alone (without nicotine), or injected with nicotine unceasingly with simultaneous pretreatment of L-theanine or DHβE. The results showed that DHβE as well as L-theanine treatment attenuated the preference of mice for the drug-paired compartment that had already been induced by nicotine. These data suggest that L-theanine can reduce the nicotine-induced rewarding effects both before and after the rewarding effects have been developed (Fig. 18-9) [Di *et al.*, 2012].

18.7.2 *L-theanine inhibits the upregulation of tyrosine hydroxylase (TH) expression and dopamine levels in the mouse midbrain*

Tyrosine hydroxylase (TH) is a monooxygenase, which is the rate-limiting enzyme that catalyzes the first step of a series of reactions in which the organism itself synthesizes DA. The increase of dopamine release is an important reward process caused by nicotine [Jiang *et al.*, 2006; Shachar & Youdim, 1990]. We next examined whether L-theanine had any effect on the DA release induced by nicotine in mice. The results showed that levels of dopamine in the midbrain increased significantly after two weeks of nicotine treatment, while pretreatment with the inhibitor of the nicotine receptor (DHβE) or L-theanine inhibited this increase ($p < 0.05$) [Di *et al.*, 2012].

TH is a key enzyme responsible for the synthesis of dopamine. Immuno-histochemical staining of TH showed that the number of TH-positive neurons was significantly increased following nicotine treatment, whereas this increase was blocked in the brains of L-theanine pretreated animals (Fig. 18-8). Meanwhile, western blotting showed that the

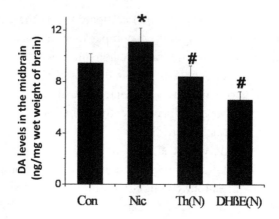

Figure 18-10.   Effects of L-theanine on nicotine-induced dopamine levels in midbrain. The levels of dopamine in the mouse midbrain were measured by HPLC with electrochemical detection. Con: mice were treated with physiological saline; Nic: mice were treated with nicotine; Th(N): mice were treated with nicotine and L-theanine (500 mg/kg); DhβE(N): mice were treated with nicotine and DHßE. The data were expressed as ratio± SEM, n = 5. *, $p < 0.05$, compared with control groups; #, $p < 0.05$, compared with nicotine groups.

expression of TH in the ventral tegmental area (VTA) increased significantly after nicotine treatment when compared with the control group, while pretreatment with L-theanine reduced the expression of TH in a dose-dependent manner. L-theanine treatment alone did not affect the production of dopamine in the midbrain and the expression of TH in the VTA, suggesting that the inhibitory effects of L-theanine on the production of DA and the expression of TH in mice only occurred when mice were stimulated by nicotine (Fig. 18-10) [Di *et al.*, 2012].

18.7.3 *L-theanine inhibits the expression of the α4, β2, and α7*

nAChRs subunits induced by nicotine in the reward circuit related areas of the mouse brain. To dissect the mechanisms by which L-theanine inhibits the rewarding effects induced by nicotine, we next examined the expression of the α4, β2, and α7 nAChR subunits in the reward circuit-related areas of the mouse brain (VTA, PFC, and NAc) following different treatments using western blotting. Our results revealed that the expression of these three forms of nAChR subunits in these three areas increased significantly after treatment with nicotine ($p < 0.05$). When mice were pretreated with L-theanine, the upregulation of the α4 and β2 nAChR subunits were inhibited in the VTA, PFC, and NAc, and the upregulation of the α7 nAChR subunit induced by nicotine was inhibited in the VTA and PFC but not in the NAc. The effects of L-theanine treatment alone on the expression of α4, β2, and α7 nAChR in the VTA, PFC, and NAc areas were examined in mice that were not treated with nicotine. L-theanine treatment alone only affected the expression of the α4 nAChR subunit, which decreased in the NAc. These data suggested that the L-theanine-induced downregulation of α4, β2, and α7 nAChR subunit expression in the three reward circuit-related areas primarily happened when mice were stimulated by nicotine (Fig. 18-11) [Di *et al.*, 2012].

18.7.4 *L-theanine inhibits nicotine-induced c-Fos expression in the reward circuit-related areas of the mouse brain*

In c-Fos proto oncogene expression, c-Fos is an immediate early gene that exists in nerve cells, and noxious stimulation can cause them to express in the nucleus of neurons related to pain transmission. The enhancement of c-Fos expression is the embodiment of vestibular function in sports. Addiction affects the expression of c-Fos protein in brain neurons and neurons in different hippocampal subregions. Addictive substance abuse may induce neuron-specific adaptation and lead to physical and psychological dependence [Mansvelder *et al.*, 2002]. Recent investigations indicate that changes in c-Fos may play a crucial role in nicotine addiction. The increase in accumbal Fos-like immunoreactivity was attenuated by pretreatment with mecamylamine (1.0 mg/kg, s.c.). Our data demonstrate that locomotor-activating effects similar to those evoked by systemically-administered nicotine, including behavioral sensitization, can be produced by intra-tegmental nicotine administration [Panagis *et al.*, 1996]. Our study showed that the expression of c-Fos was upregulated in the three

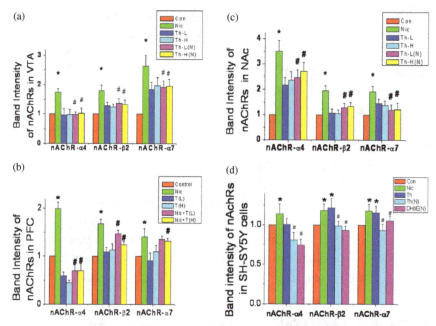

Figure 18-11.   Effects of L-theanine on the expression of nicotine receptor (nAChR) in mouse brain and neuronal cells. Protein extracts prepared from different parts of mouse brain or SH-SY5Y cells were analyzed using western blotting. The expression of nAChR was examined in (a) ventral tegmental area (VTA), (b) prefrontal cortex (PFC), (c) nuclear accumbence core (NAc), and (d) SH-SY5Y cells. Cells were treated with different chemicals as indicated. The expression of nAChR was examined by western blotting. Con: mice were treated with physiological saline; Nic: mice were treated with nicotine; Th-L: mice were treated with theanine (250 mg/kg); Th-H: mice were treated with theanine (500 mg/kg); Th-L(N): mice were pretreated with theanine (250 mg/kg); Th-H(N): mice were pretreated with theanine (500 mg/kg); Con: cells were treated with PBS; nic: cells were treated with nicotine; Th: cells were treated with theanine alone; Th(N): cells were treated with nicotine and theanine; DhβE(N): cells were treated with nicotine and inhibitor of nAChR, DHβE. The data were expressed as ratio ± SEM, n = 5. *, $p < 0.05$, compared with control groups; #, $p < 0.05$, compared with nicotine groups.

reward circuit-related areas of mice after treatment with nicotine, while this phenomenon was inhibited after pretreatment with L-theanine. Meanwhile, double-immunofluorescence staining showed that both TH and c-Fos expression were co-localized and upregulated in SH-SY5Y cells after treatment with nicotine, while these nicotine-stimulated effects were inhibited by L-theanine pretreatment (Fig. 18-12) [Di *et al.*, 2012].

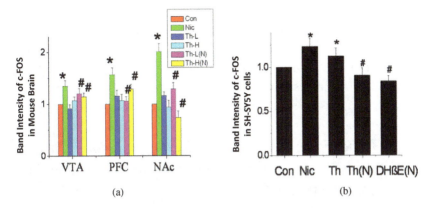

(a)　　　　　　　　　　　　　　　(b)

Figure 18-12. Effects of L-theanine on the expression of c-FOS in mouse brain and SH-SY5Y cells. Protein extracts prepared from different parts of mouse brain or SH-SY5Y cells were analyzed using western blotting. The expression of c-FOS was examined in (a) ventral tegmental area (VTA) and (b) prefrontal SH-SY5Y cells. Cells were treated with different chemicals as indicated. The expression of c-FOS was examined by western blotting. Con: mice were treated with physiological saline; Nic: mice were treated with nicotine; Th-L: mice were treated with theanine (250 mg/kg); Th-H: mice were treated with theanine (500 mg/kg); Th-L(N): mice were pretreated with theanine 250 mg/kg); Th-H(N): mice were pretreated with theanine (500 mg/kg); Con: cells were treated with PBS; Nic: cells were treated with nicotine; Th: cells were treated with theanine alone; Th(N): cells were treated with nicotine and theanine; DHβE(N): cells were treated with nicotine and inhibitor of nAChR, DHβE. The data were expressed as ratio $\pm$ SEM, n = 5. *, $p < 0.05$, compared with control groups; #, $p < 0.05$, compared with nicotine groups.

18.7.5 *L-theanine inhibits the excitatory status of SH-SY5Y cells induced by nicotine*

The sympathetic nerve is in an excited state, causing the secretion of adrenaline to increase, and promoting the intestinal absorption of glucose. Nicotine has been shown to activate cells and lead to their excitatory status. The excitability of dopaminergic neurons induced by nicotine is considered part of the mechanism for the rewarding effects of nicotine [Peng *et al.*, 1994; Sylow *et al.*, 2017]. Glucose is the key molecule of energy metabolism in cells and its uptake is an indicator of the excitatory status of the cells. Therefore, we next studied the effect of L-theanine on the nicotine-induced excitatory status of SH-SY5Y cells by assessing the glucose uptake of cells using the fluorescent probe 2-NBDG. The results showed that the fluorescent signal intensity in SH-SY5Y cells increased significantly after nicotine treatment, indicating that nicotine-induced excitatory status and glucose uptake were increased in cells. When cells were pre-treated with L-theanine or the inhibitor of the nicotine receptor (DHβE), the fluorescence intensity of cells reduced significantly when compared with cells treated with nicotine alone ($p < 0.01$) [Di *et al.*, 2012].

18.7.6 *Knockdown of c-Fos inhibits the excitatory status of the cell but not the upregulation of TH induced by nicotine in SH-SY5Y cells*

To study the importance of c-Fos in the nicotine-induced excitatory status of cells, we performed c-Fos siRNA experiments in SH-SY5Y cells and measured glucose intake using the 2-NBDG assay. The expression of c-Fos decreased by 47% following c-Fos siRNA treatment. Meanwhile, glucose uptake induced by nicotine decreased significantly in cells treated with c-Fos siRNA ($p < 0.05$). This indicated that knockdown of c-Fos inhibited the excitatory status of cells. However, the expression of TH induced by nicotine was not changed by c-Fos siRNA treatment [Di *et al.*, 2012].

To sum up the above results, the mechanism of theanine-inhibiting tobacco addiction can be shown in Figure 18-13. L-theanine, a non-proteinaceous amino acid component of green and black teas, has received growing attention in recent years due to its reported effects on the central nervous system. In this study, we used the conditioned place preference model of mice to analyze the effects of tea polyphenols, L-theanine, and caffeine, three components of tea, on nicotine-induced reinforcement. We found that only L-theanine in the three tea components could significantly affect nicotine-induced conditioned place preference. This effect of L-theanine may be due to the inhibition of nicotine-induced increase on the expression of three nicotinic acetylcholine receptor subunits in brain regions related to reward pathways, thus reducing the number of cells that can produce effects on nicotine stimulation, inhibiting the increase of nicotine induced TH protein expression, midbrain dopamine level, and c-FOS expression, and ultimately inhibiting the enhancement effect and cell excitation caused by nicotine. These data suggest that L-theanine, a natural ingredient, may safely and effectively help people solve the

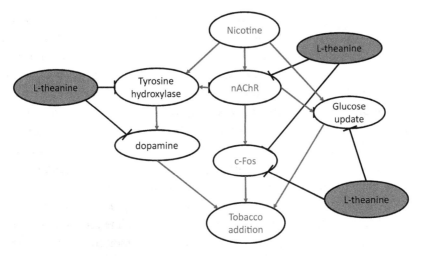

Figure 18-13. The schematic diagram of the mechanism of L-theanine inhibiting nicotine dependence through nAChRs and dopamine reward pathway.

problems of smoking addiction, public health, and environmental pollution caused by smoking. This also study shows that L-theanine reduces nicotine-induced reward effect by inhibiting the nAChR dopamine reward pathway. In this study, a tea filter and L-theanine were proposed to inhibit tobacco addiction by inhibiting nAChR, providing an effective method for treating tobacco addiction (Fig. 18-13). The implementation and promotion of this work will protect contemporary and future generations from the destructive impact of cigarette smoking on health, society, environment, and economy, and is of great significance to the construction of a harmonious human civilized society and the sustainable development of the national economy.

18.8 Conclusion

In this chapter, we discussed the effects of smoking and oxidative stress injury and various diseases caused by smoking, the inhibition of tea polyphenols and tea filters on smoking and oxidation, and the prevention of diseases caused by smoking. This chapter also discussed the effect of tea and theanine on smoking addiction. Through discussion, we can see that smoking can produce a large number of ROS stable RNS free radicals that causes oxidative stress damage, and can lead to various diseases, which has caused significant harm to human beings. It is imperative to eliminate the harm of smoking and smoking addiction. Through discussion, we can see that tea and tea polyphenols can clear ROS, stabilize RNS free radicals, inhibit oxidative stress damage caused by them, and prevent various diseases caused by smoking. Tea and L-theanine can reduce the number of cells that can produce effects on nicotine stimulation, inhibit the increase of TH protein expression, midbrain dopamine level and c-FOS expression caused by nicotine, and finally inhibit the enhancement effect and cell excitation caused by nicotine by inhibiting the increase of nicotine-induced expression of three nicotinic acetylcholine receptor subunits in brain regions related to reward pathway. It can inhibit the withdrawal of smoking addiction.

If the tourism method realizes that the abstinence of tobacco addiction not only greatly reduces the toxic effect of tobacco on human beings but also reduces the environmental pollution caused by smoking, it will be a beneficial strategy to human beings.

References

Balbani APS, Montovani JC. (2005) Methods for smoking cessation and treatment of nicotine dependence. *Rev Bras Otorrinolaringol*, **71**, 820–826.

Benwell MEM, Balfour DJK. (1992) The effects of acute and repeated nicotine treatment on nucleus accumbens dopamine and locomotor activity. *Br J Pharmacol*, **105**, 849–856.

Cardoso WV, Saldiva PHN, Criado PMP, *et al.* (1991) A comparison between the isovolume and the end-inflation occlusion methods for measurement of lung mechanics in rats. *J Applied Toxicol*, **11**, 79–84.

Cendon SP, Battlehner C, Lorenzi-Filho G, *et al.* (1997) Pulmonary emphysema induced by passive smoking: an experimental study in rats. *Braz J Med Biol Res*, **30**, 1241–1247.

Chan KH, Chan SC, Yeung SC, Man RY, Ip MS, Mak JC. (2012) Inhibitory effect of Chinese green tea on cigarette smoke-induced up-regulation of airway neutrophil elastase and matrix metalloproteinase-12 via antioxidant activity. *Free Radic Res*, **46**(9), 1123–1129.

Chow CK. (1993) Cigarette smoking and oxidative damage in the lung. *Ann N Y Acad Sci*, **686**, 289–298.

Chung FL, Morse MA, Eklind KI, Xu Y. (1993) Inhibition of tobacco-specific nitrosamine-induced lung tumorigenesis by compounds derived from cruciferous vegetables and green tea. *Ann N Y Acad Sci*, **686**, 186–201.

Chung FL. (1999) The prevention of lung cancer induced by a tobacco-specific carcinogen in rodents by green and black tea. *Proc Soc Exp Biol Med*, **220**(4), 244–248.

Chung YF, Khoo ML, Heng MK, *et al.* (1999) Epidemiology of Warthin's tumor of the paroted gland in an Asian population. *Br J Surg*, **86**, 661–664.

Church DF, Pryor WA. (1985) Free-radical chemistry of cigarette smoke and its toxicological implications. *Environ Health Perspect*, **64**, 111–123.

Diana JN, Pryor WP. (1993) Tobacco smoking and nutrition: influence of nutrition on tobacco-associated health risks. *Ann N Y Acad Sci*, **686**, 66–128.

Di X, Yan J, Zhao Y, Chang Y, Zhao B. (2012) L-theanine inhibits nicotine-induced dependence via regulation of the nicotine acetylcholine receptor-dopamine reward pathway. *Sci China Life Sci*, **55**, 1064–1074.

Egawa M, Kohno Y, Kumano Y. (1999) Oxidative effects of cigarette smoke on the human skin. *Int J Cosmet Sci*, **21**(2), 83–98.

Gu DF, Kelly TN, Wu X, *et al.* (2009) Mortality attributable to smoking. *N Engl J Med*, **360**, 150–159.

Guo SH, Yan J, Bezard E, *et al.* (2007) Protective effects of green tea polyphenols in the 6-OHDA rat model of Parkinson's disease through inhibition of ROS-NO pathway. *Biol Psychia*, **62**, 1353–1362.

Gao J, Tang H, Zhao B. (2001) Toxicologic damage of gas phase cigarette smoke on cells and protective effect of green tea polyphenols. *Res Chem Intermed*, **29**, 269–279.

Halpern A, Knieper, J. (1985) Spin trapping of radicals in gas-phase tobacco smoke. In *Proceedings of the seventeenth International Symposium on Free Radicals*. Colorado, 306.

Huang CC, Lai CY, Lin IH, Tsai CH, Tsai SM, Lam KL, Wang JY, Chen CC, Wong RH. (2022) Joint effects of cigarette smoking and green tea consumption with miR-29b and *DNMT3B* mRNA expression in the development of lung cancer. *Genes (Basel)*, **13**(5), 836.

Hudlikar RR, Pai V, Kumar R, Thorat RA, Kannan S, Ingle AD, Maru GB, Mahimkar MB. (2019) Dose-related modulatory effects of polymeric black tea polyphenols (PBPs) on initiation and promotion events in B(a)P and NNK-induced lung carcinogenesis. *Nutr Cancer*, **71**(3), 508–523.

Jankan J, Selman S H, Swiecz R. (1997) Why drinking green tea could prevent cancer. *Nature*, **387**, 661–662.

Lape R, Colquhoun D, Sivilotti LG. (2008) On the nature of partial agonism in the nicotinic receptor superfamily. *Nature*, **454**, 722–728.

Jiang H, Luan Z, Wang J, *et al.* (2006) Neuroprotective effects of iron chelator desferal on dopaminergic neurons in the substantia nigra of rats with iron-overload. *Neurochem Int*, **49**, 605–609.

Li W, Wu JX, Tu YY. (2010) Synergistic effects of tea polyphenols and ascorbic acid on human lung adenocarcinoma SPC-A-1 cells. *J Zhejiang Univ Sci B*, **11**(6), 458–464.

Mackenzie T, Leary L, Brooks W B. (2007) The effect of an extract of green and black tea on glucose control in adults with type 2 diabetes mellitus: double-blind randomized study. *Metabolism*, **56**, 1340–1344.

Mansvelder HD, Keath JR, McGehee DS. (2002) Synaptic mechanisms underlie nicotine-induced excitability of brain reward areas. *Neuron*, **33**, 905–919.

Nguyen HN, Rasmussen BA, Perry DC. (2003) Subtypeselective up-regulation by chronic nicotine of high-affinity nicotinic receptors in rat brain demonstrated by receptor autoradiography. *J Pharmacol Experim Therap*, **307**, 1090–1095.

Panagis G, Nisell M, Nomikos GG. (1996) Nicotine injections into the Ventral tegmental area increase locomotion and Fos-like immunoreactivity in the nucleus ac-cumbens of the rat. *Brain Res*, **730**, 133–142.

Peng X, Gerzanich V, Anand R, *et al.* (1994) Nicotine-induced increase in neuronal nicotinic receptors results from a decrease in the rate of receptor turnover. *Mol Pharmacol*, **46**, 523–530.

Pryor WA. (1983) An electron spin resonance study of mainstream and side-stream cigarette smoke: nature of the free radicals in gas-phase smoke. *Environ Health Perspect*, **47**, 345–355.

Pryor WA. (1984) An ESR spin trapping study of the radicals produced in NO/olefin reactions: a mechanism for the production of the apparently long-lived radicals in gasphase smoke. *J Am Chem Soc*, **106**, 5073–5086.

Ray R, Schnoll R A, Lerman C. (2009) Nicotine dependence: biology, behavior, and treatment. *Annu Rev Med*, **60**, 247–260.

Shachar BD, Youdim MB. (1990) Selectivity of melaninized nigra-striatal dopamine neurons to degeneration in Parkinson's disease may depend on iron±melanin interaction. *J Neural Transm*, **29**, 251–258.

Sylow L, Kleinert M, Richter EA, Jensen TE. (2017) Exercise-stimulated glucose uptake — regulation and implications for glycaemic control. *Nat Rev Endocrinol*, **13**(3), 133–148.

Syed D N, Afaq F, Kweon M-H, Hadi N, Bhatia N, Spiegelman V S, Mukhtar H. (2007) Green tea polyphenol EGCG suppresses cigarette smoke condensate-induced NF-kappaB activation in normal human bronchial epithelial cells. *Oncogene*, **26**(5), 673–682.

Tsuchiya M, Thompson DF, Suzuki YJ, Cross CE, Packer L. (1992) Superoxide formed from cigarette smoke impairs polymorphonuclear leukocyte active oxygen generation activity. *Arch Biochem Biophys*, **299**(1), 30–37.

Tsuchimy M, Suzuki Y, Cross CE, Packer L. (1993) Superoxide generation by cigarette smoke damages the respiratory bust and induces physical change in the membrane order and water organization of inflammatory cells. *Ann N Y Acad Sci*, **686**, 39–52.

Weisburger JH, Chung FL. (2002) Mechanisms of chronic disease causation by nutritional factors and tobacco products and their prevention by tea polyphenols. *Food Chem Toxicol*, **40**(8), 1145–1154.

Xiu X, Puskar NL, Shanata JAP, *et al.* (2009) Nicotine binding to brain receptors requires a strongcation–p interaction. *Nature*, **458**, 534–538.

Yan LJ, Zhao BL, Xin WJ. (1991a) Experimental studies on smoke aspects of toxicological effects of gas phase cigarette smoke. *Research Chem Interm*, **16**, 15–24.

Yan LJ, Zhao BL, Li X-J, Xin WJ. (1991b) ESR was used to study the effect of cigarette smoke on polymorphonuclear leukocyte respiratory burst. *J Environ Sci*, **11**, 79–82.

Yan LJ, Zhao BL, Xin WJ. (1991c) Study on the physical properties of biofilm caused by smoking. *J Biophys*, **7**, 5–9.

Yan LJ, Zhao BL, Guo X-J, Xin WJ. (1992) Study on lipid peroxidation caused by smoking gaseous substances. *Environ Chem*, **11**, 58–65.

Yan J-Qi, Di X-J, Liu C-Y, *et al.* (2010) The cessation and detoxification effect of tea filters on cigarette smoke. *Sci China Life Sci*, **53**, 533–541.

Yang FJ, Zhao BL, Xin WJ. (1992) Studies on toxicological mechanisms of gas-phase cigarette smoke and protective effects of GTP. *Res Chem Interm*, **17**, 39–57.

Yang FJ, Zhao BL, Xin WJ. (1992a) ESR spin trapping method was used to study the effect of liposomes treated with smoking smoke on O_2 production by rat granulocytes. *J Biophys*, **8**, 659–663.

Yang FJ, Zhao BL, Xin WJ. (1993a) ESR study on the relationship between lipid peroxidation induced by cigarette gaseous substances and granulocyte respiratory burst. *J Environ Sci*, **13**, 355–359.

Yang FJ, Zhao BL, Xin WJ. (1993b) ESR spectroscopic study on lipid peroxidation of rat liver microsomes induced by smoking smoke. *Environ Chem*, **12**, 117–125.

Yang FJ, Zhao, BL, Ren X-J, Xin, WJ. (1993c) ESR study of tea polyphenols inhibiting lipid free radical production in rat liver microsomes stimulated by smoking gaseous substances. *J Biophys*, **9**, 468–471.

Yang FJ, Zhao, BL, Xin, WJ. (1992b) The protective effect of tea polyphenols on the damage of lung cell membrane in rats. *Environ Chem*, **11**, 50–56.

Yang FJ, Ren S-R, Yang X-Q, Zhao, BL, Xin, WJ. (1992c) Inhibitory effect of tea polyphenol monomer L-EGCG on lung cell injury induced by gaseous smoke in rats. *J Biophys*, **8**, 450–456.

Zhao B-L. (1994) Free radical lecture. *Beijing Tobacco*, **4**, 21–25.

Zhao B-L. (1988) Smoking, free radicals and cancer. *Nature*, **12**, 453–456.

Zhao B-L. (1989b) Smoking, free radicals and heart disease. *Nature*, **12**, 655–657.

Zhao B, Li X, He R, *et al.* (1989a) Scavenging effect of extracts of green tea and natural antioxidants on active oxygen radicals. *Cell Biophys*, **14**, 175–184.

Zhao B-L, Zhang C-A, Xin W-J. (1989b) Myocardial ischemia-reperfusion injury and reactive oxygen species. *Psychol Sci*, 193–197.

Zhao B-L. (1989a) Oxygen free radicals and aging. *Nature*, **13**, 511–514.

Zhao B-L, Zhang C-A, Xin W-J. (1995) Smoking, free radicals and skin aging. *J Appl Gerontol*, **41**, 43–45.

Zhao B-L. (1996) Effect of no free radical in cigarette smoke on human body. *Beijing Tobacco*, **4**, 17–20.

Zhao B-L, Yan L-J, Hou J-W, Xin W-J. (1990) Electron spin resonance spin trapping of gaseous free radicals in smoking. *Chin J Med*, 70, 386–391.

Zhao B-L, Zhang C-A. (1994) Reduction of harmful free radicals in smoking. *Beijing Tobacco*, **3**, 16–17.

Zhang S-L, Zhao, BL, Xin, WJ. (1996) Study on the cytotoxicity of smoking smoke and the protective effect of tea polyphenols. *Chin Environ Sci*, **16**, 386–390.

Zhang WZ, Butler JJ, Cloonan SM. (2019) Smoking-induced iron dysregulation in the lung. *Free Radic Biol Med*, **133**, 238–247.

Chapter 19

Precautions for Drinking Tea and Using Tea Polyphenols

Baolu Zhao

Institute of Biophysics, Chinese Academy of Sciences, Beijing, China

19.1 Introduction

Tea is the most frequently consumed beverage worldwide besides water. Generally, there are five most popular types of tea: green, white, black, Pu'er, and oolong. Green tea and white tea are non-fermented, oolong tea is semi-fermented, and black tea and Pu'er tea are fully fermented. At present, black tea (fermented tea) and green tea (non-fermented tea) are the most popular in world. Black tea accounts for over 90% of all teas sold in western countries. The world's top-grade black teas include Qi Men black in China, Darjeeling and Assam black tea in India, and Uva black tea in Sri Lanka. However, all top 10 famous green teas in the world are produced in China, and Xi Hu Long Jing tea is the most famous among all green teas. More than 700 different kinds of components and 27 mineral elements can be found in tea. [Pan, 1995; Yang *et al.*, 2003]. Tea possesses significant antioxidant, anti-inflammatory, antimicrobial, anti-carcinogenic, antihypertensive, neuroprotective, and cholesterol-lowering properties. Several research investigations, epidemiological studies, and meta-analyses suggest that tea and its bioactive polyphenolic constituents have numerous beneficial effects on health, including the prevention of

Figure 19-1.   The possible beneficial effects of tea drinking [Zhao, 2020].

many diseases, such as cancer, diabetes, arthritis, cardiovascular disease, stroke, and obesity. Recently, there are many reports about the resistance of tea to COVID-19 virus on the Internet, which has attracted a lot of attention to tea drinking and the discussion about the pros and cons of tea drinking. Based on our research results and relevant reports from literatures, and highlighting the beneficial effects and possible side effects associated with tea consumption in my article, I have summarized the eight functions of drinking tea (Fig. 19-1) [Zhao, 2020]. Yet throughout this book, a total of 18 functions of drinking tea have already been summarized in the first 18 chapters of this book. The efficacy of drinking tea lies in its effective ingredients.

Although tea contains hundreds of chemicals, it can only drink and absorb the substances in its aqueous solution. The water extract of tea is about 40–50% of the weight of tea, and there are four main effective components: tea polyphenol, caffeine, L-theanine, and amino acid. Drinking tea can allow people to acquire a peaceful, relaxed, refreshed and cheerful enjoyment, and even longevity. According to the meridian theory of traditional Chinese medicine, different kinds of tea can activate different meridian systems in the human body. Tea polyphenols and theaflavin/thearubigins are considered to be the major bioactive components of black tea and green tea, respectively. Overly strong or overheated tea liquid should be avoided when drinking tea [Pan *et al.*, 2022; Yang *et al.*, 2003].

Tea polyphenols are the main natural antioxidants of tea. There are mainly four kinds of catechins in green tea and white tea, namely (-)-epicatein-3-gallate (EGCG), (-)-epicatechin gallate (ECG), (-)-epicatechin (EGC), and (-)-epicatechin (EC), accounting for about 30% of the dry weight of green tea. Tea polyphenols in oolong tea are partially oxidized, and tea polyphenols in black tea and Pu'er tea are completely oxidized into theaflavins and thearubins. These are all very good natural antioxidants and are the main ingredients for tea to play a series of biological functions. Caffeine is a central nerve stimulant in tea, which can temporarily drive away drowsiness and restore energy. It is clinically used to treat neurasthenia and coma recovery. All tea contains caffeine, about 3–5% of the dry weight of tea [Pan, 1995; Yang *et al.*, 2003].

Drinking tea can make people enjoy peace, relaxation, spirit, and pleasure, and even live a long life. According to the meridian theory of traditional Chinese medicine, different kinds of tea can activate different meridian systems of the human body. The antioxidant effect of tea polyphenols and their oxidation products mainly refers to their free radical scavenging effect. The antioxidant effect of tea polyphenols and their oxidation products mainly refers to their free radical scavenging effect. Tea polyphenols can prevent the oxidation of some proteins, lipids, and low molecular compounds by free radicals. Tea polyphenols can keep the free radicals in the organism in the balance between the biological production system and the biological protection system. Tea polyphenols can protect cardiovascular diseases through a variety of mechanisms, such as regulating lipid metabolism, anticoagulation, inhibiting platelet aggregation, inhibiting the proliferation of arterial smooth muscle cells, and affecting hemorheology. Tea polyphenols have anti-mutagenic activity and inhibit carcinogenic tumors of rodent skin, lung, stomach, esophagus, duodenum, and colon. Tea polyphenols can also be used as antidote for alkaloids and heavy metal salt poisoning and reduce the toxicity of these heavy metal ions. Tea polyphenols can prevent the effects of various radiation, including solar radiation burns and radiation diseases such as X-rays.

Although drinking tea and tea polyphenols have many benefits to the human body, it is necessary to drink tea and use tea polyphenols correctly to achieve the expected effect, otherwise it may cause harm to the body.

This chapter discusses how to drink tea correctly and how to use tea polyphenols and some precautions.

19.2 Matters needing attention in tea drinking

Different countries and regions have different tea drinking habits. Even different people in the same country and region also have different tea drinking habits. For example, people in China, Japan, and other countries are accustomed to drinking green tea, while people in Britain and European countries like drinking black tea. In addition, some people like to drink light tea, some people are used to drinking strong tea, and other people like to drink Kungfu tea. In China, most people in Beijing like scented tea, while most areas in the south of China are used to green tea, but Fujian Province and Taiwan Province of China prefer oolong tea, while Guangdong Province, Hong Kong, and Macao of China prefer morning tea. Inner Mongolia and Tibet of China are used to milk tea. No matter what kind of tea drinking habits you adopt, as long as you feel comfortable, drinking tea is good for your health [Lu, 1974; Pan, 1995].

In order to prevent the incorrect way of drinking tea from affecting health and make tea serve health better, there are still many things to pay attention to when drinking tea.

1. It is better not to drink tea on an empty stomach. Tea contains caffeine. If you drink tea on an empty stomach, it may stimulate the body and increase the burden on the kidney. Especially, patients with gastric ulcer need to pay more attention not to drink tea on an empty stomach. If they do not pay attention, they may get worse.
2. Avoid drinking tea that is too hot. Some people have the habit of drinking hot tea, but in fact, the maximum temperature that people's oral cavity, esophagus and gastric mucosa can tolerate is about 50–60°C. If the temperature exceeds this temperature, it is easy to cause mucosal damage. If the esophageal mucosa is repeatedly stimulated for a long time, it will increase the risk of esophageal cancer, so it is better not to drink tea more than 60°C. In addition, tea contains tea oil, vitamins, theophylline, and other ingredients. If the

temperature is too high, the vitamins in tea will be damaged, reducing the nutritional value of tea.

3. Caffeine in tea is a stimulant of nervous system. Some people drink too much tea, especially at night, which may affect sleep. If you are afraid of affecting sleep, it is advised to drink less or no tea at night. Caffeine in tea has the effect of stimulating the nerve center. Drinking strong tea, especially in the afternoon and evening, will cause insomnia and aggravate the disease. You can drink tea once in the morning and afternoon during the day, but it is not suitable to drink tea at night. In this way, patients will be refreshed during the day, while being calm and comfortable at night, and can fall asleep early.

4. In order to prevent the impact of drinking tea on iron absorption, it is better to have an hour between drinking tea and eating. In addition, it is better to eat some ascorbic acid (vitamin C) when supplementing iron with inorganic iron food.

5. Caffeine in tea may interfere with some drugs. It is not advisable to drink tea when taking medicine.

6. Patients with ulcers should drink tea carefully. Tea is a stimulant for gastric acid secretion. Drinking tea can increase the amount of gastric acid secretion and increase the stimulation of ulcer surface. Drinking strong tea often will worsen the condition. However, for mild patients, you can drink some light tea after two hours of taking medicine. Sugar black tea and milk black tea can help to diminish inflammation and protect the gastric mucosa, and have a certain effect on ulcers.

7. Chinese medicine believes that green tea is cool and black tea is warm. Drink tea according to your physical condition, whether green tea or black tea is more suitable for your health. It is easy to catch fire in spring and summer. You can drink more green tea with cool nature. It is suitable for drinking oolong tea in autumn tea in autumn. Oolong tea is gentle in nature, not cold or warm, and makes people feel refreshed. During cold winter, it is recommended to drink more warm black tea to warm your stomach.

Above precautions for drinking tea are the author's experience of drinking tea for many years and some tea drinking experience collected, which is only for the reference of tea lovers.

19.3 Precautions for using tea polyphenols

Under normal circumstances, biological individuals can always maintain the balance of oxidation and anti-oxygen and their stability in cells, which can ensure the health of the body. If the balance between oxidation and anti-oxygen is gradually broken, this may accelerate the damage and aging of the organism. If oxidation and anti-oxidation are difficult to maintain balance, it may lead to sub-health. If disease and aging occur, the balance between oxidation and anti-oxidation will be broken and oxidative stress damage will occur, leading to disease and aging (Fig. 19-1). However, it depends on the severity and duration of oxidative stress; too low or too high reactive oxygen species (ROS) and reactive nitrogen species (RNS) and the change of its cell location will affect the normal physiological function of organisms, which in turn will affect individual life span. ROS and RNS play a dual role since they can be either beneficial or harmful to living systems. With increasing ROS and RNS concentrations, the roles of ROS change from advantageous to detrimental. As redox imbalance is closely related to the occurrence and development of a variety of diseases, antioxidant therapies are an attractive option. We emphasize the precise nature of redox regulation and elucidate the importance and necessity of precision redox strategies from three aspects: differences in redox status, differences in redox function, and differences in the effects of redox therapy. We then propose the "5R" principle of precision redox in antioxidant pharmacology — "Right species, Right place, Right time, Right level, and Right target." Critical issues to highlight are: redox status must be considered in the context of species, time, place, level, and target. Precision redox is the key for antioxidant pharmacology [Meng *et al.*, 2021]. More intriguingly, Meng *et al.* found there seems to be a concentration threshold that determines the transition from their advantageous to detrimental effects. If the threshold is purposefully increased, that is, increase the range of ROS and RNS that plays an advantageous role, it should be beneficial for individuals. They found that there is a maximum level below which redox stress has benefits and named this threshold as "Redox-stress Signaling Threshold (RST)". More intriguingly, we found that increasing RST could improve Redox-stress Response Capacity (RRC) and health span, suggesting that increasing the RST value through early stimulation will be an effective strategy to delay aging

[Meng *et al.*, 2022]. Tea polyphenols are the typical antioxidant, so the use of tea polyphenols should also follow the principles of antioxidant use. Future studies to develop more accurate methods for detecting redox status and accurately evaluate the redox state of different physiological and pathological processes are needed.

19.3.1 *Attention should be paid to maintaining the balance between oxidation and anti-oxidation in the body when replenishing tea polyphenols*

Maintaining the balance of production and elimination of free radicals ROS and RNS in the body can improve immunity and ensure the health and longevity of the body. If the balance between oxidation and anti-oxidation cannot be maintained, it may cause harm to the body and lead to sub-health. If the balance between oxidation and anti-oxidation is damaged, it will lead to some diseases and aging. When there is too much oxidation caused by ROS and RNS free radicals in the body or the antioxidant system fails, the body itself cannot maintain the balance between oxidation and antioxidant, and the body will produce oxidative stress, resulting in cell death, body damage, diseases, and aging. It is necessary to supplement some antioxidants, and natural antioxidants are the best choice. On the contrary, if the anti-oxidation is too strong or insufficient, the balance between oxidation and anti-oxidation will be also destroyed, leading to cell growth injury, and immune defense system injury [Zhao *et al.*, 1989; Zhao, 1999, 2006, 2007, 2020]. Therefore, no matter which antioxidant is added, this principle must be followed. Blind supplement of antioxidants may lead to the destruction of the balance between oxidation and anti-oxidation in the body, leading to health injuries. Tea polyphenols supplementation need also follow this rule.

19.3.2 *Pay attention to the synergistic effect of various antioxidants*

The ROS and RNS free radicals produced in the body are diverse and the damage caused is also diverse. The organism needs a variety of antioxidants to promote health through synergy. Therefore, when supplementing the antioxidant tea polyphenols, we should also pay attention to synergy with other antioxidants to achieve the desired effect as there are many kinds of ROS and RNS free radicals that cause damage in the body, and the free radicals cleared by each antioxidant are also different. If an appropriate combination of antioxidant formula is used, it may be better than a single variety of antioxidant habits [Zhao *et al.*, 1989; Zhao, 1999, 2006, 2007, 2020]. Although tea polyphenols can remove a variety of free radicals, if they are combined with water-soluble and fat-soluble antioxidants, they can achieve better results than the unit use of tea polyphenols alone.

19.3.3 *Pay attention to the physical condition and needs of individuals when supplementing tea polyphenols*

Traditional Chinese medicine pays attention to the treatment based on syndrome differentiation. One person has one prescription. Only the symptomatic treatment can be effective, which is the best prescription. Similarly, everyone's physical condition is different, whether they lack tea polyphenols and to what extent. Such information needs to be clear before supplementing tea polyphenols, otherwise wrong supplementation may occur, leading to adverse reactions. Who needs to be supplemented with tea polyphenols and to what extent they need to be supplemented, etc., has to be done according to people's needs and their respective human conditions.

19.4 Conclusion

Although this chapter only briefly discusses the precautions for drinking tea and using tea polyphenols, it is very important to drink tea and using tea polyphenols correctly, because only by drinking tea and using tea polyphenols correctly can we achieve the best results and benefit for our health.

References

Lu Y. (1974) *The Classic of Tea.* Boston, MA, Little, Brown & Co.
Meng J, Lv Z, Zhang Y, Wang Y, Qiao X, Sun C, Chen Y, Guo M, Han W, Ye A, Xie T, Chu B, Shi C, Yang S, Chen C. (2021) Precision redox: the key for antioxidant pharmacology. *Antioxid Redox Signal,* **34**(14), 1069–1082.

Meng J, Lv Z, Wang Y, Chen C. (2022) Identification of the redox-stress signaling threshold (RST): increased RST helps to delay aging in C. elegans. *Free Radic Biol Med,* **178**, 54–58.

Pan G. (1995) Tea collection. China Agriculture Press, Beijing.

Pan SY, Nie Q, Tai HC, *et al.* (2022) Tea and tea drinking: China's outstanding contributions to the mankind. *Chin Med,* **17**(1), 27.

Yang X, Wang Y, Chen L. (2003) Tea polyphenol chemistry. Shanghai Science and Technology Press.

Zhao B-L, Li XJ, He RG, *et al.* (1989) Scavenging effect of extracts of green tea and natural antioxidants on active oxygen radicals. *Cell Biophys,* **14**(2), 175–185.

Zhao B-L. (1999) Oxygen free radicals and natural antioxidants (First Edition). Beijing Science Press.

Zhao B-L. (2006) The health effects of tea polyphenols and their antioxidant mechanism. *J Chin Biochem Nutr,* **38**, 59–68.

Zhao B-L. (2007) Free radicals, natural antioxidants and health. Hong Kong China Science and Culture Press.

Zhao B-L. (2020) The pros and cons of drinking tea. *Tradit Med Mod Med,* **3**(3), 1–12.

Index

Printed in the USA
CPSIA information can be obtained
at www.ICGtesting.com
JSHW010425231023
50599JS00001B/3